Nanotechnology in the Life Sciences

Series Editor

Ram Prasad, Department of Botany, Mahatma Gandhi Central University, Motihari, India

Nano and biotechnology are two of the 21st century's most promising technologies. Nanotechnology is demarcated as the design, development, and application of materials and devices whose least functional make up is on a nanometer scale (1 to 100 nm). Meanwhile, biotechnology deals with metabolic and other physiological developments of biological subjects including microorganisms. These microbial processes have opened up new opportunities to explore novel applications, for example, the biosynthesis of metal nanomaterials, with the implication that these two technologies (i.e., thus nanobiotechnology) can play a vital role in developing and executing many valuable tools in the study of life. Nanotechnology is very diverse, ranging from extensions of conventional device physics to completely new approaches based upon molecular self-assembly, from developing new materials with dimensions on the nanoscale, to investigating whether we can directly control matters on/in the atomic scale level. This idea entails its application to diverse fields of science such as plant biology, organic chemistry, agriculture, the food industry, and more.

Nanobiotechnology offers a wide range of uses in medicine, agriculture, and the environment. Many diseases that do not have cures today may be cured by nanotechnology in the future. Use of nanotechnology in medical therapeutics needs adequate evaluation of its risk and safety factors. Scientists who are against the use of nanotechnology also agree that advancement in nanotechnology should continue because this field promises great benefits, but testing should be carried out to ensure its safety in people. It is possible that nanomedicine in the future will play a crucial role in the treatment of human and plant diseases, and also in the enhancement of normal human physiology and plant systems, respectively. If everything proceeds as expected, nanobiotechnology will, one day, become an inevitable part of our everyday life and will help save many lives.

Yugal Kishore Mohanta • Kunal Biswas
Hemen Sarma
Editors

Bioresource Waste to Carbon Nanomaterial

Food, Agriculture and Circular Bioeconomy Paradigm

Springer

Editors
Yugal Kishore Mohanta
Nano-biotechnology and Translational Knowledge Laboratory
Department of Applied Biology
University of Science and Technology Meghalaya
Ri-Bhoi, Meghalaya, India

Kunal Biswas
Centre for Nanoscience & Nanotechnology, International Research Centre
Sathyabama Institute of Science and Technology (Deemed to be University)
Chennai, Tamil Nadu, India

Hemen Sarma
Department of Botany
Bodoland University
Kokrajhar (BTR), Assam, India

ISSN 2523-8027 ISSN 2523-8035 (electronic)
Nanotechnology in the Life Sciences
ISBN 978-981-92-2404-3 ISBN 978-981-92-2405-0 (eBook)
https://doi.org/10.1007/978-981-92-2405-0

This Springer imprint is published by the registered company Springer Nature Singapore Pte Ltd.
The registered company address is: 152 Beach Road, #21-01/04 Gateway East, Singapore 189721, Singapore

Preface

The book mainly addresses the economic importance of achieving useful carbon nanomaterials from diverse waste materials sources. The paradigm of biocircular economy has been achieved through this book in the process of valorizations principles. In order to illustrate diverse area of waste-based valorizations, the book has been divided into several portions, which encompass the synthesis of carbon-based nanomaterials from diverse waste-sourced origins, allied by eminent and relevant contributors.

Primary features of the book are as follows:

In a nutshell, the primary sections deal with the diverse utility of top-down and bottom-up technologies of carbonaceous nanomaterials, which have been discussed using bioresource waste recycling aspects. The myriad carbon sequestration and recycling of wastes have also been thoroughly discussed in the subsequent chapters. Further sections deal with how the conversion of agricultural biomass waste to biochar formation has been achieved by different studies and endeavors. Implications of Artificial Intelligence and Machine Learning in the process of conversion of wastes to useful products have also been discussed in the following sections with detailed examples and evidence.

The subsequent section deals with the conversion of "*waste-rice-waste*" and "*husk-derived*" synthesis of carbon nanomaterials, expressing the merits of organic-sourced conversion of rice-wastes to carbon nanomaterials conversion. The next series of sections comprises the utility of antioxidant, anticancerous, and associated biomedical attributes of waste-recycled carbon nanomaterials.

The supportive and accompanying sections comprise the food-waste biorefinery implications of carbon nanostructure synthesis and its further fabrication aspects; removal of polyethylene pollution aspects; sensor development; fabrication in determining the agricultural and food-waste determinations; dairy-based effluents recycling in carbon nanomaterial synthesis; and molecular toxicity of carbon-based nanomaterials and its environmental effects of safety and ethical issues. Altogether, the book aids in uncovering the deeper insights of the diverse waste-sourced derived synthesis of carbon-based nanomaterials and its applications in the context of food and agriculture and is allied to achieve an effective sustainable development paradigm through biocircular economic perspectives.

Overall, the book would be helpful for nanotechnologists, molecular biologists, environmentalists, policymakers, and educationists in shaping and achieving the process of waste-derived valorization principles. The different themes were written and discussed by the eminent contributors in the relevant field of waste materials, biocircular economy, and environmental management, which provided tailor-made aspects of understanding comprehensively the subjects of waste valorizations and nanotechnology at large.

Ri-Bhoi, Meghalaya, India — Yugal Kishore Mohanta
Chennai, Tamil Nadu, India — Kunal Biswas
Kokrajhar (BTR), Assam, India — Hemen Sarma

Acknowledgments

We begin by expressing our gratitude to our Creator, Almighty God, the Most Benevolent and the Most Merciful. From Him we begin, and to Him we shall ultimately return.

We are also sincerely thankful to our family members—our parents, spouses, and relatives—whose constant support, patience, and encouragement have meant a great deal to us throughout the journey of completing this book.

Without the contributing authors, this book would not have been possible. We would like to express our gratitude to each one of them for their attention and important contributions, which allowed this book to be finished on schedule. At the same time, we are immensely grateful to the publisher Springer Nature, for choosing to publish our book. We express our gratitude to Dr. Sanchi Bhimrajka, Associate Editor at Springer Nature, for her unwavering editorial support and valuable suggestions in refining the book to its current form.

Lastly, the three editors thank each other—the coordination, communication, and cooperation have been instrumental for realizing the idea of this book and making it a success.

Yugal Kishore Mohanta
Kunal Biswas
Hemen Sarma

Contents

Editors and Contributors

About the Editors

Yugal Kishore Mohanta is currently serving as Assistant Professor in the Department of Applied Biology at University of Science and Technology Meghalaya (USTM), India. With an M.Sc., Ph.D., and post-doctoral training, his specializations lie in nanobiotechnology, microbiology, and sustainable-development oriented research. As a prolific researcher, he has authored over 134 peer-reviewed articles in SCIE-indexed journals, 47 book chapters, 4 Indian patents, and edited 10 books. His work has earned significant recognition in 2025 that he was ranked among the "Top 2% Scientists in the World" by Stanford University (USA), underscoring his global impact in the field of strategic nanobiotechnology. Beyond lab-based research, he embraces translational science, where he translates lab discoveries into real-world solutions for sustainable development. He stands out as a researcher-educator bridging nanotechnology, microbiology, and sustainability, making significant contributions at both national and global levels.

Kunal Biswas is currently working as an Assistant Professor (Research) at the Centre for Nanoscience and Nanotechnology, International Research Centre at the Sathyabama Institute of Science and Technology, Chennai, India. He has more than 10 years of research experience in Nano-biosciences. He was awarded the "Outstanding Paper Awardee" in the "27th State Science and Technology Congress," 2020, supported by Govt. of West Bengal. His research interests are biotechnology, biomaterials at lower dimensions, waste valorizations, carbon nanotechnology, and food technology. He has to his credit more than 45 SCIE-indexed research articles, 2 edited books, 30 book chapters, and 2 Indian patents.

Hemen Sarma is a Professor and Head of the Department of Botany at Bodoland University, Assam, India. His research spans plant sciences, environmental pollution, nanobiotechnology, biochar innovation, plant–microbe interactions, and sustainable agriculture. He has authored more than 150 Scopus-indexed publications, edited several international books, and holds international patents. He was affiliated with The University of Texas at El Paso, USA (2017–2018). He contributed as co-author to multiple landmark The Lancet Global Burden of Disease (GBD 2023) studies. He is a Fellow of the Linnean Society of London (FLS, 2022), served as a UNESCO–TWAS Visiting Expert (2024), and is a Visiting Fellow at Hokkaido University, Japan (2026). His work addresses emerging challenges, including nanoplastic toxicity, pollutant biodegradation, green nanomaterials, and One Health–based environmental risk assessment.

Contributors

Dinesh Chand Agrawal Department of Biotechnology, School of Life and Basic Science, Jaipur National University, Jaipur, Rajasthan, India

Laith Khalil Tawfeeq Al-Ani School of Biology Science, University Sains Malaysia, Minden, Pulau Pinang, Malaysia

Makbul Alom Biodiversity and Plant Systematic Laboratory (BioPlaS_Lab), Department of Botany, University of Science and Technology Meghalaya (USTM), Baridua, Ri-Bhoi, Meghalaya, India

Amuthalinkam T. Asaithambi Department of Biotechnology, Faculty of Science and Humanities, SRM Institute of Science and Technology, Kattankulathur, Chengalpattu, Tamil Nadu, India

Sagar Barge Department of Medicine, Beth Israel Deaconess Medical Centre and Harvard Medical School, Boston, MA, USA

Arabinda Baruah Department of Chemistry, Gauhati University, Jalukbari, Guwahati, Assam, India

Mrinmoy Basak Faculty of Pharmaceutical Science, Assam down town University, Panikhaiti, Guwahati, Assam, India

Rajmoni Basumatary Department of Physics, Bodoland University, Kokrajhar, Assam, India

Himadri Tanaya Behera Imgenex India Pvt. Ltd., Bhubaneswar, Odisha, India

Nazir Ahmad Bhat Biodiversity and Plant Systematic Laboratory (BioPlaS_Lab), Department of Botany, University of Science and Technology Meghalaya (USTM), Baridua, Ri-Bhoi, Meghalaya, India

Pranaba Nanda Bhattacharyya Department of Botany, Nanda Nath Saikia College, Titabar, Jorhat, Assam, India
Institutional Biotech Hub (IBT-HUB), Nanda Nath Saikia College, Titabar, Jorhat, Assam, India

Member Coordinator of the Asia/Pacific Network, Common Microbial Biotechnology Platform (CMBP) for sustainable agriculture in Asia-Pacifi, Hanoi, Vietnam

Sudarshana Borah Royal School of Pharmacy, The Assam Royal Global University, Guwahati, Assam, India

Bhuyashi Boruah Department of Food Engineering &Technology, Central Institute of Technology Kokrajhar, Kokrajhar, BTR, Assam, India

Rangina Brahma Department of Food Engineering &Technology, Central Institute of Technology Kokrajhar, Kokrajhar, BTR, Assam, India

Saurav Chakraborty Department of Biotechnology, JIS University, Kolkata, West Bengal, India

Tuhina Das Department of Biotechnology, JIS University, Kolkata, West Bengal, India

Biswabhusan Dash Center for Biotechnology, Siksha 'O' Anusandhan (Deemed to be University), Bhubaneswar, Odisha, India

Muhammad Rehan Dastagir Department of Environmental Science, IUBAT – International University of Business Agriculture and Technology, Dhaka, Bangladesh

Monalisa Bora Deka Department of Agriculture, The Assam Royal Global University, Guwahati, Assam, India

Nilakshi Deka Department of Computer Science, The Assam Royal Global University, Guwahati, Assam, India

Ganesh Dey Department of Pharmaceutical Technology, University of North Bengal, Darjeeling, West Bengal, India

Samim Dullah Department of Zoology and DBT-IBT Hub, N.N. Saikia College, Titabor, Jorhat, Assam, India

Punniyakotti Elumalai Department of Biotechnology, Karpagam Academy of Higher Education, Coimbatore, Tamil Nadu, India
Centre for Energy and Environment, Karpagam Academy of Higher Education, Coimbatore, Tamil Nadu, India

Nikita Ghosh Department of Pharmaceutical Technology, University of North Bengal, Darjeeling, West Bengal, India

Jyotirmayee Giri Environmental Science Laboratory, School of Applied Sciences, Kalinga Institute of Industrial Technology (Deemed to be University), Bhubaneswar, Odisha, India

Krishna Giri Centre of Excellence on Sustainable Land Management, Indian Council of Forestry Research and Education, Dehradun, Uttarakhand, India

Ponkaj Gogoi Institutional Biotech Hub (IBT-HUB), Nanda Nath Saikia College, Titabar, Jorhat, Assam, India

Priyanga Govindarasu Department of Food science and Technology, Pondicherry University, Puducherry, India

Indira Gupta Department of Pharmaceutical Technology, University of North Bengal, Darjeeling, West Bengal, India

Tiyasha Gupta Department of Pharmaceutical Technology, University of North Bengal, Darjeeling, West Bengal, India

Liza Handique Department of Botany, Jagannath Barooah University, Jorhat, Assam, India

Rajeev Kumar Jha School of Health Sciences, The Neotia University, Kolkata, West Bengal, India

Salma A. Khanam Department of Chemistry, Gauhati University, Jalukbari, Guwahati, Assam, India

Arunagiri Santhosh Kumar Faculty of Science and Humanities, SRM Institute of Science and Technology, Kattankulathur, Chengalpattu, Tamil Nadu, India

Perraulta Lavanya Department of Food science and Technology, Pondicherry University, Puducherry, India

Veerapandi Loganathan Department of Food Technology, TKM Institute of Technology, Kollam, Kerala, India
Department of Food science and Technology, Hamelmalo Agricultural College, Keren, Eritrea

Santosh Mahanty Imgenex India Pvt. Ltd., Bhubaneswar, Odisha, India

Laxmidhar Mallick Environmental Science Laboratory, School of Applied Sciences, Kalinga Institute of Industrial Technology (Deemed to be University), Bhubaneswar, Odisha, India

Sanjeeb Kumar Mandal Department of Biotechnology, Chaitanya Bharathi Institute of Technology, Gandipet, Hyderabad, Telangana, India

Bishwambhar Mishra Department of Biotechnology, Chaitanya Bharathi Institute of Technology, Gandipet, Hyderabad, Telangana, India

Sanjeeb Mohapatra Department of Water Management, Delft University of Technology, Delft, The Netherlands

Shibani Mohapatra Centre for Biotechnology, Siksha 'O' Anusandhan (Deemed to be University), Bhubaneswar, Odisha, India

Abhik Mojumdar Biochemical Analysis Team/Center for Research Equipment, Korea Basic Science Institute (KBSI), Chungbuk, Republic of Korea

Alok Kumar Panda Environmental Science Laboratory, School of Applied Sciences, Kalinga Institute of Industrial Technology (Deemed to be University), Bhubaneswar, Odisha, India
Centre for Water Research and Climate Change, Kalinga Institute of Industrial Technology Deemed to Be University, Bhubaneswar, Odisha, India

Ashoutosh Panday Department of Biotechnology, Chaitanya Bharathi Institute of Technology, Gandipet, Hyderabad, Telangana, India

Punniyakotti Parthipan Faculty of Science and Humanities, SRM Institute of Science and Technology, Kattankulathur, Chengalpattu, Tamil Nadu, India

Bappi Paul School of Engineering and Technology, National Forensic Sciences University, Sector-9, Gandhinagar, Gujarat, India

Anjalu Ramchiary Department of Physics, Bodoland University, Kokrajhar, Assam, India

Subhajit Ray Department of Food Engineering &Technology, Central Institute of Technology Kokrajhar, Kokrajhar, BTR, Assam, India

Ambika Sahoo Center for Biotechnology, Siksha 'O' Anusandhan (Deemed to be University), Bhubaneswar, Odisha, India

Sambit Sahu Environmental Science Laboratory, School of Applied Sciences, Kalinga Institute of Industrial Technology (Deemed to be University), Bhubaneswar, Odisha, India

Bapi Ray Sarkar Department of Pharmaceutical Technology, University of North Bengal, Darjeeling, West Bengal, India

Bhaskar Sarma Department of Botany, Dhemaji College, Dhemaji, Assam, India

Rupak Kumar Sarma Department of Botany, Nalbari College, Nalbari, Assam, India

Aruna Selvaraj Department of Food Technology, Manakula Vinayagar Institute of Technology, Puducherry, India

Ningthoujam Samarendra Singh Division of Post-Harvest Management (PHM), ICAR-Central Institute of Sub-Tropical Horticulture (ICAR-CISH), Rehmankhera, Kakori, Lucknow, Uttar Pradesh, India

Lakshminarayanan Sivakumar Department of Biotechnology, Faculty of Science and Humanities, SRM Institute of Science and Technology, Kattankulathur, Chengalpattu, Tamil Nadu, India

Ramya Sree Department of Biotechnology, Chaitanya Bharathi Institute of Technology, Gandipet, Hyderabad, Telangana, India

Lopamudra Subudhi Centre for Biotechnology, Siksha 'O' Anusandhan (Deemed to be University), Bhubaneswar, Odisha, India

Radhika Theagarajan Department of Food Technology, Manakula Vinayagar Institute of Technology, Puducherry, India

Angila Theengh Government Pharmacy College, Sajong, Gangtok, Sikkim, India

Durairaj Thirumurugan Faculty of Science and Humanities, SRM Institute of Science and Technology, Kattankulathur, Chengalpattu, Tamil Nadu, India

Abbreviations

°C	Degree Celsius
3R	Reduce reuse recycle
AAS	Atomic absorption spectroscopy
AB	Aerobic biostabilization
AD	Anaerobic digestion
AFB_1	Aflatoxin B_1
AFM	Atomic force microscopy
AI	Artificial intelligence
Al_2O_3	Aluminum oxide
AMF	Arbuscular mycorrhizal fungi
ANN	Artificial neural network
BECCS	Bioenergy with carbon capture and storage
BET	Braunauer-Emmett-Teller
BFRPs	Basalt fiber-reinforced polymers
BiCRS	Biomass carbon removal and storage
BOD	Biochemical oxygen demand
BOF	Basic oxygen furnace
BSA	Bovine serum albumin
CAPEX	Capital on expenditure
CAT	Chloramphenicol acetyltransferase
CB	Conduction band
CCE	Circular carbon economy
CCI	Cement and concrete industry
CCU	Carbon capture and utilization
CCUS	Carbon capture, utilization, and storage
Cd^{2+}	Cadmium ion
CDR	Carbon dioxide removal
CDs	Carbon dots
CE	Circular economy
CEC	Cation exchange capacity

CFU	Colony-forming unit
CH_4	Methane
CMSs	Carbon molecular sieves
CNFs	Carbon nanofibers
CNG	Compressed natural gas
CNMs	Carbon nanomaterials
CNNs	Convolutional neural networks
CNPs	Carbon nanoparticles
CNTs	Carbon nanotubes
COD	Chemical oxygen demand
CP/MAS	Cross-polarization magic-angle spinning
CPDs	Carbonized polymer dots
CQDs	Carbon quantum dots
cRAC	Carbonated RA concrete
CVD	Catalytic chemical vapor deposition
DNA	Deoxyribonucleic acid
eCO_2R	Electrochemical CO_2 reduction
EDX	Energy-dispersive X-ray spectroscopy
EFSA	European Food Safety Authority
EIS	Electrochemical impedance spectroscopy
EPA	Environmental Protection Agency
EPI	Environmental policy integration
EPR	Extended producer responsibility
ESEM	Environmental scanning electron microscopy
eV	Electron volt
FAO	Food and Agricultural Organization
Fe	Iron
FETs	Field-effect transistors
FRET	Fluorescence resonance energy transfer
FSSAI	Food Safety and Standards Authority of India
FTIR	Fourier transform infrared spectroscopy
GC-MS	Gas chromatography–mass spectrometry
GGBFS	Ground granulated blast-furnace slag
GHG	Greenhouse gas
GO	Graphene oxide
GOBAR-DHAN	Galvanizing Organic Bio-Agro Resources Dhan
GPP	Green public procurement
GPx	Glutathione peroxidase
GQDs	Graphene quantum dots
GWP	Global warming potential
H_2	Hydrogen
Hg^{2+}	Mercury ion
HO1	Heme oxygenase-1
HPLC	Liquid chromatography

HRTEM	High-resolution TEM
HTC	Hydrothermal carbonization
HTL	Hydrothermal liquefaction
I D /I G	Intensity of the D band to the Intensity of the G band
IFE	Inner filter effect
IL-1β	Interleukin-1β
IL-6	Interleukin-6
IoT	Internet of Things
ISO	International Organization for Standardization
LCA	Life-cycle assessment
LCOE	Levelized cost of electricity
LEDs	Light-emitting devices
LFG	Landfill gas
LFGTE	Landfill gas-to-energy
LMICs	Low- and middle-income countries
LODs	Limit of detections
MB	Methylene blue
MCA	Multiple correspondence analysis
MDA	Malondialdehyde
MEA	Monoethanolamine
MHz	Megahertz
ML	Machine learning
MMMs	Mixed-matrix membranes
MO	Methyl orange
MOFs	Metal-organic frameworks
MPa	Megapascal
MRV	Monitoring, reporting, and verification
MSBW	Municipal solid biowaste
MSFT	Multi-sensor fusion technology
MSW	Municipal solid waste
MWCNTs	Multiwalled carbon nanotubes
N_2O	Nitrous oxide
NAC	Natural aggregate concrete
NFC	Near-field communication
NGOs	Nongovernment organizations
Ni	Nickel
NLBM	Non-lignocellulosic biomass
nM	Nanomolar
NMR	Nuclear magnetic resonance
Nrf2	Nuclear factor erythroid 2-related factor 2
nSOC	Native soil organic carbon
OECD	Organisation for Economic Co-operation and Development
OTA	Ochratoxin A
Pb^{2+}	Lead ion

PCSD	Policy coherence-based sustainable development
PET	Photoinduced electron transfer
PET	Polyethylene terephthalate
PGPR	Plant-growth-promoting rhizobacteria
PL	Photoluminescence
POPs	Persistent organic pollutants
ppb	Parts-per-billion
PPPs	Public-private partnerships
RBF	Radial basis function
RCA	Recycled concrete aggregates
RF	Random forest
RFA	Recycled fine aggregates
RFID	Radio frequency identification
rGO	Reduced graphene oxide
RhB	Rhodamine B
RNG	Renewable natural gas
ROS	Reactive oxygen species
SATAT	Sustainable Alternative Towards Affordable Transportation
SBRs	Sequencing batch reactors
SDG1	Sustainable Development Goal 1
SDG10	Sustainable Development Goal 10
SDG12	Sustainable Development Goal 12
SDG13	Sustainable Development Goal 10
SDG2	Sustainable Development Goal 2
SDG6	Sustainable Development Goal 6
SDG7	Sustainable Development Goal 7
SDG8	Sustainable Development Goal 8
SDGs	Sustainable Development Goals
SEM	Scanning electron microscopy
SERS	Surface enhanced Raman scattering
SiO_2	Silicon dioxide
SLG	Single-layer graphene
SOC	Soil organic carbon
SOD	Superoxide dismutase
SRFs	Slow-release fertilizers
SSbD	Safe and sustainable by design
SSF	Solid-state fermentation
SWCNTs	Single-walled carbon nanotubes
TAN	Total ammonia nitrogen
TEA	Techno-economic analysis
TEM	Transmission electron microscopy
TGA	Thermogravimetric analysis
TiO_2	Titanium dioxide
TNF α	Tumor necrosis factor-alpha

TRL	Technology readiness level
TRS	Time-resolved reflectance spectroscopy
UV	Ultraviolet
VOCs	Volatile organic compounds
W	Watt
WRA	Water retention abilities
WRCNMs	Waste-recycled carbon nanomaterials
WRL	Waste reverse logistics
XPS	X-ray photoelectron spectroscopy
XRD	X-ray diffraction

Chapter 1
An Introduction to Bioresource Waste Recycling and Circular Bioeconomy

Bhuyashi Boruah, Rangina Brahma, and Subhajit Ray

1.1 Introduction

1.1.1 Concept of Bioresources

Bioresources refer to renewable and biodegradable materials obtained from plants, animals, and microorganisms that can be used either directly or after processing for a wide range of human needs (Lakshmi et al. 2023). Unlike finite fossil-based resources, bioresources are naturally replenished and are generally more environmentally compatible, making them increasingly important in sustainable development strategies.

Their importance lies in their ability to replace conventional materials and chemicals that often harm ecosystems, reduce biodiversity, and pose risks to human health (Fritsche et al. 2020). By shifting toward bio-based alternatives, industries and communities can reduce environmental degradation, while supporting more balanced and regenerative production systems. In agriculture, the role of bioresources is particularly significant. Organic farming systems, for example, rely heavily on bio-based inputs to reduce or eliminate synthetic chemicals. This approach helps maintain soil fertility, protect beneficial microorganisms, and preserve ecological balance, ultimately promoting long-term agricultural sustainability (Ashokkumar et al. 2022).

B. Boruah · R. Brahma · S. Ray (✉)
Department of Food Engineering &Technology, Central Institute of Technology Kokrajhar, Kokrajhar, BTR, Assam, India
e-mail: subhajit@cit.ac.in

Y. K. Mohanta et al. (eds.), *Bioresource Waste to Carbon Nanomaterial*, Nanotechnology in the Life Sciences, https://doi.org/10.1007/978-981-92-2405-0_1

1.1.2 Waste Recycling

Waste recycling refers to the organized process of collecting, recovering, and transforming discarded materials into new, usable products (Chiang et al. 2024). Rather than allowing waste to accumulate in landfills or pollute natural ecosystems, recycling gives these materials a second life, helping to close resource loops within production systems.

At its core, recycling is about recognizing value in what would otherwise be thrown away. By extracting useful components from waste, societies can reduce environmental pollution, conserve raw materials, save energy, and move toward more sustainable patterns of resource use (Durak 2023). In this way, recycling not only addresses waste management challenges but also contributes to long-term environmental and economic resilience.

1.1.3 Principles of Circular Bioeconomy

A key aspect of this approach is the careful and renewable sourcing of biomass. This means harvesting biological resources in ways that respect ecological limits and allow natural systems sufficient time and capacity to regenerate (Muscat et al. 2021). By aligning resource use with nature's regenerative cycles, the circular bioeconomy seeks to balance economic development with long-term environmental stewardship.

A central principle of the circular bioeconomy is resource circularity ensuring that biological materials are continuously reused, recycled, or upcycled within closed-loop systems so that waste generation is kept to a minimum (Panahi 2025). Instead of discarding materials after a single use, this approach seeks to keep them in productive circulation for as long as possible.

Another important concept is value maximization through cascading use. Biomass is utilized in stages, where residues and by-products from one process become valuable inputs for another, leading to the production of high-value chemicals, advanced materials, biofuels, and renewable energy (Srivastav et al. 2025). This layered utilization strategy helps extract the maximum possible benefit from limited biological resources.

Eco-efficiency also lies at the heart of the circular bioeconomy. It calls for processing technologies that consume less energy, emit fewer pollutants, and operate in environmentally responsible ways (Wang et al. 2022). At the product level, materials are designed with biodegradability and renewability in mind, allowing them to safely return to natural biological cycles at the end of their life span (Duc and Dinh 2024).

Importantly, the circular bioeconomy adopts a systems-thinking perspective. It evaluates environmental, social, and economic impacts across the entire value chain from resource extraction to production, distribution, use, and recovery. Advances in biotechnology, bio-based materials, and digital technologies further enable the development of scalable and resilient bio-based industries (Holland and Shapira 2024).

Collectively, these principles aim to support climate neutrality, reduce reliance on fossil resources, revitalize rural economies, and foster a regenerative model of production and consumption that aligns economic progress with ecological sustainability (Stephenson and Damerell 2022).

1.1.4 Importance of Sustainable Resource Management

Sustainable resource management is essential for ensuring the long-term availability of natural materials, while maintaining ecological balance and economic stability (Koval et al. 2021). As global populations and consumption patterns grow, the pressure on water, energy, soil, biodiversity, forests, and mineral resources is gradually increasing (Maja and Ayano 2021). Sustainable management helps prevent depletion by promoting efficient use, conservation, and regeneration of these resources (Aiguobarueghian et al. 2024). It minimizes environmental degradation through reduced pollution, lower greenhouse gas (GHG) emissions, and protection of ecosystems and biodiversity (Kolawole and Iyiola 2023).

Economically, it supports stable supply chains, reduces costs associated with waste and inefficiency, and encourages innovation in resource-efficient technologies and circular systems; socially, sustainable resource management enhances food and water security, supports livelihoods especially in rural and resource-dependent communities, and ensures equitable access to essential resources (Fajinmolu et al. 2025). The diagrammatic representation of bioresource waste recycling and circular bioeconomy is shown in Fig. 1.1.

1.2 Types of Bioresources and Associated Wastes

Based on their biological origin and practical applications, bioresources are typically classified into major categories.

1.2.1 Agricultural Residues

Plant-based bioresources form one of the most critical categories of biological materials used in sustainable and organic agriculture. These bioresources contribute not only to soil health and crop productivity but also to pest and disease management, biodiversity conservation, and climate resilience (Priya et al. 2023). They can be broadly classified into three main groups for agricultural use: medicinal plants, crop residues, and agroforestry resources (Ingle et al. 2020).

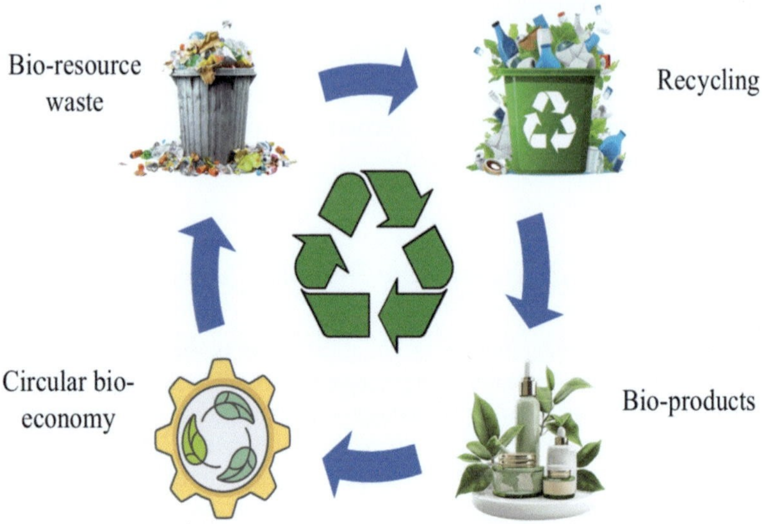

Fig. 1.1 Bioresource waste recycling and circular bioeconomy

1.2.1.1 Medicinal and Aromatic Plants

Medicinal and aromatic plants have been valued for centuries in traditional and indigenous agricultural systems, not only for their therapeutic uses but also for their role in maintaining ecological balance. Today, they are gaining renewed attention for their multifunctional benefits in organic farming systems (Mahato et al. 2025).

These plants are rich in bioactive compounds such as alkaloids, flavonoids, essential oils, phenolics, and terpenes. Such compounds naturally exhibit antimicrobial, insecticidal, fungicidal, and nematicidal properties, making them powerful allies in sustainable crop management. Because of these characteristics, medicinal and aromatic plants are increasingly used as natural sources for botanical pesticides, bio-stimulants, and plant-based growth enhancers (Mahato et al. 2025). By reducing dependence on synthetic agrochemicals, they contribute to safer, more environmentally friendly agricultural practices, while supporting long-term soil and ecosystem health.

1.2.1.2 Crop Residues and Green Manures

Crop residues such as leaves, stems, husks, and stalks that remain in the field after harvest are often viewed as agricultural waste, yet they represent a vast and largely underutilized bioresource (Andrews et al. 2021). Instead of being burned or discarded, these materials can play a significant role in restoring and maintaining soil health.

Rich in carbon and essential nutrients, crop residues contribute to nutrient cycling and help replenish organic matter in the soil. In organic farming systems, they are particularly valuable for improving moisture retention, reducing soil erosion, and stimulating beneficial microbial activity (Nair et al. 2022). By enhancing soil structure and fertility, residues support long-term productivity without the need for synthetic inputs.

Common examples of crop residues include rice straw, wheat stubble, maize husks, and legume haulms such as those from chickpea and pigeon pea crops (Gatkal et al. 2024). When managed effectively, these materials can be transformed from post-harvest leftovers into powerful tools for sustainable agriculture.

1.2.1.3 Agroforestry Resources

Agroforestry refers to the deliberate integration of trees and shrubs with crops and/or livestock to create more diverse, productive, and resilient farming systems (Chappa et al. 2024). Rather than separating forestry and agriculture, this approach combines them in ways that enhance ecological balance and farm sustainability. It is widely recognized as a key pillar of organic and sustainable agriculture because it delivers both valuable bioresources and essential ecosystem services.

Tree-based systems provide multiple forms of biomass that can be directly utilized on farms. For instance, species such as *Leucaena leucocephala* and *Gliricidia sepium* produce abundant leaf biomass that can be used for mulching, green manuring, and compost formulation. Woody residues from trees can be converted into biochar or used as renewable fuel sources. Trees also supply fodder for livestock and, in some cases, nutrient-rich inputs that indirectly benefit crops. Additionally, certain species such as *Azadirachta indica* and *Melia azedarach* are valued for their natural pest-repellent properties, supporting eco-friendly crop protection strategies (Dissanayaka et al. 2024).

Beyond these tangible products, agroforestry systems significantly enhance soil health. Leaf litter and organic residues from trees improve soil structure, stimulate beneficial soil organisms, and increase carbon sequestration. By storing carbon while boosting farm resilience and productivity, agroforestry stands out as a powerful climate-smart agricultural strategy (Singh et al. 2025a).

1.2.2 Food Processing Waste

Food processing wastes are an important category of bioresources because they are rich in valuable compounds such as fibers, proteins, starches, pectins, pigments, oils, and bioactive molecules (Sorrenti et al. 2023). These wastes can be effectively utilized for producing biofuels, biochemicals, biopolymers, enzymes, and materials like aerogels (Nguyen et al. 2022). The major types of food processing bioresource wastes include fruit and vegetable processing wastes (peels, rinds, seeds, pomace, cores, and pulp residues); cereal and grain processing wastes (bran, husk, hulls, straw, germ residues, brewer's spent grain); oilseed and nut processing wastes (press cakes, seed shells, husks, de-oiled meals); dairy industry wastes (whey, buttermilk residue, cheese whey permeate, skim milk residue); meat, poultry, and fish processing wastes (bones, skin, trimmings, blood, viscera, shells (from shrimp/crab)); sugar and starch industry wastes (molasses, bagasse, press mud, potato wastewater, rice starch water); beverage and fermentation industry wastes (spent yeast, fruit pulp residues, coffee pulp, tea waste), etc. (Yadav et al. 2020; Pandey and Dwivedi 2020; Akanbi et al. 2019; Kumar et al. 2022). These components are rich in polyphenols, proteins, pigments, and organic matter, pectin, cellulose, hemicellulose, etc. (Sorrenti et al. 2023).

Animal-based bioresources encompass livestock manure, animal compost, biodynamic preparations, bone meal, fish meal, and blood meal (Sameera et al. 2024). These bioresources are extensively utilized for nutrient cycling and soil fertility enhancement and as organic fertilizers in sustainable agricultural systems (Sameera et al. 2024).

Microbial-based bioresources include beneficial microorganisms such as bacteria (*Rhizobium*, *Azotobacter*, *Azospirillum*), fungus (*Trichoderma*, mycorrhizae), algae, and viruses, which are generally used as biofertilizers, biopesticides, biofungicides, and plant growth-promoting agents (Seenivasagan and Babalola 2021). These microorganisms contribute significantly to sustainable agriculture by enhancing nutrient availability in the soil, strengthening plant health, and improving resistance to pests and diseases (Suman et al. 2022). By supporting natural nutrient cycling and stimulating plant growth, they help reduce reliance on synthetic fertilizers, while improving overall crop resilience and productivity. Their role is especially important in organic systems, where maintaining soil biological activity is central to long-term sustainability.

However, despite the considerable promise of bioresources in sustainable farming, their widespread adoption remains limited in many regions. Technical challenges, economic constraints, limited farmer awareness, and regulatory barriers often hinder their integration into mainstream agricultural practices. Therefore, a comprehensive evaluation of existing approaches, documented benefits, practical limitations, and unresolved research gaps is essential. Such an assessment would provide clearer pathways for scaling up bioresource utilization and enabling its broader implementation within organic agriculture systems (Akanmu et al. 2023).

1.2.3 Municipal Solid Biowaste

Municipal solid biowaste (MSBW) refers to the biodegradable organic fraction of municipal solid waste generated from households, commercial areas, institutions, markets, and urban green spaces (Ram and Kumar 2021). It primarily consists of food scraps, vegetable and fruit residues, yard trimmings, garden waste, paper, cardboard, and other organic materials that can naturally decompose (Herath et al.

2024). As cities expand and consumer lifestyles change, MSBW constitutes a significant portion often 40–60% of urban waste streams, especially in developing countries (Valavanidis 2023).

Municipal solid biowaste represents a significant urban bioresource with substantial potential for valorization. Through biological processing pathways such as composting, anaerobic digestion, fermentation, and bioconversion, MSBW can be transformed into value-added products including compost, biofertilizers, biogas, bioethanol, organic acids, enzymes, and bio-based materials (Kavya et al. 2024). Beyond resource recovery, efficient MSBW management plays a critical role in mitigating methane emissions from landfills, reducing soil and water contamination, and facilitating nutrient recirculation back into agricultural systems (Gueboudji et al. 2024). In this context, sustainable MSBW utilization operationalizes circular bioeconomy principles by closing organic nutrient loops, strengthening urban–rural resource linkages, enhancing waste governance, and contributing to climate resilience (Möslinger et al. 2023).

1.2.4 Industrial Biomass Residues

Industrial biomass residues are the organic by-products and leftover materials generated from industrial activities that process agricultural, forestry, marine, or microbial biomass and hold significant potential for conversion into value-added products, energy, and bio-based materials (Kalak 2023).

These residues can be utilized to manufacture biofuels (bioethanol, biodiesel, biogas), biopolymers, biofertilizers, enzymes, organic acids, activated carbon, and advanced biomaterials such as aerogels and nanocellulose (Tran et al. 2024).

1.3 Current Challenges in Bioresource Waste Management

Technologies such as anaerobic digestion, composting, pyrolysis, and biological conversion may fail under non-optimal operating conditions (El-Sheikh et al. 2025). The absence of digital monitoring technologies limits optimization, traceability, and process performance evaluation, particularly in developing regions where waste management systems are mostly manual (Singh et al. 2024). Uncontrolled decomposition of organic waste emits methane and nitrous oxide, contributing significantly to climate change. Improper composting and landfilling exacerbate emissions (Singh et al. 2025a, b).

Improper disposal produces contaminated leachate that infiltrates soil and groundwater, risking ecosystems and human health (Kamboj et al. 2020). Open dumping, unregulated composting, and burning of waste release foul odors, particulate matter, and toxic gases affecting air quality (Bamotra et al. 2025).

1.3.1 Policy and Regulatory Gaps

Policy and regulatory gaps in bioresource waste management arise when existing frameworks fail to effectively guide, enforce, or support sustainable waste utilization (Beguedou et al. 2023). Many countries lack comprehensive and unified legislation that covers the entire value chain from waste generation and segregation to transportation, processing, and valorization (Tawo and Mbamalu 2025). As a result, responsibilities between municipal bodies, industries, and environmental agencies often overlap or remain unclear (Angouria-Tsorochidou et al. 2021). There is also inadequate enforcement of existing rules, leading to poor compliance, illegal dumping, and continued reliance on landfilling (Ntlangula et al. 2025).

A significant gap lies in the absence of standards and quality guidelines for products derived from bioresources, such as compost, digestate, biofertilizers, and bio-based materials, reducing market confidence and limiting commercialization (Mihajloska et al. 2025). Regulatory frameworks often fail to incentivize industries to adopt circular and bio-based solutions due to a lack of subsidies, tax benefits, or extended producer responsibility mechanisms for organic waste (Enriquez et al. 2020).

Additionally, insufficient monitoring and data systems hinder evidence-based policymaking, while outdated waste definitions do not account for modern biorefinery technologies (Kapoor 2025). Finally, fragmented policies on agriculture, energy, environment, and industry create inconsistencies, slowing innovation and cross-sector collaboration (Raihan 2025). Closing these policy and regulatory gaps is crucial for enabling efficient, safe, and economically viable bioresource waste utilization (Mak et al. 2020).

1.4 Recycling and Valorization Pathways

1.4.1 Biological Conversion Pathways

Composting and Vermicomposting: Organic wastes are converted into nutrient-rich compost through controlled aerobic decomposition. Products: organic fertilizers and soil conditioners (Okolie et al. 2023).

Anaerobic Digestion: Organic waste is biologically broken down without oxygen to produce biogas and digestate. Products: biogas (energy) and digestate (biofertilizer) (Faheem et al. 2021).

Fermentation and Bioconversion: Microbes convert carbohydrate-rich waste into chemicals and bio-products. Products: organic acids, enzymes, bioethanol, vitamins, pigments, and probiotics (Faheem et al. 2021).

Insect-Assisted Bioconversion: Black soldier fly larvae convert food waste into protein-rich biomass and frass. Products: animal feed and biofertilizer (Okolie et al. 2023).

1.4.2 Physicochemical Pathways

Extraction of High-Value Bioactives: Bioactive compounds are extracted from fruit pomace, peels, seeds, and shells. Products: polyphenols, flavonoids, essential oils, and pectin (Oliveira et al. 2025).

Transesterification and Chemical Upgrading: Waste oils and fats are converted into biodiesel and green chemicals (Hájek et al. 2021).

Adsorption and Bio-based Materials: Agricultural residues transformed into biosorbents and packaging materials (Carnaval et al. 2024).

1.4.3 Thermochemical Pathways

Pyrolysis and Gasification: High-temperature conversion in limited or no oxygen. Products: biochar, syngas, and bio-oil (Jahangiri et al. 2021).

Hydrothermal Carbonization and Liquefaction: Wet biomass treated under pressure. Products: hydrochar and bio-crude (Hu and Kang 2024).

1.4.4 Materials and Product Recovery Pathways

Production of Bio-based Materials: Biocomposites from fibers (Akter et al. 2022), packaging from starch and cellulose (Hassani et al. 2022), mycelium-based building materials (Khan 2025), and textiles from agricultural residues (Shafqat et al. 2023).

1.5 Technological Innovations in Waste Recycling

1.5.1 Biological and Biotechnological Innovations

Genetically engineered microbes and enzyme systems improve the degradation of complex biomass such as lignocellulose and proteins (Rasool and Irfan 2024). Tailor-made enzymatic cocktails enhance efficiency in the production of bioethanol, organic acids, and platform chemicals (Moya et al. 2024). Synthetic biology allows the design of microbial strains for specific waste streams, enhancing productivity and tolerance to inhibitors found in industrial residues (Yadav et al. 2025). The use of black soldier fly larvae is emerging as a fast, low-carbon method for converting food waste into protein-rich biomass and organic fertilizer (Tariq et al. 2025). Innovations such as microbial fuel cells enable electricity generation directly from organic matter during microbial metabolism (Ali et al. 2025).

1.5.2 Modular and Decentralized Biorefineries

Contained and modular biorefineries increase accessibility for rural and remote areas, allowing local processing and reduced transportation costs (Asghar et al. 2022). Gasification, pyrolysis, and hydrothermal liquefaction are now being designed for integration with biological systems in hybrid biorefineries (Dutta et al. 2022).

1.5.3 Digitalization and Smart Waste Management Technologies

Artificial intelligence (AI)-driven waste sorting systems use vision technology for high-accuracy separation of organics and contaminants (Lakhouit 2025). Smart bins and sensor networks facilitate real-time monitoring of waste generation, enabling efficient logistics and improved recycling rates. Blockchain improves transparency in waste-to-value chains, ensuring traceability and compliance with regulations (Gaikwad and Anerao 2025).

1.5.4 Material and Product Innovations

Waste-derived cellulose, starch, and lignin are being converted into biodegradable plastics and composites (Dejene and Geletaw 2024). Advanced pyrolysis systems generate biochar for soil carbon sequestration (Luo et al. 2023). Agricultural residues are being explored as feedstock for additive manufacturing (Carmona-Cabello et al. 2025).

1.5.5 Energy and Resource Recovery Innovations

Advanced digesters increase throughput with reduced retention time (Peces et al. 2021). Struvite precipitation and membrane filtration enable recovery of phosphorus and nitrogen (Motshekga 2025). Technological innovations in waste recycling are shown diagrammatically in Fig. 1.2.

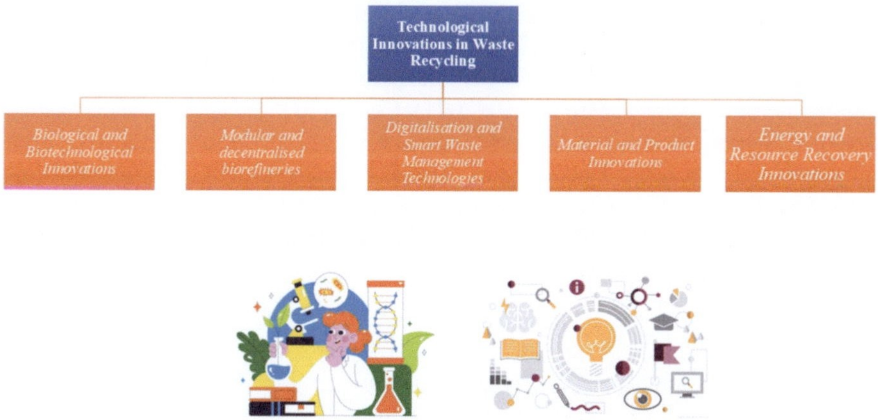

Fig. 1.2 Technological innovations in waste recycling

1.6 The Circular Bioeconomy Framework

The circular bioeconomy framework integrates biological resource management with circular economy principles to create a regenerative system that minimizes waste, maximizes resource efficiency, and supports sustainable production (Srivastav et al. 2025). It seeks to replace linear "take–make–dispose" models with biobased, closed-loop systems that renew natural capital, while generating economic and societal benefits (Has 2025). This framework emphasizes the conversion of biomass particularly wastes, residues, and renewable feedstocks into high-value materials, chemicals, and energy through environmentally responsible pathways (Has 2025). By aligning technological innovation, sustainable resource use, and policy support, the circular bioeconomy enables a transition toward low-carbon, resilient, and resource-efficient industrial ecosystems (Rasool and Irfan 2024).

1.6.1 Conceptual Models and Strategies

Conceptual models of the circular bioeconomy are grounded in the principle of maintaining biological resources within regenerative and restorative cycles. These models integrate ecological regeneration, cascading biomass utilization, and closed-loop material flows to optimize resource efficiency and minimize waste generation (Schipfer et al. 2024). Foundational frameworks such as the 3R (reduce–reuse–recycle) hierarchy, the biological cycle model proposed by the Ellen MacArthur Foundation, and cascading biomass strategies provide structured pathways for systematically recovering organic materials and converting them into higher-value products (Hsiao and Hu 2024).

Beyond resource recovery, circular bioeconomy strategies emphasize sustainable biomass sourcing, waste prevention at the design stage, technological innovation, eco-design principles, life cycle assessment, and the development of integrated bio-based value chains (Gatto and Re 2021). Central to these approaches is systems thinking, which recognizes the interdependence of agriculture, industry, urban infrastructure, and natural ecosystems. These models promote cross-sectoral integration to ensure that biological waste streams are valorized as feeds for downstream processes, enabling cascading use, nutrient recirculation, and long-term resource sustainability (Iacovidou et al. 2021).

1.6.2 Role of Bio-based Products

Bio-based products play a central role in achieving circularity by replacing fossil-derived materials with renewable, biodegradable, and lower-carbon alternatives (Alamandi 2025). These products include biofuels, bioplastics, biochemicals, natural fibers, enzymes, and emerging materials such as aerogels and nanocellulose derived from food or agricultural wastes (Ganguly et al. 2024). Their use supports circularity by reducing dependence on non-renewable resources, enhancing biodegradability, and enabling cycling of carbon within natural systems rather than adding new carbon to the atmosphere (Schirmeister and Mülhaupt 2022). Moreover, many bio-based products can be reprocessed, composted, or reintegrated into the biological cycle at end of life, reducing landfill burdens by promoting renewable feedstocks, enhancing recyclability, and enabling new sustainable industries; bio-based products serve as critical components in transforming linear supply chains into regenerative, circular systems (Tsakalova et al. 2024).

1.6.3 Industrial Symbiosis and Biorefineries

Industrial symbiosis and biorefineries constitute the technological backbone of the circular bioeconomy by operationalizing cross-sectoral resource optimization and biomass valorization. Industrial symbiosis extends beyond traditional waste exchange by integrating digital platforms, real-time material flow monitoring, and data-driven decision systems that enhance inter-firm collaboration and resource efficiency (Branca et al. 2021). Emerging models incorporate industrial metabolism mapping, AI-assisted waste matching, and blockchain-enabled traceability to improve transparency and scalability of symbiotic networks. Landmark examples such as the Kalundborg symbiosis demonstrate how industrial clusters can function as interconnected ecosystems in which waste heat, CO_2, water, and by-products are continuously exchanged, reducing emissions and enhancing economic resilience (Al-Btoush 2025).

Biorefineries, meanwhile, have evolved from conventional biomass conversion facilities into integrated, multi-product platforms that combine biochemical, thermochemical, and hybrid processing routes. Modern biorefinery concepts emphasize cascading valorization, fractionation technologies, and zero-waste design to maximize product diversity and value recovery (Pandey and Sharma 2024). Recent advancements include lignocellulosic biorefineries employing advanced pretreatment technologies (e.g., ionic liquids and deep eutectic solvents), precision fermentation for high-value biochemicals, microbial cell factories engineered through synthetic biology, and integrated carbon capture and utilization systems that convert biogenic CO_2 into fuels and chemicals (John et al. 2025). Additionally, next-generation biorefineries are increasingly linked with green hydrogen production, bioelectrochemical systems, and power-to-X technologies, enabling carbon-negative or climate-positive processing pathways.

The convergence of industrial symbiosis and advanced biorefinery systems supports regional bio-based innovation clusters, promotes distributed bio-manufacturing, and enhances supply chain resilience. By integrating renewable energy, digital monitoring tools, and circular design principles, these systems contribute not only to emission reduction and cost efficiency but also to the creation of new bio-based markets, rural industrialization opportunities, and strengthened local bioeconomic ecosystems (Lange et al. 2021).

1.6.4 Case Studies of Circular Bioeconomy in Practice

Several global case studies demonstrate the practical implementation of circular bioeconomy principles. Within the European Union, countries such as Finland and the Netherlands have advanced national bioeconomy roadmaps aligned with the European Commission Bioeconomy Strategy and the European Green Deal. Large-scale integrated biorefinery clusters such as those developed by Neste and Avantium convert forestry residues, lignocellulosic biomass, and agricultural side streams into advanced biofuels, bio-based polymers, and high-value specialty chemicals. Recently, the emergence of carbon-negative bioenergy with carbon capture and storage systems within Scandinavian bioindustrial hubs, strengthening the climate mitigation role of circular bioeconomy models, has also been seen.

In developing economies, policy-driven circular bioeconomy transitions are gaining momentum. In India, initiatives under the National Biofuel Policy (2018, amended 2022), the SATAT scheme for compressed biogas, and large-scale agro-residue valorization programs promote the conversion of crop stubble, food waste, and municipal biowaste into bio-CNG, second-generation ethanol, compost, and biofertilizers (Geetha et al. 2022). Recent reports emphasize decentralized biomethanation units and integrated rural biorefineries as mechanisms for reducing open-field burning, lowering methane emissions, and enhancing farmer income security.

Sector-specific industrial examples help to reinforce circular implementation methods. Whey, formerly considered an environmental hazard in the dairy sector, is now segregated by membrane technologies and precision fermentation into lactose derivatives, whey protein isolates, bioactive peptides, and nutraceutical components (El-Aidie and Khalifa 2024). Fruit and vegetable processing industries use green extraction technologies, supercritical CO_2 processing, and biopolymer fabrication to convert peels, pomace, and seeds into pectin, dietary fibers, essential oils, polyphenol-rich extracts, and advanced biomaterials like cellulose-based aerogels (Taneja et al. 2023). Emerging researches describe the combination of microalgae cultivation and agro-industrial effluents to produce bioactive chemicals, feed additives, and biodegradable packaging precursors.

Furthermore, urban circular bioeconomy models are gaining traction in smart cities, where organic municipal waste is integrated into anaerobic digestion systems linked with district heating networks and biofertilizer return loops. Digital monitoring tools, AI-based waste sorting systems, and blockchain-enabled biomass traceability platforms are increasingly being used as enablers of transparency and efficiency in circular value chains.

Collectively, these global case studies demonstrate that circular bioeconomy practices extend beyond waste minimization to encompass industrial symbiosis, climate-positive technologies, rural bio-industrialization, and sustainable value chain innovation. By aligning technological advancement with supportive policy frameworks and market mechanisms, these models contribute to emission reduction targets, resource security, and inclusive green growth (Stephenson and Damerell 2022).

1.7 Environmental and Socioeconomic Benefits

1.7.1 Environmental Benefits

The transition toward a circular bioeconomy offers significant environmental and socioeconomic advantages by promoting the sustainable use of biological resources, reducing pollution, and stimulating inclusive economic growth (Stephenson and Damerell 2022). Through resource-efficient technologies, waste valorization, and renewable biomass-based production, the circular bioeconomy helps address climate change, conserve ecosystems, and reduce dependence on finite fossil resources; at the same time, it creates new economic opportunities for rural and industrial communities by establishing bio-based value chains that support local livelihoods; these combined benefits make the circular bioeconomy an essential pathway toward resilient, low-carbon, and sustainable development (Ogunlade et al. 2025).

1.7.2 Reduction of Greenhouse Gas Emissions

Circular bioeconomy practices play a pivotal role in mitigating GHG emissions by diverting organic waste from landfills, enabling controlled biological degradation, and substituting fossil-derived materials with renewable bio-based alternatives (Toplicean and Datcu 2024). The conversion of biomass residues into biofuels, biogas, compost, and other bio-based products facilitates a largely closed carbon cycle, thereby limiting methane emissions that would otherwise arise from unmanaged anaerobic decomposition in landfills (Gogoi et al. 2025). Moreover, bio-based materials contribute to temporary biogenic carbon storage and often demonstrate lower life cycle emissions compared to petroleum-based counterparts. The deployment of energy-efficient and low-carbon technologies including integrated biorefinery systems, anaerobic digestion platforms, and enzyme-assisted bioprocessing further enhances carbon mitigation potential (Theodoropoulou et al. 2025).

Collectively, these interventions strengthen climate change mitigation efforts and align circular bioeconomy strategies with national decarbonization pathways and international emission reduction commitments (Liobikienė and Miceikienė 2023).

1.7.3 Soil Health and Resource Efficiency

The circular bioeconomy enhances soil health and resource efficiency by returning valuable nutrients and organic matter to agricultural systems (Srivastav et al. 2025). Compost, digestate, and biofertilizers derived from food waste, crop residues, and municipal biowaste improve soil structure, water retention, aeration, and microbial activity (Thadiyan et al. 2024). This reduces the need for synthetic fertilizers, lowers soil degradation, and enhances long-term soil fertility (Abbasi 2025). By closing nutrient loops and optimizing resource flows, the circular bioeconomy promotes sustainable agriculture and reduces environmental impacts across the production chain (Srivastav et al. 2025).

1.7.4 Job Creation and Rural Development

Bioeconomy-based industries contribute to rural development by generating employment opportunities in agriculture, biomass collection, processing, biotechnology, and manufacturing sectors (Wei et al. 2022). The establishment of decentralized biorefineries, composting plants, and biomass processing units creates value-added economic activities in rural areas, reducing migration to urban centers (Vasileiadou 2024). Additionally, small and medium enterprises involved in bio-based products such as natural fibers, bioplastics, biofertilizers, and bioenergy enable local entrepreneurship and income diversification (Ogwu et al. 2025). The circular bioeconomy strengthens farmer–industry linkages by creating markets for agricultural residues and food processing by-products, improving the economic resilience of rural communities and fostering inclusive growth (Putri 2024).

1.7.5 Contribution to Sustainable Development Goals

The circular bioeconomy supports multiple Sustainable Development Goals (SDGs) by encouraging sustainable consumption and production (SDG 12), promoting affordable and clean energy (SDG 7), and enhancing climate action (SDG 13) (United Nations 2015). Its emphasis on waste reduction, renewable resources, and low-carbon technologies contributes directly to responsible resource management and environmental protection. By creating jobs and value-added industries in rural areas, it advances goals related to poverty reduction (SDG 1), decent work (SDG 8), and reduced inequalities (SDG 10) (Food and Agriculture Organization 2018). Improvements in soil health, food system resilience, and nutrient recycling support zero hunger (SDG 2) and clean water and sanitation (SDG 6) (United Nations 2015). Overall, the circular bioeconomy provides a systemic approach that integrates ecological sustainability with socioeconomic growth, making it a key driver for achieving the SDGs (Food and Agriculture Organization 2018).

1.8 Policy, Governance, and International Perspectives

Policy and governance play a central role in advancing the circular bioeconomy by providing direction, regulatory frameworks, incentives, and collaborative platforms that support sustainable bio-based development (Lange et al. 2021). At the global level, bioeconomy strategies increasingly align with climate commitments, circular economy policies, and resource efficiency goals (Srivastav et al. 2025). Countries and international organizations are promoting integrated approaches that combine waste valorization, renewable energy, sustainable agriculture, and innovation-driven industries (Soni and Soni 2025). Effective governance requires coordinated action among governments, industries, researchers, and communities to create enabling environments for investment, capacity building, and technology deployment; however, implementation varies widely across regions due to socioeconomic disparities, regulatory limitations, and differences in technological readiness (Magliocca et al. 2024).

1.8.1 Global Policies and Strategies for Bioeconomy

Globally, the shift toward a circular bioeconomy is guided by strategic frameworks developed by international institutions and regional alliances (European Commission 2018). The European Union leads with its Bioeconomy Strategy (2018) and Circular Economy Action Plan, which emphasize sustainable biomass use, innovation, and waste reduction. OECD countries have also outlined policy roadmaps focusing on biorefineries, green chemistry, and bio-based industries (OECD 2018). The United Nations promotes circular bioeconomy principles through the SDGs, particularly those related to clean energy, sustainable production, and climate action (OECD 2018). Countries such as the United States, Canada, Japan, and Brazil have formulated national bioeconomy policies emphasizing biotechnology, renewable materials, and rural bio-industries (OECD 2018). Collectively, these global strategies highlight the need for cross-sector collaboration, responsible biomass management, and the integration of circular practices into existing economic systems.

1.8.2 National Initiatives and Case Studies

Many countries have adopted national initiatives that demonstrate practical applications of circular bioeconomy models (Marcone et al. 2022). India promotes biofuel production, organic waste biomethanation, and agricultural residue valorization through the National Biofuel Policy, Swachh Bharat Mission, and GOBAR-DHAN scheme (Meena et al. 2024). Finland's bioeconomy framework emphasizes forest-based industries and advanced biorefineries, creating one of the world's most successful circular bio-based economies (Korhonen et al. 2021). In Brazil, sugarcane bagasse and vinasse are utilized for bioethanol, biogas, and biochemicals, forming a highly integrated biomass value chain (De Almeida and Colombo 2023). The United States supports bio-based manufacturing through the BioPreferred Program, encouraging the use of renewable materials in commercial products (Daystar et al. 2021). These national case studies illustrate how targeted policy interventions, combined with strong industrial participation, can successfully accelerate the transition toward circular bioeconomy models.

1.8.3 Public–Private Partnerships

Public–private partnerships (PPPs) are essential for scaling bioeconomy initiatives by bridging gaps between research, industry, and government (Taron et al. 2023). Through shared investment, risk-sharing mechanisms, and joint innovation platforms, PPPs facilitate the development of biorefineries, waste-to-energy plants, and bio-based product manufacturing facilities (Soni and Soni 2025). Governments often support these partnerships by providing subsidies, tax incentives, or research grants, while industries contribute technical expertise, infrastructure, and market access (Zhou and Wang 2023). Public–private partnerships also play a vital role in building supply chains for biomass collection, improving waste management infrastructure, and deploying community-level bioresource technologies; successful PPP models demonstrate the importance of collaboration in transforming laboratory innovations into commercially viable, sustainable solutions (Ndlovu 2023).

1.8.4 Challenges in Global Implementation

Despite growing interest, global implementation of the circular bioeconomy faces considerable challenges; differences in economic capacity, technological development, policy stability, and institutional coordination lead to uneven progress across regions (Nijkamp et al. 2024). Many developing countries lack advanced waste processing infrastructure, biomass supply chain logistics, and financial incentives needed to promote bio-based industries (Mihajloska et al. 2025). Regulatory fragmentation and inconsistent standards for bio-based products hinder international trade and market expansion; competition for biomass between food, fuel, and industrial uses can create socioeconomic tensions if not properly managed; additionally, limited public awareness and inadequate education on bio-based solutions slow adoption at community and industry levels (Dace et al. 2024). Overcoming these challenges requires harmonized policies, inclusive governance, capacity building, and stronger global cooperation (Siddiqui 2024).

1.9 Future Perspectives and Research Directions

The future of the circular bioeconomy lies in integrating advanced technologies, digital tools, and sustainable practices that enable more efficient and resilient resource utilization (Srivastav et al. 2025). As global demand for renewable materials and climate-friendly solutions increases, research must focus on developing cost-effective bioprocesses, improving biomass conversion efficiency, and creating scalable value chains; emerging scientific innovations, combined with supportive policy frameworks and cross-sector collaboration, will be essential to overcoming existing barriers and accelerating the transition toward a regenerative bio-based economy (Ali and Russell 2025). These future directions highlight the potential for transformative change in how societies manage biological resources, reduce waste, and create sustainable economic opportunities.

1.9.1 Emerging Technologies

Emerging technologies are poised to significantly strengthen the circular bioeconomy by enhancing biomass valorization and enabling the development of novel bio-based products (Holland and Shapira 2024). Innovations such as advanced biorefineries, synthetic biology, precision fermentation, and enzyme engineering are improving the efficiency of converting complex biomass into fuels, chemicals, and high-value materials (Asghar et al. 2022). New materials like nanocellulose, bio-based aerogels, biodegradable polymers, and biochar offer sustainable alternatives to petroleum-based products (Tran et al. 2024). Technologies including pulverization-free extraction methods, pulsed electric field processing, microwave-assisted hydrolysis, and supercritical fluid extraction provide more energy-efficient routes to recover valuable compounds from organic waste (Roselli et al. 2024). Continued advancements in these technologies will help lower costs, increase product consistency, and expand commercial viability (Asghar et al. 2022).

1.9.2 Integration with Artificial Intelligence and IoT

Artificial intelligence and the Internet of Things (IoT) are critical enablers for modernizing bioresource management and circular systems; AI-driven models can optimize waste collection routes, predict biomass availability, and enhance bioprocess efficiency through real-time data analytics (Campana et al. 2025). Machine learning enables improved fermentation control, product quality prediction, and metabolic engineering design, while digital twins can simulate biorefinery performance before implementation (Campana et al. 2025; Tsui et al. 2023). IoT sensors support automated monitoring of waste generation, soil health, composting processes, and industrial symbiosis networks (Senadheera et al. 2024). When combined, AI and IoT create intelligent, data-driven systems that improve resource efficiency, reduce operational costs, and increase the scalability of circular bioeconomy solutions across urban and rural settings (Mishra and Pandey 2026).

1.9.3 Prospects of Circular Bioeconomy in Developing Countries

Developing countries have significant potential to benefit from circular bioeconomy models due to their large biomass availability, agricultural base, and expanding urban organic waste streams; the circular bioeconomy can promote sustainable rural development by generating new income sources, expanding local bio-based industries, and reducing reliance on imported fossil-derived products (Panchal and Singh 2025). However, realizing this potential requires overcoming challenges related to infrastructure, technology access, financing, and policy implementation; strengthening local innovation ecosystems, enhancing skills and capacity building, and fostering small-scale biorefineries can help bridge these gaps (Soni and Soni 2025). With appropriate support, developing countries can become global leaders in low-cost, resource-efficient, and socially inclusive circular bioeconomy practices (Ogwu et al. 2025).

1.9.4 Roadmap for a Sustainable Future

A sustainable future for the circular bioeconomy requires a comprehensive roadmap that integrates technological innovation, strong governance, and societal engagement (Brandão and Santos 2024). Key steps include scaling up green technologies, establishing standardized quality frameworks for bio-based products, and improving infrastructure for waste segregation, collection, and processing (Lange et al. 2021). Policies should emphasize incentives for renewable materials, carbon-neutral production, and industrial symbiosis (Wang et al. 2023). Collaboration between academia, industry, and government is essential to accelerate commercialization and ensure long-term investment (O'Dwyer et al. 2023). Public awareness campaigns and educational initiatives can encourage community participation and responsible consumption; ultimately, a sustainable roadmap must balance environmental protection, economic growth, and social well-being, guiding nations toward a resilient and regenerative bio-based future (Lange et al. 2021).

1.10 Conclusion

The conversion of biowaste into value-added resources represents a transformative strategy for addressing global food security while advancing environmental sustainability. Transitioning from a conventional linear model of production and disposal toward a circular bioeconomy framework ensures that food systems are designed not only for efficient production and consumption but also for resource recovery, regeneration, and reintegration. Technologies such as composting, anaerobic digestion, and upcycling illustrate how organic residues can be redirected into nutrient recycling, renewable energy generation, and novel bio-based products, thereby strengthening system resilience and reducing dependency on finite resources. By embedding circular economy principles within food systems, it becomes possible to enhance productivity, reduce food loss and waste, promote equitable resource distribution, and minimize environmental externalities such as GHG emissions, soil degradation, and water pollution. However, technological advancement alone is insufficient. The successful transition toward circular biowaste valorization requires coherent policy frameworks, targeted investments, regulatory support, and cross-sector collaboration among governments, industry stakeholders, researchers, and local communities. Strategic PPPs, economic incentives, and awareness initiatives are essential to accelerate adoption and scale sustainable practices. In essence, the pathway from food waste to food security reflects more than a technical shift; it represents a systemic transformation toward regenerative, inclusive, and climate-resilient food systems. Embracing biowaste valorization within a circular bioeconomy paradigm offers a viable and necessary route toward long-term sustainability and global food system resilience.

References

Abbasi NM (2025) Organic farming and soil health: strategies for long term agricultural sustainability. Agric Innov Sustain Ability J E-ISSN 3051-0325 1(01):25–32

Aiguobarueghian I, Adanma UM, Ogunbiyi EO, Solomon NO (2024) An overview of initiatives and best practices in resource management and sustainability. World J Adv Res Rev 22(2): 1734–1745

Akanbi TO, Dare KO, Aryee AN (2019) High-value products from cereal, nuts, fruits, and vegetables wastes. In: Byproducts from agriculture and fisheries: adding value for food, feed, pharma, and fuels, pp 347–368

Akanmu AO, Olowe OM, Phiri AT, Nirere D, Odebode AJ, Karemera Umuhoza NJ et al (2023) Bioresources in organic farming: implications for sustainable agricultural systems. Horticulturae 9(6):659

Akter M, Uddin MH, Tania IS (2022) Biocomposites based on natural fibers and polymers: a review on properties and potential applications. J Reinf Plast Compos 41(17–18):705–742

Alamandi M (2025) From fossil to function: designing next generation materials for a low carbon economy. Sustainability (2071–1050) 17(22)

Al-Btoush B (2025) The industrial Symbiosis potential in Swedish farms: exploring integrated biosystems as sector-specific industrial Symbiosis-oriented sustainable business models in agriculture

Ali A, Russell JD (2025) Accelerating the transition to wood-based circular bioeconomy: a literature review of current state, trends, opportunities, and priorities for future research. Curr For Rep 11(1):1–20

Ali HE, Hemdan BA, El-Naggar ME, El-Liethy MA, Jadhav DA, El-Hendawy HH et al (2025) Harnessing the power of microbial fuel cells as pioneering green technology: advancing sustainable energy and wastewater treatment through innovative nanotechnology. Bioprocess Biosyst Eng 48(3):343–366

Andrews EM, Kassama S, Smith EE, Brown PH, Khalsa SDS (2021) A review of potassium-rich crop residues used as organic matter amendments in tree crop agroecosystems. Agriculture 11(7):580

Angouria-Tsorochidou E, Teigiserova DA, Thomsen M (2021) Limits to circular bioeconomy in the transition towards decentralized biowaste management systems. Resour Conserv Recycl 164:105207

Asghar A, Sairash S, Hussain N, Baqar Z, Sumrin A, Bilal M (2022) Current challenges of biomass refinery and prospects of emerging technologies for sustainable bioproducts and bioeconomy. Biofuels Bioprod Biorefin 16(6):1478–1494

Ashokkumar V, Flora G, Venkatkarthick R, SenthilKannan K, Kuppam C, Stephy GM, Kamyab H, Chen W-H, Thomas J, Ngamcharussrivichai C (2022) Advanced technologies on the sustainable approaches for conversion of organic waste to valuable bioproducts: emerging circular bioeconomy perspective. Fuel 324:124313

Bamotra S, Yadav S, Tandon A (2025) Air pollution from dumping sites in India: implications on environmental health. Water Air Soil Pollut 236(10):624

Beguedou E, Narra S, Agboka K, Kongnine DM, Afrakoma Armoo E (2023) Review of Togolese policies and institutional framework for industrial and sustainable waste management. Waste 1(3):654–671. MDPI

Branca TA, Fornai B, Colla V, Pistelli MI, Faraci EL, Cirilli F, Schröder AJ (2021) Industrial symbiosis and energy efficiency in European process industries: a review. Sustainability 13(16): 9159

Brandão AS, Santos JM (2024) Sustainability from policy to practice: assessing the impact of European research and innovation frameworks on circular bioeconomy. Sustainability 16(6): 2355

Campana P, Censi R, Tarola AM, Ruggieri R (2025) Artificial intelligence and digital twins for sustainable waste management: a bibliometric and thematic review. Appl Sci 15(11):6337

Carmona-Cabello M, Herreros JM, Molero E, Lopez-Uceda A, Pinzi S, Dorado MP, Romero PE (2025) Upcycling of agricultural residues for additive manufacturing: corn straw waste as reinforcing agent in acrylonitrile-butadiene-styrene composite matrix. Biomass Convers Biorefinery 15(5):7869–7880

Carnaval LDS, Jaiswal AK, Jaiswal S (2024) Agro-food waste valorization for sustainable bio-based packaging. J Compos Sci 8(2):41

Chappa LR, Nungula EZ, Makwinja YH, Ranjan S, Sow S, Alnemari AM et al (2024) Outlooks on major agroforestry systems. Agroforestry:21–48

Chiang PC, Ma HW, Wen L, Lin CH (2024) Reduction, reuse, recycling, recovery, redesign. In: Introduction to green science and Technology for Green Economy: principles and applications. Springer Nature Singapore, Singapore, pp 255–285

Dace E, Cascavilla A, Bianchi M, Chioatto E, Zecca E, Ladu L, Yilan G (2024) Barriers to transitioning to a circular bio-based economy: findings from an industrial perspective. Sustainable Prod Consumption 48:407–418

Daystar J, Handfield R, Golden JS, McConnell E, Pascual-Gonzalez J (2021) An economic impact analysis of the US biobased products industry. Ind Biotechnol 17(5):259–270

De Almeida MA, Colombo R (2023) Production chain of first-generation sugarcane bioethanol: characterization and value-added application of wastes. Bioenergy Res 16(2):924–939

Dejene BK, Geletaw TM (2024) Development of fully green composites utilizing thermoplastic starch and cellulosic fibers from agro-waste: a critical review. Polymer-Plastics Technol Mater 63(5):540–569

Dissanayaka DMNS, Udumann SS, Atapattu AJ (2024) Synergies between tree crops and ecosystems in tropical agroforestry. Agroforestry:49–87

Duc KN, Dinh NC (2024) Circular economy. Navigating the circular age of a sustainable digital revolution. IGI Global, Hershey, p 1

Durak H (2023) Comprehensive assessment of thermochemical processes for sustainable waste management and resource recovery. PRO 11(7):2092

Dutta N, Usman M, Luo G, Zhang S (2022) An insight into valorization of lignocellulosic biomass by optimization with the combination of hydrothermal (HT) and biological techniques: a review. Sustain Chem 3(1):35–55

El-Aidie SA, Khalifa GS (2024) Innovative applications of whey protein for sustainable dairy industry: environmental and technological perspectives—a comprehensive review. Compr Rev Food Sci Food Saf 23(2):e13319

El-Sheikh HM, Gammal YOE, Ahmad SS (2025) Optimizing food waste digestion: a review of parameters and pre-treatment techniques. Egypt J Pet 34(2):4

Enriquez S, Sánchez-Triana E, Guerra López MG (2020) Economic instruments and financial mechanisms for the adoption of a circular economy. In: An introduction to circular economy. Springer Singapore, Singapore, pp 455–482

European Commission (2018) A sustainable bioeconomy for Europe: strengthening the connection between economy, society and the environment

Faheem M, Hassan MA, Mehmood T, Hashim S, Ashraf MA (2021) Waste valorization: physical, chemical, and biological routes. In: Sustainable solutions for environmental pollution: waste management and value-added products, vol 1, pp 229–255
Fajinmolu AO, Sanuade AO, Akinsorotan OA, Owolabi BA (2025) Status of biological resources in the global south: baseline assessment strategies. In: Sustainable bioeconomy development in the global south: volume I status and perspectives. Springer Nature Singapore, Singapore, pp 1–36
Food and Agriculture Organization (2018) Sustainable bioeconomy guidelines
Fritsche U, Brunori G, Chiaramonti D, Galanakis C, Hellweg S, Matthews R, Panoutsou C (2020) Future transitions for the bioeconomy towards sustainable development and a climate-neutral economy—knowledge synthesis final report, vol 10. Publications Office of the European Union, Luxembourg, p 667966
Gaikwad S, Anerao K (2025) Industry 5.0 innovations for waste management and circular economy in sustainable food systems: a systematic review. In: BIO web of conferences, vol 178. EDP Sciences, p 01004
Ganguly A, Nag S, Gayen K (2024) Synthesis of cellulosic and nano-cellulosic aerogel from lignocellulosic materials for diverse sustainable applications: a review. Prep Biochem Biotechnol 54(3):419–434
Gatkal NR, Nalawade SM, Sahni RK, Walunj AA, Kadam PB, Bhanage GB, Datta R (2024) Present trends, sustainable strategies and energy potentials of crop residue management in India: a review. Heliyon 10(21):e39815
Gatto F, Re I (2021) Circular bioeconomy business models to overcome the valley of death. A systematic statistical analysis of studies and projects in emerging bio-based technologies and trends linked to the SME instrument support. Sustainability 13(4):1899
Geetha TK, Ramesh D, Iniyakumar M, Rakesh S, Shivakumar KM, Karthikeyan S (2022) Integrated waste biorefinery for biofuels and biochemicals. In: Microbial biotechnology for renewable and sustainable energy. Springer Nature Singapore, Singapore, pp 1–34
Gogoi R, Borah MP, Daimari FS, Debbarma P, Rath A, Bora SS (2025) Bioenergy and circular economy: closing the loop. In: Bioenergy: sustainable solutions for a greener planet. American Chemical Society, pp 131–165
Gueboudji Z, Mahmoudi M, Kadi K, Nagaz K (2024) Characteristics and impacts of municipal solid waste (MSW): a review. In: Technical landfills and waste management: volume 1: landfill impacts, characterization and valorisation, pp 115–134
Hájek M, Vávra A, de Paz Carmona H, Kocík J (2021) The catalysed transformation of vegetable oils or animal fats to biofuels and bio-lubricants: a review. Catalysts 11(9):1118
Has M (2025) Bio-based resources: rethinking sustainable supply: social metabolism, resource scarcity, raw material substitution, circular economy. Walter de Gruyter GmbH & Co KG
Hassani FZSA, Salim MH, Kassab Z, Sehaqui H, Ablouh EH, Bouhfid R et al (2022) Crosslinked starch-coated cellulosic papers as alternative food-packaging materials. RSC Adv 12(14):8536–8546
Herath HMTC, Udayanga WDC, Dunusinghe DNL, Vidusanka KD (2024) Sustainable solid waste management in developing countries: converting urban wood, garden, food, and cardboard waste into biochar using pyrolysis. J Desk Res Rev Anal 2(1):92
Holland C, Shapira P (2024) Building the bioeconomy: a targeted assessment approach to identifying biobased technologies, challenges and opportunities. Eng Biol 8(1):1–15
Hsiao CJ, Hu JL (2024) Biomass and circular economy: now and the future. Biomass 4(3):720–739
Hu Y, Kang K (2024) Hydrothermal liquefaction of biomass to produce bio-crude oil and hydrochar. In: Biomass to bioenergy. Woodhead Publishing, pp 127–153
Iacovidou E, Hahladakis JN, Purnell P (2021) A systems thinking approach to understanding the challenges of achieving the circular economy. Environ Sci Pollut Res 28(19):24785–24806
Ingle AP, Philippini RR, Martiniano S, Marcelino PRF, Gupta I, Prasad S, da Silva SS (2020) Bioresources and their significance: prospects and obstacles. In: Current developments in biotechnology and bioengineering. Elsevier, pp 3–40

Jahangiri H, Santos J, Hornung A, Ouadi M (2021) Thermochemical conversion of biomass and upgrading of bio-products to produce fuels and chemicals. In: Catalysis for clean energy and environmental sustainability: biomass conversion and green chemistry, vol 1, pp 1–47

John SU, Onu CE, Ezechukwu CMJ, Nwokedi IC, Onyenanu CN (2025) Multi-product biorefineries for biofuels and value-added products: advances and future perspectives. Acad Green Energy 2(1)

Kalak T (2023) Potential use of industrial biomass waste as a sustainable energy source in the future. Energies 16(4):1783

Kamboj N, Bisht A, Kamboj V, Bisht A (2020) Leachate disposal induced groundwater pollution: a threat to drinking water scarcity and its management. In: Advances in environmental pollution management: wastewater impacts and treatment technologies, vol 1, pp 54–76

Kapoor K (2025) Regulatory and policy frameworks. In: Algal bioengineering and microbial synergy to green remediation. Springer Nature Singapore, Singapore, pp 315–338

Kavya, Vashisht M, Jain B, Shrivastava S (2024) Transforming waste into wealth: a review on microbial conversion of organic municipal wastes to value-added products. Discov Environ 2(1):112

Khan R (2025) Mycelium-based bioproducts: a novel material for a sustainable economy-a comprehensive review. Microbe 8:100439

Kolawole AS, Iyiola AO (2023) Environmental pollution: threats, impact on biodiversity, and protection strategies. In: Sustainable utilization and conservation of Africa's biological resources and environment. Springer Nature Singapore, Singapore, pp 377–409

Korhonen J, Miettinen J, Kylkilahti E, Tuppura A, Autio M, Lähtinen K et al (2021) Development of a forest-based bioeconomy in Finland: insights on three value networks through expert views. J Clean Prod 299:126867

Koval V, Mikhno I, Udovychenko I, Gordiichuk Y, Kalina I (2021) Sustainable natural resource management to ensure strategic environmental development

Kumar V, Yadav SK, Patel AK, Mishra BB, Ahluwalia V, Thakur LK, Kumar J (2022) Bioprocessing of agri-food processing residues into nutraceuticals and bioproducts. In: Current developments in biotechnology and bioengineering. Elsevier, pp 301–322

Lakhouit A (2025) Revolutionizing urban solid waste management with AI and IoT: a review of smart solutions for waste collection, sorting, and recycling. Results Eng 25:104018

Lakshmi M, Lekshmi S, Castillo-Godina R, Campos-Muzquiz L, Rodriguez-Herrera R, Biju S, Krishnan R, Swapna T (2023) Bioresources and diversity. In: Conservation and sustainable utilization of bioresources. Springer, Berlin/Heidelberg, pp 1–37

Lange L, Connor KO, Arason S, Bundgård-Jørgensen U, Canalis A, Carrez D et al (2021) Developing a sustainable and circular bio-based economy in EU: by partnering across sectors, upscaling and using new knowledge faster, and for the benefit of climate, environment & biodiversity, and people & business. Front Bioeng Biotechnol 8:619066

Liobikienė G, Miceikienė A (2023) Contribution of the European bioeconomy strategy to the green deal policy: challenges and opportunities in implementing these policies. Sustainability 15(9): 7139

Luo L, Wang J, Lv J, Liu Z, Sun T, Yang Y, Zhu YG (2023) Carbon sequestration strategies in soil using biochar: advances, challenges, and opportunities. Environ Sci Technol 57(31): 11357–11372

Magliocca P, Faggioni F, Muto V, Caputo F (2024) Technology readiness and digital gap for depicting socio-economic dynamics in society 5.0: a meso-level observation. J Technol Transf:1–17

Mahato D, Mahto H, Kumari S (2025) Medicinal and aromatic plant cultivation and sustainable development. In: Industrial crops improvement: biotechnological approaches for sustainable agricultural development. Springer Nature Switzerland, Cham, pp 135–153

Maja MM, Ayano SF (2021) The impact of population growth on natural resources and farmers' capacity to adapt to climate change in low-income countries. Earth Syst Environ 5(2):271–283

Mak TM, Xiong X, Tsang DC, Yu IK, Poon CS (2020) Sustainable food waste management towards circular bioeconomy: policy review, limitations and opportunities. Bioresour Technol 297:122497

Marcone RD, Schmid M, Meylan G (2022) Closing the gap between EU-wide national bioeconomy monitoring frameworks and urban circular bioeconomy development. J Clean Prod 379:134563

Meena PK, Pal A, Gautam S (2024) Zone-wise biogas potential in India: fundamentals, challenges, and policy considerations. Environ Sci Pollut Res 31(2):1841–1862

Mihajloska E, Zdraveva P, Gjorgievski V, Markovska N, Abazi A (2025) Unlocking bio-based potential: assessing biomass and nutrient availability for sustainable development. J Sustain Dev Energy Water Environ Syst 13(2):1–17

Mishra R, Pandey DK (2026) Application of artificial intelligence in the development of circular bioeconomy. In: Applications of artificial intelligence in removal of emerging contaminants. Elsevier, pp 221–245

Möslinger M, Ulpiani G, Vetters N (2023) Circular economy and waste management to empower a climate-neutral urban future. J Clean Prod 421:138454

Motshekga SC (2025) Nutrient recovery from wastewater using bioelectrochemical systems: a review on nitrogen and phosphorus removal. J Water Process Eng 79:109031

Moya EB, Syhler B, Dragone G, Mussatto SI (2024) Tailoring a cellulolytic enzyme cocktail for efficient hydrolysis of mildly pretreated lignocellulosic biomass. Enzym Microb Technol 175: 110403

Muscat A, de Olde EM, Ripoll-Bosch R, Van Zanten HH, Metze TA, Termeer CJ et al (2021) Principles, drivers and opportunities of a circular bioeconomy. Nat Food 2(8):561–566

Nair PR, Kumar BM, Nair VD (2022) Soil organic matter (SOM) and nutrient cycling. In: An introduction to agroforestry: four decades of scientific developments. Springer International Publishing, Cham, pp 383–411

Ndlovu D (2023) The role of public-private-partnerships (PPPs) in the creation of a sustainable circular economy within the City of Ekurhuleni. University of Johannesburg (South Africa)

Nguyen PT, Do NH, Goh XY, Goh CJ, Ong RH, Le PK et al (2022) Recent progresses in eco-friendly fabrication and applications of sustainable aerogels from various waste materials. Waste Biomass Valoriz 13(4):1825–1847

Nijkamp P, Ţigănaşu R, Bănică A, Pascariu GC (2024) Institutional adaptability: its relevance for enhancing resilience and smart specialization capacity of the European Union's lagging regions. Eurasian Geogr Econ 65(1):1–33

Ntlangula L, Xelelo Z, Chitongo L (2025) Evaluating enforcement strategies for curbing illegal waste dumping: a case study of King Sabata Dalindyebo local municipality, South Africa. IRASD J Manag 7:41–57

O'Dwyer M, Filieri R, O'Malley L (2023) Establishing successful university–industry collaborations: barriers and enablers deconstructed. J Technol Transf 48(3):900–931

OECD (2018) Meeting policy challenges for a sustainable bioeconomy

Ogunlade CA, Olaniyan AR, Babalola RT, Oyefeso BO, Jaiyeoba KF (2025) Processing techniques for bio-based products in the global south. In: Sustainable bioeconomy development in the global south: volume II bioeconomy techniques. Springer Nature Singapore, Singapore, pp 39–65

Ogwu MC, Kosoe EA, Osawaru ME (2025) Raw materials for sustainable bioeconomy development in the global south. In: Sustainable bioeconomy development in the global south: volume I status and perspectives. Springer Nature Singapore, Singapore, pp 37–63

Okolie JA, Jimoh T, Akande O, Okoye PU, Ogbaga CC, Adeleke AA et al (2023) Pathways for the valorization of animal and human waste to biofuels, sustainable materials, and value-added chemicals. Environments 10(3):46

Oliveira MRD, Cantorani JRH, Pilatti LA (2025) Sustainable extraction of bioactive compounds from food processing by-products: strategies and circular economy insights. PRO 13(11):3611

Panahi O (2025) Innovative biomaterials for sustainable medical implants: a circular economy approach. Europ J Innov Stud Sustainab 1(2):20–29

Panchal K, Singh V (2025) An introduction to sustainable circular bioeconomy. In: Sustainable waste management towards circular bioeconomy: components, design innovation and impact. Springer Nature Singapore, Singapore, pp 1–15

Pandey S, Dwivedi N (2020) Utilisation and management of agriculture and food processing waste. In: Innovations in food technology: current perspectives and future goals. Springer Singapore, Singapore, pp 269–288

Pandey A, Sharma YC (2024) Advancements in biomass valorization in integrated biorefinery systems. Biofuels Bioprod Biorefin 18(6):2078–2090

Peces M, Astals S, Jensen PD, Clarke WP (2021) Transition of microbial communities and degradation pathways in anaerobic digestion at decreasing retention time. New Biotechnol 60: 52–61

Priya AK, Alagumalai A, Balaji D, Song H (2023) Bio-based agricultural products: a sustainable alternative to agrochemicals for promoting a circular economy. RSC Sustainab 1(4):746–762

Putri NS (2024) Circular economy on agricultural product: assessing food security resilience and sustainable growth. Sukuk Int Jo Banking Finance Manag Business 3(I):16–28

Raihan A (2025) Frontiers in public policy: integrating sustainability, governance, and innovation for a changing world. Frontiers 13:14

Ram C, Kumar A (2021) 2 municipal solid waste. In: Nanobiotechnology for green environment, p 37

Rasool G, Irfan M (2024) The role of microbial diversity in lignocellulosic biomass degradation: a biotechnological perspective. ChemBioEng Rev 11(3):613–635

Roselli V, Pugliese G, Leuci R, Brunetti L, Gambacorta L, Tufarelli V, Piemontese L (2024) Green methods to recover bioactive compounds from food industry waste: a sustainable practice from the perspective of the circular economy. Molecules 29(11):2682

Sameera VS, Rao AR, Srividya R, Boppena K, Rani RP (2024) From farm to fuel: animal-based bioenergy and bioproducts. In: From waste to wealth. Springer Nature Singapore, Singapore, pp 247–279

Schipfer F, Burli P, Fritsche U, Hennig C, Stricker F, Wirth M et al (2024) The circular bioeconomy: a driver for system integration. Energy Sustain Soc 14(1):34

Schirmeister CG, Mülhaupt R (2022) Closing the carbon loop in the circular plastics economy. Macromol Rapid Commun 43(13):2200247

Seenivasagan R, Babalola OO (2021) Utilization of microbial consortia as biofertilizers and biopesticides for the production of feasible agricultural product. Biology 10(11):1111

Senadheera UE, Jayasanka J, Udayanga D, Hewawasam C, Amila B, Takimoto Y et al (2024) Beyond composting basics: a sustainable development goals—oriented strategic guidance to IoT integration for composting in modern urban ecosystems. Sustainability 16(23):10332

Shafqat AR, Hussain M, Nawab Y, Ashraf M, Ahmad S, Batool G (2023) Circularity in materials: a review on polymer composites made from agriculture and textile waste. International Journal of Polymer Science 2023(1):5872605

Siddiqui A (2024) Climate change and global governance: challenges in policy coordination. J Future Build 1(1):15–30

Singh S, Chhabra R, Arora J (2024) A systematic review of waste management solutions using machine learning, Internet of Things and blockchain technologies: state-of-art, methodologies, and challenges. Arch Computat Methods Eng 31(3):1255–1276

Singh AK, Shukla S, Dey A, Singh A, Tantuway G (2025a) Agroforestry: climate-smart agriculture technique for semi-arid climate. In: Sustainable agriculture management in semi-arid climates: volume 1. Springer Nature Switzerland, Cham, pp 51–68

Singh A, Tirpude A, Mathew N, Arfin T (2025b) Gaseous emissions from solid waste disposal. In: Waste derived carbon nanomaterials, vol 1. American Chemical Society, pp 95–114

Soni SK, Soni R (2025) Policy and regulatory frameworks for green biorefinery: driving sustainable waste valorization. In: Green biorefinery solutions: transforming biodegradable waste into resources. Springer Nature Singapore, Singapore, pp 315–349

Sorrenti V, Burò I, Consoli V, Vanella L (2023) Recent advances in health benefits of bioactive compounds from food wastes and by-products: biochemical aspects. Int J Mol Sci 24(3):2019
Srivastav AL, Bagherian A, Ghosh D (2025) The circular bioeconomy: pathways to sustainability and resource optimization. Clean Techn Environ Policy 27(2):815–832
Stephenson PJ, Damerell A (2022) Bioeconomy and circular economy approaches need to enhance the focus on biodiversity to achieve sustainability. Sustainability 14(17):10643
Suman J, Rakshit A, Ogireddy SD, Singh S, Gupta C, Chandrakala J (2022) Microbiome as a key player in sustainable agriculture and human health. Front Soil Sci 2:821589
Taneja A, Sharma R, Khetrapal S, Sharma A, Nagraik R, Venkidasamy B et al (2023) Value addition employing waste bio-materials in environmental remedies and food sector. Meta 13(5): 624
Tariq MR, Liu S, Wang F, Wang H, Mo Q, Zhuang Z et al (2025) Black soldier Fly: a keystone species for the future of sustainable waste management and nutritional resource development: a review. Insects 16(8):750
Taron A, Majumder A, Bodach S, Agbefu D (2023) Public-private partnerships for the circular bio-economy in the global south: lessons learned
Tawo OE, Mbamalu MI (2025) Advancing waste valorization techniques for sustainable industrial operations and improved environmental safety. Int J Sci Res Arch 14(02):127–149
Thadiyan V, Kaur M, Gupta R (2024) Agricultural waste for biofertilizers and soil amendments. In: Transforming agriculture residues for sustainable development: from waste to wealth. Springer Nature Switzerland, Cham, pp 183–204
Theodoropoulou A, Bagaki DA, Gaspari M, Kougias P, Treu L, Campanaro S et al (2025) Technological approaches for the capture and reuse of biogenic carbon dioxide towards sustainable anaerobic digestion. Sustainability 17(22):10385
Toplicean IM, Datcu AD (2024) An overview on bioeconomy in agricultural sector, biomass production, recycling methods, and circular economy considerations. Agriculture 14(7):1143
Tran MH, Paramasivam P, Le HC, Nguyen DT (2024) Biomass: a versatile resource for biofuel, industrial, and environmental solution. Int J Adv Sci Eng Inform Technol 14(1):268
Tsakalova M, Onorato C, Karathanasis AZ, Deligkiozi I (2024) A circular economy framework for the assessment of bio-based value chains. Sustain Chem Environ 6:100099
Tsui TH, van Loosdrecht MC, Dai Y, Tong YW (2023) Machine learning and circular bioeconomy: building new resource efficiency from diverse waste streams. Bioresour Technol 369:128445
United Nations (2015) Transforming our world: the 2030 Agenda for Sustainable Development
Valavanidis A (2023) Global municipal solid waste (MSW) in crisis. In: Two billion tonnes of MSW every year, a worrying worldwide environmental problem, vol 1, pp 1–28
Vasileiadou A (2024) From organic wastes to bioenergy, biofuels, and value-added products for urban sustainability and circular economy: a review. Urban Sci 8(3):121
Wang G, Shi R, Mi L, Hu J (2022) Agricultural eco-efficiency: challenges and progress. Sustainability 14(3):1051
Wang L, Zhang Q, Zhang G, Wang D, Liu C (2023) Can industrial symbiosis policies be effective? Evidence from the nationwide industrial symbiosis system in China. J Environ Manag 331: 117346
Wei X, Luo J, Pu A, Liu Q, Zhang L, Wu S et al (2022) From biotechnology to bioeconomy: a review of development dynamics and pathways. Sustainability 14(16):10413
Yadav SK, Kauldhar BS, Sandhu PP, Thakur K, Sucheta, Sharma TR (2020) Retrospect and prospects of secondary agriculture and bioprocessing. J Plant Biochem Biotechnol 29(1):1–14
Yadav J, Marwah H, Kumar C (2025) Synthetic biology and metabolic engineering paving the way for sustainable next-gen biofuels: a comprehensive review. Energy Adv 4(10):1209–1228
Zhou J, Wang M (2023) The role of government-industry-academia partnership in business incubation: evidence from new R&D institutions in China. Technol Soc 72:102194

Chapter 2
Biosynthesis and Characterization of Carbon Nanoscaled Materials: Top-Down and Bottom-Up Approaches

Ningthoujam Samarendra Singh

2.1 Introduction

Carbon nanoscaled materials have emerged as one of the most versatile and intensively investigated classes of nanomaterials, primarily due to their remarkable physicochemical properties, structural diversity, and exceptional application potential (Bai et al. 2024; Jesmin et al. 2026). When carbon is engineered at the nanometer scale, it exhibits phenomena that are fundamentally different from its bulk form, including quantum confinement effects, high surface-to-volume ratios, tunable electronic band structures, enhanced charge transfer, and increased chemical reactivity (John et al. 2022; Sharma and Das 2019). These size-dependent characteristics have given rise to a broad spectrum of carbon nanostructures such as graphene and its derivatives, carbon nanotubes, carbon quantum dots (CQDs), nanodiamonds, fullerenes, and amorphous carbon nanoparticles, each distinguished by unique dimensionality, crystallinity, and functional behavior (Atchudan et al. 2021). Collectively, these materials have demonstrated transformative potential across diverse sectors, including energy conversion and storage, optoelectronics, catalysis, environmental remediation, biomedical diagnostics and therapeutics, food systems, and sustainable agriculture. The growing interest in carbon nanoscaled materials has also highlighted the importance of adopting sustainable and biologically inspired synthesis strategies (Sharma and Das 2019). Conventional fabrication methods often rely on extreme temperatures, toxic solvents, and energy-intensive processes, which pose limitations related to environmental impact, scalability, and biocompatibility (Bressi et al. 2023). In response, biosynthesis and green synthesis approaches have gained considerable attention, utilizing renewable biomass, plant extracts, microorganisms, enzymes, and other bio-derived precursors to generate carbon nanomaterials under

N. S. Singh (✉)
Division of Post-Harvest Management (PHM), ICAR-Central Institute of Sub-Tropical Horticulture (ICAR-CISH), Rehmankhera, Kakori, Lucknow, Uttar Pradesh, India

Y. K. Mohanta et al. (eds.), *Bioresource Waste to Carbon Nanomaterial*, Nanotechnology in the Life Sciences, https://doi.org/10.1007/978-981-92-2405-0_2

mild and environmentally benign conditions (Chahal et al. 2021; Gedda et al. 2023). These approaches not only minimize ecological burden but also introduce inherent surface functionalities, improved aqueous dispersibility, and enhanced biological interactions, which are critical for applications involving living systems or agro-environmental interfaces.

In fact, the carbon nanoscaled materials are synthesized through two overarching methodological frameworks—top-down and bottom-up routes—each distinguished by its underlying mechanistic principles (Sohail 2025). In the top-down paradigm, bulk carbon sources are progressively reduced to nanoscale dimensions using physical or chemical processes such as laser ablation, chemical oxidation, electrochemical exfoliation, and mechanical milling (Dananjaya et al. 2024). While these methods enable effective size reduction and structural modification, they may introduce defects or require additional purification steps (Ojiako et al. 2026). In contrast, bottom-up approaches rely on the controlled assembly of carbon atoms or molecular precursors into nanostructures via processes such as hydrothermal carbonization (HTC), thermal decomposition, microwave-assisted synthesis, and bio-templated self-assembly (Cai et al. 2025). These methods typically provide enhanced control over particle size distribution, morphology, crystallinity, and surface chemistry, thereby facilitating precise modulation of functional and physicochemical properties (Jesmin et al. 2026). By integrating characterization data with synthetic parameters, researchers can systematically optimize material design, enhance reproducibility, and ensure reliability for targeted technological applications (Targhan et al. 2024). In this context, the present chapter offers an expanded and cohesive overview of the biosynthesis and characterization of carbon nanoscaled materials, with particular emphasis on top-down and bottom-up approaches, their advantages and limitations, and their role in advancing sustainable nanotechnology.

2.2 Carbon Quantum Dots: Structure and Properties

The CQD represents a significant advancement in the field of nanotechnology, acting as a class of zero-dimensional carbon-based nanomaterials that possess distinctive optical, electronic, and surface-driven features (Sen and Sarma 2026). These unique characteristics arise primarily from their ultrasmall size—typically less than 10 nm—and their intricate chemical architecture (Yang et al. 2023). Because of their restricted dimensions, CQDs experience pronounced quantum confinement effects, which allow for the precise tuning of their electronic bandgaps and optical properties through structural manipulation (Sheshmani et al. 2025). From a structural perspective, CQDs are composed of a central core containing graphitic or amorphous carbon domains, which are frequently interspersed with complex sp^2- and sp^3-hybridized carbon networks (Daood et al. 2025). This carbonaceous core provides a robust framework for electron mobility, while the periphery of the nanoparticle is heavily decorated with an array of oxygen- and nitrogen-rich functional groups. The presence of hydroxyl, carboxyl, carbonyl, and amine moieties on the surface serves a dual purpose; these groups ensure high solubility and stability in aqueous environments through electrostatic or steric hindrance, while also providing reactive sites that facilitate the conjugation of the dots with various biomolecules, polymers, or specific metal ions (Santos et al. 2025). The intrinsic performance metrics of CQDs, including their high resistance to photobleaching, remarkably low cytotoxicity, and highly tunable photoluminescence (PL), distinguish them from traditional semiconductor quantum dots. These properties, combined with an exceptional level of biocompatibility, have fuelled their expanding application across diverse scientific disciplines such as high-resolution bioimaging, molecular sensing, heterogeneous catalysis, and advanced energy storage systems (Seo et al. 2026). Furthermore, their role in sustainable agriculture is becoming increasingly prominent, as they can be engineered to enhance nutrient uptake or act as sensitive nanosensors for environmental monitoring (Sharma and Das 2019). Ultimately, the synergy between their internal carbon morphology and their external surface chemistry establishes CQDs as a versatile and indispensable platform in modern nanoscience, positioning them as a critical tool for the next generation of technological and biological innovations.

2.2.1 Size, Morphology, and Surface Chemistry

The CQDs are ultrasmall, carbon-based nanomaterials typically measuring below 10 nm in diameter (Tavan et al. 2025). Their nanoscale dimensions impart unique quantum confinement effects, leading to size-dependent optical and electronic behaviors (Akter et al. 2025). The morphology of CQDs generally ranges from nearly spherical to quasi-amorphous structures, although their appearance can vary depending on the carbon source and synthesis conditions (Afdhal et al. 2024). High-resolution imaging often reveals well-defined lattice fringes, indicating partial graphitic domains within an otherwise disordered carbon matrix (Toth 2021). Equally important is the surface chemistry of CQDs, which is characterized by abundant functional groups—including hydroxyl, carboxyl, amine, and carbonyl moieties (Elugoke et al. 2024). These surface functionalities govern water dispersibility, PL, chemical stability, and the ability to interact with biomolecules or metal ions (Litowczenko et al. 2021). Tailoring the surface through passivation or heteroatom doping further enhances their optical performance, biocompatibility, and catalytic activity (Yousaf et al. 2025).

2.2.2 Optical and Electronic Characteristics

The CQDs represent a sophisticated class of zero-dimensional carbon-based nanomaterials defined by their ultrasmall size—typically below 10 nm—and a rich chemical architecture (Jesmin et al. 2026). Structurally, these nanoparticles consist of a core composed of graphitic or amorphous carbon domains, characterized by a complex network of sp^2- and sp^3-hybridized carbon atoms (Daood et al. 2025). This carbonaceous framework is heavily decorated with an array of oxygen- and nitrogen-rich functional groups, including hydroxyl, carboxyl, carbonyl, and amine moieties (Tewari et al. 2026). These surface groups serve a dual purpose: they provide excellent colloidal stability in aqueous environments and offer versatile reactive sites for conjugation with biomolecules, polymers, or metal ions (Mu et al. 2024). The distinctive optical and electronic characteristics of CQDs emerge from the synergy between these surface states, quantum confinement effects, and the heterogeneous nature of the carbon core (Giordano et al. 2023). A hallmark of these materials is their highly tunable PL, which can be precisely modulated by manipulating the particle size, implementing surface passivation, or introducing heteroatom dopants (Bhattacharya et al. 2017). Most CQDs exhibit a characteristic excitation-dependent emission, a phenomenon that allows for the generation of a broad spectrum of fluorescence colors from a single material (Ren et al. 2024). This spectroscopic behavior is fundamentally attributed to the distribution of varying energy levels within the carbon core and the presence of multiple emissive traps created by surface defects. In terms of light absorption, CQDs demonstrate robust absorbance within the UV–visible range, featuring prominent π–π^* and n–π^* transitions that arise from the aromatic carbon domains and surface functional groups, respectively (You et al. 2024). These optical properties are complemented by exceptional electronic capabilities, including efficient charge-transfer and potent electron-donating or electron-accepting behaviors (Sead et al. 2025). The integration of sp^2 clusters within sp^3 matrix creates localized electronic states that facilitate rapid electron mobility (Vejpravov 2021). Furthermore, by doping the lattice with elements such as nitrogen, sulfur, or phosphorus, the electronic bandgap can be strategically engineered to optimize conductivity and catalytic activity (Hareesh 2024). Owing to this combination of high photostability, low toxicity, and exceptional biocompatibility, CQDs have emerged as a versatile platform in modern nanoscience. Their unique structural and electronic attributes continue to drive innovation across diverse fields, including high-resolution bioimaging, molecular sensing, heterogeneous catalysis, energy storage, and sustainable agriculture (Sharma and Das 2019). Consequently, CQDs are no longer viewed merely as laboratory curiosities but as essential tools for the next generation of technological and biological systems.

2.2.3 Functionalization and Passivation Strategies

Functionalization and passivation represent crucial approaches for tailoring the properties and performance of CQDs (Azmat and Perviaz 2024). These strategies modify the surface chemistry, enhance fluorescence efficiency, and expand the compatibility of CQDs across biological, environmental, and technological applications. Surface functionalization involves the deliberate attachment or incorporation of chemical groups, polymers, biomolecules, or heteroatoms onto the CQD surface

(Serrano-lotina et al. 2023). Through covalent bonding, electrostatic interactions, or coordination chemistry, a wide range of functional groups—such as amines, thiols, carboxyls, and aromatic moieties—can be introduced (Sim et al. 2019). This process improves solubility, biorecognition, catalytic activity, and selective interaction with target molecules. Functionalization also allows CQDs to act as carriers for drugs, nutrients, or pesticides in agricultural and biomedical systems, enabling controlled release and enhanced bioavailability. Surface passivation, on the other hand, focuses on reducing surface defects that often act as non-radiative recombination centers. By coating the CQDs with polymers (e.g., polyethylene glycol, polyethyleneimine), small organic molecules, or inorganic shells, the fluorescence emission becomes more stable and intense (Redondo-fernandez et al. 2023). Passivation effectively smooths out energy traps, boosts quantum yield (QY), and improves photostability, thereby enhancing their performance in imaging and sensing applications. Furthermore, heteroatom doping, a specialized form of functionalization, incorporates elements such as nitrogen, sulfur, phosphorus, or boron into the carbon matrix (Gao et al. 2021). This strategy fine-tunes the electronic structure, modulates bandgap energy, and strengthens optical responsiveness. Doped CQDs often show improved electron mobility, superior catalytic performance, and distinctive emission characteristics.

2.3 Synthesis Approaches for CQDs

The synthesis of CQDs has evolved into a diverse and rapidly expanding field, offering multiple routes to tailor their structure, size distribution, surface chemistry, and optical behavior (Tavan et al. 2025). Broadly, CQD synthesis can be categorized into top-down and bottom-up strategies, each with distinct advantages and controllable parameters suited for specific applications (Elugoke et al. 2024). Top-down approaches involve breaking down larger carbonaceous materials—such as graphite, carbon fibers, soot, coal, or activated carbon—into nanoscale domains (Sharma and Das 2019). Techniques such as laser ablation, arc discharge, electrochemical oxidation, and chemical exfoliation are commonly used. These methods typically generate CQDs with graphitic cores and abundant defect sites, offering high photostability and strong fluorescence. However, top-down methods may involve harsh conditions and require post-synthesis purification to obtain uniform particle sizes. Bottom-up approaches, in contrast, construct CQDs from molecular precursors through controlled carbonization, pyrolysis, or hydrothermal/solvothermal treatment (Bressi et al. 2023). Organic molecules, polymers, biomass wastes, and natural products are widely used as precursors due to their rich carbon content (Guerrero-Gonzalez et al. 2023). These methods allow precise control over particle nucleation and growth, making it easier to adjust CQD size, surface groups, and heteroatom doping. Bottom-up routes are generally cost-effective, environmentally friendly, and suitable for large-scale production (Mao et al. 2021). The use of biomass-derived resources, such as fruit peels, agricultural residues, and food waste, has further advanced the

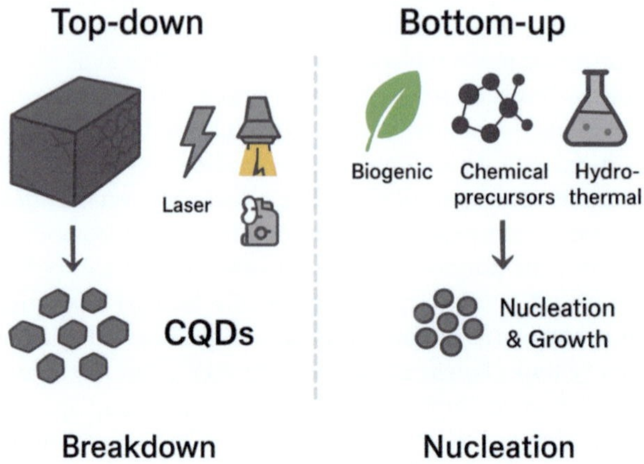

Fig. 2.1 Schematic illustration of top-down and bottom-up approaches for the synthesis of CQDs

development of sustainable "green" CQDs (Sahu et al. 2012). Modern synthesis strategies increasingly integrate passivation agents, heteroatom dopants, and surface modifiers directly into the reaction environment, enabling simultaneous formation and functionalization. These combined methods enhance QY, improve dispersibility, and create application-specific CQDs for sensing, imaging, catalysis, and agricultural uses (Fig. 2.1).

2.3.1 Top-Down Approaches

2.3.1.1 Arc Discharge and Laser Ablation

Arc discharge and laser ablation are two of the earliest and most established top-down physical methods used to produce CQDs from bulk carbon materials (Azam et al. 2021). Both techniques rely on high-energy inputs to fragment larger carbon structures into nanoscale particles with unique optical and electronic characteristics (Sim et al. 2019). Arc discharge involves generating an electric arc between graphite electrodes under controlled atmospheres such as helium, argon, or a mixture of inert gases (Iravani and Varma 2020). The intense localized heat produced by the arc (often exceeding several thousand degrees Celsius) vaporizes the carbon source (Manikandan and Lee 2022). As the vapor cools and condenses, nanosized carbon fragments, including CQDs, are formed. This method typically yields CQDs with crystalline graphitic cores, stable emission properties, and relatively narrow size distributions (Targhan et al. 2024). However, arc discharge may require extensive purification steps to remove amorphous carbon and metal impurities, and controlling particle uniformity can be challenging (Bai et al. 2024).

Laser ablation, in contrast, employs high-powered pulsed lasers to target carbon-containing solids (Azam et al. 2021). When the laser beam strikes the target material, it induces rapid heating, vaporization, and plasma formation. The subsequent cooling of the ablated carbon plume leads to the formation of CQDs (Iravani and Varma 2020). Parameters such as laser wavelength, pulse duration, power intensity, and solvent environment allow fine-tuning of particle size, surface chemistry, and emission characteristics (Sim et al. 2019). Laser ablation often produces CQDs with fewer impurities and well-defined structures compared to arc discharge, and the process allows facile incorporation of heteroatoms or functional groups by altering the surrounding medium (Guerrero-Gonzalez et al. 2023). Together, arc discharge and laser ablation provide robust physical routes to high-quality CQDs with strong photostability and graphitic features (Mehta et al. 2014). While they can be more resource-intensive than chemical or green synthesis methods, these approaches remain important for producing structurally uniform CQDs and for studying fundamental properties derived from pristine carbon frameworks (Tadesse et al. 2018).

2.3.1.2 Electrochemical Exfoliation and Chemical Oxidation

Electrochemical exfoliation and chemical oxidation are widely used top-down chemical approaches for producing CQDs by breaking down bulk carbon precursors into nanoscale fragments (Sahu et al. 2012). These methods offer greater control over surface chemistry and functional groups compared to physical techniques, making them highly attractive for generating CQDs tailored for sensing, catalysis, and biological applications (Gedda et al. 2023). Electrochemical exfoliation utilizes an external electric potential to induce the disintegration of carbon-based electrodes such as graphite rods, carbon fibers, or graphene sheets (Azam et al. 2021). When a voltage is applied in an electrolyte solution—commonly containing salts, acids, or ionic liquids—various oxidation and reduction reactions occur on the electrode surface (Rani et al. 2020). These reactions weaken the carbon lattice, promote interlayer spacing, and ultimately cleave the material into nanosized carbon domains. The resulting CQDs often possess abundant oxygen- or nitrogen-containing groups, excellent water dispersibility, and strong PL (Bressi et al. 2023). By adjusting parameters such as applied voltage, electrolyte composition, and reaction time, the size distribution and surface chemistry of the CQDs can be effectively tuned. Chemical oxidation, on the other hand, relies on strong oxidizing agents—such as nitric acid, sulfuric acid, permanganate, or hydrogen peroxide—to chemically break down carbon precursors (Chahal et al. 2021). The oxidation process introduces numerous surface defects and functional groups, which not only reduce the particle size to the quantum dot scale but also significantly enhance solubility and reactivity (Afdhal et al. 2024). Chemical oxidation methods are relatively simple and scalable, producing CQDs with high yields and rich functional surfaces (Daood et al. 2025). However, they may require careful purification to remove residual acids, by-products, and metal ions, particularly when used for biological applications.

2.3.2 Bottom-Up Approaches

Bottom-up synthesis routes construct CQDs from molecular or polymeric building blocks, allowing precise control over their formation at the atomic and nanoscale levels (Azam et al. 2021). In these methods, small carbon-rich precursors—such as citric acid, sugars, amino acids, natural extracts, or biodegradable polymers—undergo chemical transformations that gradually assemble into nanodots with well-defined structures and functional surfaces (Bian et al. 2025). This family of techniques has gained significant attention due to its simplicity, tunability, and compatibility with sustainable, low-cost materials. During bottom-up synthesis, precursor molecules typically experience dehydration, polymerization, aromatization, and carbonization, ultimately forming nucleation centers that grow into CQDs (Manikandan and Lee 2022; Sim et al. 2019). Approaches such as hydrothermal and solvothermal treatment, thermal pyrolysis, microwave irradiation, and ultrasonic-assisted reactions provide the energy needed to drive these transformations. Because reaction conditions—temperature, solvent composition, pH, precursor ratio, and reaction time—directly influence nucleation and growth, bottom-up methods offer exceptional control over particle size, crystallinity, surface groups, and PL behavior (Sharma and Das 2019). A distinctive advantage of this strategy is the ability to incorporate passivating agents or heteroatom dopants directly during synthesis. Introducing nitrogen-, sulfur-, phosphorus-, or boron-containing compounds can significantly enhance emission intensity, tune the bandgap, and improve catalytic and electronic performance (Iravani and Varma 2020). Similarly, polymers and organic ligands added in situ can smooth surface defects, leading to higher QYs and better stability under physiological or environmental conditions. Bottom-up approaches have also enabled a strong shift toward green and sustainable nanomaterial production (Manikandan and Lee 2022). Plant extracts, fruit peels, agricultural residues, and food processing wastes serve as renewable carbon sources that naturally provide functional groups beneficial for biocompatibility and targeted applications. This eco-friendly trend reduces environmental burden and strengthens the feasibility of CQD production for biomedical imaging, food safety monitoring, water purification, energy devices, and precision agriculture.

2.3.2.1 Hydrothermal Carbonization

Hydrothermal carbonization is one of the most widely adopted bottom-up techniques for synthesizing CQDs, particularly valued for its simplicity, environmental friendliness, and ability to process a broad range of organic precursors (Azam et al. 2021). In this approach, carbon-rich materials—such as sugars, amino acids, plant extracts, polymers, or biomass wastes—are dispersed in water and subjected to elevated temperatures (typically 150–250 °C) within a sealed autoclave of 3–12 h (Bressi et al. 2023). Under these conditions, the solution experiences high pressure, which promotes a series of thermochemical reactions that convert the precursor molecules into nanoscale carbonized domains. During HTC, precursor molecules undergo dehydration, condensation, polymerization, and aromatization, gradually forming nucleation centers that evolve into CQDs. The hydrothermal environment facilitates uniform heat transfer and efficient carbonization, producing dots with controlled size, high water dispersibility, and rich surface functionalities. The presence of oxygen-, nitrogen-, or sulfur-containing groups generated during the reaction enhances fluorescence behavior and provides active sites for further chemical modification (Sharma and Das 2019). One of the major strengths of HTC is its compatibility with natural and renewable feedstocks. Biomass residues—such as fruit peels, leaves, agricultural by-products, and food wastes—serve as excellent precursors due to their intrinsic carbon content and built-in functional groups (Bai et al. 2024). Using such materials not only reduces synthesis costs but also supports sustainable nanomaterial production. Reaction parameters, including temperature, pH, precursor concentration, and reaction duration, directly influence the structural and optical features of CQDs (Bressi et al. 2023). Lower temperatures often produce dots with abundant surface defects and stronger excitation-dependent emission, whereas higher temperatures tend to improve graphitization and enhance PL stability. Additional reagents introduced during the reaction, such as passivating molecules or heteroatom dopants, can further tailor the QY, emission wavelength, and catalytic properties (de Yro et al. 2019). Overall, HTC offers a versatile and eco-friendly route for generating high-quality CQDs with tunable characteristics (Chahal et al. 2021). Its scalability, mild reaction conditions, and suitability for green precursors make HTC a preferred technique for CQD production in applications spanning bioimaging, sensing, environmental remediation, and sustainable agriculture (Atchudan et al. 2021).

2.3.2.2 Microwave and Ultrasonic Treatment

Microwave and ultrasonic treatments are rapid and energy-efficient bottom-up strategies for producing CQDs, offering significant advantages in reaction speed, uniform heating, and structural control (Bai et al. 2024). These techniques use electromagnetic or acoustic energy to initiate and accelerate the carbonization of molecular precursors, making them attractive for laboratories seeking fast synthesis routes and scalable production (Azam et al. 2021). Microwave-assisted synthesis employs microwave irradiation to heat precursor solutions directly and uniformly (Iravani and Varma 2020). Unlike conventional heating—which relies on conduction and convection—microwave energy interacts with polar molecules, generating instantaneous and homogeneous thermal effects (Manikandan and Lee 2022). This leads to rapid dehydration, polymerization, and carbon nucleation, allowing CQDs to form within minutes (Bian et al. 2025). The resulting nanodots often display narrow size distributions, bright fluorescence, and abundant surface functionalities due to the controlled, high-rate heating (Ishak et al. 2025). Parameters such as microwave power, exposure time, and solvent polarity can be fine-tuned to adjust particle size, graphitization degree, and emission characteristics (Bressi et al. 2023).

Microwave methods are also compatible with green precursors, enabling the conversion of biomass and natural extracts into CQDs with minimal energy consumption (Dager et al. 2019). Ultrasonic treatment, also known as sonochemical synthesis, utilizes high-frequency sound waves to generate acoustic cavitation—the rapid formation, growth, and collapse of microbubbles in solution (Phuc et al. 2025). These collapsing bubbles create localized hotspots with extreme temperatures and pressures, producing unique reaction environments that promote carbonization and nanodot formation (Serrano-lotina et al. 2023). Ultrasonic methods enhance mixing and mass transfer, which helps generate CQDs with uniform properties and well-defined surface groups (Hamid et al. 2026). The technique is particularly effective in breaking molecular aggregates and facilitating nucleation, resulting in nanodots with high dispersibility and strong luminescent behavior (Tavan et al. 2025). Both microwave and ultrasonic approaches share common advantages—short reaction times, lower energy demand, and minimization of harsh chemicals—making them ideal for sustainable nanomaterial production. They also permit easy integration of dopants or passivating agents directly into the reaction mixture, offering additional modifiability of optical and electronic features (Afdhal et al. 2024).

2.3.2.3 Thermal Decomposition

In this method, small molecules such as citric acid, urea, polyethylene glycol, amino acids, or natural polymers are subjected to elevated temperatures—typically between 180 and 300 °C—under an inert or ambient atmosphere (Akbar et al. 2021). As the precursor is heated, it undergoes a sequence of transformations that convert it into nanoscale carbon-rich domains (Rezania et al. 2024). During the decomposition process, the precursor molecules experience melting, dehydration, polymer formation, and eventual aromatization, which create nucleation centers that gradually evolve into CQDs (Dern et al. 2025). The thermal environment promotes efficient carbonization, while simultaneously generating functional groups—such as hydroxyl, amine, and carboxyl moieties—on the CQD surface. These groups improve water solubility and enable further chemical modification or doping. One of the key strengths of thermal decomposition is its simplicity and minimal equipment requirement (Sahu et al. 2012). The reaction can be performed in conventional furnaces, hot plates, or simple heating mantles, making it accessible for both research laboratories and scalable industrial processes (Gedda et al. 2023). Additionally, the method allows rapid synthesis with high product yields, especially when using low-melting organic solids that readily undergo carbonization (Rani et al. 2020). Thermal decomposition also facilitates in situ heteroatom doping, as precursors containing nitrogen, sulfur, phosphorus, or boron naturally introduce these elements into the carbon matrix (Sharma and Das 2019). Such dopants significantly influence the electronic structure, enhancing fluorescence intensity, catalytic activity, and charge-transfer efficiency.

2.4 Green Synthesis of CQDs

Green synthesis of CQDs offers an eco-friendly alternative to conventional methods that depend on toxic chemicals and energy-intensive processes (Hamid et al. 2026). By using renewable, carbon-rich biowaste such as fruit peels, crop residues, and natural biopolymers, this approach minimizes environmental impact, while supporting circular bioeconomy goals (Gedda et al. 2023). Energy-efficient techniques like hydrothermal treatment, microwave or ultrasonic carbonization, and mild pyrolysis enable the conversion of biomass into photoluminescent CQDs, often in a single step (Deepika et al. 2023). Water is commonly used as the only solvent, further enhancing sustainability. Natural heteroatoms in biomass precursors promote self-doping, reducing the need for chemical additives. Green-derived CQDs typically possess abundant surface functional groups—such as hydroxyl, carboxyl, and amine groups—that enhance water solubility, biocompatibility, and interaction with biological systems (Chahal et al. 2021). This makes them highly suitable for applications in bioimaging, antioxidant delivery, pollutant sensing, and sustainable agriculture, including seed priming and stress-resilient crop development. Economically, green synthesis transforms waste into high-value nanomaterials, reducing disposal demands and enabling broader technological applications across agriculture, food preservation, and environmental remediation (Tadesse et al. 2018). Key challenges include batch variability, size uniformity, and scalable purification, although ongoing process optimization continues to improve production reliability. Green synthesis is a promising pathway for producing functional, cost-effective, and environmentally safe CQDs, advancing sustainable nanotechnology solutions for food systems, agriculture, health, and the environment (Guerrero-Gonzalez et al. 2023).

2.4.1 Renewable and Biomass-Derived Precursors

Renewable and biomass-derived precursors serve as the foundation for green synthesis of CQDs, offering a sustainable alternative to fossil-based materials. These precursors are sourced from naturally abundant and continuously regenerating resources such as agricultural residues, fruit and vegetable peels, forestry by-products, and food processing wastes. They are rich in essential organic components like carbohydrates, amino acids, lipids, and phenolic compounds, which readily undergo carbonization to form nanoscale carbon domains. In addition to waste materials, biopolymers such as cellulose, starch, chitosan, and gelatin are also frequently utilized. Their inherent heteroatoms—such as nitrogen, sulfur, and phosphorus—promote in situ doping of CQDs, enhancing optical and surface properties without the need for external chemical dopants. The use of these precursors aligns with waste valorization strategies by transforming low-value or discard-worthy materials into high-performance nanomaterials. Moreover, renewable feedstocks

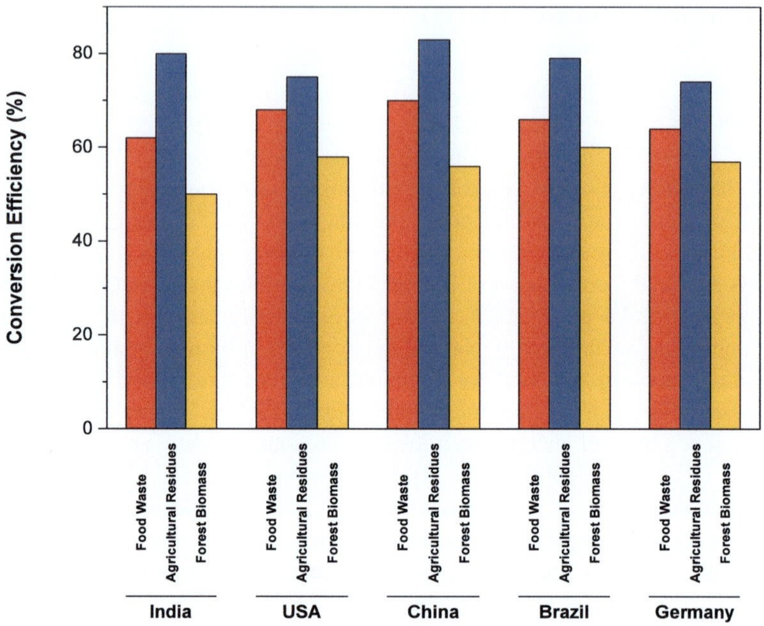

Fig. 2.2 Country-wise comparative performance analysis of carbon nanoscaled materials synthesized via different approaches

support circular bioeconomy principles, as they reduce reliance on non-renewable resources and mitigate environmental burdens associated with waste accumulation. By integrating waste recovery with nanomaterial production, biomass-derived precursors not only contribute to cost-effective synthesis but also enable environmentally responsible innovations for agriculture, food systems, medicine, and environmental remediation (Fig. 2.2).

2.4.2 Sustainable Synthesis Methods and Scalability

Sustainable synthesis methods are central to the development of environmentally responsible CQDs. Green synthesis emphasizes low-energy processing, reduced chemical hazards, and minimal waste generation, while maintaining high product quality. Techniques such as hydrothermal treatment, microwave irradiation, ultrasonic-assisted carbonization, and mild pyrolysis are widely favored because they efficiently convert biomass into CQDs under milder reaction conditions compared to conventional chemical methods. Water often serves as the primary reaction medium, eliminating the need for harsh organic solvents and enhancing the biocompatibility of the final product. These methods enable one-pot synthesis with fewer processing steps, lowering operational costs and energy consumption. Importantly, the reaction mechanisms naturally incorporate heteroatom functionalization from biomass feedstocks, reducing or eliminating external dopant requirements. Scalability remains a critical objective in translating laboratory-scale green CQD production to industrial deployment. Variability in biomass composition can influence CQD size, fluorescence, and yield, making process standardization essential. Efforts to overcome these challenges include the development of continuous-flow reactors, improved precursor pretreatment strategies, and advanced purification technologies that avoid solvent-intensive steps. Integration of automated control systems and optimization of heat and mass transfer dynamics further enhance consistency at larger scales. Adopting scalable green synthesis not only supports global sustainability goals but also offers significant economic advantages. By valorizing agricultural and food wastes into high-value nanoproducts, this approach aligns with circular economy practices and encourages industrial investment. As technological refinements progress, sustainable CQD manufacturing is poised to meet growing market demands in agriculture, environmental monitoring, food safety, and biomedicine—demonstrating that ecological responsibility and commercial feasibility can be achieved in tandem.

2.5 Characterization Techniques

Comprehensive characterization of CQDs is essential to evaluate their physicochemical properties, structural integrity, surface chemistry, and functional performance. Since the characteristics of CQDs—such as size, morphology, optical behavior, and chemical composition—directly influence their applicability in biological, agricultural, and environmental systems, multiple analytical tools are employed to ensure quality and reproducibility.

2.5.1 Structural Analysis (TEM, XRD, Raman)

Structural characterization plays a crucial role in understanding the nanoscale organization, crystallinity, and graphitic nature of CQDs. Advanced analytical techniques such as transmission electron microscopy (TEM), X-ray diffraction (XRD), and Raman spectroscopy provide complementary insights that help confirm successful synthesis and evaluate structural quality.

(i) TEM: It enables direct visualization of CQDs with high spatial resolution, allowing precise determination of particle size, shape, and distribution (Kudilatt et al. 2020). High-resolution TEM further reveals lattice fringes that indicate the presence of graphitic crystalline domains formed during carbonization (Bian et al. 2025). These lattice spacings provide evidence of quantum confinement effects, which are directly associated with the photoluminescent behavior of CQDs. Additionally, TEM can assess aggregation tendencies, which are critical for determining colloidal stability in biological and agricultural systems.

(ii) XRD: It is employed to analyze the crystalline phase and structural ordering of synthesized CQDs (Bian et al. 2025). Typically, green-derived CQDs exhibit broad diffraction peaks around $2\theta \approx 20–25°$, characteristic of partially graphitized amorphous carbon structures (Phuc et al. 2025). The presence and sharpness of these peaks help distinguish between well-organized sp^2-hybridized carbon networks and more disordered carbon arrangements. X-ray diffraction results are essential for evaluating the degree of carbonization achieved during synthesis and for optimizing process parameters to enhance optical performance.

(iii) Raman Spectroscopy: Raman spectroscopy provides deeper insight into the graphitic quality and defect structure of CQDs (Ishak et al. 2025). Two primary peaks are observed: the D band ($\approx 1350\ cm^{-1}$), related to structural defects and disorder, and the G band ($\approx 1580\ cm^{-1}$), associated with the in-plane vibration of sp^2 carbon atoms. The intensity ratio of ID/IG serves as a key indicator of defect levels and the extent of graphitization. Green-synthesized CQDs often show higher ID/IG ratios due to the presence of functional groups originating from biomass precursors (Phuc et al. 2025), which can be advantageous for enhancing surface reactivity and fluorescence properties.

Thus, TEM, XRD, and Raman provide a comprehensive structural profile—linking nanoscopic morphology, crystallinity, and defect states to functional attributes. This structural validation is essential for tailoring CQDs toward specific applications such as bioimaging, seed priming, environmental sensing, and crop stress mitigation.

2.5.2 *Optical Characterization (UV–Vis, PL, QY)*

The optical characterization is fundamental for assessing the functional performance of CQDs, particularly for applications in sensing, bioimaging, and agricultural diagnostics. These properties arise from quantum confinement effects, surface defects, and heteroatom dopants, making optical measurements essential for understanding both electronic structure and application potential. The UV–visible spectroscopy provides primary insight into the electronic transitions occurring within CQDs (Dager et al. 2019). Typically, absorption spectra display prominent features in the ultraviolet region that correspond to $\pi–\pi^*$ transitions of aromatic sp^2-hybridized carbon frameworks and $n–\pi^*$ transitions associated with surface-bound oxygen- and nitrogen-containing functional groups (Wang et al. 2023). Variations in peak position, spectral width, and absorption intensity reflect the degree of carbonization, extent of π-conjugation, and effectiveness of surface passivation (Sohail 2025). Consequently, UV–Vis analysis serves as a rapid and reliable indicator of synthesis progression, structural evolution, and overall material quality. The PL spectroscopy represents one of the most distinctive and informative analytical techniques for CQDs (Yang et al. 2023). A characteristic feature of these nanomaterials is excitation-dependent fluorescence, wherein altering the excitation wavelength produces tunable emission profiles across a broad spectral range (Wang et al. 2017). This phenomenon arises from the coexistence of multiple emissive surface states, particle size heterogeneity, and defect-mediated electron–hole recombination pathways (Castaing et al. 2024). Through PL measurements, researchers can systematically evaluate emission wavelength distribution and spectral stability, fluorescence intensity and decay lifetime, as well as responsiveness to environmental stimuli such as pH fluctuations, metal ion presence, or temperature variations (Hamid et al. 2026). Such environmentally sensitive luminescent behavior underpins the deployment of CQDs in smart sensing devices, responsive diagnostic probes, and real-time biological imaging systems.

2.5.3 Surface and Chemical Analysis (FTIR, XPS, Zeta Potential)

Understanding the surface chemistry and interfacial properties of CQDs is vital, as these features directly influence their solubility, biocompatibility, reactivity, and overall functional performance in biological and environmental systems. Techniques such as Fourier transform infrared spectroscopy (FTIR), X-ray photoelectron spectroscopy (XPS), and zeta potential measurements offer complementary information on functional groups, elemental composition, and colloidal stability.

(a) FTIR: It is employed to identify characteristic vibrational bands associated with the chemical functionalities present on the CQD surface (Azam et al. 2021). Peaks related to hydroxyl (–OH), carboxyl (–COOH), carbonyl (C=O), amine (–NH_2), and aromatic C=C stretching are typically detected in green-synthesized CQDs due to incomplete carbonization of biomass precursors. These functional groups provide hydrophilicity, facilitate bonding with biomolecules, and enable metal ion sensing, enhancing CQDs' versatility in agricultural and environmental applications.
(b) XPS: The instrument provides a detailed assessment of surface elemental composition and chemical bonding states. It detects carbon, oxygen, nitrogen, and other heteroatoms inherent in renewable feedstocks, confirming the presence of natural self-doping that improves fluorescence and electronic properties. Analysis of high-resolution C1s, N1s, and O1s spectra allows deconvolution of functional moieties, supporting structure–property correlations and validating successful heteroatom incorporation without toxic additives.
(c) Zeta Potential Analysis: Zeta potential measures the electrostatic charge on the CQD surface and reflects colloidal stability in aqueous media (Azam et al. 2021). High negative or positive zeta potential values indicate repulsive forces

sufficient to prevent agglomeration, ensuring uniform distribution in biological assays and plant systems (Serrano-lotina et al. 2023). Surface charge also influences cellular uptake, interaction with root tissues, and transport efficiency in foliar applications, making it a key parameter for agricultural nanotechnology deployment.

2.6 Applications of CQDs

The CQDs have emerged as promising nanomaterials for bioimaging and biosensing due to their exceptional fluorescence, high biocompatibility, and chemical stability. Their tunable PL enables real-time visualization of cells, tissues, and biological processes without causing significant toxicity, making them suitable alternatives to traditional semiconductor quantum dots that often contain heavy metals.

2.6.1 Bioimaging and Biosensing

In bioimaging, CQDs can penetrate cellular membranes and distribute uniformly within biological environments, providing strong and stable emission under different excitation wavelengths. This allows multicolor imaging, improved signal clarity, and compatibility with advanced imaging techniques (Azam et al. 2021). In biosensing applications, surface functional groups on CQDs facilitate selective interaction with biomolecules, ions, and metabolic indicators (Tavan et al. 2025). These interactions induce measurable fluorescence changes such as enhancement, quenching, or wavelength shifts, enabling sensitive detection of analytes including glucose, metal ions, pathogens, and biomarkers of stress or disease (Pouri et al. 2025). The rapid responsiveness and high detection accuracy support early diagnostics and environmental monitoring.

2.6.2 Catalysis and Photocatalytic Pollutant Degradation

The CQDs have emerged as highly efficient nanocatalysts owing to their remarkable electron transfer ability, tunable bandgap, and strong absorption of visible and ultraviolet light (Azmat and Perviaz 2024). These properties make CQDs excellent materials for catalytic conversions and photocatalytic degradation of environmental pollutants. In catalytic applications, CQDs can accelerate redox reactions by acting as electron shuttles or providing active surface sites for reactant adsorption (Afdhal et al. 2024). Their abundant functional groups and heteroatom doping enhance catalytic selectivity and diffusion of reactants at the nanoscale interface. For photocatalytic pollutant degradation, CQDs generate reactive oxygen species

(ROS) such as hydroxyl radicals and superoxide ions upon light irradiation (Akbar et al. 2021). These ROS effectively break down toxic organic substances including synthetic dyes, antimicrobial residues, pesticides, and emerging contaminants in water systems. The CQDs can also serve as co-catalysts by coupling with traditional semiconductors like TiO_2, ZnO, and g-C_3N_4, improving charge separation efficiency and extending light absorption into the visible region—allowing pollutant degradation under natural sunlight (Dern et al. 2025). Green-synthesized CQDs further enhance sustainability by replacing hazardous catalysts with environmentally benign materials derived from renewable waste (Saini et al. 2022). This integration of nanotechnology with ecological remediation strategies supports cleaner industrial practices and aligns with circular economy goals. The CQDs offer a promising route toward low-cost, energy-efficient, and eco-safe catalytic systems capable of addressing global challenges in water purification and environmental protection.

2.6.3 Environmental Remediation

The CQDs have emerged as highly promising nanomaterials in environmental remediation owing to their exceptional surface reactivity, robust photostability, and remarkable dispersibility in aqueous systems (Targhan et al. 2024). Their nanoscale dimensions, coupled with abundant oxygen-, nitrogen-, and sulfur-containing functional groups, provide numerous active sites that facilitate strong physicochemical interactions with diverse environmental contaminants (Sen and Sarma 2026). This multifunctional surface chemistry enables CQDs to participate in pollutant removal through mechanisms such as adsorption, fluorescence-based sensing, redox reactions, and photocatalytic degradation (Aziz et al. 2025). The affinity of CQDs toward hazardous substances extends across a wide spectrum of pollutants, including heavy metal ions (e.g., lead, cadmium, and mercury), synthetic organic dyes, pesticide residues, and emerging pharmaceutical contaminants (Rezania et al. 2024). The electrostatic attraction, surface complexation, π–π stacking interactions, and hydrogen bonding collectively contribute to their binding efficiency (Sheshmani et al. 2025). In addition to passive adsorption, CQDs can actively neutralize toxic species by facilitating electron transfer processes, thereby transforming pollutants into less harmful forms. Their intrinsic fluorescence properties further enable sensitive and selective detection of trace contaminants, allowing real-time monitoring of pollutant concentrations and supporting early warning systems for environmental protection (Cai et al. 2025). As photocatalysts, CQDs exhibit the ability to harvest visible and solar light, promoting the generation of ROS such as hydroxyl and superoxide radicals (Bressi et al. 2023). These reactive intermediates drive the oxidative breakdown of persistent organic pollutants into smaller, less toxic, or biodegradable compounds (Seo et al. 2026). When integrated into composite systems with semiconductors or metal oxides, CQDs enhance charge separation efficiency and extend light absorption ranges, thereby significantly improving photocatalytic performance under environmentally relevant conditions

(Azam et al. 2021). Importantly, green-synthesized CQDs offer additional sustainability advantages. The use of renewable biomass, agricultural residues, or food waste as carbon precursors reduces reliance on hazardous chemicals and lowers overall environmental impact during production (Bai et al. 2024). Such eco-friendly synthesis strategies align with circular economy principles and promote safer wastewater treatment and soil restoration technologies (Bai et al. 2024). Furthermore, CQDs demonstrate strong compatibility with hybrid nanostructures, enabling their incorporation into advanced filtration membranes, high-capacity adsorption matrices, catalytic coatings, and multifunctional nanocomposites (Dananjaya et al. 2024). Collectively, the versatile physicochemical characteristics of CQDs position them as eco-conscious and technologically adaptable materials for mitigating water pollution, remediating contaminated soils, and safeguarding ecosystem health. Their multifunctionality, combined with sustainable production pathways, supports the development of next-generation environmental management strategies tailored to the escalating global challenges of pollution and resource degradation.

2.7 Challenges and Future Perspectives

Although green-synthesized CQDs demonstrate considerable promise across environmental, biomedical, and agricultural applications, several technical and operational constraints must be resolved to enable their large-scale commercial adoption. The long-term viability of these nanomaterials within sustainable nanotechnology frameworks depends largely on improved synthesis precision, enhanced performance characteristics, and the development of scalable production systems. One of the primary challenges lies in achieving consistent control over particle size distribution, morphology, and crystallinity. Because biomass precursors naturally vary in biochemical composition, inconsistencies often arise in optical behavior and functional efficiency. Addressing this limitation requires careful precursor selection, standardized pretreatment and purification protocols, and strict control of reaction parameters such as temperature, pH, and residence time. The implementation of continuous-flow reactors and automated synthesis platforms is increasingly viewed as a promising pathway to enhance reproducibility and batch-to-batch uniformity. Another critical area for improvement involves QY enhancement and advanced surface engineering. While biomass-derived CQDs inherently possess diverse functional groups that confer biocompatibility and environmental responsiveness, their fluorescence efficiency may be lower than that of conventionally synthesized analogues. Strategic surface passivation, controlled heteroatom doping, and targeted post-synthetic functionalization can significantly improve radiative recombination efficiency and overall luminescence intensity. Moreover, rational tailoring of surface chemistry can optimize charge transfer behavior, selective molecular recognition, and biological targeting capabilities, thereby broadening the applicability of CQDs in imaging diagnostics, pollutant sensing, and photocatalytic systems.

Scaling up production remains another major barrier, as most green synthesis routes are still confined to laboratory settings. The challenges such as limited product yield, variability between batches, energy-intensive carbonization steps, and solvent-dependent purification processes hinder industrial translation. Overcoming these obstacles will require the adoption of continuous and automated manufacturing systems, cost-effective biomass preprocessing strategies, energy-efficient heating technologies, and environmentally benign purification approaches that minimize or eliminate organic solvents. In future, research trends are increasingly aligned with sustainability-driven innovation. Integrating CQD production into circular bioeconomy models—where agricultural or agro-industrial residues are transformed into high-value nanomaterials—can enhance resource efficiency, while reducing waste streams. The development of hybrid nanostructures, achieved by coupling CQDs with metals or semiconductor materials, offers opportunities to create multifunctional systems with synergistic optical and catalytic properties. Emerging artificial intelligence and data-driven optimization tools are also enabling predictive control over structural and photophysical attributes, accelerating material discovery and process refinement. Equally important is the establishment of comprehensive regulatory and safety frameworks to ensure environmental compatibility and biological safety, particularly for agricultural and medical applications. In parallel, carbon-negative production concepts that sequester atmospheric carbon during CQD synthesis are being explored to align nanomaterial fabrication with climate mitigation strategies.

2.8 Conclusion

The CQDs have emerged as versatile nanomaterials with remarkable optical, electronic, and catalytic characteristics, enabling their application across fields as diverse as bioimaging, sensing, energy conversion, and environmental remediation. Their fabrication generally follows either top-down strategies, which reduce bulk carbon sources into nanoscale domains, or bottom-up approaches, which construct quantum dots from molecular or polymeric precursors. In recent years, increasing attention has been directed toward green synthesis pathways that utilize biomass-derived feedstocks, underscoring the feasibility of scalable, environmentally benign, and economically viable production systems. Among the available techniques, HTC stands out for its operational simplicity, moderate energy requirements, and ability to generate functionalized CQDs with desirable physicochemical attributes. Nevertheless, persistent challenges—including control over particle size uniformity, enhancement of QY, and optimization of energy efficiency—continue to limit full industrial translation. Addressing these constraints through refined reaction engineering, targeted surface modification, and integrated sustainable processing frameworks will be critical for advancing CQD technology. Collectively, such progress is expected to expedite the incorporation of CQDs into next-generation technological platforms, strengthening innovation in nanoscience, while contributing to environmentally responsible development. Beyond technical refinement, the broader impact of CQDs will depend on interdisciplinary collaboration that bridges materials science, environmental engineering, biotechnology, and data-driven optimization. Integrating life-cycle assessment, techno-economic analysis, and regulatory evaluation into research pipelines can ensure that large-scale implementation remains both sustainable and socially responsible. As global priorities increasingly emphasize green innovation and resource efficiency, CQDs are well positioned to serve as a model nanomaterial that aligns high-performance functionality with ecological stewardship, reinforcing their role in shaping a resilient and sustainable technological future.

References

Afdhal AM, Fazlur A, Piliang R, Goei R, Risky Ramadhan HTS, Gea S (2024) Synthesis, properties, and utilization of carbon quantum dots as photocatalysts on degradation of organic dyes: a mini review. Catal Commun 187:106914. https://doi.org/10.1016/j.catcom.2024.106914

Akbar K, Moretti E, Vomiero A (2021) Carbon dots for photocatalytic degradation of aqueous pollutants: recent advancements. Adv Opt Mater:2100532. https://doi.org/10.1002/adom.202100532

Akter M, Hossain I, Tuz F (2025) Quantum dot nanomaterials: empowering advances in optoelectronic devices. Chem Eng J Adv 21:100704. https://doi.org/10.1016/j.ceja.2025.100704

Atchudan R, Jebakumar Immanuel Edison TN, Shanmugam M, Perumal S, Somanathan T, Lee YR (2021) Sustainable synthesis of carbon quantum dots from banana peel waste using hydrothermal process for in vivo bioimaging. Phys E Low-Dimens Syst Nanostruct 126:114417. https://doi.org/10.1016/j.physe.2020.114417

Azam N, Ali MN, Khan TJ (2021) Carbon quantum dots for biomedical applications: review and analysis. Front Mater 8:1–21. https://doi.org/10.3389/fmats.2021.700403

Aziz DM, Hassan SA, Aziz SB, Kader DA (2025) Efficient adsorption and photocatalytic degradation of methylene blue using HKUST-1: a novel approach for dye removal and wastewater treatment under sunlight. Next Mater 9:101015. https://doi.org/10.1016/j.nxmate.2025.101015

Azmat E, Perviaz M (2024) Degradation efficiency of organic dyes on CQDs as photocatalysts: a review. ACS Omega. https://doi.org/10.1021/acsomega.3c09547

Bai S, Lv T, Chen M, Li C, Wang Z, Yang X, Xia T (2024) Carbon quantum dots assisted BiFeO3@BiOBr S-scheme heterojunction enhanced peroxymonosulfate activation for the photocatalytic degradation of imidacloprid under visible light: performance, mechanism and biotoxicity. Sci Total Environ 915:170029. https://doi.org/10.1016/j.scitotenv.2024.170029

Bhattacharya D, Mishra MK, De G (2017) Carbon dots from a single source exhibiting tunable luminescent colors through the modification of surface functional groups in ORMOSIL films. J Phys Chem C. https://doi.org/10.1021/acs.jpcc.7b08039

Bian Z, Gomez E, Gruebele M, Levine BG, Link S, Mehmood A, Nie S (2025) Bottom-up carbon dots: purification, single-particle dynamics, and electronic structure. Chem Sci:4195–4212. https://doi.org/10.1039/d4sc05843g

Bressi V, Balu AM, Iannazzo D, Espro C (2023) Recent advances in the synthesis of carbon dots from renewable biomass by high-efficient hydrothermal and microwave green approaches. Curr Opin Green Sustain Chem 40:100742. https://doi.org/10.1016/j.cogsc.2022.100742

Cai D, Zhong X, Xu L, Xiong Y, Deng W, Zou G, Hou H, Ji X (2025) Biomass-derived carbon dots: synthesis, modification and application in batteries. Chem Sci 16:4937–4970. https://doi.org/10.1039/d4sc08659g

Castaing V, Romero M, Rytz D, Lozano G, Míguez H (2024) Quantification of emission efficiency in persistent luminescent materials. Adv Opt Mater 2401638:1–8. https://doi.org/10.1002/adom.202401638

Chahal S, Macairan JR, Yousefi N, Tufenkji N, Naccache R (2021) Green synthesis of carbon dots and their applications. RSC Adv 11:25354–25363. https://doi.org/10.1039/d1ra04718c

Dager A, Uchida T, Maekawa T, Tachibana M (2019) Synthesis and characterization of mono-disperse carbon quantum dots from fennel seeds: photoluminescence analysis using machine learning. Sci Rep:1–10. https://doi.org/10.1038/s41598-019-50397-5

Dananjaya V, Marimuthu S, Yang RC, Grace AN, Abeykoon C (2024) Synthesis, properties, applications, 3D printing and machine learning of graphene quantum dots in polymer nanocomposites. Prog Mater Sci 144:101282. https://doi.org/10.1016/j.pmatsci.2024.101282

Daood U, Daood S, Mei N, Zain E, Lin L, Matinlinna J, Sauro S, Peters OA, Tezvergil-mutluay A, Yiu C, Blum IR (2025) Carbon quantum dots as modulators of hydroxyapatite c-axis orientation and mechanical reinforcement in dentin: Unlocking quantum entanglement and coherence on biomineralization. J Dent 163:106198. https://doi.org/10.1016/j.jdent.2025.106198

de Yro PAN, Quaichon GMO, Cruz RAT, Emolaga CS, Que MCO, Magdaluyo ER Jr, Basilia BA (2019) Hydrothermal synthesis of carbon quantum dots from biowaste for bio-imaging. AIP Conf Proc 2083:020007. https://doi.org/10.37934/arfmts.101.1.192206

Deepika, Kumar L, Gaikwad KK (2023) Carbon dots for food packaging applications. Sustain Food Technol 1:185–199. https://doi.org/10.1039/d2fb00020b

Dern E, Kong H, Lai CW, Juan JC, Pang YL, Khe CS, Badruddin IA, Gapsari F, Anam K (2025) Recent advances in titanium dioxide bio-derived carbon photocatalysts for organic pollutant degradation in wastewater. iScience 28:112368. https://doi.org/10.1016/j.isci.2025.112368

Elugoke SE, Uwaya GE, Quadri TW, Ebenso EE (2024) Part 1 – Carbon quantum dots: basics, properties, and fundamentals. In: Carbon dots: recent developments and future perspectives. https://doi.org/10.1021/bk-2024-1465.ch001

Gao Y, Wang Q, Ji G, Li A, Niu J (2021) Doping strategy, properties and application of heteroatom-doped ordered mesoporous carbon. RSC Adv 11:5361–5383. https://doi.org/10.1039/d0ra08993a

Gedda G, Sankaranarayanan SA, Putta CL, Gudimella KK, Rengan AK, Girma WM (2023) Green synthesis of multi-functional carbon dots from medicinal plant leaves for antimicrobial, antioxidant, and bioimaging applications. Sci Rep. https://doi.org/10.1038/s41598-023-33652-8

Giordano MG, Seganti G, Bartoli M (2023) An overview on carbon quantum dots optical and chemical features. Molecules 28:2772

Guerrero-Gonzalez R, Vázquez-Dávila F, Saucedo-Flores E, Ruelas R, Ceballos-Sánchez O, Pelayo JE (2023) Green approach synthesis of carbon quantum dots from agave bagasse and their use to boost seed germination and plant growth. SN Appl Sci 5. https://doi.org/10.1007/s42452-023-05428-2

Hamid A, Sohal N, Singh P, Kolte J (2026) Green synthesis of carbon quantum dots from biomass: a sustainable approach to nanotechnology. Biomass Bioenergy 204:108447. https://doi.org/10.1016/j.biombioe.2025.108447

Hareesh K (2024) A review on carbon quantum dot/semiconductor-based nanocomposites as hydrogen production photocatalysts. RSC Adv:23404–23422. https://doi.org/10.1039/d4ra04149f

Iravani S, Varma RS (2020) Green synthesis, biomedical and biotechnological applications of carbon and graphene quantum dots. A review. Environ Chem Lett 18:703–727. https://doi.org/10.1007/s10311-020-00984-0

Ishak N, Galář P, Mekkat R, Grandcolas M, Šoóš M (2025) Fine-tuning photoluminescence and photocatalysis: Exploring the effects of carbon quantum dots synthesis and purification on g-C3N4. Colloids Surf A Physicochem Eng Asp. https://doi.org/10.1016/j.colsurfa.2024.135789

Jesmin F, Mitu JF, Khandaker T, Ibrahim ABM, Md Al Amin Mia Anik KIA, Suwaylih WAA, Hossain Hasan MK, Sarwar M (2026) Recent advances in nanoparticle-modified zeolites:

functionalization strategies and diverse applications. RSC Adv 16:8828–8860. https://doi.org/10.1039/D5RA07712E

John BK, Abraham T, Mathew B (2022) A review on characterization techniques for carbon quantum dots and their applications in agrochemical residue detection. J Fluoresc 32:449–471. https://doi.org/10.1007/s10895-021-02852-8

Kudilatt H, Hou B, Welland ME (2020) Quantum dots microstructural metrology from time-resolved spectroscopy to spatially resolved electron microscopy. Part Part Syst Charact 37: 2000192

Litowczenko J, Wieszczycka K, Staszak K, Woz MJ, Maciejewska BM, Jurga S (2021) Surface functionalization – the way for advanced applications of smart materials. Coord Chem Rev 436. https://doi.org/10.1016/j.ccr.2021.213846

Manikandan V, Lee NY (2022) Green synthesis of carbon quantum dots and their environmental applications. Environ Res 212:113283. https://doi.org/10.1016/j.envres.2022.113283

Mao H, Xue M, Han G (2021) Synthesis, properties and applications of fluorescent carbon dots. Gongneng Cailiao/J Funct Mater 52:01053–01063. https://doi.org/10.3969/j.issn.1001-9731.2021.01.009

Mehta VN, Jha S, Kailasa SK (2014) One-pot green synthesis of carbon dots by using Saccharum officinarum juice for fluorescent imaging of bacteria (Escherichia coli) and yeast (Saccharomyces cerevisiae) cells. Mater Sci Eng C 38:20–27. https://doi.org/10.1016/j.msec.2014.01.038

Mu L, Wu W, Huang J, Peng H, Song Y, Tao J, Chen G (2024) Simultaneous synthesis of carbon quantum dots and hydrothermal biochar from corn straw through hydrothermal treatment. Ind Crop Prod 219:119026. https://doi.org/10.1016/j.indcrop.2024.119026

Ojiako TP, Osei R, Buchely M, Wen H, Lekakh S, Malley RO (2026) The effects of mold flux contamination on oxide scale formation and hydro – descaling efficiency during steel processing. J Mater Sci. https://doi.org/10.1007/s10853-026-12420-9

Phuc PH, Khanh NK, Thuy HT (2025) Synthesis and characterization of carbon dots derived from biomass sources. Dong Thap Univ J Sci 14:20–25

Pouri H, Panta R, Bharathan P, Fang J (2025) Advances in nanostructured fluorescence sensors for H2O2 detection: current status and future direction. Micro 5:1–30

Rani UA, Ng LY, Ng CY, Mahmoudi E (2020) A review of carbon quantum dots and their applications in wastewater treatment. Adv Colloid Interf Sci 278:102124. https://doi.org/10.1016/j.cis.2020.102124

Redondo-fernandez G, Canga JC, Soldado A, Encinar JR, Costa-fernandez JM (2023) Functionalized heteroatom-doped carbon dots for biomedical applications: a review. Anal Chim Acta 1284:341874. https://doi.org/10.1016/j.aca.2023.341874

Ren H, Qi F, Feng X, Liu J, Zhao Y (2024) Facile synthesis of fluorescent carbon quantum dots with high product yield using a solid-phase strategy. Molecules 29:1–12

Rezania S, Darajeh N, Rupani PF, Mojiri A, Kamyab H (2024) Recent advances in the adsorption of different pollutants from wastewater using carbon-based and metal-oxide nanoparticles. Appl Sci 14:11492

Sahu S, Behera B, Maiti TK, Mohapatra S (2012) Simple one-step synthesis of highly luminescent carbon dots from orange juice: application as excellent bio-imaging agents. Chem Commun 48: 8835–8837. https://doi.org/10.1039/c2cc33796g

Saini S, Kumar K, Saini P, Mahawar DK, Rathore KS, Kumar S, Dandia A, Parewa V (2022) Sustainable synthesis of biomass-derived carbon quantum dots and their catalytic application for the assessment of α, β-unsaturated compounds. RSC Adv:32619–32629. https://doi.org/10.1039/d2ra05201f

Santos CIM, Almeida AC, Martins ALF, Araújo AR, Lourenço LMO, Gonçalves G, Neves MGPMS (2025) Carbon dot-phthalocyanine hybrids: synergistic effects that boost their multifaceted applications. Nanoscale 17:12554–12579. https://doi.org/10.1039/d5nr00466g

Sead FF, Jadeja Y, Kumar A, Rekha MM (2025) Nanoscale advances carbon quantum dots for sustainable energy: enhancing electrocatalytic reactions through. Nanoscale Adv 7:3961–3998. https://doi.org/10.1039/D5NA00205B

Sen B, Sarma H (2026) Carbon quantum dots as multifunctional nanomaterials for sustainable optoelectronic biosensing and green photonics. Commun Mater:1–14. https://doi.org/10.1038/s43246-025-01056-y

Seo G, Kim BS, Lim H, Choi J, Kim M, Lee H, Kim HO (2026) Biomedical applications and future perspectives of carbon dots and their hybrid nanomaterials. Mater Adv 7:157–174. https://doi.org/10.1039/d5ma00816f

Serrano-lotina A, Portela R, Baeza P, Alcolea-Rodriguez V, Villarroel M, Avila P (2023) Zeta potential as a tool for functional materials development. Catal Today 423. https://doi.org/10.1016/j.cattod.2022.08.004

Sharma A, Das J (2019) Small molecules derived carbon dots: synthesis and applications in sensing, catalysis, imaging, and biomedicine. J Nanobiotechnol 17:1–24. https://doi.org/10.1186/s12951-019-0525-8

Sheshmani S, Mardali M, Shokrollahzadeh S, Bide Y, Tarlani R (2025) Synthesis, optical, and photocatalytic properties of cellulose-derived carbon quantum dots. Sci Rep 15:1–13. https://doi.org/10.1038/s41598-025-04453-y

Sim LC, Tai JY, Khor JM, Wong JL, Lee JY, Leong KH, Saravanan P, Aziz AA (2019) Carbon dots synthesized from green precursors with an amplified photoluminescence: synthesis, characterization, and its application, nanotechnology in the life sciences. https://doi.org/10.1007/978-3-030-16379-2_1

Sohail A (2025) Progress in nanomaterials: synthesis, characterization, and applications. Next Nanotechnol 8:100263. https://doi.org/10.1016/j.nxnano.2025.100263

Tadesse A, RamaDevi D, Battu G, Basavaiah K (2018) Facile green synthesis of fluorescent carbon quantum dots from citrus lemon juice for live cell imaging. Asian J Nanosci Mater 1:36–46. https://doi.org/10.26655/ajnanomat.2018.1.5

Targhan H, Rezaei A, Aliabadi A, Zheng H, Cheng H, Aminabhavi TM (2024) Adsorptive and photocatalytic degradation of imidacloprid pesticide from wastewater via the fabrication of ZIF-CdS/Tpy quantum dots. Chem Eng J 482:148983. https://doi.org/10.1016/j.cej.2024.148983

Tavan M, Yousefian Z, Bakhtiar Z, Rahmandoust M (2025) Carbon quantum dots: multifunctional fluorescent nanomaterials for sustainable advances in biomedicine and agriculture. Ind Crop Prod 231:121207. https://doi.org/10.1016/j.indcrop.2025.121207

Tewari S, Kaur J, Tandon S, Das DK (2026) Recent advances in functionalized nanocarbon adsorbents for toxic pollutant removal: challenges and future perspectives. Discov Nano 21:1–39

Toth P (2021) Nanostructure quantification of turbostratic carbon by HRTEM image analysis: state of the art, biases, sensitivity and best practices. Carbon N Y 178:688–707. https://doi.org/10.1016/j.carbon.2021.03.043

Vejpravov J (2021) Mixed sp2–sp3 nanocarbon materials: a status quo review. Nanomaterials 11:2469

Wang H, Sun C, Chen X, Zhang Y, Colvin VL, Rice Q, Seo J, Feng S, Wang S, Yu WW, Tech L (2017) Excitation wavelength independent visible color emission of carbon dots. Nanoscale 9:1909–1915. https://doi.org/10.1039/c6nr09200d

Wang Y, Liu C, Tang S, Tian J, Wang Y, Yang Y (2023) Thermodynamics, kinetics and structure-activity relationship of hydroxyanthraquinones scavenging free radicals. Food Biosci 53:102705. https://doi.org/10.1016/j.fbio.2023.102705

Yang H, Bai L, Geng Z, Chen H, Xu L, Xie Y, Wang D, Gu H, Wang X (2023) Carbon quantum dots: preparation, optical properties, and biomedical applications. Mater Today Adv 18:100376. https://doi.org/10.1016/j.mtadv.2023.100376

You W, Zou W, Jiang S, Zhang J (2024) Fluorescent carbon quantum dots with controllable physicochemical properties fantastic for emerging applications: a review. Carbon Neutralization:245–284. https://doi.org/10.1002/cnl2.120

Yousaf A, Imran M, Warsi MF, Alsafari IA, Khan FA, Parra-saldívar R, Gutiérrez-soto G, Iqbal HMN (2025) Nanomaterials as a new frontier platform: metal-doped and hybrid carbon dots as enzyme mimics for environmental applications. Front Mater. https://doi.org/10.3389/fmats.2025.1553214

Chapter 3
Carbon Sequestration and Recycling from Wastes: A Sustainable Approach

Nazir Ahmad Bhat and Makbul Alom

3.1 Introduction

The carbon cycle's sensitivity to tropical temperatures nearly doubled in the tropical world between 1959 and 2011, leading to an upward trend of carbon emission since the year 2000, reaching their highest equilibrium or peak at 34.36 billion tons in 2019, followed by a short decline during the pandemic in 2020, before they increased again shortly after the pandemic (Dong et al. 2022; Li et al. 2024). Record concentrations of CO_2 reflect the increasing influence of human activities, further reinforcing the urgency of staying within a limited licensing of cumulative emissions of greenhouse gases (GHGs) (as advocated in IPCC evaluation assessments) (Dong et al. 2022). The waste sector, which is responsible for only 3–5% of global GHG emissions, contains great mitigation potential in terms of waste reduction, recovery, and energy conversion, with landfills alone accounting for 3–4% of global human-driven discharge (Ilmas et al. 2018). In developing countries such as Myanmar, open dumping is a major way waste is disposed of, resulting in a potential of 2000 Gg CO-eq in 2018 and 3350 Gg total by 2025 (Tun and Juchelková 2018). In Rawal Town, Pakistan, landfills contributed 10.73 Gg of CH_4 over the total 417.84 Gg CO_2 equivalent emissions from the waste sector, i.e., 73% of the total CO_2 equivalent emissions from the waste sector (Ilmas et al. 2018). Similarly, the CO_2 emissions from the landfills in China were around ~165 million tons in 2020. Food waste alone adds 6.8% to global emissions and wastes 250 billion liters per year of water. Reducing it by half would have prevented 77 million metric tons of methane in 2015 (Guo et al. 2022; Krause et al. 2023).

Carbon sequestration by waste management would provide a good way to fight climate change by capturing and storing CO_2 from the atmosphere. Soil application

N. A. Bhat (✉) · M. Alom
Biodiversity and Plant Systematic Laboratory (BioPlaS_Lab), Department of Botany, University of Science and Technology, Baridua, Meghalaya, India
e-mail: nazirahmadbhat@ustm.ac.in

Y. K. Mohanta et al. (eds.), *Bioresource Waste to Carbon Nanomaterial*, Nanotechnology in the Life Sciences, https://doi.org/10.1007/978-981-92-2405-0_3

of biosolids and biochar improves carbon storage, and composting with CO_2 capture is a cost-effective carbon removal, like the biomass carbon removal and storage (BiCRS) technology, achieving 0.32–0.97 t-CO_2 removal per ton of dry biomass, which is more effective than anaerobic digestion (AD) (0.04–0.40 t-CO_2/t-biomass) and hydrothermal liquefaction (0.49–0.62 t-CO_2/t-biomass) (Wijesekara et al. 2021; Woods et al. 2025). Food waste valorization (conversion to valuable bio-based products) by fermentation, AD, and catalytic transformation reduces the environmental as well as societal and economic costs and plays a role in building a circular bioeconomy. Processes such as the food waste to bioethanol process can greatly alleviate climate impacts by 4.0–4.8 trillion USD and mitigate average warming by 0.06–0.07 °C between 2011 and the year 2050 (Ansari et al. 2024).

The circular economy (CE) is a creative model that encourages resource efficiency, waste reduction, and sustainability using the three Rs—recycle, reuse, and reduce—within a closed-loop system (Dong et al. 2022). The plans, such as the EU's "Circular Economy Package Plan" (2015) and Saudi Arabia's CCE framework, in line with the Paris Agreement, indicate global efforts toward a low-carbon future through coordinated G20 efforts (Dong et al. 2022; Al Shehri et al. 2023). The combination of carbon capture, utilization, and storage with the circular approaches will allow the countries to immensely diminish GHG emissions and enhance environmental sustainability. This chapter discusses technologies of waste valorization in the CE; the life cycle of carbon sequestration, recycling, and sustainable energy recovery; and the dual opportunities that municipal solid waste (MSW) holds in terms of energy generation and carbon capture in a circular carbon economy (CCE).

3.2 Carbon Sequestration and Principles of the Circular Economy

Carbon sequestration is the biological, geological, mineral, and product-based capturing and safe sequestration of atmospheric carbon dioxide (CO_2) to mitigate climate change. Biological methods contribute to climate change mitigation by enhancing photosynthesis and increasing carbon storage in biomass and soils (Lorenz and Lal 2015). Mineral carbonation is a process of stable carbonates being formed by CO_2 via metal cation reactions, including in situ mineralization in alkaline rock formations and ex situ processes by reactors, or applications of carbonation reactions such as carbonation of construction and demolition (C&D) waste containing abundant calcium compounds, and enhanced weathering also increases this process by spreading powdered silicate rocks onto cultivation areas for replication of CO_2 removal (Romanov et al. 2015). Product-based capturing makes use of CO_2 in low-emission materials such as carbon-negative binders or biogenic concrete in order to allow infrastructure to complement the natural sinks of a circular bioeconomy.

The CE is an economic system that minimizes waste and maximizes resource efficiency through reuse, sharing, repair, refurbishment, remanufacturing, and recycling. It also reduces resource inputs, pollution, and emissions while and aligning with the UN Sustainable Development Goal (SDG) 12, which promotes responsible production and consumption (Venkatesh 2022). Building on CE, the CCE, introduced by William McDonough in 2016, combines technology-based strategies that include better control of CO_2 emissions by reduction, reuse, recycle, and removal, especially in the domains of energy efficiency, non-bio-renewables, nuclear power, fuel switching, and carbon productivity, which is intended to reduce climate change, potentially by 70% (Dong et al. 2022). Central to both is the "4R" framework (reduce, reuse, recycle, remove): reduction of waste (e.g., food), reuse of materials (e.g., wood), recycling (e.g., oil to biodiesel), and removal of CO_2 to reduce emissions by 39% and resource use by 28% (Venkatesh 2022; Ghiat et al. 2025).

Life-cycle assessment (LCA), as described by ISO (2006), is defined as the quantification of environmental impacts throughout a product's entire life cycle, which is ideal for assessing carbon utilization and sequestration technologies. Since 2015, LCA has been developed considerably and embraced new methods such as the planetary boundaries framework to examine environmental effects in relation to a global ecological boundary. It is a vital systematic tool that can help to detect environmental issues and mitigate the corresponding effects, so it is indispensable to companies and organizations that want to enhance their sustainability (Muller et al. 2020). Developments in digitalization, regionalization, and integration of the CE contribute to the further development of the methodology, turning it into an extensive sustainability strategy (Muller et al. 2020). For carbon utilization, the system boundaries must be defined from capturing waste gas, pretreatment, purification, and CO_2 or methane conversion to products, covering the energy/material requirements, which may produce emissions. Net GHG benefits exist if products displace more emitting alternatives, e.g., CO_2-polyols vs. traditional products (Müller et al. 2020). Sequestration involves three key components viz., the accounting boundaries for biogenic CO_2 storage, the assessment of CO_2 re-emission in incinerator LCAs based on the waste composition, and the application of standardized frameworks such as the Biogenic Carbon Project, best practices for Biomass Carbon Removal and Storage (BiCRS), and guidelines offered by the European Commission to ansure comparability and policy agreement.

3.3 Organic and Biomass Wastes

The production of biochar from organic and biomass wastes by the pyrolysis process is a sustainable way of carbon sequestration and carbon recycling, which promotes sustainability by storing carbon in agricultural settings, reducing GHG emissions, and reducing carbon while reusing organic waste. The deliberate incorporation of modern practices of biomass use, including biochar production and application pretreatment, has the potential to enhance carbon sequestration improves soil fertility, and generate renewable energy that can offset fossil fuel emissions (Jalali et al. 2025).

3.3.1 Sources and Characteristics

Organic waste, including food scraps, yard waste, paper, wood, fabrics, sewage, and manure, makes up roughly three-quarters of the global waste stream, where it is decomposed in landfills and dumpsites. About 25–30% of all food in the world is wasted, and as it rots, not only CO_2 but especially powerful methane gases are released (McDonald and Krol 2026). Food waste composition varies depending on typology, source of collection, and geography (Lelicińska-Serafin et al. 2023). Agro-industrial residues are a problem worldwide, since the world population is expected to rise to 10 billion people by 2050, and the future of food is expected to cause an increase in food demand and food waste reaching 4.5 times the municipal volumes in the world. These residues (AFRs) from plant (straws, bagasse, and peels) and animal (whey and feathers) processing produce an amount of ~250 million tons per year and are particularly well developed in regions such as Latin America. As low-cost feedstocks, they allow biomedical materials through microbial valorization, i.e., solid-state fermentation, allowing for ecological and economic benefits (Sosa-Martínez et al. 2023). Physicochemical aspects are very important when it comes to the application, but this is a great challenge in the case of compositional diversity, which interferes with standardization.

Agricultural crop residues comprise inedible plant materials remaining after after harvest, including straw, stover, stubble, stalks, leaves, branches, seeds, nuts, fruits and vegetables residues. Agro-industrial residues include secondary by-products such as husks, hulls, peels, and bagasse, as well as tertiary kinds such as rice hulls, corn cobs. Exceeding 5 billion metric tons worldwide every year, these residues are primarily composed of cellulose (40–50 wt%), hemicellulose (15–25 wt %), and lignin (20–30 wt%). Carbon-rich biomasses (40–45% C, 0.6–1% N, 0.45–2% P, 14–23% K) are used as an animal feed, soil amendments, and bioenergy production. However, if not properly managed, these biomasses are easily contaminated due to their porous and heterogeneous structure. Their physical properties vary considerably among feedstocks e.g., rice straw has a surface area of c.a. 0.77 m^2 g^{-1} and a pore volume of 0.0059 m^3 g^{-1}, while bulk density differs substantially with low in wheat/maize and high in soybean/cotton (Jing et al. 2020; Fu et al. 2021). Forestry by-products (stems, branches, bark, logging/sawmill waste) represent another important biomass resource. In the United States alone, *c.a.* 132 Mt of residues are generated annually, while global forest biomass resources are estimated at about 1,194 million tons per year, with significant potential for renewable energy production (National Academies 2019).

3.3.2 Biochar-Based Carbon Sequestration

Pyrolysis is a thermochemical process in an oxygen-free environment, in which carbonaceous materials are destroyed to obtain syngas, pyrolytic oil, and biochar, a carbon-rich porous solid obtained by thermolysis of biomass at 300–700 °C (Fig. 3.1). Slow (low heating rates, residence time) and fast (high heating rates, residence time) pyrolysis maximizes pure biochar, and in certain cases, commonly higher temperatures are preferred for biochar due to lower initial temperatures (around 400 °C), whereas excess liquids are often favored by fast temperatures, as well as the ability of higher temperature values (600–700 °C) to promote solids. It enhances the biochar's carbon content, pH, and conductivity but reduces its O/C and H/C ratios. Moreover, prolonged low-temperature residence time can further increases biochar yield and solid product formation (Jayakumar et al. 2023; Jalali et al. 2025). In contrast, gasification involves the partial oxidation of biomass using gasifying agents, such as steam, air, or CO_2, to produce syngas (mainly CO and H_2). This syngas can be utilized for heat and power generation or converted into liquid fuels through downstream processing (National Academies 2019; Suparmin et al. 2024).

Biochar yield and properties depend highly on feedstock characteristics, e.g., carbon content, moisture, ash composition, lignin content, and pyrolysis conditions (Table 3.1). High levels of lignin biomass (e.g., softwood, hardwood, nut shells) produce biochar with high porosity, high aromaticity, a high surface area, high levels

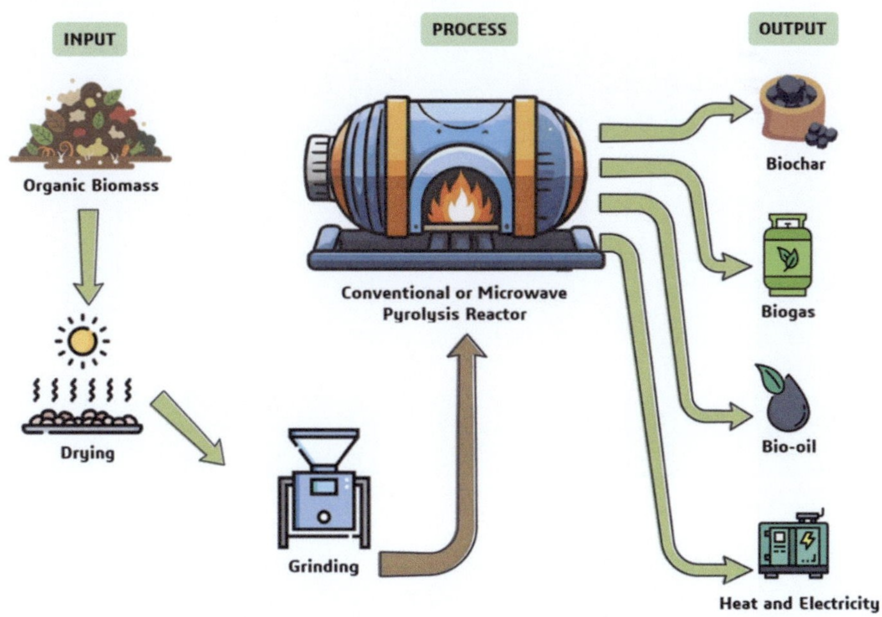

Fig. 3.1 Biochar-based carbon sequestration of organic biomass via the pyrolysis technique

Table 3.1 Detailed biochar properties and sequestration metrics by feedstock and pyrolysis conditions

Feedstock type	Pyrolysis temperature (°C)	Biochar yield (%)	Fixed carbon (%)	Surface area (m^2/g)	C/N ratio	Sequestration potential (t CO_2/t biochar)	Key applications and benefits	References
Rice straw	300–400	High (peaks ~400 °C)	Moderate (40–45%)	0.77	14–23	0.7–1.8 Gt CO_2-C(eq)/year globally	Soil amendment reduces nSOC mineralization by 13.5%	Jing et al. (2020); Fu et al. (2021)
Sugarcane filter cake	400–500	Medium	Elevated	Variable (porous)	High	Enhanced stability via negative priming	Nutrient retention, crop yield +10.8%	Eykelbosh et al. (2014); Jalali et al. (2025)
Farmyard manure	500–600	Lower	High	Moderate	Lower	Superior for recalcitrant C	GHG reduction (CH_4, −13.5%; N_2O, −21.4%)	Qi et al. (2023); Jalali et al. (2025)
Wood/softwood (high lignin)	600–700	Low	Very high (>50%)	Superior (>1.0)	High	0.32–0.97 (BiCRS tech)	Long-term soil C storage, +52.5% SOC	Weng et al. (2022); Jalali et al. (2025)
Nut shells/hardwood	600–700	Low	Very high	High porosity	High	Outperforms fecal matter	Aromaticity boost, CEC enhancement	Mukome et al. (2013); Jalali et al. (2025)
Engineered (KOH-activated)	400 (carbonization) + 600 (activation)	Optimized	N/A	Micropore-focused ($R^2 = 0.90$)	Tunable	4.06 mmol/g at 0 °C	CO_2 adsorption, ML-predicted capacity	Yuan et al. (2024)

of entire and fixed carbon, and low levels of volatile matter (Weng et al. 2022; Jalali et al. 2025). Carbon stability, which is critical for long-term sequestration in soils, is enhanced by the use of wood-based feedstock and high pyrolysis temperature, which yields more recalcitrant carbon despite low yields from high burning rates. Plant biomass with elevated values of C/N ratio performs better for sequestration potential than fecal matter compositional variations for producing biochars from agricultural wastes such as sugarcane filter cake, farmyard manure, and rice straw (Jalali et al. 2025).

Engineered biochar from biomass waste helps achieve climate mitigation and waste management through CO_2 capture; however, optimization is labor-intensive. Active learning now facilitates the synthesis of biochar by locating narrow micropore volume, which highly correlates with adsorption capacity. The main properties, surface area, microporosity (micropore area $R^2 = 0.9032$, volume $R^2 = 0.8793$), aromaticity, hydrophobicity, and basic groups induced by KOH activation are the determinants of its work with N/O combination, which enhance the electrostatic interactions, e.g., carbonization of 400 °C and KOH activation (ratio 2) at 600 °C, resulting in 4.06 and 2.54 mmol/g at 0 °C and 25 °C, respectively (Yuan et al. 2024). Prediction of adsorption and surpassing pristine biochar can be applied using machine learning.

Biochar application to soil is a potential carbon sequestration strategy and offers the potential mitigation of global warming through the reduction of atmospheric CO_2 concentration by high biochar stability and its influence on native soil organic carbon (nSOC) mineralization with global potential between 0.7 and 1.8 Gt CO_2-C(eq)/year, where biochar produced at ≥600 °C with high C/N ratios, low O/C and H/C ratios, and increased aromaticity exhibits superior sequestration efficiency by retarding mineralization of nSOC due to negative priming effects such as substrate switching, microbial shifts, and nSOC sorption. Beyond carbon benefits, biochar helps in improving soil physicochemical properties like pH and cation exchange capacity (CEC), increases gross soil organic carbon (SOC) levels by almost 27%, and stimulates microbial biomass, enzymatic activities, and biota dynamics, improving soil fertility, reducing nutrient loss, and enhancing plant growth in a variety of soils, practices, and climates (Enaime and Lübken 2021). Long-term annual applications maintain agronomic benefits, such as increasing yield of crops (e.g., +10.8% overall, up to +42.1% in sandy soils after ≥5 years), improved SOC contents (+52.5% per annum, up to +83.5% long term due to stable carbon content, rhizodeposit preservation, and microbial necromass stabilization), and GHG reductions of CH_4 (−13.5% in particular in acidic paddy soils by declining the ratio of methanogens/methanotrophs) and N_2O (−21.4% with repeated applications or − 23.3% single doses, driven by N immobilization and enhanced N_2O reductase gene expression) (Yang et al. 2025).

3.3.3 Bioenergy with Carbon Capture and Storage

The net-negative emissions can be achieved through bioenergy production with carbon capture and sequestration or BECCS. The concept behind BECCS is that it is an indicator of climate risk and it has evolved as an important tool of CO_2 removal (CDR) to maintain the global atmospheric CO_2 level at lower than 500 ppm (National Academies 2019).

The BECCS pathways broadly include:

- *Biomass combustion with carbon capture and sequestration (traditional BECCS):* This is done by burning biomass for thermal and electricity generation, followed by the CO_2 capture and geologic sequestration of CO_2.
- *Biomass thermochemical conversion to fuel with biochar soil amendment:* Thermochemical conversion of biomass into fuels (co-produces biochar, then used like soil amendments to sequester carbon).
- *Biomass fermentation to fuel with carbon capture and sequestration:* This involves the fermentation of biomass into fuels with carbon capture during the fermentation process.

Typical sources of biomass feedstocks are forest management (i.e., logging residues, sawmill waste), agricultural (i.e., dedicated energy crop, crop residues), algae growth, or municipal organic solid waste. CO_2 capture technologies in BECCS systems are mostly comparable to the conventional fossil fuel power plants and come in a variety of post-combustion, pre-combustion (gasification), oxy-combustion, and chemical looping (Table 3.2). The technology readiness level of post-combustion capture, e.g., in the case of supercritical super-fine power plants using pulverized coal, is 9, and the cost of CO_2 capture is in the range of 36–53/t. As an example, fermented production of ethanol can use carbon capture technology because carbon capture is already available, and it will capture 60% of the CO_2 by-product, which will cost less than 25/t CO_2 (National Academies 2019).

3.3.3.1 Integration of Biochar, Biogas, and CCS

The combination of biochar, biogas, and carbon capture and storage (CCS) in BECCS can offer a complex approach toward carbon sequestration and waste value generation. Biochar as produced from the pyrolysis of biomass to a stable carbon-rich material can either be applied as a long-term soil amendment or as an underground burial for direct sequestration, or co-deployment with bioenergy production and CCS will increase the overall CDR, where the study reports a CDR increase of 15–400% as a result of co-deployment, most notably in areas lacking CO_2 transport infrastructure (Jayakumar et al. 2023; Deng et al. 2025). Complementarily, the anaerobic digestion of organic residues and sewage sludge produce biogas for energy generation and a concentrated stream of CO_2 that can be captured and stored. A notable example is the South Korea's nation-wide food waste

Table 3.2 Comprehensive BECCS pathways and constraints

Pathway	CO_2 flux potential (by 2050)	Feedstocks	Technology readiness level	Economic cost ($/t CO_2)	Sustainability constraints	References
Biomass combustion + CCS	Global, 10–15 Gt/year; USA, 0.522–1.5 Gt/year	Forest residues, crop residues, MSW organics	9 (post-combustion)	36–53	Land use (380–700 Mha), life-cycle emissions	National Academies (2019); Shahbaz et al. (2021)
Pyrolysis to biofuel + biochar amendment	Enhances CDR by 15–400%	Logging waste, agro-residues	Pilot scale	37–132	Feedstock competition, nutrient depletion	National Academies (2019); Jayakumar et al. (2023)
Fermentation to fuel + CCS	60% CO_2 capture from ethanol	Algae, municipal organics	Existing (e.g., <25/t CO_2)	<25	Water use: Scalability in the global south	National Academies (2019)
Gasification	Syngas for fuels/power	Sawmill waste, energy crops	Mature	Variable	Biomass transport emissions, black carbon	National Academies (2019); Suparmin et al. (2024)
Cumulative storage potential	Global, 57–86 Gt; USA, 11 Gt	All biomass wastes	N/A	N/A	Biodiversity loss, social acceptance	National Academies (2019)

collection program, in which almost all organin wastes used for biogas production, composting, and other resource recovery applications, thereby reducing waste disposal and enhancing resource circularity (McDonald and Krol 2026). However, this competition for limited feedstocks for biomass production with those used to produce bioliquids or biopower for CCS requires LCAs to confirm net carbon benefits, while reducing sustainability trade-offs.

3.3.3.2 Removal Potentials and Sustainability Constraints

The global upper-bound BECCS potential flux of CO_2 is estimated to be 10–15 Gt/year CO_2 by 2050, according to IPCC. The economically feasible lower-bound BECCS CO_2 flux potential for the USA is 522 Mt/year CO_2, while the upper bound is estimated at 1500 Mt/year CO_2. These estimates assume all the carbon content of the biomass is absorbed and sequestered, not taking losses or other GHG emissions of the supply chain into consideration. The lower-bound cumulative CO_2 storage capacity for the USA is estimated at 11 Gt CO_2 by 2050, and the global lower bound is estimated at 57–86 Gt CO_2 by 2050 (National Academies 2019).

The BECCS has serious shortcomings limiting its application on a massive scale. Most importantly, it takes up huge portions of land, posing competition with food production and other ecosystem services. The reduction of 12 Gt/year CO_2e may require 380–700 Mha of land, 36–163 of available abandoned plus marginal lands (National Academies 2019). The availability of biomass and its productivity are vastly different for regional and feedstock origins. Life-cycle analyses must be comprehensive in nature because emissions come with the use of fossil fuels, burning of biomass or biodiesel, disturbance of the land, transportation, and short-lived climate forcers such as black carbon and also the non-CO_2 gases such as methane and nitrous oxide emissions (Xu et al. 2022). The BECCS is economically still expensive: biomass CCS power plants are more expensive than natural gas-based combined-cycle plants because they are less efficient and fuel and capital costs are higher, and biochar-based CO_2 capture is between \$37 and \$132 per ton (National Academies 2019). Among the issues that are considered environmental concerns are indirect emissions, threat to food security, loss of biodiversity, water competition, social equity concerns, and biomass harvesting of soil, which is a major part of the non-climate impacts, for which the Intergovernmental Panel on Climate Change has not given significant focus (Hansen and van den Bergh 2023). In this case, the technologies through gasification and burn, still with CCS, are at maturity, whereas pyrolysis and liquefaction at the pilot phase are at BECCS and the laboratory scale, respectively. Lastly, social acceptability and infrastructural limitations are other limiting factors to its general adoption in the mitigation of climate change (Hansen and van den Bergh 2023).

3.4 Municipal Solid Waste Systems

The proper management of MSW is a major issue to the developing economies, and the global community has to work together to create carbon neutrality by avoiding placement of waste in the disposal facilities (Fig. 3.2). The activities of separate streams of waste collection, the maximum amount of recyclable recovery, and the creation of waste-to-energy facilities with minimum landfill waste will enable the MSW industry to become a carbon sink on an international level.

3.4.1 Carbon Traffic in MSW Management

Municipal solid waste consists of different fractions that undergo varied biological degradation processes when discarded in the environment, and their organic fraction (OFMSW) that makes 30–40% (or 20–80%) of the total MSW is dominated by carbohydrates (30–60%) and contains approximately 30% of compostable wastes with variable moisture, organic content, C/N ratio, and year-to-year changes in pH (Stylianou et al. 2020). The contents of landfills shift as they mature, and those related to organic matter have a significant impact on the operations within landfills and the amounts of biodegradable organic carbon found in MSW and Subtitle D waste streams, which can be assessed using sample-and-sort composition data (EREF 2024). A thorough characterization of the OFMSW is therefore crucial in order to maximize the AD to produce methane.

In landfills and dumpsites, the organic fraction of MSW undergoes the process of AD, which significantly exposes the environment to major hazards and the population to potential health complications, especially in low- and middle-income countries (LMICs) where open dumping is frequently practiced and there is no more specific control over this process other than providing barriers or covering with a capping (Vinti et al. 2025). The global MSW generation is projected to increase by 81% rising from 2.1 billion tons in 2020 to 3.8 billion tons by 2050. In 2020, approximately 38% of MSW (about 810 million tons) was disposed of through uncontrolled dumping or open burning, a figure that is expected to increase substantially by 2050 (Maalouf and Mavropoulos 2023). Even modern landfills in high-

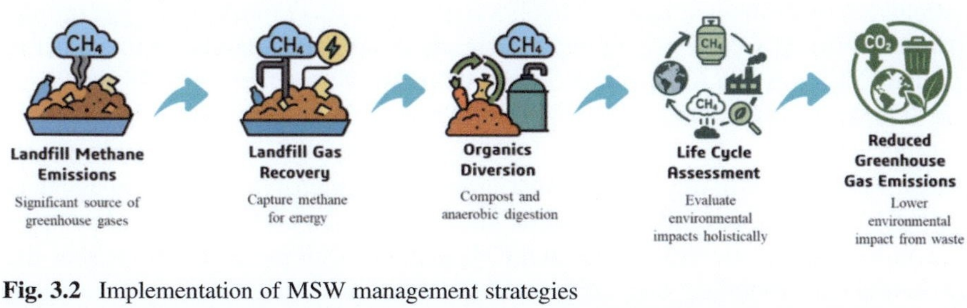

Fig. 3.2 Implementation of MSW management strategies

income countries such as Denmark (50% biogas capture), Sweden (58%), and the USA (63%) still emit large amounts of uncaptured methane. Although carbon cycling models have been developed, carbon cycle management in landfills has not yet been systematically addressed (Vinti et al. 2025).

3.4.2 *Methane Mitigation and Utilization in Landfills*

The third largest anthropogenic source, methane (CH_4), of MSW landfills is the result of anaerobic organic breakdown to landfill gas (LFG), which depends on temperature and moisture, composition of waste, and maturation and microbes through hydrolysis, acidogenesis, acetogenesis, methanogenesis, and maturation stages. Outputs of aerobic breakdown give initial CO_2, H_2, and heat, while anaerobic acidogenesis gives acids, CO_2, and H_2 (pH < 5), acetogenesis gives acetic acid, and methanogenesis gives 50–55% CH_4 and 45–50% CO_2 until depleted. LandGEM and other first-order models of decay forecast the emissions of CH_4, CO_2, and other emissions as a result of waste decomposition rates, which reach a maximum immediately following the phase of disposal and fall exponentially as a function of the primary parameters such as methane generation potential (L_0, use default of 170 m^3/t, generally 125–310 m^3/t) and rate constant (k), depending on factors including landfill conditions and the content of the organic material (Tadesse and Lee 2023).

Landfill gas, which is mostly 40–65% methane (CH_4), a GHG that is more potent than CO_2 over 100 years, is recovered through MSW dumps using wells, blowers, and flares and provides a raw material as part of best practices (range, 104–70,000; best practices, 80%), routed centrally where any flaring can be done to prevent emissions (Tadesse and Lee 2023). Technologies, including internal combustion engines (1–3 MW, cost-effective and efficient at 50% CH_4, is best used for cogeneration through waste heat recovery), turbines (>5 MW), micro-turbines (30–250 kW, approximately 30% electric yield at >35% CH_4, 95% available but requiring service every 8 years), and fuel cells, are used to generate electricity taking advantage of technologies such as adsorption, biological treatment, gasification, pyrolysis, membrane separation, and gradual oxidizers (95% availability at $\geq$1.5% CH_4, 29% yield, servicing every 9 years for low CH_4 streams) (Tadesse and Lee 2023). The cost of LFG electricity is USD 0.0604/kWh, which is high compared to hydro (0.0018/kWh) or steam (0.0415/kWh) but low compared to geothermal (0.0637/kWh), combined cycle (0.0763/kWh), diesel (0.1773/kWh), or gas turbine (0.1966/kWh), as well as where the infrastructure has been established to collect it, and the CO_2 emission (0.27×10^{-9} kg/kWh) outperforms fuel oil (0.85 kg/kWh), gas (0.85 kg/kWh), or coal (1.18 kg/kWh) (Tadesse and Lee 2023).

Methane (CH_4) emits about 400 kg CO_2 equivalent (CO_2e) of organic waste, which is why landfilling is the most greenhouse-intensive waste management approach (Nordahl et al. 2020). Landfill gas-to-energy (LFGTE) projects can reduce such emissions by 60–90% of the total CH_4 that a landfill produces throughout life by flaring or producing electricity or heat, effectively transforming the highly dangerous methane into insignificant CO_2 and water, and by replacing fossil fuels such as coal, oil, or natural gas with other activities to save or reduce indirect GHG (Tadesse and Lee 2023). Collected and demolished CH_4 in million metric tons of CO_2e is determined annually and relates to improvements in collection efficiency and gas engines, also enhancing environmental and economic conditions and carbon accounting, proving to equalize landfill emission base with that of LFGTE and consequently offsetting the base of carbon accounting (Tadesse and Lee 2023).

3.4.3 Diversion of the Organics Out of the Landfill

To minimize GHG to −41 kg CO_2e/ton and generate nutrient-rich compost that enhances soil and stores carbon and reduces the need for synthetic fertilizers with high global warming potential (GWP), biowaste such as the OFMSW is treated through composting and aerobic biostabilization (Nordahl et al. 2020). Examples are windrow, tunnel, and forced aeration composting, where inventory data include machinery, fossil fuel, and electricity consumption with air, water, and soil emissions, and the efficiency of the process depends on the time spent in the untreated material storage and innovations in the form of biochar addition or airflow optimization, whereby the mass of the waste is reduced variably and the sustainability of agricultural by-products with the help of the best practices is supported (Serafini et al. 2025).

An anaerobic digestion (AD) is a potentially viable technology in North America and Europe to treat organic fraction of municipal solid waste (OFMSW) as microbes degrade organics, producing a low-cost substitute for fossil fuel (biogas), that is rich in methane. Compared with landfilling, AD reduces greenhouse gas emissions and diverts organic waste from landfill disposal, thereby contributing to more sustainable waste management (Nordahl et al. 2020). Rapid modification of biogas into renewable natural gas (RNG) results in −36 to −2 kg CO_2e/ton of waste by eliminating impurities. The challenges are transport costs, more so when used together with incineration, but source-segregated and co-managed systems that eliminate food waste from landfills are usually known to have the lowest environmental footprint than mixed or landfilled streams (Nordahl et al. 2020).

A LCA is used to assess the environmental effects of the management of MSW, which shows the importance of biowaste in the world. Standard landfilling and incineration produce great amounts of GHGs (5 out of 1000 kg CO_2e/ton of organic waste), leachate, and land use problems (Table 3.3), whereas novel approaches, such as mechanical biological treatment (MBT), composting, and AD, alleviate the quantities and yield compost or biogas energy (Nordahl et al. 2020). The LCA indicates composting to be the least GHG emitter at −41 kg CO_2e/ton, but AD demonstrates −500 kg CO_2e/ton through energy recovery, which beats composting at −413 kg CO_2e/ton with co-digestion reducing the GHG emissions. The net environmental impacts are also integrated in an AD manner, as it encourages producing biogas electricity and revitalizing landfills through diversion (Nordahl et al. 2020; Serafini et al. 2025). Separation of sources improves the quality of

Table 3.3 MSW management and GHG emission comparison (kg CO_2e/ton)

Management method	GHG emissions (kg CO_2e/ ton organic waste)	Key factors influencing emissions	Additional benefits	Limitations	References
Landfilling	~400	Methane from anaerobic decomposition, waste composition, and moisture	Low initial cost in LMICs	High CH_4 emissions (3–4% global anthropogenic), leachate pollution	Nordahl et al. (2020)
Incineration	Variable (high without energy recovery)	Organic content, energy recovery efficiency	Volume reduction, energy generation	Non-CO_2 GHGs, ash disposal needs	Nordahl et al. (2020)
Composting/ aerobic biostabilization	−41	Process efficiency, biochar addition, airflow optimization	Nutrient-rich compost, soil carbon storage	Emissions from machinery, quality variability	Nordahl et al. (2020)
Anaerobic digestion (AD)	−36 to −2 (RNG upgrade), up to −500 with energy recovery	Source segregation, co-digestion, biogas upgrading	Biogas for energy, lowest footprint with diversion	Transport costs, contamination risks	Nordahl et al. (2020)
Mechanical biological treatment (MBT)	Lower than landfilling	Mixed waste processing	Reduces landfill burden	Heavy metals in compost	Nordahl et al. (2020)

compost; cost is coupled with emissions, whereas mixing of the waste in MBT compost adds to contamination of heavy metals. However, in the end, holistic comparisons require investigation of the net energy requirements, not just the recovery of valorization processes.

3.5 Construction, Demolition, and Mineral Wastes

The C&D waste can be regarded as a potential target of recycling into new building materials, which can boost the need to use natural resources and also decrease the proportion of the landfill use (Fig. 3.3). Significant environmental benefits across different categories of impact can also be achieved in relation to the recycling of C&D wastes as building materials or on the condition of innovative processes, in this case electrical fragmentation. It is responsive to sustainable development because it tends to conserve its resources and reduce wastage in unauthorized places.

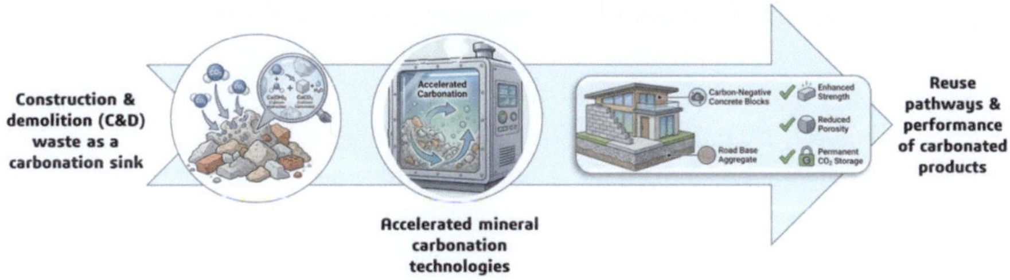

Fig. 3.3 Carbon sequestration via C&D waste recycling

3.5.1 Construction and Demolition (C&D) Waste as a Carbonation Sink

The C&D waste, especially the recycled concrete aggregates (RCAs), also has hydrates that can be carbonated, as well as alkaline metallic oxides, i.e., calcium oxide (CaO) and magnesium oxide (MgO), which react with CO_2 to create solid carbonates that can be irreversibly trapped. Stainless steel slag contains mainly merwinite, bredigite, gehlenite-akermanite, cuspidine, portlandite, periclase, and donathite, whereas basic oxygen furnace (BOF) steel slag mainly contains amorphous Ca silicates, larnite, portlandites, srebrodolskite, free lime, and wustite (Quaghebeur et al. 2015). Calcite and C-S-H gel are regarded as the major reactive components in C&D incineration, and research on siliceous and calcareous fine-grained products has shown that they are reactive to cement pastes. Quantitative changes in such mineral phases are also revealed by the activation of mines tailings and C&D-generated silt (Mencía et al. 2021).

Natural carbonation of concrete is a slow process where portlandite ($Ca(OH)_2$) reacts with CO_2 in the atmosphere to produce calcium carbonate ($CaCO_3$), which lowers porosity during years of exposure, which is worsened by the increasing CO_2 levels from 280 ppm before the industrial era to 381 ppm now (Alshaeer et al. 2022). This reinforced concrete reduces the pH of the surroundings of the concrete to between 5 and 9, affecting the passivation layer of the steel and leading to corrosion. Accelerated carbonation tests mimic this process casually by subjecting samples to high concentrations of CO_2 (e.g., 4–5%) under controlled humidity (50–60%) and temperature, and the period might take 7–15 days typically equating to 1 year of natural exposure depending on various factors such as the CO_2 concentration, time, temperature, relative humidity, and cement used (Merah 2024). Whereas this type of acceleration may reduce porosity and increase longevity, it is preferentially used in non-reinforced structures such as concrete blocks.

3.5.2 Accelerated Mineral Carbonation Technologies

Direct gas-solid CO_2 sequestration capture CO_2 in re-used materials such as steel slag and concrete waste under high pressure and temperature conditions to produce value-added construction materials. The Carbstone process enhances mineral carbonation in non-hydraulic stainless steel and basic oxygen furnace (BOF) slags, producing carbonate-based binders with concrete-like properties that can permanently trap 100–150 g CO_2/kg slag (Quaghebeur et al. 2015). The recycled concrete aggregates (RCAs) can be directly carbonated through three phases, which enhances quality by reducing porosity, lowers the carbon footprint of concrete, and captures 10–50 kg CO_2/ton RCA depending on the extent of natural carbonation, water content, temperature, and particle size. Industrial-scale tests have achieved carbonation levels of 3–4%, corresponding to approximately 30 kg CO_2 per ton of crushed concrete (Torrenti et al. 2022). The concrete contains calcium hydroxide, which reacts with atmospheric CO_2 in situ to create stable $CaCO_3$, which fills the voids and microcracks, densifies the microstructure, decreases porosity, and improves properties and sequesters CO_2 in leak-proof carbonates (Merah 2024). CO_2 is naturally adsorbed by bare concrete with time, which can be optimized through CO_2 curing at the batch plants. Factors influencing carbonation include CO_2 concentration, temperature, relative humidity (50–60% RH being optimal, as complete saturation prevents the reaction), and cement type (Alshaeer et al. 2022).

CO_2 is captured and trapped as stable carbonates using alkaline by-products such as steel slags and mining wastes, and mineral carbonation recycles by-products of steelmaking (170–250 Mt of global emissions, estimated in 2014). The CaO–(24–65%) and MgO-rich (3–20%) steel slags increase the rate of weathering by thousands of years and shorten it down to a few minutes (Quaghebeur et al. 2015). Experiments of fast carbonation show an increase in CO_2 absorption, strength formation, and environmental advantages in steel slags at higher temperatures (20–140 °C), CO_2 pressure, and concentrations, although volumetric instability limits aggregate use (76% used in Europe) (Quaghebeur et al. 2015). Technologies that use magnesium silicate-dominated mining tailings also contribute to the possible sequestration, and this would be equivalent to the reduction of approximately 1.5% of the world's yearly CO_2 emissions through the formation of carbonates (Li et al. 2018).

3.5.3 Reuse Pathways and Performance of Carbonated Products

Carbonated recycled coarse aggregates (cRAs) improve the microstructure of recycled aggregate concrete, thereby increasing its durability due to increased densities by 13–55 times relative to non-carbonated RA concrete. Carbonation treatment also also mitigates the adverse effects associated with recycled fine aggregates. They increase ordinary Portland cement-based RA compressive strength and density and reduce water uptake and permeability to as much as 90% of natural aggregate concrete strength (Zhang et al. 2025). However, the durability of carbonated RA concrete is less than that of natural aggregate concrete because RA is porous and has low interfacial transition zones. Nevertheless, carbonated RA concrete exhibits superior durability compared with untreated recycled aggregate concrete owing to its lower porosity, denser interfacial transition zones, reduced shrinkage, lower sorptivity (reduced by 26% after 7 days of carbonation), decreased water absorption, and reduced chloride penetration (Zhang et al. 2025). Nevertheless, carbonated recycled aggregate concrete exhibits superior durability compared with untreated recycled aggregate concrete owing to its lower porosity, denser interfacial transition zones, reduced shrinkage, lower sorptivity (reduced by 26% after 7 days of carbonation), decreased water absorption, and reduced chloride penetration (Zhang et al., 2025).

Beyond the limitations of physical CO_2 adsorption, mineral carbonation provides permanent and stable CO_2 fixation and Steel slags demonstrate a theoretical fixation of 460 g CO_2/kg based on their total Ca and Mg level, while basic oxygen furnace (BOF) slags can theoretically sequester approximately 360 g CO_2/kg. This capacity can be further increased with milling (to less than 125 μm particle sizes) which promotes near-complete conversion of free lime during carbonation (Quaghebeur et al. 2015). Carbonated cement paste has demonstrated CO_2 uptake of 7–15% on initial cement mass (max 55.43%). In contrast, natural carbonation results in CO_2 uptake of 0.0514% and 0.0338% in ordinary Portland cement concrete and ground granulated blast-furnace slag concrete, respectively. CO_2 uptake is typically quantified by comparing samples before and after exposure to carbonation conditions and can be further enhanced in high-ionic-strength brines, highlighting its its potential as a promising emission mitigation measure (Neves Junior et al. 2019; Yuan et al. 2025).

The carbonated products, including carbonated RCAs as developed using FastCarb projects, generated 80 tons of C25/30 and C45/55 concretes, and the carbonated materials are used in precast and cast-in-situ construction, increasing low-carbon construction without any worsening of properties (Torrenti et al. 2022). The feasible accelerated carbonation depends on the regional availability of CO_2 sources to deliver net-positive effects and CO_2 uptake, with fractions of RCA sand proving high CO_2 uptake and allowing it to be valorized within the framework of the CE. Combined with the basalt fiber-reinforced polymers, it also goes further in ensuring the sustainability of the infrastructure made of concrete (Torrenti et al. 2022).

3.6 Industrial Residues and Carbon Capture and Utilization

The use of carbon capture and utilization (CCU) technologies can serve as a viable alternative to the conventional method of generating useful chemicals by mitigating the effects of global warming and depleting the amount of fossil resources. The CCU

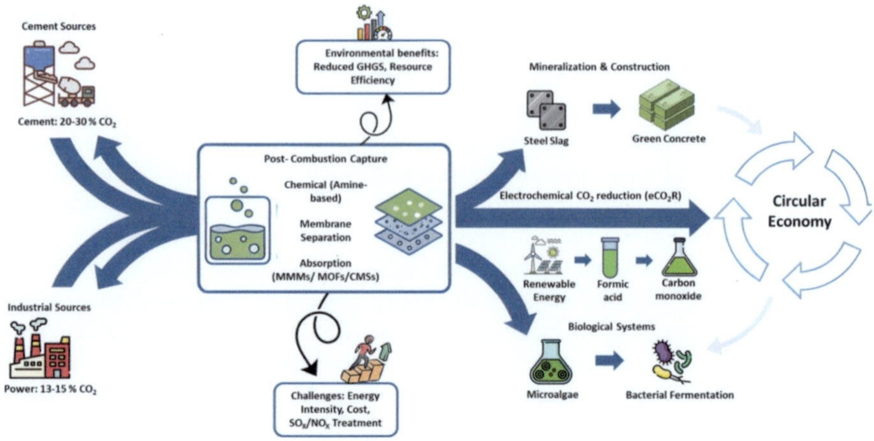

Fig. 3.4 Carbon capture and utilization and storage process

strives to power up the CO_2 that is generated in the industrial processes and uses it as a raw material in different applications without discharging it to the atmosphere (Fig. 3.4). In most cases, the environmental credits of the CCU processes can outweigh the conventional production process of the same products especially in the mitigation of global warming and climate change effects as long as renewable energy sources are used.

3.6.1 Wastes and Sources of Industrial CO_2

The cement and concrete industry is a significant contributor to CO_2 emissions, accounting for nearly 8% of global anthropogenic CO_2 emissions. The process emissions are most of the emissions that are caused by the chemical reaction of calcination during the production of cement (Cho et al. 2025). The steel industry also relies on fossil energy, making it one of the most energy-intensive and carbon-intensive industries in the world today. Chemical plants and waste incinerators are also viewed as the secondary CO_2 emitters, alongside cement and steel (Gkotsis et al. 2023). Biomass processing plants, most notably fermentation plants used to produce ethanol, are also sources of highly concentrated industrial CO_2.

The cement and steel plants typically produce a higher percentage of CO_2, ranging from 20% to 30% in the flue gas, which can be useful in various separation processes. By contrast, the gas of power plants that utilize coal includes 13–15% CO_2 (Baker et al. 2018). Post-combustion capture involves the removal or isolation of CO_2 from the flue gases generated after the combustion of fossil fuels or biomass. These flue gases are mostly composed of 73–77% of N_2, 15–16% of CO_2, 5–7% of H_2O, 3–4% of O_2, and other contaminants such as SOx and NOx. Treatment of SOx and NOx and the high levels of oxygen in the flue gases should be adapted in the process (Gkotsis et al. 2023). The supply of the CO_2-containing flue gas that can be used to carry out CO_2 mineralization is virtually inexhaustible.

3.6.2 Capture Technologies Linked to Waste Valorization

The CO_2 means of post-combustion CO_2 capture include chemical absorption, adsorption, and cryogenic and membrane separation, of which the amine-based absorption (e.g., monoethanolamine) is considered to have a great capture capacity and could achieve near-100% removal efficiency (Gkotsis et al. 2023). The limitations have been high regeneration energy, slow kinetics, and solvent toxicity by adverse solvents, which are addressed by greener adsorption on biomass-derived activated carbons with programmable surface chemistry, high surface areas, and cost performance. New mixed matrix membranes are proposed using metal-organic frameworks, carbon molecular sieves, nanocomposites, and ionic liquids, as membrane technologies offer compact alternatives with low costs and energy efficiency and do not always require energy usage in power and cement industries (Gkotsis et al. 2023).

CO_2 mineralization is an attractive carbon capture approach and method in which CO_2 is reacted with alkaline earth metals, including calcium and magnesium, to create a stable carbonate mineral, and industrial solid wastes and by-products with uncarbonated calcium are also positively used (Driver et al. 2024). Steel slag and blast furnace slag, produced during the manufacture of steel, coal fly ash, cement kiln dust, municipal incinerator ash, phosphogypsum, and wastes that contain a lot of aluminum are some of the viable feedstocks, as they allow for the generation of valuable carbonated by-products, with a global production of 3.9 Gt/year, as of 2020 (Driver et al. 2024).

3.6.3 Deliberation Pathways in the Utilized CO_2

Electrochemical CO_2 reduction (eCO_2R) is a potentially scalable, neutral carbon technology to convert waste CO_2 into useful synthetic fuels and carbon-based chemicals in mild conditions, capturing either industrial waste gas (such as cement and steel plants) or atmospheric CO_2 (Kumar et al. 2024). One of them is the production of carbon monoxide (CO) and formic acid (HCOOH), which are particularly profitable, as they provide industrial-scale rates competitive with those of commercial products, because their formation has fewer electron transfer steps than multi-electron products such as hydrocarbons (Kumar et al. 2024).

The use of CO_2 in construction materials aims to reduce the carbon footprint associated with cement and concrete production. CO_2-carbonated aggregates, including recycled concrete aggregates (RCAs) and steel slag, can be used to reduce waste generation while improving the mechanical performance of construction materials (Cho et al. 2025). In addition, equipment such as the emerging binders, like the alkali-activated and magnesium-based cements, facilitate in situ CO_2 mineralization, which contributes to the provision of a low-carbon substitute as opposed to Portland cement. Previous research has shown that carbonation curing improves strength at early ages, accelerates hydration kinetics, optimizes pore structures, and permanently sequesters CO_2 within the material matrixs (Cho et al. 2025). The application of CO_2 on cement-based products, recycled aggregate, and calciferous solid waste is synthesized and contrasted in the existing research. Building materials and composites with a carbonated structure also present a possible solution to store CO_2 for hundreds of years in infrastructure, potentially even longer than at the end of their life (Driver et al. 2024).

Biological systems offer the potential option of CO_2 conversion due to their great selectivity and benign operation conditions; microalgae exemplify an ideal CO_2 absorption capability of flue gas, and bacterial fermentation employs carbonaceous gases to grow, which are also appealing as a substitute for CO_2 fixation (Yen et al. 2015; Almomani et al. 2023). Complementary to these biological approaches is the electrochemical reduction of CO_2 (eCO_2R), which enables precise control of reaction rates and product selectivity through the application of electrical potential. This technology has significant potential for the production of clean fuels and value-added chemicals while mitigating greenhouse gas emissions. Recent large-scale laboratories have fuelled eCO_2R concepts in dusting up catalysts, mechanistic understandings, heightened product preference, and refined configurations of electrolyzers (Kumar et al. 2024).

3.7 Integrated Circular Waste-Carbon Systems

Circular waste valorization system-level design converts waste materials into valuable goods such as energy, chemicals, and raw materials and, therefore, reduces the carbon footprint of the treatment procedures and develops the principles of the CE. In the current trends, the activity demands the use of multi-criteria techniques by relying on the concept of the Safe and Sustainable by Design approach, which comprises the combination of sustainability pillars and risk analysis at the early design phases (Arias et al. 2025). With population growth and increasing demand for food and services, the generation of biotic waste is expanding, and its management becomes the key to a sustainable bioeconomy. The effective assessment and implementation of biotechnological and biorefinery valorization strategies depend on integrated urban waste management systems, which are fundamental to achieving a circular economy (Arias et al. 2025).

Circular waste-carbon systems are constructed in place of specific waste stream and process interventions to produce biochar, BECCS, carbonation, and CCU, so that the process of cycling up resources becomes more sustainable (Fidel et al. 2017). Biochar and BECCS are the two strategies of carbon removal that are complementary, but in some situations, biochar-based pyrolysis platforms could be good. Innovations comprise pyrolysis of wood ash to increase biochar and introduction of pyrogenic carbons made of biomass to cyclic CO_2 adsorption, and these biomass carbons do as well or better than synthetic sorbents (Buss et al. 2019).

Digital technologies also contribute to carbon evaluation and optimization in the CE. Customized tracing of the composition and condition of building elements with the help of material passports and digital records makes it possible to recover and reuse the material of buildings, whereas waste-to-energy systems and data-driven platforms based on Internet-of-Things (IoT) allow real-time monitoring of material flows to enhance efficiencies and reduce carbon footprint (Zhou 2025). Artificial intelligence (AI)-based digital twins (including integral digital twins) simulate projects, predict the carbon footprint of a product, and assess material products. Carbon counting, waste mapping through AI, and decentralized modular waste-to-energy systems rank among the new solutions to sustainable engineering (Zhou 2025).

3.8 Environmental, Economic, and Social Assessment

The Life cycle assessment (LCA) procedures are becoming more popular in the policymaking process to assess and rank the sustainable CDR interventions (Fig. 3.5). In addition to bioenergy with carbon capture and storage (BECCS), afforestation, reforestation, biochar application, and direct air capture and storage (DACCS), CDR technologies are required to mitigate residual emissions and support the achievement of net-zero CO_2 emissions by 2050, a key objective of the Paris Agreement aimed at limiting global temperature rise to well below 2°C (Butnar et al. 2024).

3.8.1 Life-Cycle Assessment of Waste-Based Sequestration Pathways

The research on textile waste valorization as a BECCS underlines using wastes as a source of carbon sequestration through the gate-to-grave life cycle of textile waste measurement, based on 100 metric tons of post-consumer textile waste as the standard and reporting biogenic removals and emissions under ISO 14067 (Morizet-Davis et al. 2025). For example, electricity generation from 100% cotton without carbon capture and storage (CCS) results in a net global warming potential (GWP) of 783 t CO_2e/year, primarily due to indirect emissions associated with biomass transport, material inputs, and other embodied sources. In contrast, a 50/50 blend of cotton and polyethylene terephthalate (PET) by anthropogenic combustion of polyester produces a substantially higher net GWP of 37,930 t CO_2e/year. With the inclusion of CCS, electricity generation from 100% cotton becomes a significant net carbon removal pathway, achieving a net GWP of −48,533 t CO_2e/year. By

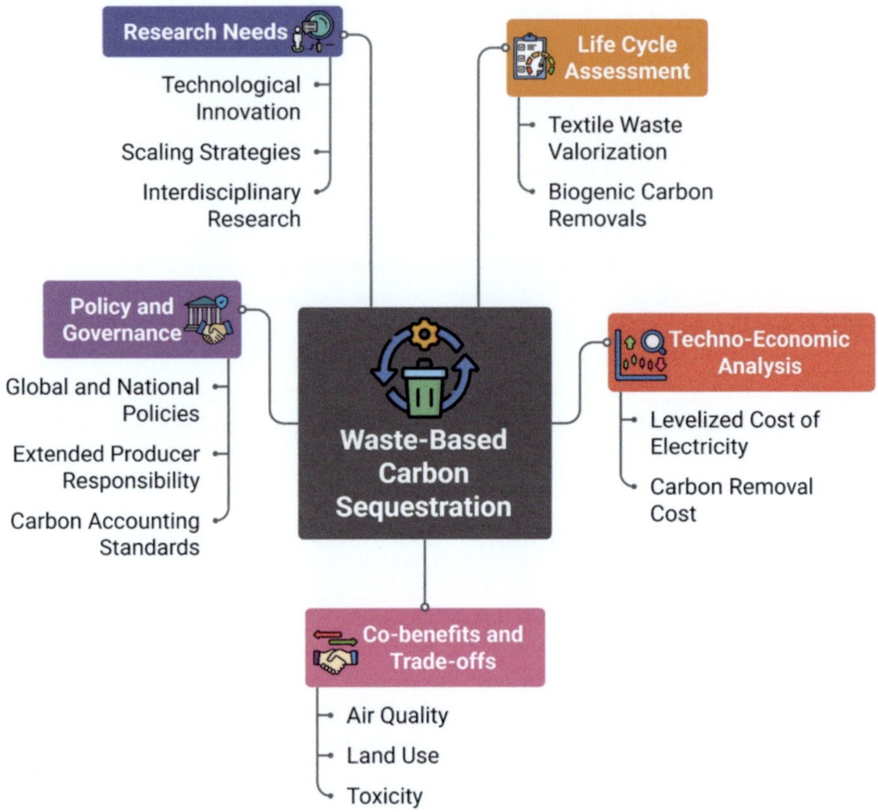

Fig. 3.5 Waste-based carbon sequestration: challenges and opportunities

comparison, the cotton–PET blend achieves a lower net removal of −22,523 t CO_2e/year because a portion of the captured carbon originates from non-atmospheric PET-derived carbon rather than biogenic carbon from cotton (Morizet-Davis et al. 2025).

3.8.2 Implications of Carbon Pricing and Techno-Economic Analysis

The techno-economic analysis of waste-based sequestration pathways evaluates the economic probability of these in terms of cost of carbon removal and levelized cost of electricity (Fig. 3.5). It is shown that 100% cotton waste is lower in terms of the cost of carbon removal of 329 USD/t CO_2e versus subsidiary 50/50 cotton-PET blends with 777 USD/t CO_2e, and the lowest levelized cost of electricity is 0.17 USD/kWh because of higher feedstock energy density versus subsidiary 50/50 cotton-PET blends at 0.29 USD/kWh owing to lower energy density, elevated capital expenditure, and reduced CO_2 sales from its lower carbon content. Policy incentives have been greatly enhanced in the recent past, such as the US 45Q tax credit that provides 85 USD/t of geologic sequestering of CO_2 and possibilities of sale of credits through cap-and-trade initiatives, along with the liquid CO_2 in the US market, which sells at 380 USD/t (Morizet-Davis et al. 2025).

3.8.3 Co-Benefits, Risks, and Trade-Offs (Air Quality, Land Use, Toxicity)

When carbon capture and storage (CCS) is incorporated within waste-based sources of energy, including PET mixtures and valorized cotton, co-benefits and trade-offs are observed in the environment for their impacts, including air quality, land use, and even toxicity (Fig. 3.5). All four CCS valorization scenarios also performed better than landfilling in overarching LCA effects, but CCS increased the risk of acidification and ecotoxicity more than non-CCS and landfilling baselines, largely due to ammonia emissions as a result of monoethanolamine degradation and upstream monoethanolamine production (greater than the amount of NaOH or activated carbon input) (Morizet-Davis et al. 2025). Although CCS scenarios may triggered respiratory impacts through ammonia-mediated PM2.5 generation as well as potentials of hazardous releases from BECCS, electricity credits from biogenic CO_2 can reduce the chances of global warming, with a result depending on the relative contributions of biogenic (cotton) and anthropogenic (PET) CO_2 (Morizet-Davis et al. 2025).

3.9 Policy, Governance, and Implementation

Carbon capture, utilization, and storage technologies and other waste-to-resource policies can only be successfully implemented with effective policy and governance (Fig. 3.5). Regarding carbon markets, a systematic simulator of carbon credit allocation and stage trading has indicated that a policy of learning in terms of carbon credit allocation can be successfully used to attain productivity, equity, and carbon emission balancing (Zhang 2021).

3.9.1 International and Domestic Waste and Climatic Policy Environments

High consideration of climate and CE in foreign policy should be adopted since neither mitigation of climate change nor the shift to a CE can be achieved without a combination of unilateral national interventions, but with the combination of such dimensions, effective international cooperation based on the 2030 Agenda of balancing economic, social, and environmental dimensions can be achieved for SDGs. This concept will represent policy coherence-based sustainable development, a concept that is best summed up by more classical policy coherence criteria that need policies sensitive to their inter-relationship aspect (Ashraf et al. 2020). Such integration can be analyzed through key factors, such as past experience, particular policy frameworks, operationalization strategies, and institutional mechanisms. This approach is closely aligned with the concept of environmental policy integration, which integrates environmental goals in non-environmental domains, such as foreign affairs (Runhaar et al. 2020).

3.9.2 Circular Economy and Extended Producer Responsibility Regulation

Extended producer responsibility (EPR) is known as one of the major policy tools that can be used to support the marketization of waste management. The EPR is perceived as a vital initiative to overcome the numerous problems in the waste management context, and its impact and efficacy have been assessed with the help of the management practice (Compagnoni et al. 2025). The relation between the CE and the EPR policies, as well as their potential to develop effective strategies in terms of waste reduction, has been discussed. The EPR concept in terms of packaging has been examined in light of its efficiency in shifting toward the CE. The effect of EPR programs on the behavior of consumers and producers is also under research on how circular they can be. The EPR policies are also involved in the system architecture of the policy tool and in the similarities between waste reverse logistics networks of various products (Mayanti and Helo 2024).

3.9.3 Carbon Accounting Standards, MRV, and Certification

Monitoring, reporting, and verification (MRV) is a set of processes, techniques, and guidelines for quantifying, reporting, and verifying carbon emissions and removals, which have become the core of credible CDR programs (WRI 2024). The quality of the MRV systems requires strict requirements that come in overriding rules, regulations, and procedures that ensure uniformity and transparency, which is proved with the help of the technical review of CDR methods and their accounting schemes, and a detailed methodological instruction to disclose GHG emissions in six categories, made public in November 2020 (PCAF 2022). To integrate available research output, such endeavors underpin the development of GHG accounting practices and international standards and focus on effective and science-oriented MRV based on institutional frameworks sensitive to the various socio-economic contexts. Finally, the last solution is to ensure that all the carbon sinks are registered in the Global C-Sink Registry so that effective MRV can be promoted and create confidence in the certification processes (Schmidt et al. 2025).

3.9.4 Financing Mechanisms and Market Instruments (Carbon Credits, Green Procurement)

Clean energy projects can provide low- and middle-income countries with important financing support through carbon credits mechanisms, while waste management facilities may generate additional revenues by accessing these markets. Green public procurement represents a prominent demand-side policy instrument that helps to reduce carbon emissions in urban environments. For example, assessments of procurement practices in the food sector have reported reductions of more than 3.1 million tons of CO_2 emissions and the creation of over 12,000 jobs following the implementation of green procurement policies (Hasanbeigi et al. 2019; Nassary et al. 2025). Complementary measures include the application of composted organic feedstocks as soil supplements to reduce near-term climatic effects and the use of carbon-enriched biomass to reduce atmospheric emission. In addition, composted biomass products can support sustainable crop production while reducing reliance on energy-intensive inputs and synthetic fertilizers.

3.10 Future Research and Requirements

The current studies in carbon sequestration and recyclability of wastes should be centered on enhancing the potential carbon sequestration of biochar, especially by determining how physicochemical characteristics and surface chemistry influence the potential carbon sequestration of biochar. Besides, to determine the long-term consequences of biochar aging in soil and its carbon trapping capabilities, large field studies are necessary (Fig. 3.5).

3.10.1 Waste Stream Technological Innovation Priorities

The example of blockchain and AI is also changing how the CE operates in that waste can be harnessed to produce sustainable energy recovery and material recirculation, and this aspect is part of the virtuous biomass-driven model, as the use of biological resources is being operated on according to the principles of the circular bioeconomy. The efforts in encouraging people to be more aware of such sustainable practices, such as IEA Bioenergy Task 36, are critical in helping to achieve SDGs by reducing the number of GHGs produced into the atmosphere and reducing the pressure at landfills as the world transitions to biofuel production processes based on the principles of the CE and green innovation (Roberts et al. 2022; Islam et al. 2024).

3.10.2 *Scaling the Waste-Based Carbon Removal in the Global South*

The Global North has largely dominated discussion on carbon removal policy evaluation, innovation, and development due to its advantages in research capacity, technological innovation, funding, and policy development across the climate, energy, and land-use sectors. However, as population growth, greenhouse gas emissions, and economic influence increasingly shift to the Global South, expanding carbon removal research, policy development, and implementation beyond the Europe, the USA, Japan, and Australia has become essential (Sovacool 2023). The varying justice, governance, and systems in the region require the South to prioritize limited technologies, which would enable slow decarbonization to take care of historical harms, overcoming limitations such as finance, IP (mostly North-owned patents and copyrights), capacity, and knowledge by sharing models such as public licensing, which would adapt to localization (Nawaz et al. 2025).

3.10.3 *The Gaps in the Knowledge and Interdisciplinary Research*

Major studies are seriously needed to define the sustainable amount of carbon removal regarding its social and ecological effects, other than the top-down estimates prevalent today, based on temperature or residual emission levels, optimized on cost. A just transition requires evaluating the extent to which carbon removal technologies can be implemented while addressing resource limitations, trade-offs, and concerns about equity and justice. This includes balancing the high energy requirements of carbon removal approaches with the availability of renewable energy needed to meet broader societal needs (Nawaz et al. 2025). The studies with the level of spatial comprehensiveness and explicit learning of resource dimensions and pathways, as well as the regional examination of the public opinion, will be crucial in providing feasibility and consistency with the principles of equity. Besides, techno-economic studies should include social science and co-research of the environmental/social effects, cooperating with communities to assess the benefits and risks of the project (Nawaz et al. 2025).

3.10.4 *Evolution and Perspectives of Waste-Based Adverse Emissions*

Carbon removal involves both technological solutions and nature-based options that help to extract the CO_2, a by-product of fossil fuel combustion out of the atmosphere. Over the past century, human use of natural resources has increased dramatically. Between 1900 and 2015, the world increased its material extraction by 12 times, reaching 89 Gt/year, and in-use stocks rose subtly than those of inputs to produce waste outflows, with about 58 Gt/year in 2015 (including 35% solid wastes and 25% emissions). After 2002, the rate material consumption has accelerated and reached 53 times higher in 2015 compared to the 2008 economic crisis. If current trends continue, the projections yet warn of doubling per capita metabolic rates to 22 tons/capita/year and 218 Gt/year extraction by 2050 (Krausmann et al. 2018). These growing material cycles require an urgent intervention to reduce them to sustainable amounts with direct air capture, emerging as the lowest-risk strategy due to its immediate, verifiable, and permanent carbon sequestration.

3.11 Conclusion

This chapter has demonstrated that how waste valorization serves as a key pillar of the circular carbon economy (CCE), and displays how different waste streams can be transformed into valuable resources. These streams include organic biomass, MSW, construction and demolition waste, and industrial residues. The valorization of these waste streams can enhance carbon sequestration and contribute to greenhouse gas (GHG) emission reductions through a range of innovative approaches, including biochar production, bioenergy with carbon capture and storage (BECCS), accelerated mineral carbonation, and the utilization of CO_2 in construction materials and fuels. Studies have shown that employing all of these strategies not only produces net-negative GHG emissions but also can remove more carbon from the atmosphere than they release. The BECCS and biochar alone have the potential to offset billions of tons of CO_2 per year—a global mitigation potential of more than 10 Gt CO_2/year. Beyond carbon removal, it also led to other valuable co-benefits, such as improved health of the soil, creation of renewable energy, recovery of resources, and financial feasibility.

However, there are still challenges, such as competition on feedstock, limited scalability in the Global South, and trade-offs related to land use or toxicity. To overcome these challenges, we need more focused and specialized research, along with innovative solutions such as digital technologies like AI-driven digital twins, which can help improve system design and efficiency. Innovative methods to finance interventions are also essential to bridge the knowledge gap and accelerate deployment. Ultimately, adopting waste as a carbon sink can support a fair and healthy transition to a lower carbon future, which is in line with the Paris Agreement objectives and with SDGs for future generations.

References

Al Shehri T, Braun JF, Howarth N, Lanza A, Luomi M (2023) Saudi Arabia's climate change policy and the circular carbon economy approach. Clim Pol 23(2):151–167. https://doi.org/10.1080/14693062.2022.2070118

Almomani F, Abdelbar A, Ghanimeh S (2023) A review of the recent advancement of bioconversion of carbon dioxide to added value products: a state of the art. Sustainability 15(13):10438. https://doi.org/10.3390/su151310438

Alshaeer HAY, Irwan JM, Alshalif AF, Al-Fakih A, Ewais DYZ, Salmi A, Alhokabi AA (2022) Review on carbonation study of reinforcement concrete incorporating bacteria as a self-healing approach. Materials 15(16):5543. https://doi.org/10.3390/ma15165543

Ansari SA, Kumar T, Sawarkar R, Gobade M, Khan D, Singh L (2024) Valorization of food waste: a comprehensive review of individual technologies for producing bio-based products. J Environ Manag 364:121439. https://doi.org/10.1016/j.jenvman.2024.121439

Arias A, Feijoo G, Moreira MT, Tukker A, Cucurachi S (2025) Advancing waste valorization and end-of-life strategies in the bioeconomy through multi-criteria approaches and the safe and sustainable by design framework. Renew Sust Energ Rev 207:114907. https://doi.org/10.1016/j.rser.2024.114907

Ashraf N, Knaepen H, Van Seters J, Mackie J (2020) ECDPM's the integration of climate change and circular economy in foreign policies. The European Centre for Development Policy Management, Maastricht. www.ecdpm.org/dp274

Baker RW, Freeman B, Kniep J, Huang YI, Merkel TC (2018) CO 2 capture from cement plants and steel Mills using membranes. Ind Eng Chem Res 57(47):15963–15970. https://doi.org/10.1021/acs.iecr.8b02574

Buss W, Jansson S, Wurzer C, Mašek O (2019) Synergies between BECCS and biochar – maximizing carbon sequestration potential by recycling wood ash. ACS Sustain Chem Eng 7(4):4204–4209. https://doi.org/10.1021/acssuschemeng.8b05871

Butnar I, Lynch J, Vetter S, Gamaralalage D, Tang Y, McKechnie J, Foteinis S, Rodway-Dyer S, Röder M, Sogbesan S, Hastings A, Renforth P, Brander M, House J (2024) A review of life cycle assessment methods to inform the scale-up of carbon dioxide removal interventions. WIREs Energy Environ 13(6):e540. https://doi.org/10.1002/wene.540

Cho S, Lee N, Kim MO (2025) A comprehensive review on CO_2 utilization in cement and concrete: from microstructural transformations to structural applications. Case Stud Constr Mater 23: e05265. https://doi.org/10.1016/j.cscm.2025.e05265

Compagnoni M, Grazzi M, Pieri F, Tomasi C (2025) Extended producer responsibility and trade flows in waste: the case of batteries. Environ Resour Econ 88(1):43–76. https://doi.org/10.1007/s10640-024-00907-5

Deng X, Teng F, Zhang X, Fan J-L, Forsell N, Reiner DM (2025) Co-deploying biochar and bioenergy with carbon capture and storage improves the cost-effectiveness and sustainability of China's carbon neutrality. One Earth 8(1):101172. https://doi.org/10.1016/j.oneear.2024.12.008

Dong Z, Zhang L, Li H, Gong Y, Jiang Y, Peng Q (2022) Knowledge mapping and institutional prospects on circular carbon economy based on scientometric analysis. Int J Environ Res Public Health 19(19):12508. https://doi.org/10.3390/ijerph191912508

Driver JG, Bernard E, Patrizio P, Fennell PS, Scrivener K, Myers RJ (2024) Global decarbonization potential of CO2 mineralization in concrete materials. Proc Natl Acad Sci USA 121(29): e2313475121. https://doi.org/10.1073/pnas.2313475121

Enaime G, Lübken M (2021) Agricultural waste-based biochar for agronomic applications. Appl Sci 11(19):8914. https://doi.org/10.3390/app11198914

EREF (2024) Analysis of waste streams entering MSW landfills. Published on the Internet. https://erefdn.org/product/analysis-of-waste-streams-entering-msw-landfills/

Eykelbosh AJ, Johnson MS, Santos de Queiroz E, Dalmagro HJ, Guimarães Couto E (2014) Biochar from sugarcane filtercake reduces soil CO2 emissions relative to raw residue and improves water retention and nutrient availability in a highly-weathered tropical soil. PLoS One 9(6):e98523. https://doi.org/10.1371/journal.pone.0098523

Fidel R, Archontoulis S, Babcock B, Brown RC, Dokoohaki H, Hayes D, Laird DA, Miguez F, Wright MM (2017) Commentary on 'current economic obstacles to biochar use in agriculture and climate change mitigation' regarding uncertainty, context-specificity and alternative value sources. Carbon Manag 8(2):215–217. https://doi.org/10.1080/17583004.2017.1306408

Fu B, Chen L, Huang H, Qu P, Wei Z (2021) Impacts of crop residues on soil health: a review. Environ Pollut Bioavailability 33(1):164–173. https://doi.org/10.1080/26395940.2021.1948354

Ghiat I, Banu A, Bicer Y, Amhamed AI, Al-Ansari T (2025) Circularity within carbon capture networks: a review of capture and utilization technologies. J CO_2 Util 95:103075. https://doi.org/10.1016/j.jcou.2025.103075

Gkotsis P, Peleka E, Zouboulis A (2023) Membrane-based technologies for post-combustion CO2 capture from flue gases: recent progress in commonly employed membrane materials. Membranes 13(12):898. https://doi.org/10.3390/membranes13120898

Guo H, Xu H, Liu J, Nie X, Li X, Shu T, Bai B, Ma X, Yao Y (2022) Greenhouse gas emissions in the process of landfill disposal in China. Energies 15(18):6711. https://doi.org/. https://doi.org/10.3390/en15186711

Hansen OW, van den Bergh J (2023) Environmental problem shifting from climate change mitigation: a mapping review. PNAS Nexus 3(1):pgad448. https://doi.org/10.1093/pnasnexus/pgad448

Hasanbeigi A, Becqué R, Springer C (2019) Curbing carbon from consumption: the role of green public procurement. Int Organ Pandemic Control 4:1–104

Ilmas B, Mir KA, Khalid S (2018) Greenhouse gas emissions from the waste sector: a case study of Rawalpindi in Pakistan. Carbon Manag 9(6):645–654. https://doi.org/10.1080/17583004.2018.1530025

Islam NF, Gogoi B, Saikia R, Yousaf B, Narayan M, Sarma H (2024) Encouraging circular economy and sustainable environmental practices by addressing waste management and biomass energy production. Reg Sustain 5(4):100174. https://doi.org/10.1016/j.regsus.2024.100174

Jalali M, Panam Z, Jalali M, Buss W (2025) The impact of feedstock type and pyrolysis parameters on the physical and chemical properties of biochars for sorption, agricultural and carbon sequestration applications: a meta-analysis. J Anal Appl Pyrolysis 192:107271. https://doi.org/10.1016/j.jaap.2025.107271

Jayakumar A, Morrisset D, Koutsomarkos V, Wurzer C, Hadden RM, Lawton L, Edwards C, Mašek O (2023) Systematic evaluation of pyrolysis processes and biochar quality in the operation of a low-cost flame curtain pyrolysis kiln for sustainable biochar production. Curr Res Environ Sustain 5:100213. https://doi.org/10.1016/j.crsust.2023.100213

Jing Y, Zhang Y, Han I, Wang P, Mei Q, Huang Y (2020) Effects of different straw biochars on soil organic carbon, nitrogen, available phosphorus, and enzyme activity in paddy soil. Sci Rep 10(1):8837. https://doi.org/10.1038/s41598-020-65796-2

Krause M, Kenny S, Stephenson J, Singleton A (2023) Quantifying methane emissions from landfilled food waste. US Environmental Protection Agency Office of Research and Development

Krausmann F, Lauk C, Haas W, Wiedenhofer D (2018) From resource extraction to outflows of wastes and emissions: the socioeconomic metabolism of the global economy, 1900–2015. Glob Environ Chang 52:131–140. https://doi.org/10.1016/j.gloenvcha.2018.07.003

Kumar B, Muchharla B, Dikshit M, Dongare S, Kumar K, Gurkan B, Spurgeon JM (2024) Electrochemical CO 2 conversion commercialization pathways: a concise review on experimental Frontiers and technoeconomic analysis. Environ Sci Technol Lett 11(11):1161–1174. https://doi.org/10.1021/acs.estlett.4c00564

Lelicińska-Serafin K, Manczarski P, Rolewicz-Kalińska A (2023) An insight into post-consumer food waste characteristics as the key to an organic recycling method selection in a circular economy. Energies 16(4):1735. https://doi.org/10.3390/en16041735

Li J, Hitch M, Power I, Pan Y (2018) Integrated mineral carbonation of ultramafic mine deposits—a review. Minerals 8(4):147. https://doi.org/10.3390/min8040147

Li N, Sippel S, Linscheid N, Rödenbeck C, Winkler AJ, Reichstein M, Mahecha MD, Bastos A (2024) Enhanced global carbon cycle sensitivity to tropical temperature linked to internal climate variability. Sci Adv 10(39):eadl6155. https://doi.org/10.1126/sciadv.adl6155

Lorenz K, Lal R (2015) Managing soil carbon stocks to enhance the resilience of urban ecosystems. Carbon Manag 6(1–2):35–50. https://doi.org/10.1080/17583004.2015.1071182

Maalouf A, Mavropoulos A (2023) Re-assessing global municipal solid waste generation. Waste Manag Res 41(4):936–947. https://doi.org/10.1177/0734242X221074116

Mayanti B, Helo P (2024) Circular economy through waste reverse logistics under extended producer responsibility in Finland. Waste Manag Res 42(1):59–73. https://doi.org/10.1177/0734242X231168801

McDonald L, Krol A (2026) Organic waste. Published on the Internet. https://climate.mit.edu/explainers/organic-waste

Mencía RV d l V, Rojas MF, Martínez-Ramírez S, Fernández-Carrasco L, Cociña EV, García-Giménez R (2021) Reactivity of binary construction and demolition waste mix as supplementary cementitious materials. *Materials* 14(21):6481. https://doi.org/10.3390/ma14216481

Merah A (2024) Methods of concrete accelerated carbonation test: a review. Discov Civ Eng 1(1): 53. https://doi.org/10.1007/s44290-024-00057-z

Morizet-Davis J, Qiu Y, Khongpatimakorn J, Daystar J, Lan K, Park S, Sagues WJ, Venditti RA (2025) Environmental life cycle assessment and techno-economic analysis of textile waste valorization via modular bioenergy with carbon capture, utilization, and storage. Bioenergy Res 19(1):4. https://doi.org/10.1007/s12155-025-10909-w

Mukome FND, Zhang X, Silva LCR, Six J, Parikh SJ (2013) Use of chemical and physical characteristics to investigate trends in biochar feedstocks. J Agric Food Chem 61(9): 2196–2204. https://doi.org/10.1021/jf3049142

Müller LJ, Kätelhön A, Bachmann M, Zimmermann A, Sternberg A, Bardow A (2020) A guideline for life cycle assessment of carbon capture and utilization. Front Energy Res 8:15. https://doi.org/10.3389/fenrg.2020.00015

Nassary EK, Magubika AJ, Mwampashi LL, Fukah FK, Kahangwa CA, Tryphone GM (2025) Transitioning to clean energy and opportunities for developing countries. Energ Strat Rev 62: 101909. https://doi.org/10.1016/j.esr.2025.101909

National Academies (2019) Chapter 4: Bioenergy with carbon capture and sequestration. In: Negative emissions technologies and reliable sequestration: a research agenda. National Academies Press, pp 137–188. https://www.nationalacademies.org/read/25259/chapter/6

Nawaz S, McLaren D, Caggiano H, Hudson AD, Scott-Buechler C (2025) Carbon removal for a just transition. Clim Pol 25(6):926–937. https://doi.org/10.1080/14693062.2024.2418305

Neves Junior A, Dweck J, Filho RDT, Ellis B, Li V (2019) Determination of CO2 capture during accelerated carbonation of engineered cementitious composite pastes by thermogravimetry. J Therm Anal Calorim 138(1):97–109. https://doi.org/. https://doi.org/10.1007/s10973-019-08210-y

Nordahl SL, Devkota JP, Amirebrahimi J, Smith SJ, Breunig HM, Preble CV, Satchwell AJ, Jin L, Brown NJ, Kirchstetter TW, Scown CD (2020) Life-cycle greenhouse gas emissions and human health trade-offs of organic waste management strategies. Environ Sci Technol 54(15): 9200–9209. https://doi.org/10.1021/acs.est.0c00364

PCAF (2022) The global GHG accounting and reporting standard part A. https://carbonaccountingfinancials.com/files/downloads/PCAF-Global-GHG-Standard.pdf

Qi JQ, Yuan HY, Zhuang QL, Zama EF, Tian XF, Tao BX, Zhang BH (2023) Effect of different types of biochar on soil properties and functional microbial communities in rhizosphere and bulk soils and their relationship with CH4 and N2O emissions. Front Microbiol 14:1292959. https://doi.org/10.3389/fmicb.2023.1292959

Quaghebeur M, Nielsen P, Horckmans L, Van Mechelen D (2015) Accelerated carbonation of steel slag compacts: development of high-strength construction materials. Front Energy Res 3:52. https://doi.org/10.3389/fenrg.2015.00052

Roberts D, Edo M, Johansson I, Hoffman B, Becidan M, Ciceri G, Murphy F, Trois C, Curran TP (2022) Material and energy valorisation of waste in a circular economy. IEA Bioenergy. Task Rep 36:1–27. https://www.ieabioenergy.com/wp-content/uploads/2022/05/T36_Waste_Circuar_Economy_final_report.pdf

Romanov V, Soong Y, Carney C, Rush GE, Nielsen B, O'Connor W (2015) Mineralization of carbon dioxide: a literature review. ChemBioEng Rev 2(4):231–256. https://doi.org/10.1002/cben.201500002

Runhaar H, Wilk B, Driessen P, Dunphy N, Persson Å, Meadowcroft J, Mullally G (2020) Policy integration. In: Architectures of earth system governance. Cambridge University Press, pp 183–206. https://doi.org/10.1017/9781108784641.009

Schmidt H, Kammann C, Hagemann N (2025) Global biochar C-sink standard 3.2. Ithaka Institute. https://www.carbon-standards.com/docs/transfer/4000039EN.pdf

Serafini LF, Praça PJGM, González-Andrés F, Gonçalves A (2025) Life cycle approach as a tool for assessing municipal biowaste treatment units: a systematic review. Waste Manag Res 43(10): 1509–1524. https://doi.org/10.1177/0734242X251326866

Shahbaz M, AlNouss A, Ghiat I, McKay G, Mackey H, Elkhalifa S, Al-Ansari T (2021) A comprehensive review of biomass-based thermochemical conversion technologies integrated with CO2 capture and utilisation within BECCS networks. Resour Conserv Recycl 173:105734. https://doi.org/10.1016/j.resconrec.2021.105734

Sosa-Martínez JD, Montañez J, Contreras-Esquivel JC, Balagurusamy N, Gadi SK, Morales-Oyervides L (2023) Agroindustrial and food processing residues valorization for solid-state fermentation processes: a case for optimizing the co-production of hydrolytic enzymes. J Environ Manag 347:119067. https://doi.org/10.1016/j.jenvman.2023.119067

Sovacool BK (2023) Expanding carbon removal to the global south: thematic concerns on systems, justice, and climate governance. Energy Clim Chang 4:100103. https://doi.org/10.1016/j.egycc.2023.100103

Stylianou E, Pateraki C, Ladakis D, Cruz-Fernández M, Latorre-Sánchez M, Coll C, Koutinas A (2020) Evaluation of organic fractions of municipal solid waste as renewable feedstock for succinic acid production. Biotechnol Biofuels 13(1):72. https://doi.org/10.1186/s13068-020-01708-w

Suparmin P, Purwanti N, Oscar Nelwan L, Halomoan Tambunan A (2024) Syngas production by biomass gasification: a meta-analysis. Renew Sust Energ Rev 206:114824. https://doi.org/10.1016/j.rser.2024.114824

Tadesse A, Lee J (2023) Utilization of methane from municipal solid waste landfills. Environ Eng Res 29(1):230166–230160. https://doi.org/10.4491/eer.2023.166

Torrenti JM, Amiri O, Barnes-Davin L, Bougrain F, Braymand S et al (2022) The FastCarb project: taking advantage of the accelerated carbonation of recycled concrete aggregates. Case Stud Constr Mater 17:e01349. https://doi.org/10.1016/j.cscm.2022.e01349

Tun MM, Juchelková D (2018) Estimation of greenhouse gas emissions: an alternative approach to waste management for reducing the environmental impacts in Myanmar. Environ Eng Res 24(4):618–629. https://doi.org/10.4491/eer.2018.364

Venkatesh G (2022) Circular bio-economy—paradigm for the future: systematic review of scientific journal publications from 2015 to 2021. Circ Econ Sustain 2(1):231–279. https://doi.org/10.1007/s43615-021-00084-3

Vinti G, Bauza V, Clasen T, Di Trapani D, Viviani G (2025) Prioritizing dumpsite risks and interventions: an overview and framework for action with a focus on LMICs. Environ Monit Assess 198(1):3. https://doi.org/10.1007/s10661-025-14842-5

Weng Z(H), Van Zwieten L, Tavakkoli E, Rose MT, Singh BP et al (2022) Microspectroscopic visualization of how biochar lifts the soil organic carbon ceiling. Nat Commun 13(1):5177. https://doi.org/10.1038/s41467-022-32819-7

Wijesekara H, Colyvas K, Rippon P, Hoang SA, Bolan NS, Manna MC, Thangavel R, Seshadri B, Vithanage M, Awad YM, Surapaneni A, Saint C, Tian G, Torri S, Ok YS, Kirkham MB (2021) Carbon sequestration value of biosolids applied to soil: a global meta-analysis. J Environ Manag 284:112008. https://doi.org/10.1016/j.jenvman.2021.112008

Woods E, Berlin P, Daystar J, Sagues WJ (2025) Gaseous carbon dioxide removal from the composting of biomass and cotton textile waste. Waste Manag 207:115108. https://doi.org/10.1016/j.wasman.2025.115108

WRI (2024) High-quality carbon removal requires credible and consistent MRV. Published on the Internet. https://www.wri.org/technical-perspectives/measurement-reporting-verification-of-carbon-removal

Xu H, Ou L, Li Y, Hawkins TR, Wang M (2022) Life cycle greenhouse gas emissions of biodiesel and renewable diesel production in the United States. Environ Sci Technol 56(12):7512–7521. https://doi.org/10.1021/acs.est.2c00289

Yang J, Xia L, van Groenigen KJ, Zhao X, Ti C, Wang W, Du Z, Fan M, Zhuang M, Smith P, Lal R, Butterbach-Bahl K, Han X, Meng J, Liu J, Cai H, Cheng Y, Liu X, Shu X et al (2025) Sustained benefits of long-term biochar application for food security and climate change mitigation. Proc Natl Acad Sci 122(33):e2509237122. https://doi.org/10.1073/pnas.2509237122

Yen H, Ho S, Chen C, Chang J (2015) CO2, NOx and SOx removal from flue gas via microalgae cultivation: a critical review. Biotechnol J 10(6):829–839. https://doi.org/10.1002/biot.201400707

Yuan X, Suvarna M, Lim JY, Pérez-Ramírez J, Wang X, Ok YS (2024) Active learning-based guided synthesis of engineered biochar for CO2 capture. Environ Sci Technol 58(15):6628–6636. https://doi.org/10.1021/acs.est.3c10922

Yuan W, Zheng Y, Yu N, Shen Y, Hao X-K (2025) Comparative analysis of time-dependent CO2 uptake in cement and geopolymer concretes via diffusion-induced carbonation. Case Stud Constr Mater 22:e04573. https://doi.org/10.1016/j.cscm.2025.e04573

Zhang H (2021) Regulations for carbon capture, utilization and storage: comparative analysis of development in Europe, China and the Middle East. Resour Conserv Recycl 173:105722. https://doi.org/10.1016/j.resconrec.2021.105722

Zhang Z, Angst U, Troian V, Guo B, Zeng Q (2025) Durability performance of concrete incorporating carbonated recycled coarse aggregates: a review. Npj Mater Sustain 3(1):27. https://doi.org/10.1038/s44296-025-00071-x

Zhou Y (2025) AI-driven digital circular economy with material and energy sustainability for industry 4.0. Energy AI 20:100508. https://doi.org/10.1016/j.egyai.2025.100508

Chapter 4
Recent Advancements in the Generation of Biochar Bioeconomy from Agricultural Biomass Waste and Sustainable Applications in Climate-Smart Agriculture

Pranaba Nanda Bhattacharyya, Bappi Paul, Bhaskar Sarma, Laith Khalil Tawfeeq Al-Ani, Liza Handique, Ponkaj Gogoi, Rupak Kumar Sarma, and Muhammad Rehan Dastagir

4.1 Introduction

Rapid advancements in biotechnological interventions have propelled global progress across multiple sectors, driving accelerated industrialization and urban expansion. As per experts, such developments and innovations are crucial in enhancing the living standards of future generations. However, alongside these multiparous benefits, the environmental burden has intensified due to the massive generation of agrowastes, agrochemicals (Bhattacharyya et al. 2023, 2024a, b), and other toxic industry materials, leading to deposition of rapid toxic molecules in soils, water, and environment (Mahanta et al. 2023; Sarma et al. 2023). According to Chen et al. (2020),

P. N. Bhattacharyya (✉)
Department of Botany, Nanda Nath Saikia College, Titabar, Jorhat, Assam, India

Institutional Biotech Hub (IBT-HUB), Nanda Nath Saikia College, Titabar, Jorhat, Assam, India

Member Coordinator of the Asia/Pacific Network, Common Microbial Biotechnology Platform (CMBP) for sustainable agriculture in Asia-Pacific, Hanoi, Vietnam
e-mail: P.Bhattacharya@nnsaikiacollege.org

B. Paul
School of Engineering and Technology, National Forensic Sciences University, Gandhinagar, Gujarat, India

B. Sarma
Department of Botany, Dhemaji College, Dhemaji, Assam, India

L. K. T. Al-Ani
School of Biology Science, University Sains Malaysia, Minden, Pulau Pinang, Malaysia

L. Handique
Department of Botany, Jagannath Barooah University, Jorhat, Assam, India

Y. K. Mohanta et al. (eds.), *Bioresource Waste to Carbon Nanomaterial*, Nanotechnology in the Life Sciences, https://doi.org/10.1007/978-981-92-2405-0_4

approximately 7–9 billion tons of wastes are produced annually, encompassing municipal, agricultural, industrial, and hazardous categories. Agro-industrial waste generates detrimental impacts on soil, water, and air ecosystems (Bhattacharyya and Sarmah 2018), adversely affecting the plant and animal health (Bhattacharyya and Jha 2012; Bhattacharyya et al. 2016); consequently, waste generation has become a serious environmental and public health concern. As the global focus shifts towards sustainable development, addressing these emerging issues of toxicity generation and management is becoming increasingly crucial, leading to significant consideration of economically viable and environmentally sustainable alternatives (Bhattacharyya et al. 2024c). The current article highlights the transformation of agricultural waste into biochar and its subsequent application in elimination of toxicity, which has garnered significant interest. Biochar contains abundant surface functional groups, mineral ash, and charged sites, which tend to bind, adsorb, or immobilize pollutants through physical adsorption, ion exchange, complexation, and precipitation mechanisms.

Biochar, a carbon-rich charcoal-like substance (Bhattacharyya et al. 2024d), is characterized by its exceptional resilience to environmental degradation and is recognized for its porous structure and high level of aromatization. It is usually generated through the thermal transformation of plant- or animal-derived biomass at elevated temperatures under oxygen-limited conditions, thereby yielding materials with unique physical and chemical properties (Amalina et al. 2022a). Recent findings have indicated certain key attributes of biochar including its elemental composition, density, porosity, and unique pH, the properties that are strongly governed by the nature of the feedstock and production technologies. These properties, in turn, determine its suitability for multiparous applications in agriculture and industry. In water and wastewater treatment systems, biochar is widely used for the removal of both organic and inorganic contaminants (Olugbenga et al. 2023), including persistent organic pollutants (POPs). In farming, biochar enhances soil quality, minimizes nutrient depletion, and improves overall soil characteristics (Kamali et al. 2022). Furthermore, due to its high carbon content, biochar has become an important component of new-generation fuel source for sustainable energy generation (Seo et al. 2022). As per Bhatia et al. (2021), biomass itself is widely recognized as a valuable resource for producing diverse fertilizers, renewable energy resources, and diversified mineral products. Ghodake et al. (2021) have identified agricultural residues, crop by-products, activated sludge, algal biomass, animal waste, and

P. Gogoi
Institutional Biotech Hub (IBT-HUB), Nanda Nath Saikia College, Titabar, Jorhat, Assam, India

R. K. Sarma
Department of Botany, Nalbari College, Nalbari, Assam, India

M. R. Dastagir
Department of Environmental Science, IUBAT – International University of Business Agriculture and Technology, Dhaka, Bangladesh

energy crops as major biomass feedstocks. However, the conversion of agricultural biomass into value-added products including the biochar conversion can be achieved through physical, biochemical, or thermochemical processes. Hydrothermal carbonization (HTC), gasification, pyrolysis, and torrefaction are some of the most typical thermal conversion techniques usually employed in biochar production (Amalina et al. 2022b).

Although unmodified biochar has shown substantial promise in environmental applications, its performance is often hindered by limiting factors such as confined surface area, comparatively low number of active sites, and inconsistent adsorption efficiency, thereby highlighting the necessity for improved biochar production. To cope with diverse issues to improve the use efficiency of biochar, methods pertaining to physical activation, chemical modification, and surface functionalization have widely been explored to improve its structural characteristics and functional performances, thereby enhancing pollutant removal efficiency and suitability for specific agro-industry applications (Gong et al. 2022). Focusing on sustainable resource utilization, improved production methodologies, and environmental implications, the current article aims to provide an integrative and comprehensive assessment of biochar production from agricultural biomass. Primarily, we intend to illustrate how biochar can play an integral role in enhancing environmental sustainability through connecting waste valorization to real-world applications. The article critically evaluates different agricultural waste biomass feedstocks, thermal conversion processes, and the effects of multifaceted production parameters in maintaining biochar quality and functionality. This analysis distinguishes itself from earlier studies by offering a comprehensive assessment of the entire biochar life cycle, encompassing feedstock selection from agro-waste, manufacturing processes, and post-production modifications. Special emphasis is given to correlate between production circumstances and biochar efficacy in diverse environmental applications. This study provides a strategic perspective by incorporating recent advancements and pinpointing research gaps to facilitate the targeted optimization and scalable implementation of biochar for effective environmental solutions.

4.2 Types, Sources, and Components of Agricultural Biomass Wastes

Agricultural biomass wastes include organic residues including the straw, husk, leaves, crop residues, plant by-products, and animal remains from livestock that can be used for generating sustainable sources of energy (Bhattacharyya et al. 2024d; Piarpuezán Enríquez et al. 2025) in agriculture and industry including biofertilizers (Bhattacharyya et al. 2025), biopesticides, biochar, and biofuels (Bairwa et al. 2023). As per experts, agricultural biomass wastes are basically composed of three macromolecules such as cellulose, hemicellulose, and lignin and fall under two separate categories like lignocellulosic and non-lignocellulosic biomass (NLBM).

Lignocellulosic biomass is primarily composed of plant by-products, and NLBM is composed of wastes generated from livestock. Pyrolysis, a thermochemical conversion process, can turn these two types of biomass into a carbon-based material—biochar (Bhat et al. 2022). Because of its large surface area, abundance, and porosity, biochar is considered as a promising agent for environmental remediation activities. When subjected to pyrolysis at optimal temperatures, biochar effectively absorbs contaminants and heavy metals, enhances soil quality, and facilitates carbon sequestration potential.

4.2.1 Lignocellulosic vs. Non-lignocellulosic Residues

Lignocellulosic biomass usually refers to cellulose (a polymer of α-D-glucose), hemicellulose (heteropolymer of C5 and C6 glucose), or lignin components (heteropolymer of phenylpropanoid units) derived from the plants (Blasi et al. 2023). As per Hemansi et al. (2019), different lignocellulosic compounds have varied percentages of dry weight of hexosans, pentosans, and lignin. These products have originated from the by-products of agriculture and forestry activities, as well as solid waste from paper, wood, and pulp recycling facilities (Singh et al. 2022). It is noteworthy to mention that biomass, referred to as lignocellulosic biomass, is usually plant-based materials. According to agricultural experts, most of the lignocellulosic biomasses are usually obtained during the post-harvest seasons of various crops. Agricultural residues obtained from wheat, maize, paddy, cotton, sugarcane, olive pits, etc.; forestry residues like wood, wood chips, sawdust, etc.; energy crops including the switchgrass, elephant grass, and timothy grass; cellulosic wastes; herbal biomass obtained from forage plants; and specific types of municipal wastes are classified as certain key examples of lignocellulosic biomass. Fajobi et al. (2022) have investigated the physicochemical properties of selected lignocellulosic biomasses of Nigerian origin obtained from cow dung, leaves of *Chromolaena odorata*, and mango pulp. They recorded the percentages of fixed carbon in between 6.22% and 7.25%, volatile matter 5.02% and 7.79%, and ash content 1.14% and 1.91% and calorific values between 13.77 MJ/kg and 16.16 MJ/kg, respectively. The final results of the analysis showed carbon (C) content of *C. odorata* leaves, cow dung, and mango pulp to be 41.69%, 43.08%, and 39.98%; hydrogen content 9.86%, 7.87%, and 6.74%; nitrogen (N) content 1.51%, 1.53%, and 1.34%; sulphur (S) content 0.25%, 0.46%, and 0.12%; and oxygen (O) content 46.69%, 47.06%, and 51.82%, respectively. The range of percentage composition for lignin was examined as 0.22–6.33, for cellulose 3.71–12.03, and for hemicellulose 7.47–11.37, respectively. Cow dung was found to be rich in key microbial population dynamics including the species of *Bacillus cereus*, *Escherichia coli*, *Micrococcus* spp., *Proteus morganii*, *Pseudomonas aeruginosa*, and *Staphylococcus aureus*. According to Jeswani et al. (2020), utilizing waste materials for biofuel production can minimize the production costs, thereby contributing to environmental cleanliness through reducing greenhouse gas (GHG) emissions. On the other hand, biomasses obtained from animals, microorganisms, and certain plant derivatives are basically categorized as NLBM. The primary components of NLBM are lipids, inorganics, saccharides, proteins, and a small portion of plant derivatives including lignin and cellulose. Ayaz et al. (2021) have revealed on the common similarities within the lignocellulosic biomass and NLBM that includes C, N, S, phosphorus (P), calcium (Ca), potassium (K), and magnesium (Mg) in varying amounts. Besides, NLBM also contains greater quantities of heavy metals and elements like N, S, P, O, and halogens. According to Li and Jiang (2017), the presence of these metals and other components makes NLBM quite challenging materials for their disposal and utilization. However, several studies have focused on preparation of biochar from NLBM due to their unique structural organization (Liu et al. 2015). NLBM contains metals typically absent in lignocellulosic biomass, which, when subjected to heat treatment, catalyse pyrogenic decomposition. This process results in the creation of porous carbon compounds with structural anomalies, making them suitable for applications in adsorption, electrochemistry, and catalysis. Because agriculture provides income for 50–90% of the people in developing countries, processed biochar from various agricultural residues assists in waste management, reduces GHG emissions, and supports with carbon sequestration abilities.

4.3 Emerging Risks Associated with Dumping of Agricultural Wastes

According to the Food and Agriculture Organization, agricultural waste is defined as any change in food quality that makes it unsafe for human consumption. Agricultural experts argue if food waste analysis should start at harvesting or cover the entire production cycle. In middle- and high-income countries, food waste mostly happens at the consumption stage due to consumer behaviour, while in low-income countries, it primarily occurs early in the food supply chain before reaching the consumers. Agricultural waste poses considerable environmental, economic, and social challenges worldwide, with estimates indicating that 10–50% of products are squandered annually. Inadequate waste disposal techniques, such as open incineration and interment, cause significant health risks and severe ecological problems. The production of agricultural waste results in significant economic losses, including resource misallocation, diminished product quality, squandered subsidies, heightened import dependence, and lower income, thereby jeopardizing livelihoods. Besides, the management of agricultural waste-derived amendments also poses serious challenges beyond production losses. Inefficiencies in agricultural practices result in waste and economic losses, due to inadequate harvesting, improper pesticide use, and issues in transportation, leading to quality deterioration. Excessive compost application negatively impacts the soil and crop health, while animal manure contributes to potent water contaminations. Composted wastes may contain toxic heavy metals from various agricultural inputs, leading to major soil contamination, degradation, and accumulation of phytotoxicity in agriculture. Thus, sustainable management of agricultural sector is seemed to be essential to maintain the economic integrity in emerging and developed countries, emphasizing the sustainable development through safeguarding the natural resources for larger benefits of mankind. This section explores how various agricultural waste management and disposal strategies lead to considerable environmental and public health issues.

4.3.1 Burning Agricultural Residues Led to Global Warming and Air Pollution

A prevalent method employed by farmers to clean agricultural fields post-harvest for subsequent crop production is the open incineration of rice straw (Kaur et al. 2022). This combustion process subsequently generates substantial quantities of toxic residues including the polycyclic aromatic hydrocarbons, particulate matter, volatile organic compounds (VOCs), sulphur dioxide (SO_2), nitrogen dioxide (NO_2), carbon monoxide (CO), and other air pollutants. According to Maximilian et al. (2022), burning crop residues is one way that agricultural and food wastes contribute to climate crises. Another way is that they release hazardous compounds like methane (CH_4) and N_2O. Patel and Panwar (2023) have addressed that these pollutants pose significant threats to public health and contaminate natural air. Specifically, NO_2 and SO_2 are key building blocks for generation of smog, ground-level ozone, particulate matter of nitrate (NO_3^-) and sulphate, and other toxic pollutants (Ravindra et al. 2019).

4.3.2 Biosolids as a Sink of Emerging Pollutants

Biosolids generated from municipal sewage sludge are recognized as significant pollutants to soil and environment. Biosolids are typically disposed in soil through landfilling and are extensively utilized as soil amendments in farming. Borthakur et al. (2022) have addressed the benefits of biosolids in land restorations, as these substances are rich resources of essential nutrients including N, phosphate, organic carbon (C_{org}), and other minerals that are desirable for maintaining sustenance in soil quality and enhancing crop production. While the advantages of recycling of essential nutrients for crop production and minimizing the reliance on energy-intensive synthetic fertilizers are significant, biosolids can function as a repository for emerging pollutants. Mitigating unsafe exposure to these emerging pollutants during land application poses a significant challenge (Liu et al. 2021). The debate regarding the safe utilization of biosolids produced from human waste as soil amendments, thus, persists. The US Environmental Protection Agency standards for assessing biosolid quality are delineated in Title 40 of the Code of Federal

Regulations, Part 503, with a specific emphasis on the presence of ten inorganic metals including arsenic (As), cadmium (Cd), copper (Cu), mercury (Hg), molybdenum (Mo), nickel (Ni), lead (Pb), chromium (Cr), selenium (Se), and zinc (Zn), as well as microbial pathogenicity and other vector attractions. As per Black et al. (2021), agricultural soils have becoming major reservoirs of natural microplastics due to rapid exposure of biosolids in soil.

4.3.3 Release of Pathogens from Farming Wastes

Cattle and other farm animals act as major carriers of a wide range of pathogens, including bacteria, fungi, viruses, and parasites. When animal wastes containing these pathogens are directly applied to soil, they can contaminate the crops and nearby water bodies, facilitating their easy transfer to human beings. Increasing negative consequences of pathogenic microorganisms of animal origin (zoonoses) have emerged as a significant public health concern, particularly in relation to the rising incidence of water-borne diseases driven by faecal contamination of organic wastes. Farm animal wastes, including faeces, urine, and respiratory secretions, contain very high levels of pathogens, potentially millions to billions per gram. Notably, even healthy animals harbour numerous potential pathogens including germs, viruses, and other protozoa capable of causing severe infections in humans. Intensive livestock production leads to deposition of concentrated waste, thereby increasing the risks of environmental contamination and health hazards. The principal pathways through which these microbial infections are transmitted to humans are illustrated in Fig. 4.1.

Animal wastes can harbour infectious agents, including viruses like swine hepatitis E virus, bacterial pathogens including the pathogenic species of *Salmonella*, and parasites like *Cryptosporidium parvum*, which threatens both animals and human health. The widespread presence of these pathogens in livestock complicates their elimination from farming. Therefore, proper treatment and management of animal-derived agricultural wastes are crucial to reduce the environmental contamination and associated health risks.

4.3.4 Ammonia Toxicity in Waterways

Ammonia occurs in aquatic environments primarily in two forms like ammonium ions (NH_4^+) and unionized ammonia (NH_3). It is naturally produced during the decomposition of organic matter and is also excreted by fish as a nitrogenous waste product. In aquatic organisms, NH_3 is mainly eliminated through the gill membranes as a by-product of protein metabolism, with a small fraction excreted via urine. Microbial activity subsequently converts ammonia into nitrite (NO_2^-) and NO_3^- through nitrification processes (Fig. 4.2). The toxic nature of ammonia in aquatic

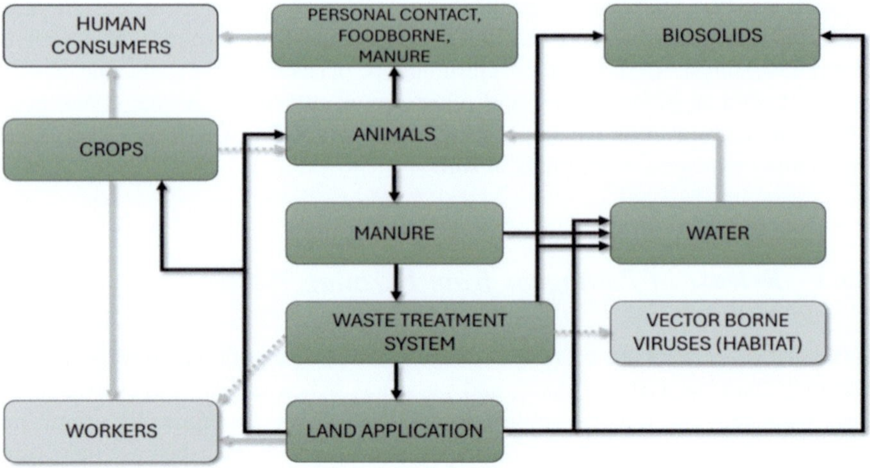

Fig. 4.1 The transmission pathways of pathogens to humans from veterinary and animal agriculture

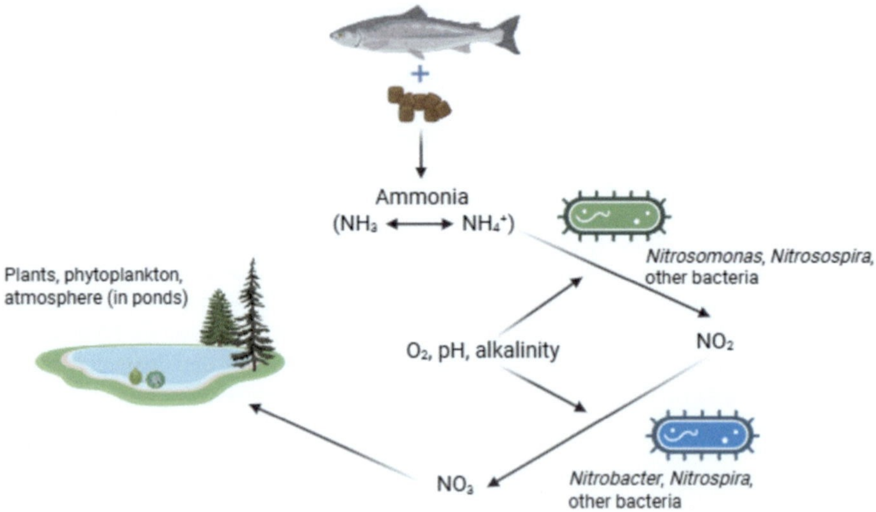

Fig. 4.2 Ammonia transformation in aquatic environments

systems is largely attributed to its unionized form, NH_3. As water pH increases, the proportion of NH_3 also rises, thereby enhancing ammonia toxicity, as illustrated in Fig. 4.2. Although NH_3 is an essential nutrient at low concentrations, excessive accumulation can disrupt metabolic processes in aquatic organisms and alter internal physiological balance, including the changes in body pH.

Exposure to heightened NH_3 concentrations can trigger hyperexcitability, loss of balance, increased oxygen intake and respiratory rates, and raised heart rate in fish. Agricultural wastes abundant in organic matter frequently include elevated levels of NH_3, representing an increased pollution threat to fish and freshwater invertebrates. NH_3 contamination impacts groundwater systems in addition to surface waters. Moreover, elevated concentrations of free NH_3 in rivers can impede nitrification in sediments, resulting in NO_2^- build-up, which poses an additional risk to aquatic ecosystems (Kim et al. 2008).

4.4 Thermochemical Conversion of Agro-waste and Production of Biochar

Agricultural biomass contains cellulose (40–50%), hemicellulose (25–35%), and lignin (15–30%) and possesses various thermal behavioural characteristics in conversion. For instance, hemicellulose breaks down at 220–315 °C, cellulose breaks down at 315–400 °C, and lignin breaks down at 160–900 °C due to its complex aromatic backbone. Thermochemical conversion process converts the agricultural biomass waste into biochar by transforming these key organic substances through controlled degradation of complex lignocellulosic structures in oxygen-limited environments (Lehmann and Joseph 2015). Nonetheless, four main technologies have emerged in thermochemical conversion of agricultural biomass waste to biochar including pyrolysis, gasification, HTC, and torrefaction. These processes differ in the operating temperature, product distribution, and biochar properties; each offers its own advantages in valorization of agricultural waste, thereby contributing to the strengthening of the circular economy. A comparative analysis to depict diverse process conditions related to thermochemical conversion has been represented in Table 4.1.

4.4.1 Diverse Pyrolysis Technologies and Process Conditions

Slow pyrolysis is the most commonly used biochar production process, operating at moderate heating rates (5–80 °C/min), higher residence times (hours to days), and temperatures of 300–700 °C, which maximize the yield of biochar (30–40 °C by mass) (Ronsse et al. 2013). Its operation involves the steps of moisture evaporation (<200 °C), active pyrolysis (200–500 °C), and passive pyrolysis, which involves stabilizing the char component (>500 °C). The temperature strongly affects the production of biochar; at 300 °C, yield decreases to 28%, and fixed carbon content decreases to 78%, but at higher temperature (600 °C), yield falls to 28%, and fixed carbon content falls to 45%. Biochar produced from pyrolysing the rice husks at 500 °C has indicated a carbon content of 65–70%, surface area of 250–350 m^2/g, and

Table 4.1 Comparative overview of thermochemical conversion technologies

Different technologies	Temperature ranges	Heating rate	Residence time	Primary products	Biochar yield (%)	Oxygen requirement
Slow pyrolysis	300–700 °C	5–80 °C/min	Hours to days	Biochar, bio-oil, gas	30–40	Oxygen-free
Fast pyrolysis	400–650 °C	200–1000 °C/s	<5 s	Bio-oil, biochar, gas	10–20	Oxygen-free
Flash pyrolysis	800–1000 °C	>1000 °C/s	<1 s	Gas, bio-oil, biochar	5–15	Oxygen-free
Gasification	700–1500 °C	Variable	Minutes to hours	Syngas, biochar	5–10	Limited oxygen
HTC	180–350 °C	Variable	Minutes to hours	Hydrochar, organics	50–80	Aqueous medium
Torrefaction	200–300 °C	Slow	30–120 min	Torrefied biomass	70–90	Inert atmosphere

cation exchange capacity (CEC) of 25–40 cmol/kg (Zhang et al. 2015), respectively. In contrast, fast pyrolysis operates at higher temperatures of 400–650 °C with rapid heating rates (200–1000 °C/s) and short vapour residence times (<5 s), thereby yielding bio-oil (60–75%) as the primary product and biochar as a by-product (10–20% yield) (Bridgwater 2012). Fast pyrolysis produces biochar with 65–70% of carbon content, surface area of 250–350 m^2/g, and CEC of 25–40 cmol/kg. Fast pyrolysis biochar contains 40–60% carbon and exhibits higher volatile matter and oxygen functional groups, which contribute to its increased reactivity and nutrient retention abilities, making it more challenging to dry.

Furthermore, flash pyrolysis represents an advanced thermochemical conversion process with a residence period of less than 1 s and operating temperatures ranging from 800 °C to 1000 °C. This process may produce as much gas as possible with an operating efficiency of 70–80% and biochar that is both highly aromatic and graphitic in nature, yielding 5–15% and exhibiting outstanding stability (Qian et al. 2013). The elevated temperature promotes significant aromatic condensation, resulting in biochar with a substantial surface area (400–800 m^2/g) and the capacity for carbon sequestration exceeding 1000 years in soil.

4.4.2 Hydrothermal Carbonization and Gasification

Gasification: It is a partial oxidation reaction mechanism conducted at temperatures of 700–1500 °C with a controlled amount of oxygen added into the feedstock with a sustained amount of oxygenation reaction producing syngas and biochar (5–10% yield) (Basu 2010). Biochar produced through gasification has a large carbon content (>80%), could undergo a great deal of aromatization, and bears small quantities of volatile matter with a median residence time in the soil of more

Table 4.2 Biochar properties of various thermochemical conversion technologies

Biochar properties	Slow pyrolysis	Fast pyrolysis	Gasification	HTC	Torrefaction + pyrolysis
Carbon content (%)	60–80	45–65	80–95	50–70	70–85
Oxygen content (%)	5–15	15–30	<5	15–30	8–18
H/C molar ratio	0.3–0.6	0.6–1.0	<0.3	0.6–1.2	0.4–0.7
O/C molar ratio	0.1–0.3	0.2–0.5	<0.1	0.3–0.6	0.15–0.35
Surface area (m^2/g)	200–450	50–200	300–800	5–50	250–500
pH	8–11	6–9	9–12	4–7	7–10
CEC (cmol/kg)	25–50	35–65	15–35	40–80	30–55
Ash content (%)	10–30	8–25	15–40	5–15	12–28
Stability (years in soil)	100–500	50–200	>1000	10–100	200–700

than 1000 years. The equivalence ratio is maintained between 0.2 and 0.4, which dictates the extent of partial oxidation; besides, the interactions between the gasifying agents (air, oxygen, steam, or CO_2) significantly influence the physicochemical characteristics and recovery rate of the resulting char.

Hydrothermal carbonization: HTC is an aqueous process performed at 180–350 °C and 2–10 Mpa with the key benefits of handling high-moisture agricultural wastes without undertaking the energy-intensive drying process. Subcritical water hydrolysis, dehydration, decarboxylation, and aromatization processes yield 50–80% hydrochar. Elevating the temperature from 200–280 °C enhances the carbon content from 52–68%, while diminishing the yield from 72–58%. Hydrochar possesses higher oxygen content (15–30%), an elevated concentration of surface functional groups, and a lower pH (4.7) compared to pyrolysis biochar, rendering it advantageous for cation exchange and nutrient retention. Table 4.2 represents diverse biochar properties generated from different thermochemical conversion technologies.

4.4.3 Comparative Analysis and Technology Selection

The choice of biochar production technology varies based on the properties of the feedstock, desired biochar properties, and application criteria. Slow pyrolysis and HTC result in maximum production of biochar, compared to gasification and high-temperature pyrolysis (>600 °C), which make biochar maximally stable for sequestration of carbon (Bhattacharyya et al. 2024d). Faster pyrolysis and HTC biochar with larger oxygen content give better nutrient retention abilities and CEC to soil to improve its fertility. Pyrolysis and gasification biochar at high temperature, offering large surface area, has been shown to have a high adsorption potential in the remediation of contaminants. Refined reactor technologies, such as microwave-assisted pyrolysis, auger reactor, and rotary kiln, have enhanced the production capacity and product characteristics of biochar. Thermochemical conversion, a thermochemical approach, integrates both thermochemical and biological processes to maximize the use of resources and attain very high thermal efficiencies, typically above 75% using heat integration. Life-cycle assessment (LCA) indicates that biochar systems serve net carbon-negative technology. The reduction in GHGs is achieved by three mechanisms: firstly, by long-term sequestration of stable carbon in the soil, followed by the utilization of syngas and bio-oil that leads to the displacement of fossil fuels, and finally by reduction in the synthetic carbon requirements that further mitigates the carbon footprint during chemical manufacturing (Roberts et al. 2010). Specifically, the process of slow pyrolysis exhibits a high carbon trapping efficiency of 40–60%. Although the above-mentioned techniques are environmentally friendly, the mass adoption of thermochemical conversion is governed by maintaining a balance between production costs and market value. Presently, the cost of producing biochar typically ranges from $200 to $800 per ton, which is mainly dictated by the scale of production and the specific type of reactor technology. For the operation scaling, industries have identified a sweet spot for viability, which is 5–10 tons per hour. At this capacity, the capital on expenditure efficiency is relatively optimized with respect to logistic challenges and biomass feedstock collection and availability.

4.5 Sustainable Chemistry and Biochar Formation

Lignocellulosic biomass is a biogenic material composed of both organic and inorganic fractions and is widely being recognized as a renewable resource for the production of value-added products (Rangabhashiyam and Balasubramanian 2019). Owing to its inherent surface chemistry characterized by abundant hydroxyl, carboxyl, and phenolic functional groups, this biomass serves as an excellent precursor material for biochar generation with high reactivity. Additionally, its elevated carbon content, structural diversity, widespread availability, and low cost render lignocellulosic biomass particularly effective for producing high-quality biochar with improved adsorption and pollutant-binding capabilities. It exhibits a structurally complex matrix mainly consisting of cellulose, hemicellulose, and lignin, along with minor quantities of extractives (such as hydrocarbons and lipids) and inorganic ash in the form of oxides and carbonates. Consequently, the type and composition of lignocellulosic biomass significantly influence the pyrolytic degradation pathways and the nature of the resulting products.

However, the suitability of lignocellulosic biomass for biochar production is largely determined by its intrinsic surface chemistry, which originates from the distinct thermal degradation behaviours of its major structural components including cellulose, hemicellulose, and lignin. The thermal degradation mechanisms of these components vary considerably due to differences in their molecular structure and composition, thereby directly influencing the surface functional groups and reactivity of the resulting biochar. During pyrolysis, cellulose primarily undergoes depolymerization activities followed by intramolecular rearrangement, dehydration, decarboxylation, aromatization, and intermolecular condensation between 315 °C and 400 °C, leading to biochar surfaces enriched with the most abundant functional groups, O–H and C–O. Hemicellulose follows a broadly similar degradation pattern but decomposes at lower temperatures (220–315 °C), contributing to the formation of carboxylic and phenolic groups, thereby promoting pore development, along with the release of low-molecular-weight compounds such as bio-oil and synthetic gas. In contrast, lignin, owing to its highly aromatic and cross-linked structure, degrades over a much wider temperature range (160–900 °C) through complex free radical-driven mechanisms, resulting in biochar with highly aromatic, graphitic-like surfaces and π–π electron systems that enhance structural stability and affinity towards organic pollutants. Collectively, these component-specific degradation pathways explain how variations in lignocellulosic biomass composition shape biochar surface chemistry, thereby determining its adsorption efficiency and environmental performances (Zhang et al. 2019).

Cellulose and hemicellulose mainly contribute to the formation of volatile products during pyrolysis, whereas lignin acts as the principal precursor of biochar formation, resulting in higher biochar yields and distinct physicochemical characteristics. For example, woody biomass produces approximately 19 wt.% biochar at 900 °C after 64 min, whereas lignin-derived biochar yields up to 45.69% (Farrokh et al. 2018). Therefore, biomass feedstocks rich in lignin are particularly examined as suitable for thermochemical conversion processes to obtain biochar with high fixed carbon content, enhanced surface area, and well-developed aromatic structures (Wang et al. 2020). Thus, feedstock selection plays a decisive role in determining the biochar yield, quality, and overall economic feasibility of the pyrolysis process.

4.5.1 Biochar Formation: An Integrative Approach

In recent years, multiple biomass conversion and valorization strategies have been developed, including thermochemical conversion processes (torrefaction, pyrolysis, hydrothermal liquefaction, hydrothermal combustion, and gasification) and biochemical (anaerobic and aerobic digestion, sugar fermentation) and physicochemical methodologies such as lipid extraction (Wang et al. 2018). Among all of these technologies, pyrolysis has emerged as a prominent thermochemical route due to its effectiveness in valorizing residual biomass from both the environmental and economic perspectives. Slow pyrolysis specifically yields biochar, a cost-effective, carbon-rich solid by-product characterized by a porous and heterogeneous structure. This technique has garnered significant interest due to its efficacy in transforming waste biomass into stable carbon compounds. The following subsections will highlight different standard protocols for biochar formation for improved sustainability in agro-industries.

4.5.1.1 Thermochemical Conversion of Lignocellulosic Biomass: An Effective Way for Improved Biochar Formation

Thermochemical conversion is often regarded carbon-neutral, as the CO_2 released during the process is approximately equivalent to the CO_2 absorbed during biomass growth (Jafri et al. 2018). Thermochemical conversion of lignocellulosic biomass involves techniques such as torrefaction, hydrothermal liquefaction, HTC, pyrolysis, and gasification. These methods utilize controlled temperature and pressure under inert or oxygen-limited conditions to thermally cleave biomass bonds, yielding biochar along with liquid, gaseous, and aqueous by-products. This conversion process results in irreversible alterations to the biomass's physical and chemical structure due to the breakdown and reorganization of lignocellulosic components. Among available techniques, slow pyrolysis has gained significant prominence in producing biochar with tailored properties and for serving as a sustainable waste management strategy through utilizing residual biomass, otherwise disposed of in landfills.

The generalized pyrolysis reaction can be represented in as follows:

$$\text{Lignocellulosic biomass} + \text{heat} \rightarrow \text{Biochar} + \text{pyrolytic liquid} + \text{gaseous fraction}$$

4.5.1.2 Effects of Pyrolytic Process Parameters on Quality Biochar Yield

Biochar formation occurs in three distinct stages, i.e., pre-pyrolysis (<200 °C), main pyrolysis (200–500 °C), and carbonaceous product formation (>500 °C). Increasing pyrolysis temperature strongly affects the biochar structure, surface chemistry, and diverse physicochemical properties. Higher temperatures generally increase surface area, aromaticity, pH, fixed carbon, and hydrophobicity, while reducing the CEC and oxygen-containing functional groups. Elevated temperatures promote pore development through the release of volatiles and organic matter degradation activities. Above 650 °C, biochar becomes thermally stable and highly hydrophobic. Micropore formation and aromatic condensation enhance resistance to microbial decomposition and sorption capacity (Xie et al. 2015). Conversely, low-temperature biochars retain more labile functional groups, influencing the nutrient availability and plant growth patterns.

Variations in pyrolysis temperature also cause volatile accumulation and changes in ash and carbon content and pH variations. Increasing temperature reduces volatile matter due to enhanced cracking reactions (Zhao et al. 2017), while carbon and ash contents increase due to concentrations of inorganic residues and polymerized carbon structures (Domingues et al. 2017). Biochar pH increases with temperature owing to carbonate formation, alkali metal concentration, and loss of acidic functional groups (Ding et al. 2014). According to Tripathi et al. (2016), the yield and qualities of biochar are significantly impacted by the biomass feedstock characteristics including moisture, lignin, cellulose, hemicellulose, and ash content. Woody biomass generally exhibits lower moisture and ash content and higher calorific value than non-woody biomass (Jafri et al. 2018). Low moisture content is desirable, as it reduces energy consumption and enhances process efficiency (Tripathi et al. 2016). High lignin content in biomass enhances char formation, while cellulose and hemicellulose are more prone to decompose into volatile products. Biochar produced from woody biomass is characterized by lower ash content and greater surface area compared to biochars derived from manure or sludge. Variations in carbon content, aromaticity, surface area, and porosity are influenced by feedstock type and their respective thermal degradation behaviours and inorganic compositions. Besides, the CEC of biochar also varies with feedstock type and ash composition. Manure- and algae-derived biochars generally exhibit higher CEC due to increased inorganic content and oxygen-containing functional groups, whereas woody biochars typically show lower CEC (Tag et al. 2016). Carboxylic functional groups contribute most significantly to CEC, along with specific surface area.

Biochars are generally alkaline, with pH values ranging from 7.1 to 10.5, depending on feedstock and pyrolysis conditions. The pH of biochar is closely associated with its lignin, hemicellulose, and cellulose composition, as well as the presence of oxygen-containing functional groups. The alkaline nature of biochars mainly arises from –COO– and –O– functional groups along with their carbonate content. Mukome et al. (2013) demonstrated that biochar pH showed the strongest correlation with oxygen content ($R^2 = 0.7$), supporting earlier reports that biochar basicity originates from oxygen-rich functional moieties such as γ-pyrone-type, chromene, diketone, and quinone groups. During the pyrolysis of cellulose-rich materials, compounds such as levoglucosan and its derivatives including levoglucosenone, furfural, 2,3-butanedione, and 5-methylfurfural are generated, leading to the formation of oxygen-based functional groups. This phenomenon is associated with polymerization and condensation reactions of aliphatic compounds, as well as dehydration processes occurring within the feedstock. Furthermore, Li et al. (2013) observed that the pH of biochars produced from rice straw and rice bran exhibited a negative relationship with aliphatic O-alkyl carbons and anomeric O–C–O carbons, while showing a positive association with fused aromatic ring structures and aromatic C–O groups. As a result, a higher proportion of carboxyl groups reduced during pyrolysis and/or acidic groups that undergo deprotonation to form conjugate bases, thereby contributing to an increase in the alkalinity of biochar (Tag et al. 2016).

4.6 Surface Chemistry of Biochar and Functional Group Analysis

Understanding biochar surface chemistry is crucial for knowing its interactions with soil minerals, nutrients, organic matter, water, and contaminants, thereby playing key roles in agricultural sustainability. The heterogeneity of biochar surface arises from the aromatic carbon structure, oxygen-based functional groups, inorganic minerals, and porous structures formed during the thermochemical conversion process. Characterization employs multiple techniques including spectroscopy, chromatographic, and surface area analyses. This section highlights on diverse techniques usually employed during the thermochemical conversions of agro-waste into quality biochar production.

4.6.1 *Fourier Transform Infrared Spectroscopy*

Fourier transform infrared spectroscopy (FTIR) is used to determine the functional groups and chemical bonds, which quantify infrared absorptions caused by molecular vibrations in the mid-infrared (4000–400 cm^{-1}) region. The most salient bands are identified as hydroxyl groups, aliphatic C–H, aromatic C=C, carboxyl/carbonyl groups, and C–O stretching.

Hydroxyl groups (3600–3200 cm^{-1}): The presence of the wide peak implies OH vibrations associated with phenolic groups, carboxylic acids, and adsorbed water, which are linked to the hydrophilicity and CEC of biochar (Yuan et al. 2011). The intensity decreases with an increase in pyrolysis temperature, which can be explained by the progressive dehydration.

Aliphatic C–H (2920–2850 cm^{-1}): The sharp peaks typical of the methylene groups of residual cellulose and hemicellulose reduce significantly with temperature as the aliphatic structures are aromatized (Chen and Chen 2009).

Aromatic C=C (1600–1580 cm^{-1}): The high absorption ability is a characteristic, indicating stretching of aromatic rings; the signal also increases with an increase in pyrolysis temperature; hence, it is considered as a marker of biochar stability (Keiluweit et al. 2010).

Carboxyl/Carbonyl Groups (1700–1600 cm^{-1}): The absorption is associated with C=O stretching frequency (carboxyl acids, ketones, and quinones) and plays an essential role in cation exchange and nutrient retention.

C–O Stretching (1300–1000 cm^{-1}): There are several bands that can be ascribed to phenolic C–O, ether linkage, and polysaccharide residues, which brings out hydrophilicity and nutrient sorption. The FTIR analysis of rice straw biochar indicates that increasing the pyrolysis temperatures from 300 °C to 700 °C eventually leads to an 85% decrease of aliphatic C–H intensities, a 70% drop in hydroxyl absorption intensity, and a 120% increase in aromatic C=C intensity (Zhang et al. 2015).

4.6.2 *X-Ray Photoelectron Spectroscopy*

X-ray photoelectron spectroscopy provides quantitative elemental composition and chemical state data analysis at biochar surfaces by measuring photoelectron kinetic energies and providing high surface sensitivity (2–10 nm analysis depth).

Carbon 1s Analysis: High-resolution carbon 1s (C1s) spectra indicate for many component analyses including the graphitic/aromatic carbon (284.5–284.8 eV, >70% in high temperature biochar), C–O bonds (286.0–286.5 eV), C=O bonds (287.5–288.5 eV), and carboxyl groups (289.0–290 eV). The intensity of the aromatic carbon peak increases as the temperature of pyrolysis rises, which is considered as an indicator of increased graphitization.

Oxygen 1s Analysis: Oxygen 1s (O1s) spectra show effects of carbonyl/quinone oxygen (530–531 eV), hydroxyl oxygen/ether (532–533 eV), and adsorbed water/carboxyl oxygen (534–535 eV). Surface O/C ratios decrease between 0.4 and 0.6 at 300 °C and to <0.1 at 700 °C, indicating progressive deoxygenation (Cheng et al. 2008).

Nitrogen, Sulphur, and Phosphorus: Nitrogen 1s (N1s) spectra show the presence of pyridinic N (398–399 eV), pyrrolic N (400–401 eV), and quaternary N (401–402 eV) and transition towards stable heterocyclic forms as temperature increases. The phosphorus spectrum (133–134 eV) is dominated by inorganic phosphates providing information about nutrient abundance (Wang et al. 2015). A ratio of (O + N)/C <0.2 indicates that biochar is extremely stable and may remain stationary over centuries or millennia in soils.

4.6.3 Nuclear Magnetic Resonance Spectroscopy

Solid-state ^{13}C nuclear magnetic resonance utilizing the cross-polarization magic-angle spinning technology provides molecular-scale data on carbon structures. The following structural definition is used to determine the chemical shift regions such as alkyl carbon (0–50 ppm) for aliphatic groups, O-alkyl carbon (50–110 ppm) for carbohydrates and alcohols/ethers, aromatic carbon (110–165 ppm) for benzenoid structures, and carboxyl/carbonyl carbon (165–220 ppm) for highly oxidized functional groups such as organic acids, ketones, aldehydes, and esters (Cao et al. 2011). The aromaticity index (aromatic carbon/total carbon) increases between 20% and 40% at 300 °C to 70% and 90% at 700 °C, which is directly related to soil stability. A factor (ratio of aromatic/O-alkyl carbon) is a delicate chart of thermal change, rising between <0.5 in natural biomass and >5 in biochar formed above 500 °C (Knicker et al. 2005).

4.6.4 Surface Area and Porosity

The Brunauer-Emmett-Teller (BET) analysis is a standard analytical method used to determine the surface area and porosity of the biochar. This approach analyses the gas adsorption isotherms by nitrogen (N_2) at 77 K to quantify the total area available for physical and chemical interactions. The surface area of the agricultural biochar varies significantly between 5 m^2/g and 800 m^2/g, based on the input material and pyrolysis optimizations (Downie et al. 2012). Surface area development is non-linear with high sensitivity to peak temperature. Intermediate temperature (400–600 °C) records optimum growth for porosity development at 200–400 m^2/g. In this stage, there is a rapid discharge of VOCs, which clears the micropores that in turn increases the accessible internal surface. High temperatures above 600 °C led to a decline in surface area growth due to pore debris and sintering, which reduces overall porosity. Rice husk exhibits an outstanding surface area growth (250–450 m^2/g of silica) due to its higher silica content, which allows silica to form silica-carbon frameworks. Besides, pore size selection seems to be critical to functionality: small pores (<2 nm) raise surface area but may be inaccessible to large molecules; middle pores (2–50 nm) enable nutrient adsorption and contribute to the development of microbial colonizations; and large pores (>50 nm) enable mass transfer of quality microbes and consortia.

4.6.5 Other Key Characterization Techniques

Raman Spectroscopy: It is a non-destructive technique, which is used to determine the microstructural evolution of biochar, specifically focusing on the degree of order in its carbon matrix. The carbon structure is probed in Raman spectroscopy with the G-band (~1580 cm^{-1}) signifying the presence of graphitic structures and the D-band (~1360 cm^{-1}) indicating the presence of disordered carbon, and the within ratio of I_D/I_G to it (around 0.8–1.2 for agricultural biochar) indicates structural ordering. An increase in I_D/I_G ratio indicated the formation of higher number or small aromatic clusters in the early stages of carbonization. At extremely high temperatures, a decrease in this ratio infers the coalescence of the clusters into larger graphitic sheets.

Scanning Electron Microscopy and Energy-Dispersive X-Ray Spectroscopy: The cellular structure, surface texture, and porous architecture derived from biomass feedstock are characterized by scanning electron microscopy (SEM), while elemental content and distribution mapping are analysed using energy-dispersive X-ray spectroscopy (Lee et al. 2013).

Thermogravimetric Analysis: The ability to obtain oxidation resistance is indicated by the temperature of the transition to oxidation, which is indicated to be higher by high-temperature biochar (400 °C) compared to low-temperature biochar (<300 °C) (Crombie et al. 2013).

4.6.6 Biochar Surface Functionalization for Smart Applications in Agriculture

Deliberate surface functionalization is performed to increase the agricultural functionality and sustainability. Several functionalization techniques are explained as follows:

Oxidative Functionalization: This technique involves the exposure of biochar to hydrogen peroxide (H_2O_2), nitric acid (HNO_3), or ozone, which adds oxygen-based functional groups and significantly elevates the CEC. The density of carboxyl groups can be increased by 200–400% through nitric acid oxidation, which enhances nutrient retention abilities (Xue et al. 2012).

Nitrogen Functionalization: Treatment of NH_3 at temperatures of 300–600 °C adds nitrogen to aromatic structures to produce a slow-release nitrogen fertilizer with 30–50% increased nitrogen levels (Shafeeyan et al. 2011). Urea impregnation forms nitrogen-functionalized biochar, which has both organic and inorganic nitrogen species.

Mineral Loading: Impregnation of phosphate salts to biochar produces biochar-based phosphorus fertilizers that could bring longevity to release upon crops, thereby assisting in raising crop P uptake by 25–45% as compared to standard biofertilizers (Yao et al. 2013). Combined soil conditioning and nutrient provision is achieved through K, Ca, and micronutrient loading.

Microbial Inoculation: The incorporation of biochar with plant growth-promoting (PGP) rhizobacteria or mycorrhizal fungi, particularly the arbuscular mycorrhizal fungi (AMF), produces value-added microbial biochar formulations, thereby indicating crop yield up to 15–30% (Hale et al. 2015). The soil's permeability offers better shelter to microbial inhabitants, thus enhancing their survivability within the microhabitat.

Polymer Composites: Biochar surface functions are altered through coating with chitosan or polyacrylamide, resulting in enhanced mechanical strength and precise release of nutrients. Chitosan-modified biochar exhibits enhanced CEC and antibacterial properties (Xiao et al. 2018). However, to confirm successful functionalization, fine characterization with FTIR (recognizing new functional groups), X-ray photoelectron spectroscopy (determining the surface elemental changes), and statistical analysis of surface elements (BET) and functional group quantification (chemical titration) is required.

4.7 Biochar and Climate-Smart Agriculture

4.7.1 Biochar as a Carrier Material for Slow-Release Fertilizers and Microbial Inoculants

Biochar is a stable, carbon-rich material produced via the pyrolysis of organic wastes in limited-oxygen conditions. Utilized primarily as a soil additive in agriculture and for environmental management, it can be derived from diverse biomass sources (Bhattacharyya et al. 2024d). Biochars in agriculture play key roles in boosting soil fertility, improving soil water retention abilities, and minimizing GHG emissions. As mentioned earlier, although biochar can be obtained from both the lignocellulosic biomass and NLBM, the biochar derived from lignocellulosic biomass is found to be more suitable for applying as soil amendments (Nzediegwu et al. 2021). A study by Suo et al. (2021) demonstrated that pyrolysing corn straw alongside *Enteromorpha prolifera* leads to a synergistic effect, producing biochar with improved water solubility and higher N/P ratios as compared to corn straw alone. Biochar requires modification for key agricultural use, leveraging its physicochemical properties as a platform for slow-release fertilizers (SRFs). As per agricultural experts, different processes can be used to make biochar-based SRFs, such as impregnation (An et al. 2020a), encapsulation (Sim et al. 2021), granulation (Zheng et al. 2017), and co-pyrolysis (An et al. 2020b). It is worth mentioning that SRFs help with environmental clean-up since they release nutrients gradually, which reduces the risk of leaching and makes fertilizer work more efficiently (Kim et al. 2014). Biochar, a porous material with a large surface area, is highly effective in precision agriculture and thereby assists in mitigating climate crises. Its microstructure, high surface area, and substantial charge improve nutrient utilization abilities in plants, making it a suitable carrier material for essential nutrients like NPK. Field experiment demonstrated that applying low amounts (<0.1%) of biochar-based fertilizers could significantly improve crop yield and soil fertility and reduce Cd levels in pulses/grains (Chen et al. 2021). Furthermore, biochar application could enhance native soil quality, facilitate the growth and proliferation of beneficial microbial communities, and increase both the quality and quantity of pluckable tea shoots (Zhou et al. 2021). Biochar application also enhanced bacterial community dynamics in brown soil, thereby positively influencing crop development, soil fertility, and total soil microbiota. Additionally, biochar can also be used in the removal of toxic substances from soils, water, and environments as well as can be considered as sustainable key sources of energy and fuel (Varkolu et al. 2025). The absorbent abilities of biochar make them a potent material for removal of pollutants from air and water.

4.7.2 The Role of Biochar in Soil Fertility Enhancement and Land Restoration

Soil nutrient retention is an essential characteristic of biochar applications, which can adsorb various nutrients and prevent their leaching into the soil, making them available for plant roots. Soil amendment with biochar-based fertilizers can improve the fertility of soil, thereby playing a vital role in land restoration. The CEC of biochar aids in retaining positively charged nutrient ions like Ca, K, and Mg, which is vital for replenishing nutrients in nutrient-poor soils. Yan et al. (2021) conducted a field experiment in the Karst mountains of China, revealing that biochar-based fertilizers effectively restored soil and enhanced the populations of beneficial microbial communities. This resulted in increased participation of key microbial species in soil restoration, demonstrating a positive correlation between key soil nutrients (N, P, K) and potent microbial activity. Biochar, owing to its alkaline characteristics, can reduce soil acidity and stabilize pH. This modulation of pH solubilizes important nutrients in soil, rendering them accessible to plants (Amini et al. 2016). Biochar can mitigate soil pollution through absorbing organic and inorganic pollutants due to its porous nature, thereby immobilizing soil contaminants and lowering soil toxicity. The toxicity removal abilities of biochar substantially assist in rehabilitating soil microbial communities, maintaining adequate moisture, and facilitating nutrient cycling, consequently improving overall soil health and prosperity.

4.7.3 Biochar in Fostering Beneficial Microbial Populations in Soil

Biochar acts as a substrate for soil microbial populations, offering a carbon source that supports significant microbial growth and prosperity. The use of biochar in soil in optimal quantities considerably enhances rhizosphere engineering (Kolb et al. 2009). As per observations of Yu et al. (2018), biochar directly or indirectly influences the populations of rhizospheric microbiota, including the species of *Trichoderma*, *Bacillus*, and *Pseudomonas* that were found to be higher in biochar-applied soils. Similarly, Yan et al. (2022) have observed that biochar application significantly increased the α-diversity of bacterial communities in rhizosphere, thereby playing an essential role in releasing inorganic nutrients in soil and making it available for growing plants. Biochar-applied soils are naturally rich in beneficial microbial communities, which eventually suppress pest infestations in agriculture, through competing for nutrients and resources, thereby enhancing crop resistance against major pests and pathogens (Zhu et al. 2017). Biochar application alters soil physicochemical properties, influencing the natural growth, development, and metabolism of beneficial microbiota, and alters microbial community structure, thereby playing an essential role in climate-smart agriculture.

4.7.4 The Role of Biochar in Carbon Sequestration and Climate Mitigation

The addition of biochar to soil increases the carbon pool through creating stable complexes with soil organic matter, thereby significantly contributing to long-term carbon storage. The process substantially mitigates climate change through reducing atmospheric carbon that would otherwise be released, if it were not sequestered by biochar application. Some estimates state that biochar when used on a global scale can mitigate GHG emissions up to 12% (Wang et al. 2021). Biochar is recognized for its crucial role in mitigating methane emissions from rice fields by fostering the proliferation of methanotrophs that consume methane, consequently diminishing the populations of methane-producing bacteria (Singh et al. 2017). The production of biochar from lignocellulosic biomass and NLMB would facilitate long-term carbon sequestration and enhance soil health and environmental conditions. On a global scale, agricultural waste can produce approximately 373 million tons of biochar annually, capable of sequestering 0.55 petagrams of CO_2. Biochar can exceed the release of gaseous CO_2 from completely degraded biomass. Stabilization of carbon through biochar is regarded as a carbon-negative pathway, significantly reducing GHG emissions. A market-driven approach to combatting GHG emissions involves carbon markets, including carbon credit and offset systems, which financially incentivize emissions reduction and aid climate change mitigation (Kishwan 2023). Participants such as entrepreneurs, governments, and non-governmental organizations (NGOs) engage in activities like afforestation and renewable energy use, verified by third-party organizations. This has led to carbon removal markets with platforms focusing on carbon dioxide removal units, exemplified by organizations like Puro Earth (Finland), Nori (USA), MoorFutures (Germany), and Switzerland (Kujanpää et al. 2023). Thus, the role of biochar in sustainability improvements and climate change mitigation demonstrates its significant potential, particularly in enhancing crop resilience, boosting soil fertility, enhancing water retention, and facilitating carbon sequestration and reductions of GHG emissions.

4.8 Biochar as a Circular Economy Tool

Globally, agricultural activities generate an estimated 3758 million tons of waste annually, thereby exerting increasing pressure on ecosystems, accelerating soil degradation, reducing soil organic matter (SOM), and increasing pollution emissions. This challenge is particularly evident in sectors such as viticulture, which rely on specific climatic conditions, while generating substantial amounts of lignocellulosic waste. Improper management of these wastes not only poses environmental risks but also leads to the loss of valuable resources that could otherwise be locally recovered through circular approaches. The circular economy provides a framework for converting biological waste streams into regenerative resources. Biochar is a solid carbonaceous material produced through the thermal decomposition of organic biomass. In the pursuit of sustainable agriculture, biochar production has emerged as a key component of closed-loop and circular agricultural systems. This approach simultaneously addresses waste management, soil fertility enhancement, and environmental protection, while supporting carbon sequestration and efficient resource use potential in GHG mitigation. Biochar is increasingly recognized as an important tool in the circular bioeconomy, with research focused on its environmental management potential and versatility across various sectors. It is to be noted that, since the adoption of Paris Agreement in 2015, global climate initiatives have aimed to reduce GHG emissions and limit global temperature rise to 1.5 °C above preindustrial levels. The integration of biochar into agricultural practices improves soil health, enhances nutrient availability, and supports long-term sustainability (Ayaz et al. 2021). Thus, its emerging importance within the circular bioeconomy has positioned biochar as a valuable environment management tool, with ongoing research focused on the better understanding of its mitigation potential.

4.8.1 Biochar: Techno-economic Feasibility and Life-Cycle Assessment

Biochar serves as a soil amendment to enhance soil structure, improve water retention abilities, increase nutrient availability, and boost carbon sequestration potential (Bhattacharyya et al. 2024d). It can be mixed into the soil, spread on the surface, or added to compost to improve nutrient levels and facilitate beneficial microbial interactions. Circular agricultural systems are commonly conceptualized through two interconnected cycles: the technical cycle and the biological cycle. The technical cycle focuses on reducing waste from agricultural components and products through improved design, reuse, and efficient utilization, whereas the biological cycle emphasizes the reintegration of organic residues into natural ecosystems to support soil fertility and ecological balance (Rodríguez-Espinosa et al. 2023). Closed-loop agriculture focuses on recycling resources to minimize waste and enhance resource efficiency, reducing dependence on external inputs. Integrating biochar into these systems offers a scalable approach to promote circular agriculture, vital for maintenance of sustainable and resilient agricultural production. Within this framework, biochar plays a central role by transforming agricultural and agro-industrial residues such as crop residues, animal manure, and food processing wastes into a stable, value-added soil amendment. Once applied to agricultural soils, biochar effectively closes the nutrient and carbon loop through returning essential elements into the soil while stabilizing the soil organic carbon. Thermochemical conversion processes transform waste biomass into a stable carbon-rich material, while its application to soil enhances biological nutrient cycling, improves soil structure, and supports beneficial microbial activity. Biochar in closed-loop agriculture offers benefits such as transforming organic residues into soil amendments and improving soil health and fertility through enhanced structure, aeration, water retention, and nutrient availability. It contributes to carbon sequestration, decreases GHG emissions, and promotes resource efficiency through recycling nutrients. Although economic factors like costs and labour are important, a cost-benefit analysis demonstrates that the benefits surpass the expenses, highlighting biochar's significance in sustainable agriculture. From a broader perspective, closed-loop agriculture aligns closely to the principles of circular economy, which emphasize efficient management of materials and resources throughout their life cycle.

Blending biochar with compost significantly enhances nutrient conservation abilities, prominent microbial activity, compost stability, and overall agronomic value beyond what either amendment provides alone. Its porous structure reduces leaching and nutrient volatilizations, particularly N, thereby decreasing its losses by more than 50% and favouring NO_3^- over ammonia/ammonium. Biochar also mitigates GHG emissions and odours during composting, with reported reductions of over 80% for CH_4, 50% for N_2O, and up to 60% for NH_3 (Yin et al. 2021), respectively. By creating microhabitats, biochar supports beneficial microbial population dynamics, including the populations of fungi, nitrifying bacteria, and methanotrophic populations, enhancing nitrogen retention abilities, improving

NH_4^+/NO_3^- ratios, and reducing CH_4 production (Godlewska et al. 2017). It expedites compost maturation, enhances humification, diminishes bulk density, stabilizes moisture, mitigates anaerobic zones, elevates humic extractable carbon, optimizes humic to fulvic acid ratios, and lowers water-soluble carbon. Co-composting with biochar extends the thermophilic phase and raises compost temperature without additional carbon loss (Godlewska et al. 2017). Furthermore, biochar is also known to enhance the aquaculture sustainability through improving feed efficiency, nutrient digestibility, immune response, and water quality parameters (Amjad et al. 2024). It adsorbs excess nutrients like NH_3, NO_2^-, NO_3^-, and phosphate, reduces eutrophication risks, and promotes beneficial microbial populations for improved bioremediation abilities (Mosa et al. 2023). Besides, biochar's aromatic C-structure is resistant to microbial degradation, enabling long-term soil carbon storage abilities. It improves soil pH, promotes ferric ion (Fe^{3+})/ aluminium ion (Al^{3+})-associated soil organic carbon stabilization, enhances microbial biomass, and supports nutrient cycling (Zhu et al. 2025). High C/N ratio of biochars can temporarily immobilize nitrogen, unless supplemented with fertilizers/ manures (Tsai and Chang 2021). Field studies have demonstrated the potential of biochar in improving soil structure, organic matter, nutrient retention abilities, and sustainable plant growth and protection (Martinez-Fernandez et al. 2024). Its interaction with plant growth-promoting rhizobacteria enhances soil microbial activity, nutrient-use efficiency, and ecosystem resilience.

The techno-economic feasibility and LCA of biochar production are critical for evaluating its long-term viability as a sustainable agricultural and environmental tool. Despite extensive scientific evidence highlighting improvements in crop yield, soil health, and environmental benefits, the global biochar market remains in its infancy, characterized by limited transactions and slow adoption (Campbell et al. 2018). High fixed and variable production costs, often associated with characteristics of a natural monopoly, have constrained the profitability of biochar enterprises. Moreover, insufficient market demands further limit cost recovery and prevent the full monetization of biochar's environmental benefits, such as carbon sequestration and GHG mitigation. Currently, biochar adoption is more prevalent at household or small-scale levels, particularly in rural regions of Asia and Europe, where it is sometimes used as an alternative to charcoal (Vochozka et al. 2016). However, studies indicate that the immediate economic returns from biochar use for energy purposes often outweigh the long-term benefits derived from soil application, which typically requires extended payback periods. Biochar has a number of agronomic and economic benefits; however, large-scale farmers around the world are either sceptical or ignorant of these benefits (Bezerra et al. 2019). Despite an extensive patent base and 20 years of supporting scientific literature, biochar use is still limited or categorized as experimental in numerous countries due to persistent regulatory hurdles (Peiris et al. 2017). Production cost estimates for biochar technology exhibit considerable variance depending on the technology and scale, ranging from 1674 USD to 1909 USD per ton for biochar solutions and from 528 USD to 1051 USD per ton for air curtain burner technologies, highlighting the need for technological optimizations. Comprehensive techno-economic feasibility and LCA are crucial to identify the cost reduction pathways and minimize environmental impacts while aligning with SDGs (Bergman et al. 2022). The integration of biochar production within a circular bioeconomy framework offers significant economic and environmental benefits. Conventional linear economic models prioritize cost-effective practices like open burning of crop residues, overlooking nutrient loss and environmental toxicity. As global food demand rises and GHG mitigation becomes urgent, these practices are considered as increasingly unsustainable. Policy initiatives, such as the European Commission's Circular Economy Strategy, recognize the importance of bio-based resources like biochar. A circular bioeconomy integrates the ideas of circular economy and bioeconomy to enhance the utilization of biomass and waste. Biochar promotes nutrient recycling, carbon sequestration, and waste valorization, supporting the SDGs of the United Nations. However, successful large-scale agricultural implementation demands market competitiveness, economic viability, and supportive policies. Technological innovation enhances feasibility, particularly through mobile pyrolysis units, which causes lower transportation costs and increases rural production efficiency. The biochar industry is in its market introduction stage, facing high production costs, low sale volumes, and minimal competition. Future profitability hinges on scaling production for economies of scale, potentially leading to competitive pressures during the growth stage of the product life cycle. Despite its potential, the biochar industry encounters significant challenges, including high initial costs, lack of scientific consensus on benefits, insufficient commercial operations, and risk of feedstock overexploitations. To facilitate the sector's transition to sustainability, it is crucial to conduct thorough techno-economic analyses and LCAs and ensure regulatory clarity and financial investments. Despite ongoing uncertainties regarding the market's potential, research indicates that with proper management, biochar has the potential to make significant strides in sustainable agriculture and popularization of circular bioeconomy standards.

4.9 Trends in Biochar Bioeconomy and Climate-Smart Agriculture

Figure 4.3a illustrates the worldwide distribution of research concerning the development of biochar bioeconomy from agricultural biomass waste, as determined by the authors' affiliations. China, Poland, India, Italy, the USA, and Norway are among the top contributors, showing intense intellectual effort while developing this field of research. Different shades of blue colour represent varying productivity levels, with darker shades indicating higher research output, while grey colour indicates regions with nil publications. This approach outlines a global geography of research participations in the biochar bioeconomy from agricultural biomass waste, highlighting the productivity and significant collaborations among the authors (Fig. 4.3b).

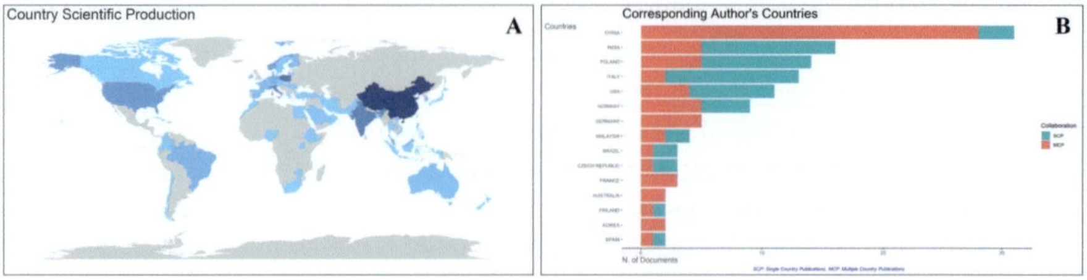

Fig. 4.3 Global distribution of biochar research. (**a**) The dispersion of research publications across different regions related to the generation of biochar bioeconomy from agricultural biomass waste. (**b**) Representation of top 15 countries by corresponding author

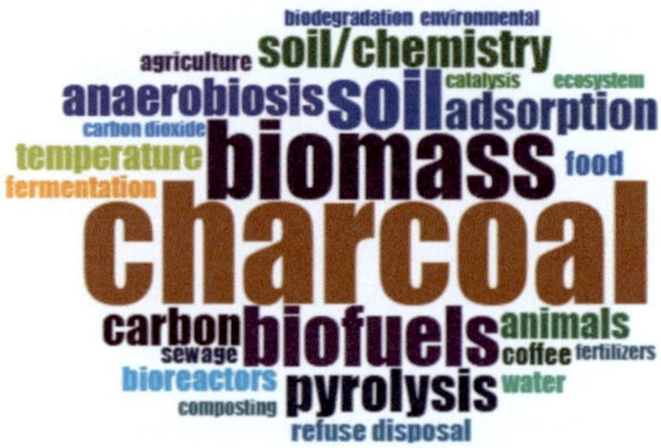

Fig. 4.4 Word cloud of top 25 prominent keywords

Figure 4.4 depicts a word cloud visualization map produced using Biblioshiny R (version 4.5.2), which offers an alternative perspective on prominent keywords, emphasizing the significance of rapidly used terms such as "charcoal", "biomass", "biofuels", "soil", and "sewage", while emphasizing the development of a biochar bioeconomy.

The co-occurrence network of keywords was analysed using multiple correspondence analysis (Mahanta et al. 2025), as depicted in Fig. 4.5. Two clusters of frequently used keyword combinations are identified—a large blue cluster and a smaller red cluster—which rarely appear together in publications. Anaerobiosis, methane, and bioreactors are close to the cluster's centre of gravity (red cluster), which means they frequently appear together in publications. Greater peripherality (as in the case of fermentation, food loss, and waste) suggests a lower frequency of co-occurrence with central keywords. The larger and complex blue cluster comprises the major focus of the emergence of biochar bioeconomy from agricultural biomass waste perspectives. The red cluster has a few highly central keywords and relatively few peripheral ones, whereas the blue cluster contains numerous reasonably central keywords and a lower number of peripheral ones. This shows that the red cluster is more concentrated but less consolidated, whereas the blue cluster is more diverse and interconnected.

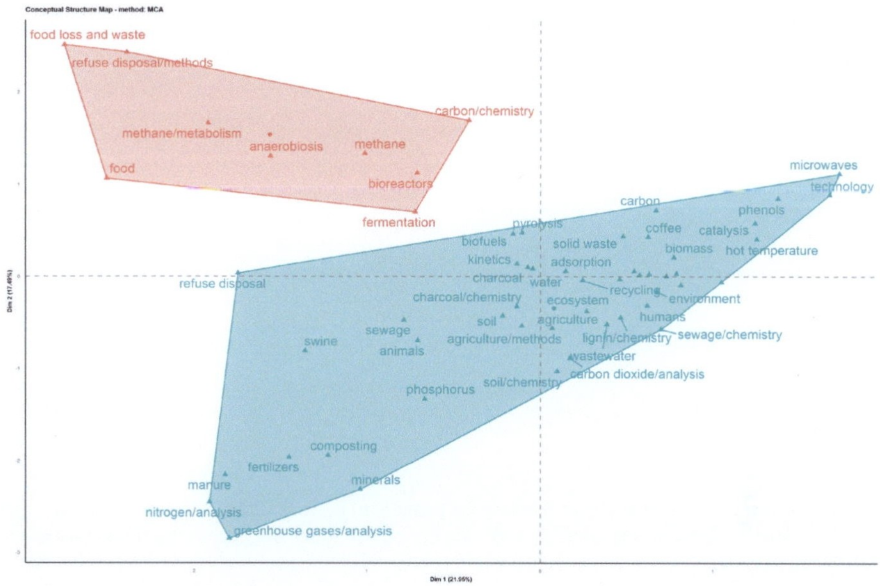

Fig. 4.5 Factorial analysis of keywords related to the generation of biochar bioeconomy from agricultural biomass waste using correspondence analysis

The three-field plot analysis or Sankey diagram, depicting the relationships among different journals, keywords, and authors in articles about generation of biochar bioeconomy from agricultural biomass waste, is illustrated in Fig. 4.6. Rectangles of different colours were used to highlight the key components (Mahanta et al. 2025), with their height corresponding to the number of links among the authors, journals, and keywords. The largest rectangle indicated the element with the most numbers of relationships (Koo 2021). The representation of Sankey diagram indicates the most productive journals (e.g., *Bioresource Technology*, *Chemosphere*, *Scientific Reports*, *Journal of Environmental Management*) and authors (e.g., Chen WH, Li Y), emphasizing their significant contributions to research on generation of biochar bioeconomy from agricultural biomass waste.

Figure 4.7 depicts the network's cluster structure, which consists of eight distinct clusters. Each cluster represents a thematic concentration, with varied levels of interconnections. The keywords in a common theme are represented by an identical colour. The circles represent a certain keyword, called nodes (Bhattacharyya et al. 2024b). Of the 9753 total keywords, 386 meet the threshold, if the minimum number of occurrences is 10. In keyword network analysis, the most frequently occurring keywords are biochar (1575 occurrences, total link strength 4487), charcoal (1543 occurrences, total link strength 7670), biomass (269 occurrences, total link strength 1316), agriculture (218 occurrences, total link strength 1226), soil (887 occurrences, total link strength 4770), and biofuels (65 occurrences, total link strength 270), indicating the research focus on generating biochar bioeconomy from agricultural biomass waste.

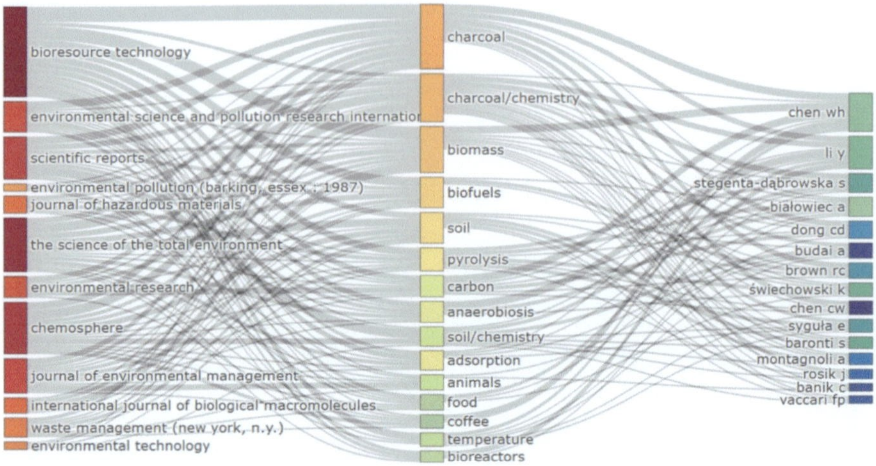

Fig. 4.6 A three-field plot analysis between the prominent journals (left), keywords (middle), and authors (right) to highlight the generation of biochar bioeconomy from agricultural biomass waste

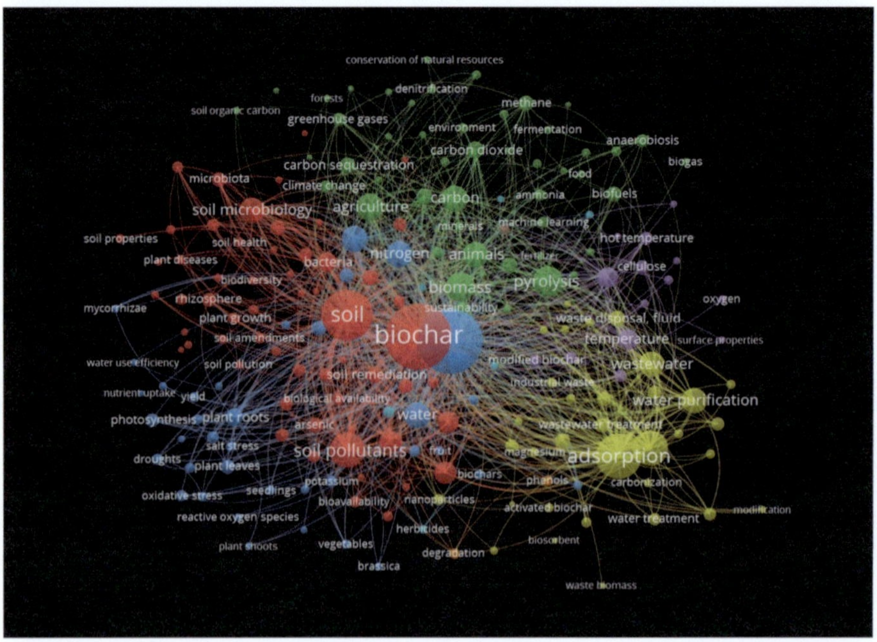

Fig. 4.7 Cluster-based network visualization of keyword co-occurrence using VOSviewer (version 1.6.20)

4.10 Conclusions, Limitation, and Future Perspectives

Transforming commonly available agricultural wastes into biochar represents an environmentally sustainable approach facilitating waste valorization. The practical utility of biochar is enhanced by its simple production process and dependence on inexpensive feedstocks, such as agro-industrial waste materials, which can become sources of harmful pollutants, if managed improperly. Thermochemical conversion of biomass results in notable differences in both biochar yield and quality, primarily governed by parameters such as heating rate, reaction temperature, and oxygen availability. Generally, higher heating rates or increased oxygen presence led to reduced biochar yields. Owing to its multifunctional nature, biochar has relevant applications across multiple sectors, functioning as an economical adsorbent tool in environmental remediation, besides supporting rural economic development and empowerment opportunities. In recent years, biochar research has also been highlighted in several emerging application pathways, including its use efficiency in nutrient recovery, carbon sequestration strategies, soil health improvement, and integration into circular bioeconomy frameworks. When used as an electrode material, modified biochar exhibits high surface area and enhanced electrical conductivity, making it a strong candidate for bioelectricity generation, particularly in microbial fuel cells where energy recovery from waste is achieved with. In catalytic applications, both unmodified and modified biochars promote pollutant degradation and redox reactions, demonstrating significant potential in biohydrogen production through facilitating essential microbial and chemical processes. As a filtration medium, its porous architecture and surface functional groups enable efficient treatment of water and wastewater. In agricultural systems, biochar acts as a soil amendment through enhancing soil fertility, water-holding capacity, and microbial activity, thereby supporting crop productivity and carbon sequestration abilities.

Biochar from agricultural wastes is a sustainable material with potential in environmental, agricultural, and energy applications. However, its adoption faces challenges due to variability in feedstock, optimizations in production conditions, and properties. With process optimization and regulatory support, biochar could play a significant role in circular bioeconomy strategies, aiding in waste valorization, resource efficiency, and climate change mitigation. Notwithstanding the aforementioned benefits of converting agro-waste into enhanced biochar, a more profound comprehension is necessary to optimize biochar efficacy across various disciplines in agro-industries. Additional study is required to delineate the explicit connections between biochar properties and their functional qualities across various sectors. A comprehensive assessment of the heterogeneity in characteristics and impacts linked to various biochar types is crucial, especially indicating the existing absence of process standardizations. Despite its extensive applicability, biochar faces hurdles stemming from property discrepancies and the lack of standardized production protocols. Furthermore, a significant portion of the current evidence originates from laboratory-scale tests, with a limited number of research confirming performance under field conditions.

Future research should prioritize the integration of production techniques, such as coupling HTC with pyrolysis, to improve carbon retention and customize biochar attributes for targeted uses and delivery. Developing application-specific quality standards will be crucial to ensure uniformity and reliability across different biochar applications. Additionally, implementing field-scale evaluations and LCAs is necessary to assess the long-term effectiveness, sustainability, and real-world impacts. Further exploration of the links between biochar properties and application-specific performance will also enable targeted modification strategies to enhance functionality. There is a great potential for zero-waste initiatives and cost-efficiency improvements in environmental management and sustainability to be achieved through ongoing research into the usage, modification, and enhancement of biochar. The environmental, agricultural, and energy sectors will embrace biochar on a wide scale if these fields continue to advance, which will strengthen its role in circular bioeconomy frameworks, thereby mitigating climate crises. The sustained and prospective growth of the biochar market relies on the strategic engagement and the concrete establishment of standardized frameworks by government entities, non-governmental organizations, policymakers, stakeholders, and entrepreneurs.

Acknowledgements The principal of Nanda Nath Saikia College, Titabar, Jorhat, Assam, India, is heartily thanked by the corresponding author for giving the logistics and support. The corresponding author appreciated the Department of Biotechnology (DBT), Government of India, for granting research funds for the Institutional Biotech Hub (BT/NER/143/SP44344/2021) at NNS College, under the NER Biotech Hub Programme. The corresponding author is thankful to Parismita Sarma, Department of Botany, Rangia College, for her assistance in reference menagement.

References

Amalina F, Razak ASA, Krishnan S, Zularisam A, Nasrullah M (2022a) A comprehensive assessment of the method for producing biochar, its characterization, stability, and potential applications in regenerative economic sustainability—a review. Cleaner Mater 3:100045. https://doi.org/10.1016/j.clema.2022.100045

Amalina F, Razak ASA, Krishnan S, Sulaiman H, Zularisam A, Nasrullah M (2022b) Biochar production techniques utilizing biomass waste-derived materials and environmental applications—a review. J Hazard Mater Adv 7:100134. https://doi.org/10.1016/j.hazadv.2022.100134

Amini S, Ghadiri H, Chen C, Marschner P (2016) Salt-affected soils, reclamation, carbon dynamics, and biochar: a review. J Soils Sediments 16:939–953. https://doi.org/10.1007/s11368-015-1293-1

Amjad M, Hussain SM, Ali S, Rizwan M, Al-Ghanim KA, Yong JWH (2024) Effectiveness of feeding different biochars on growth, digestibility, body composition, hematology and mineral status of the Nile tilapia, Oreochromis niloticus. Sci Rep 14:13526. https://doi.org/10.1038/s41598-024-63463-4

An X, Wu Z, Yu J, Ge L, Li T, Liu X, Yu B (2020a) High-efficiency reclaiming phosphate from an aqueous solution by bentonite modified biochars: a slow-release fertilizer with a precise rate regulation. ACS Sustain Chem Eng 8(15):6090–6099. https://doi.org/10.1021/acssuschemeng.0c01112

An X, Wu Z, Yu J, Cravotto G, Liu X, Li Q, Yu B (2020b) Copyrolysis of biomass, bentonite, and nutrients as a new strategy for the synthesis of improved biochar-based slow-release fertilizers. ACS Sustain Chem Eng 8(8):3181–3190. https://doi.org/10.1021/acssuschemeng.9b06483

Ayaz M, Feizienė D, Tilvikienė V, Akhtar K, Stulpinaitė U, Iqbal R (2021) Biochar role in the sustainability of agriculture and environment. Sustainability 13(3):1330. https://doi.org/10.3390/su13031330

Bairwa P, Kumar N, Devra V, Abd-Elsalam KA (2023) Nano-biofertilizers synthesis and applications in agroecosystems. Agrochemicals 2:118–134. https://doi.org/10.3390/agrochemicals2010009

Basu P (2010) Biomass gasification and pyrolysis: practical design and theory. Academic Press. https://doi.org/10.1016/C2009-0-20099-7

Bergman R, Sahoo K, Englund K, Mousavi-Avval SH (2022) Lifecycle assessment and techno-economic analysis of biochar pellet production from forest residues and field application. Energies 15(4):1559. https://doi.org/10.3390/en15041559

Bezerra J, Turnhout E, Vasquez IM, Rittl TF, Arts B, Kuyper TW (2019) The promises of the Amazonian soil: shifts in discourses of Terra Preta and biochar. J Environ Policy Plan 21(5): 623–635. https://doi.org/10.1080/1523908X.2016.1269644

Bhat AS, Kuriqi A, Dar MUD, Bhat O, Sammen SS, Towfiqul Islam ARM, Elbeltagi A, Shah O, Al-Ansari N, Ali R, Heddam S (2022) Application of biochar for improving physical, chemical, and hydrological soil properties: a systematic review. Sustainability 14(17):11104. https://doi.org/10.3390/su141711104

Bhatia SK, Palai AK, Kumar A, Bhatia RK, Patel AK, Thakur VK, Yang YH (2021) Trends in renewable energy production employing biomass-based biochar. Bioresour Technol 340: 125644. https://doi.org/10.1016/j.biortech.2021.125644

Bhattacharyya PN, Jha DK (2012) Plant growth-promoting rhizobacteria (PGPR): emergence in agriculture. World J Microbiol Biotechnol 28:1327–1350. https://doi.org/10.1007/s11274-011-0979-9

Bhattacharyya PN, Sarmah SR (2018) The role of microbes in tea cultivation. In: Sharma VS, Gunasekare MTK (eds) Global tea science: current status and future needs. Burleigh Dodds Science Publishing, Cambridge. https://doi.org/10.1201/9781351114516

Bhattacharyya PN, Goswami MP, Bhattacharyya LH (2016) Perspective of beneficial microbes in agriculture under changing climatic scenario: a review. Journal of Phytology 8:26–41. https://doi.org/10.19071/jp.2016.v8.3022

Bhattacharyya PN, Sarmah SR, Roy S, Sarma B, Nath BC, Bhattacharyya LH (2023) Perspectives of *Beauveria bassiana*, an entomopathogenic fungus for the control of insect-pests in tea [*Camellia sinensis* (L.) O. Kuntze]: opportunities and challenges. Int J Trop Insect Sci 43(1): 1–19. https://doi.org/10.1007/s42690-022-00932-1

Bhattacharyya PN, Baishya B, Borgohain DJ, Sarma B, Bhattacharyya LH, Garganese F, Barman TS (2024a) Unveiling the virulence and stress tolerance of *Beauveria bassiana* against major insect pests in tea with an emphasis on Bibliometrics using data mining tools. In: Deshmukh SK, Sridhar KR (eds) Entomopathogenic fungi: prospects and challenges. Springer, Singapore. https://doi.org/10.1007/978-981-97-5991-0_11

Bhattacharyya PN, Nath BC, Sarma B, Al-Ani LK, Borgohain DJ, Garganese F, Ahmed S, Batsya S, Mudoi A, Kumari R (2024b) Interaction of *Metarhizium anisopliae* against emergent insect pest problems in the north-eastern tea industry. In: Deshmukh SK, Sridhar KR (eds) Entomopathogenic fungi. Springer, Singapore. https://doi.org/10.1007/978-981-97-5991-0_13

Bhattacharyya PN, Islam NF, Sarma B, Nath BC, Al-Ani LKT, Lesueur D (2024c) Frankia-actinorhizal symbiosis: a non-chemical biological assemblage for enhanced plant growth, nodulation and reclamation of degraded soils. Symbiosis 92:1–26. https://doi.org/10.1007/s13199-023-00956-2

Bhattacharyya PN, Sandilya SP, Sarma B, Pandey AK, Dutta J, Mahanta K, Lesueur D, Nath BC, Borah D, Borgohain DJ (2024d) Biochar as soil amendment in climate-smart agriculture: opportunities, future prospects, and challenges. J Soil Sci Plant Nutr 24:135–158. https://doi.org/10.1007/s42729-024-01629-9

Bhattacharyya PN, Sarma B, Bhattacharyya LH (2025) A textbook of biofertilizers. As per FYUGP (NEP) syllabus, 1st edn. Publisher Mahaveer Publications, pp 1–174
Black GP, He G, Denison MS, Young TM (2021) Using estrogenic activity and nontargeted chemical analysis to identify contaminants in sewage sludge. Environ Sci Technol 55(10): 6729–6739. https://doi.org/10.1021/acs.est.0c07846
Blasi A, Verardi A, Lopresto CV, Siciliano S, Sangiorgio P (2023) Lignocellulosic agricultural waste valorization to obtain valuable products: an overview. Recycling 8(4):61. https://doi.org/10.3390/recycling8040061
Borthakur A, Leonard J, Koutnik VS, Ravi S, Mohanty SK (2022) Inhalation risks of wind-blown dust from biosolid-applied agricultural lands: are they enriched with microplastics and PFAS? Curr Opin Environ Sci Health 25:100309. https://doi.org/10.1016/j.coesh.2021.100309
Bridgwater AV (2012) Review of fast pyrolysis of biomass and product upgrading. Biomass Bioenergy 38:68–94. https://doi.org/10.1016/j.biombioe.2011.01.048
Campbell RM, Anderson NM, Daugaard DE, Naughton HT (2018) Financial viability of biofuel and biochar production from forest biomass in the face of market price volatility and uncertainty. Appl Energy 230:330–343. https://doi.org/10.1016/j.apenergy.2018.08.085
Cao X, Ro KS, Chappell M, Li Y, Mao J (2011) Chemical structures of swine-manure chars different carbonization conditions investigated by advanced solid-state ^{13}C nuclear magnetic resonance (NMR) spectroscopy. Energy Fuel 25(1):388–397. https://doi.org/10.1021/ef101342v
Chen B, Chen Z (2009) Sorption of naphthalene and 1-naphthol by biochars of orange peels with different pyrolytic temperatures. Chemosphere 76(1):127–133. https://doi.org/10.1016/j.chemosphere.2009.02.004
Chen DMC, Bodirsky BL, Krueger T, Mishra A, Popp A (2020) The world's growing municipal solid waste: trends and impacts. Environ Res Lett 15(7):074021. https://doi.org/10.1088/1748-9326/ab8659
Chen Z, Pei J, Wei Z, Ruan X, Hua Y, Xu W et al (2021) A novel maize biochar-based compound fertilizer for immobilizing cadmium and improving soil quality and maize growth. Environ Pollut 277:116455. https://doi.org/10.1016/j.envpol.2021.116455
Cheng CH, Lehmann J, Engelhard MH (2008) Natural oxidation of black carbon in soils: changes in molecular form and surface charge along a climosequence. Geochim Cosmochim Acta 72(6): 1598–1610. https://doi.org/10.1016/j.gca.2008.01.010
Crombie K, Mašek O, Sohi SP, Brownsort P, Cross A (2013) The effect of pyrolysis conditions on biochar stability as determined by three methods. Glob Change Biol Bioenergy 5(2):122–131. https://doi.org/10.1111/gcbb.12030
Ding W, Dong X, Ime IM, Gao B, Ma LQ (2014) Pyrolytic temperatures impact lead sorption mechanisms by bagasse biochars. Chemosphere 105:68–74. https://doi.org/10.1016/j.chemosphere.2013.12.042
Domingues RR, Trugilho PF, Silva CA, de Melo ICNA, Melo LCA, Magriotis ZM, Sánchez-Monedero MA (2017) Properties of biochar derived from wood and high-nutrient biomasses with the aim of agronomic and environmental benefits. PLoS One 12(5):e0176884. https://doi.org/10.1371/journal.pone.0176884
Downie A, Crosky A, Munroe P (2012) Physical properties of biochar. In: Biochar for environmental management. Routledge, pp 45–64
Fajobi MO, Lasode OA, Adeleke AA, Ikubanni PP, Balogun AO (2022) Investigation of physicochemical characteristics of selected lignocellulose biomass. Sci Rep 12:2918. https://doi.org/10.1038/s41598-022-07061-2
Farrokh NT, Suopajärvi H, Mattila O, Umeki K, Phounglamcheik A, Romar H et al (2018) Slow pyrolysis of by-product lignin from wood-based ethanol production–a detailed analysis of the produced chars. Energy 164:112–123. https://doi.org/10.1016/j.energy.2018.08.161

Ghodake GS, Shinde SK, Kadam AA, Saratale RG, Saratale GD, Kumar M, Palem RR, Al-Shwaiman HA, Elgorban AM, Syed A et al (2021) Review on biomass feedstocks, pyrolysis mechanism and physicochemical properties of biochar: state-of-the-art framework to speed up vision of circular bioeconomy. J Clean Prod 297:126645. https://doi.org/10.1016/j.jclepro.2021.126645

Godlewska P, Schmidt HP, Ok YS, Oleszczuk P (2017) Biochar for composting improvement and contaminants reduction: a review. Bioresour Technol 246:193–202. https://doi.org/10.1016/j.biortech.2017.07.095

Gong H, Zhao L, Rui X, Hu J, Zhu N (2022) A review of pristine and modified biochar immobilizing typical heavy metals in soil: applications and challenges. J Hazard Mater 432:128668. https://doi.org/10.1016/j.jhazmat.2022.128668

Hale L, Luth M, Crowley D (2015) Biochar characteristics relate to its utility as an alternative soil inoculum carrier to peat and vermiculite. Soil Biol Biochem 81:228–235. https://doi.org/10.1016/j.soilbio.2014.11.023

Hemansi, Gupta R, Yadav G, Kumar G, Yadav A, Saini JK, Kuhad RC (2019) Sustainable approaches for biofuels production technologies. In: Srivastava N et al (eds) Biofuel and biorefinery technologies. Springer, pp 121–146. https://doi.org/10.1007/978-3-319-94797-6

Jafri N, Wong WY, Doshi V, Yoon LW, Cheah KH (2018) A review on production and characterization of biochars for application in direct carbon fuel cells. Process Saf Environ Prot 118:152–166. https://doi.org/10.1016/j.psep.2018.06.036

Jeswani HK, Chilvers A, Azapagic A (2020) Environmental sustainability of biofuels: a review. Proc R Soc A 476:20200351. https://doi.org/10.1098/rspa.2020.0351

Kamali M, Sweygers N, Al-Salem S, Appels L, Aminabhavi TM, Dewil R (2022) Biochar for soil applications: sustainability aspects, challenges and future prospects. Chem Eng J 428:131189. https://doi.org/10.1016/j.cej.2021.131189

Kaur M, Malik DP, Malhi GS, Singh G, Singh S, Singh J, Singh P, Singh R, Singh A, Singh V et al (2022) Rice residue management in the Indo-Gangetic Plains for climate and food security: a review. Agron Sustain Dev 42:92. https://doi.org/10.1007/s13593-022-00817-0

Keiluweit M, Nico PS, Johnson MG, Kleber M (2010) Dynamic molecular structure of plant biomass-derived black carbon (biochar). Environ Sci Technol 44(4):1247–1253. https://doi.org/10.1021/es9031419

Kim DJ, Lee DI, Cha GC, Keller J (2008) Analysis of free ammonia inhibition of nitrite oxidizing bacteria using a dissolved oxygen respirometer. Environ Eng Res 13(3):125–130. https://doi.org/10.4491/eer.2008.13.3.125

Kim Y, Ullah S, Moore TR et al (2014) Dissolved organic carbon and total dissolved nitrogen production by boreal soils and litter: the role of flooding, oxygen concentration, and temperature. Biogeochemistry 118:35–48. https://doi.org/10.1007/s10533-013-9903-8

Kishwan J (2023) REDD+ and its implementation in the Himalayan region: policy issues. In: Kumar A, Jong WD, Kumar M, Pandey R (eds) Climate change in the Himalayas. Academic Press, pp 121–139. https://doi.org/10.1016/B978-0-443-19415-3.00006-2

Knicker H, González-Vila FJ, Polvillo O, González JA, Almendros G (2005) Fire-induced transformation of C-and N-forms in different organic soil fractions from a dystric cambisol under a Mediterranean pine forest (*Pinus pinaster*). Soil Biol Biochem 37(4):701–718. https://doi.org/10.1016/j.soilbio.2004.09.008

Kolb SE, Fermanich KJ, Dornbush ME (2009) Effect of charcoal quantity on microbial biomass and activity in temperate soils. Soil Sci Soc Am J 73:1173–1181. https://doi.org/10.2136/sssaj2008.0232

Koo M (2021) Systemic lupus erythematosus research: a bibliometric analysis over a 50-year period. Int J Environ Res Public Health 18(13):7095. https://doi.org/10.3390/ijerph18137095

Kujanpää L, Reznichenko A, Saastamoinen H, Mäkikouri S, Soimakallio S, Tynkkynen O et al (2023) Carbon dioxide use and removal: prospects and policies. https://julkaisut.valtioneuvosto.fi/handle/10024/164795

Lee Y, Park J, Ryu C, Gang KS, Yang W, Park YK, Jung J, Hyun S (2013) Comparison of biochar properties from biomass residues produced by slow pyrolysis at 500 °C. Bioresour Technol 148: 196–201. https://doi.org/10.1016/j.biortech.2013.08.135

Lehmann J, Joseph S (2015) Biochar for environmental management: an introduction. In: Biochar for environmental management. Routledge, pp 1–13

Li DC, Jiang H (2017) The thermochemical conversion of non-lignocellulosic biomass to form biochar: a review on characterizations and mechanism elucidation. Bioresour Technol 246:57–68. https://doi.org/10.1016/j.biortech.2017.07.029

Li X, Shen Q, Zhang D, Mei X, Ran W, Xu Y, Yu G (2013) Functional groups determine biochar properties (pH and EC) as studied by two-dimensional (13)C NMR correlation spectroscopy. PLoS One 8(6):e65949. https://doi.org/10.1371/journal.pone.0065949

Liu N, Zhang Y, Liu C, Wang W, Liu Z (2015) Characterization of biochars derived from agricultural wastes and their adsorptive removal of atrazine from aqueous solution: a comparative study. Bioresour Technol 198:55–62. https://doi.org/10.1016/j.biortech.2015.09.023

Liu H, Wang Z, Nghiem LD, Gao L, Zamyadi A, Zhang Z, Sun J, Wang Q (2021) Solid embedded microplastics from sewage sludge to agricultural soils: detection, occurrence, and impacts. ACS EST Water 1(6):1322–1333. https://doi.org/10.1021/acsestwater.0c00218

Mahanta K, Bhattacharyya PN, Sharma AK et al (2023) Residue and soil dissipation kinetics of chloroacetanilide herbicides on rice (*Oryzae sativa* L.) and assessing the impact on soil microbial parameters and enzyme activity. Environ Monit Assess 195:910. https://doi.org/10.1007/s10661-023-11513-1

Mahanta DK, Krishnappa C, Bhoi TK (2025) Microbial control of forest insect pests over 60 years (1964–2024): network analysis and bibliometric mapping. J Nat Pestic Res 12:12100132. https://doi.org/10.1016/j.napere.2025.100132

Martinez-Fernandez G, Kinley RD, Smith WJ, Simington J, Joseph S, Tahery S, Durmic Z, Vercoe P (2024) Effect of fit-for-purpose biochars on rumen fermentation, microbial communities, and methane production in cattle. Front Microbiol 15:1463817. https://doi.org/10.3389/fmicb.2024.1463817

Maximilian L, Baharak S, Chen WY (2022) Handbook of climate change mitigation and adaptation, 3rd edn. Springer, Cham. https://doi.org/10.1007/978-3-030-72579-2

Mosa A, Mansour MM, Soliman E, El-Ghamry A, El Alfy M, El Kenawy AM (2023) Biochar as a soil amendment for restraining greenhouse gases emission and improving soil carbon sink: current situation and ways forward. Sustainability 15(2):1206. https://doi.org/10.3390/su15021206

Mukome FND, Zhang X, Lucas CRS, Six J, Parikh SJ (2013) Use of chemical and physical characteristics to investigate trends in biochar feedstocks. J Agric Food Chem 61:2196–2204. https://doi.org/10.1021/jf3049142

Nzediegwu C, Arshad M, Ulah A, Naeth MA, Chang SX (2021) Fuel, thermal and surface properties of microwave-pyrolyzed biochars depend on feedstock type and pyrolysis temperature. Bioresour Technol 320:124282. https://doi.org/10.1016/j.biortech.2020.124282

Olugbenga OS, Adeleye PG, Oladipupo SB, Adeleye AT, John KI (2023) Biomass-derived biochar in wastewater treatment: a circular economy approach. Waste Manag Bull 1:1–14. https://doi.org/10.1016/j.wmb.2023.07.007

Patel MR, Panwar NL (2023) Biochar from agricultural crop residues: environmental, production, and life cycle assessment overview. Resour Conserv Recycl Adv 19:200173. https://doi.org/10.1016/j.rcradv.2023.200173

Peiris C, Gunatilake SR, Mlsna TE, Mohan D, Vithanage M (2017) Biochar based removal of antibiotic sulfonamides and tetracyclines in aquatic environments: a critical review. Bioresour Technol 246:150–159. https://doi.org/10.1016/j.biortech.2017.07.150

Piarpuezán Enríquez MA, Zapata Hidalgo DR, Pantoja-Suárez F (2025) Characterization of briquettes from potato stalk residues for sustainable solid biofuel production. Processes 13(6): 1851. https://doi.org/10.3390/pr13061851

Qian K, Kumar A, Patil K, Bellmer D, Wang D, Yuan W, Huhnke RL (2013) Effects of biomass feedstocks and gasification conditions on char properties. Energies 6(8):3972–3986. https://doi.org/10.3390/en6083972

Rangabhashiyam S, Balasubramanian P (2019) The potential of lignocellulosic biomass precursors for biochar production: performance, mechanism and wastewater application—a review. Ind Crop Prod 128:405–423. https://doi.org/10.1016/j.indcrop.2018.11.041

Ravindra K, Singh T, Mor S (2019) Emissions of air pollutants from primary crop residue burning in India and their mitigation strategies for cleaner emissions. J Clean Prod 208:261–273. https://doi.org/10.1016/j.jclepro.2018.10.031

Roberts KG, Gloy BA, Joseph S, Scott NR, Lehmann J (2010) Life cycle assessment of biochar systems: estimating the energetic, economic, and climate change potential. Environ Sci Technol 44(2):827–833. https://doi.org/10.1021/es902266r

Rodríguez-Espinosa T, Papamichael I, Voukkali I, Gimeno AP, Candel MBA, Navarro-Pedreño J, Zorpas AA, Lucas IG (2023) Nitrogen management in farming systems under the use of agricultural wastes and circular economy. Sci Total Environ 876:162666. https://doi.org/10.1016/j.scitotenv.2023.162666

Ronsse F, Van Hecke S, Dickinson D, Prins W (2013) Production and characterization of slow pyrolysis biochar: influence of feedstock type and pyrolysis conditions. Glob Change Biol Bioenergy 5(2):104–115. https://doi.org/10.1111/gcbb.12018

Sarma B, Kashtoh H, Lama Tamang T, Bhattacharyya PN, Mohanta YK, Baek KH (2023) Abiotic stress in rice: visiting the physiological response and its tolerance mechanisms. Plants (Basel, Switz) 12(23):3948. https://doi.org/10.3390/plants12233948

Seo JY, Tokmurzin D, Lee D, Lee SH, Seo MW, Park YK (2022) Production of biochar from crop residues and its application for biofuel production processes—an overview. Bioresour Technol 361:127740. https://doi.org/10.1016/j.biortech.2022.127740

Shafeeyan MS, Daud WMAW, Houshmand A, Arami-Niya A (2011) Ammonia modification of activated carbon to enhance carbon dioxide adsorption: effect of pre-oxidation. Appl Surf Sci 257(9):3936–3942. https://doi.org/10.1016/j.apsusc.2010.11.127

Sim DHH, Tan IAW, Lim LLP, Hameed BH (2021) Encapsulated biochar-based sustained release fertilizer for precision agriculture: a review. J Clean Prod 303:127018. https://doi.org/10.1016/j.jclepro.2021.127018

Singh C, Tiwari S, Boudh S, Singh JS (2017) Biochar application in management of paddy crop production and methane mitigation. In: Singh J, Seneviratne G (eds) Agro-environmental sustainability. Springer, Cham. https://doi.org/10.1007/978-3-319-49727-3_7

Singh N, Singhania RR, Nigam PS, Dong CD, Patel AK, Puri M (2022) Global status of lignocellulosic biorefinery: challenges and perspectives. Bioresour Technol 344:126415. https://doi.org/10.1016/j.biortech.2021.126415

Suo F, You X, Yin S, Wu H, Zhang C, Yu X, Sun R, Li Y (2021) Preparation and characterization of biochar derived from co-pyrolysis of *Enteromorpha prolifera* and corn straw and its potential as a soil amendment. Sci Total Environ 798:149167. https://doi.org/10.1016/j.scitotenv.2021.149167

Tag AT, Duman G, Ucar S, Yanik J (2016) Effects of feedstock type and pyrolysis temperature on potential applications of biochar. J Anal Appl Pyrolysis 120:200–206. https://doi.org/10.1016/j.jaap.2016.05.006

Tripathi M, Sahu JN, Ganesan P (2016) Effect of process parameters on production of biochar from biomass waste through pyrolysis: a review. Renew Sust Energ Rev 55:467–481. https://doi.org/10.1016/j.rser.2015.10.122

Tsai CC, Chang YF (2021) Higher biochar rate can be efficient in reducing nitrogen mineralization and nitrification in the excessive compost-fertilized soils. Agronomy 11(4):617. https://doi.org/10.3390/agronomy11040617

Varkolu M, Gundekari S, Omvesh P, Palla VCS, Kumar P, Bhattacharjee S, Vinodkumar T (2025) Recent advances in biochar production, characterization, and environmental applications. Catalysts 15(3):243. https://doi.org/10.3390/catal15030243

Vochozka M, Maroušková A, Váchal J, Straková J (2016) Biochar pricing hampers biochar farming. Clean Techn Environ Policy 18:1225–1231. https://doi.org/10.1007/s10098-016-1113-3

Wang Y, Lin Y, Chiu PC, Imhoff PT, Guo M (2015) Phosphorus release behaviors of poultry litter biochar as a soil amendment. Sci Total Environ 512-513:454–463. https://doi.org/10.1016/j.scitotenv.2015.01.093

Wang H, Wang X, Cui Y, Xue Z, Ba Y (2018) Slow pyrolysis polygeneration of bamboo (*Phyllostachys pubescens*): product yield prediction and biochar formation mechanism. Bioresour Technol 263:444–449. https://doi.org/10.1016/j.biortech.2018.05.040

Wang D, Jiang P, Zhang H, Yuan W (2020) Biochar production and applications in agro and forestry systems: a review. Sci Total Environ 723:137775. https://doi.org/10.1016/j.scitotenv.2020.137775

Wang L, Ok YS, Tsang DC, Alessi DS et al (2021) Biochar composites: emerging trends, field successes and sustainability implications. Soil Use Manag 38(1):14–38. https://doi.org/10.1111/sum.12731

Xiao R, Wang JJ, Gaston LA, Zhou B, Park JH, Li R, Dodla SK, Zhang Z (2018) Biochar produced from mineral salt-impregnated chicken manure: fertility properties and potential for carbon sequestration. Waste Manag 78:802–810. https://doi.org/10.1016/j.wasman.2018.06.047

Xie T, Sadasivam BY, Reddy KR, Wang C, Spokas K (2015) Review of the effects of biochar amendment on soil properties and carbon sequestration. J Hazard Toxic Radioact Waste 20: 04015013. https://doi.org/10.1061/(ASCE)HZ.2153-5515.0000293

Xue Y, Gao B, Yao Y, Inyang M, Zhang M, Zimmerman AR, Ro KS (2012) Hydrogen peroxide modification enhances the ability of biochar (hydrochar) produced from hydrothermal carbonization of peanut hull to remove aqueous heavy metals: batch and column tests. Chem Eng J 200:673–680. https://doi.org/10.1016/j.cej.2012.06.116

Yan T, Xue J, Zhou Z, Wu Y (2021) Effects of biochar-based fertilizer on soil bacterial network structure in a karst mountainous area. Catena 206:105535. https://doi.org/10.1016/j.catena.2021.105535

Yan H, Cong M, Hu Y, Qiu C, Yang Z, Tang G, Xu W, Zhu X, Sun X, Jia H (2022) Biochar-mediated changes in the microbial communities of rhizosphere soil alter the architecture of maize roots. Front Microbiol 13:1023444. https://doi.org/10.3389/fmicb.2022.1023444

Yao Y, Gao B, Chen J, Yang L (2013) Engineered biochar reclaiming phosphate from aqueous solutions: mechanisms and potential application as a slow-release fertilizer. Environ Sci Technol 47(15):8700–8708. https://doi.org/10.1021/es4012977

Yin Y, Yang C, Li M, Zheng Y, Ge C, Gu J et al (2021) Research progress and prospects for using biochar to mitigate greenhouse gas emissions during composting: a review. Sci Total Environ 798:149294. https://doi.org/10.1016/j.scitotenv.2021.149294

Yu L, Yu M, Lu X, Tang C, Liu X, Brookes PC, Xu J (2018) Combined application of biochar and nitrogen fertilizer benefits nitrogen retention in the rhizosphere of soybean by increasing microbial biomass but not altering microbial community structure. Sci Total Environ 640–641:1221–1230. https://doi.org/10.1016/j.scitotenv.2018.06.018

Yuan JH, Xu RK, Zhang H (2011) The forms of alkalis in the biochar produced from crop residues at different temperatures. Bioresour Technol 102(3):3488–3497. https://doi.org/10.1016/j.biortech.2010.11.018

Zhang J, Liu J, Liu R (2015) Effects of pyrolysis temperature and heating time on biochar obtained from the pyrolysis of straw and lignosulfonate. Bioresour Technol 176:288–291. https://doi.org/10.1016/j.biortech.2014.11.011

Zhang Z, Zhu Z, Shen B, Liu L (2019) Insights into biochar and hydrochar production and applications: a review. Energy 171:581–598. https://doi.org/10.1016/j.energy.2019.01.035

Zhao SX, Na T, Wang XD (2017) Effect of temperature on the structural and physicochemical properties of biochar with apple tree branches as feedstock material. Energies 10:1293. https://doi.org/10.3390/en10091293

Zheng J, Han J, Liu Z, Xia W, Zhang X, Li L et al (2017) Biochar compound fertilizer increases nitrogen productivity and economic benefits but decreases carbon emission of maize production. Agric Ecosyst Environ 241:70–78. https://doi.org/10.1016/j.agee.2017.02.034

Zhou M, Ying S, Chen J, Jiang P, Teng Y (2021) Effects of biochar-based fertilizer on nitrogen use efficiency and nitrogen losses via leaching and ammonia volatilization from an open vegetable field. Environ Sci Pollut Res Int 28(46):65188–65199. https://doi.org/10.1007/s11356-021-15210-9

Zhu X, Chen B, Zhu L, Xing B (2017) Effects and mechanisms of biochar-microbe interactions in soil improvement and pollution remediation: a review. Environ Pollut 227:98–115. https://doi.org/10.1016/j.envpol.2017.04.032

Zhu S, Guo Y, Zhou H, Luo W, Yi X, Zhou Y, Wu Y, Petticord DF, Song X (2025) Biochar efficacy in enhancing soil carbon fractions is mediated by parent soil type in grazing karst grassland. Carbon Res 4(1):52. https://doi.org/10.1007/s44246-025-00222-8

Chapter 5
Artificial Intelligence in Recycling Wastes to Carbon Nanomaterials

Monalisa Bora Deka, Nilakshi Deka, Sudarshana Borah, and Sagar Barge

5.1 Introduction

With the global boom in populations, ever-increasing complexity in food systems, and mounting levels of waste piling up quicker than ever, there is an urgent demand for smart ways through which trash would be transformed into treasure (FAO 2019). The agri-food supply chain generates huge leftovers ranging from crop scraps and household wastes to factory by-products and postharvest biomass. If neglected, these materials further deteriorate the environment and squander valuable resources (Chen et al. 2020). However, in today's circular bioeconomy, the scenario has been flipped; biowastes are not problems to be dumped—they form a renewable gold mine for high-value products (Geissdoerfer et al. 2017; Kirchherr et al. 2018).

Within this framework, the transformation of bioresource wastes into carbon-based nanomaterials has thus generated tremendous interest as a renewable and highly promising technological pathway (Sevilla and Fuertes 2009; Wang et al. 2023). Such CNMs include graphene, CNTs, CQDs, and designed porous carbons, which exhibit an astonishing combination of properties: extremely large specific surface areas, tunable pore architectures, excellent electrical conductivity, and very strong chemical stability (Novoselov et al. 2004; Iijima 1991; Zhu et al. 2015).

M. B. Deka
Department of Agriculture, The Assam Royal Global University, Guwahati, Assam, India

N. Deka
Department of Computer Science, The Assam Royal Global University, Guwahati, Assam, India

S. Borah (✉)
Royal School of Pharmacy, The Assam Royal Global University, Guwahati, Assam, India

S. Barge
Department of Medicine, Beth Israel Deaconess Medical Centre and Harvard Medical School, Boston, MA, USA

Y. K. Mohanta et al. (eds.), *Bioresource Waste to Carbon Nanomaterial*, Nanotechnology in the Life Sciences, https://doi.org/10.1007/978-981-92-2405-0_5

These features make them highly versatile for next-generation applications such as energy storage, catalysis, environmental remediation, chemical sensing, and the development of advanced functional materials (Wang et al. 2023). By using waste-derived precursors to produce such materials, the approach not only reduces environmental burden but also adds value to materials that would otherwise contribute to pollution, thus completing an elegant loop of resource recovery.

Producing carbon nanomaterials from agricultural and biomass waste resonates strongly with the ethos of the circular economy, where the goal is to convert "waste" into "wealth" while reducing reliance on fossil-based carbon sources. This approach not only mitigates the environmental burden of waste disposal but also unlocks new material streams for advanced technologies. A growing body of research has shown that an impressive diversity of biowastes such as rice husk, coconut shell, fruit peels, and other lignocellulosic residues can be effectively transformed into high-value CNMs and composite materials with finely tunable structural and electrochemical characteristics (Wang et al. 2023).

The promise of waste-derived CNMs has been underscored by several pivotal studies. Lähde et al. (2020), for instance, demonstrated that agricultural residues can be engineered into graphene-like carbon structures, offering a low-cost and sustainable alternative to conventional precursors. Galanakis (2012) further expanded this perspective by illustrating how food waste streams can be valorized not only for bioactive compounds but also for carbon-based functional materials. Recent comprehensive reviews have continued to advance the field by elucidating the intricate links between synthesis strategies, structural evolution, and functional performance in biomass-derived carbons, particularly for applications in energy storage, environmental remediation, and catalysis (Yan et al. 2025; Liu et al. 2025).

When viewed through the lens of agricultural systems, this paradigm reveals a range of mutually reinforcing benefits. First, agricultural residues that would otherwise be burned or landfilled can be redirected as valuable feedstocks for advanced material production. Second, converting these low-cost wastes into nanocarbon materials opens new avenues for value addition, supporting regional bioeconomies and creating local income opportunities. Third, the resulting biomass-derived CNMs can be reintroduced into the agricultural sector itself for instance, in smart sensors, water purification units, or sustainable energy storage devices creating a closed-loop model in which waste valorization directly supports agri-technological progress (Liu et al. 2025).

However, while the field has advanced rapidly, several challenges still need to be addressed. Key issues include refining green and energy-efficient synthesis routes, managing the inherent variability of biomass feedstocks, achieving cost-effective scale-up, and ensuring that these materials meet safety and life-cycle sustainability standards. Additionally, successful deployment will require seamless integration with existing agricultural and industrial infrastructures an area where continued innovation and policy support remain essential (Wang et al. 2023).

5.2 Waste Streams as Feedstock for Carbon Nanomaterials (CNMS)

5.2.1 *Categories of Waste Streams*

Modern human activities driven by rapid industrialization, urbanization, and consumer lifestyles generate an enormous diversity of waste streams. Many of these wastes are inherently rich in carbon, making them attractive precursors for the sustainable production of carbon nanomaterials. Broadly, these feedstocks can be grouped into several categories: municipal and household organic waste, including kitchen scraps, yard trimmings, paper waste, and sewage sludge; industrial waste, such as residues from food processing; pulp and paper mill sludge; textile effluents; and other organic-rich by-products, plastic and polymeric waste particularly polyolefins and packaging films that can be thermochemically converted into carbon through pyrolysis, electronic waste (e-waste) containing polymeric fractions that can serve as carbon sources; and biomass and agro-residues, including rice husk, wheat straw, corn stover, bagasse, coconut shells, nut shells, fruit peels, and vegetable skins.

Among all these streams, biomass and agricultural residues stand out as especially promising. Their natural abundance, renewability, low cost, and high lignocellulosic content make them ideal candidates for sustainable CNM synthesis. With appropriate pre-treatment and conversion such as carbonization, activation, hydrothermal processing, or catalytic transformation these materials can be engineered into a broad spectrum of CNMs, including activated porous carbons, carbon quantum dots (CQDs), carbon nanosheets, graphene-like carbons, and carbon nanotubes (CNTs) (Yan et al. 2025).

5.2.2 *Environmental Benefits and Circular Bioeconomy Context*

The use of waste to produce carbon nanomaterials (CNMs) is at the very center of the circular bioeconomy, as it provides a route that is both environmentally sustainable and economically viable. In this regard, this technology has the effect of preventing organic, agricultural, and polymeric wastes from being disposed of in landfills and through open burning, thus reducing environmental pollution and greenhouse gas emissions (Chen et al. 2020). Biomass and organic waste are used as renewable carbon sources, thus providing an environmentally sustainable alternative to fossil carbon precursors in the global decarbonization of material production (Wang et al. 2023). This is not only environmentally sustainable but also aligns with the principles of the circular economy, as it closes the loop between waste generation and resource recovery (Geissdoerfer et al. 2017; Kirchherr et al. 2018).

In addition to the environmental advantages, the processing of low-value waste to high-value CNMs opens up new economic avenues, especially in rural and agro-industrial economies, where agricultural residues and industrial by-products are processed into high-value nanomaterials (Deng et al. 2022; Li and Chen 2023). Waste-derived CNMs have been shown to possess diverse applications in environmental sensors, catalysis, water purification, and energy storage devices owing to their high surface area, superior electrical conductivity, and adjustable surface chemistry (Novoselov et al. 2004; Zhu et al. 2015). These applications help in enhancing agricultural productivity, environmental monitoring, and clean energy technologies, thus reiterating their importance as sustainable materials that facilitate economic development and a healthy planet. Overall, the waste-to-CNM process is a revolutionary technological approach that synergistically combines waste management, renewable resource utilization, and high-value material synthesis, thereby promoting the principles of sustainable development and circular bioeconomy.

5.2.3 Evidence from Recent Studies: Biowaste as a CNM Source

Recent peer-reviewed articles have emerged with convincing evidence that various biowastes can be effectively utilized to produce high-performance carbon nanomaterials with superior functional properties. Biowaste-based precursor carbons and carbon-metal oxide composites have shown great promise in catalysis, supercapacitor electrodes, and environmental applications due to their high surface area, improved charge transport, and catalytic properties. In a similar manner, carbon quantum dots (CQDs) produced from biomass have been found to possess highly tunable photoluminescence, excellent water solubility, and natural biocompatibility, making them highly attractive for applications in sensing, bioimaging, and novel agricultural monitoring technologies (Liu et al. 2025). Moreover, carbon nanomaterials of zero-dimensional, one-dimensional, two-dimensional, and three-dimensional morphologies produced from agricultural residues have been found to possess electrochemical properties comparable to, and in some cases superior to, those of carbon nanomaterials produced through conventional methods, especially in energy storage and electrochemical devices (Yan et al. 2025).

It is pertinent to note that recent studies have shown the excellent electrocatalytic activity of transition-metal borides supported by carbon derived from onion skin in water-splitting reactions, thus emphasizing the versatility and superiority of biowaste-derived carbon precursors. In light of these developments, the maturity and applicability of biowaste conversion to carbon nanomaterials have been reinforced as a sustainable, economically feasible, and environmentally responsible approach to advanced material synthesis.

5.3 Carbon Nanomaterials: Properties and Synthesis

Carbon nanomaterials (CNMs) constitute a promising class of advanced materials, distinguished by the unique physicochemical properties arising from carbon atoms with sp^2 and sp^3 hybridization. Owing to these distinctive characteristics, CNMs particularly carbon nanotubes and graphene-based materials exhibit a wide range of interdisciplinary applications, including energy storage, catalysis, environmental remediation, healthcare, and nanoelectronics (Iijima 1991; Geim and Novoselov 2007). Their high specific surface area, tunable surface chemistry, and excellent electrical conductivity make carbon nanomaterials highly suitable for next-generation functional materials.

5.3.1 Types of Carbon Nanomaterials

5.3.1.1 Carbon Nanotubes (CNTs)

Carbon nanotubes (CNTs) are nanoscale cylindrical materials made up of rolled graphene sheets and are generally categorized into single-walled carbon nanotubes (SWCNTs) and multi-walled carbon nanotubes (MWCNTs), as described by Iijima for the first time in 1991. From a material science perspective, CNTs have an exceptionally high aspect ratio, with nanometer-scale diameters and several micrometers-long lengths, which is primarily responsible for their outstanding mechanical, electrical, and thermal properties. CNTs have exceptional tensile strength, which is almost two orders of magnitude higher than that of steel, along with high thermal stability and outstanding electrical conductivity due to their highly ordered sp^2-hybridized carbon framework (Avouris et al. 2007). Notably, the electronic properties of CNTs are highly dependent on their chirality and diameter, which enables CNTs to be either metallic or semiconducting in nature, making them extremely useful for a wide range of electronic and optoelectronic applications.

These distinct physicochemical properties have made it possible for CNTs to be applied extensively in conductive composites, high-performance sensors, drug delivery systems, nanoelectronics, and advanced energy storage devices such as supercapacitors and lithium-ion batteries, where they enhance charge transport, energy density, and mechanical robustness. The combination of high aspect ratio, high mechanical strength, controllable electrical properties, and chemical stability makes CNTs one of the most valuable and intensively investigated carbon nanomaterials for future technological and sustainable applications.

5.3.1.2 Graphene and Graphene-Based Materials

Graphene is a two-dimensional monolayer of carbon atoms arranged in a hexagonal lattice. It exhibits remarkably high electron mobility, mechanical flexibility, and optical transparency (Novoselov et al. 2004). Graphene derivatives such as graphene oxide (GO) and reduced graphene oxide (rGO) contain oxygen-functional groups that enhance dispersibility and chemical reactivity, thereby expanding their applicability in catalysis and biological systems (Geim and Novoselov 2007). These materials are characterized by high carrier mobility, large specific surface area, and excellent mechanical flexibility, which collectively contribute to their outstanding performance in applications such as nanoelectronics, energy storage, sensors, composites, and biomedical systems (Geim and Novoselov 2007; Novoselov et al. 2004).

5.3.1.3 Carbon Dots (CDs)

Carbon dots are quasi-spherical, fluorescent carbon nanoparticles typically less than 10 nm in size. They are well known for their strong and tunable photoluminescence, excellent biocompatibility, and low toxicity (Lim et al. 2015). Due to easy surface functionalization, carbon dots have emerged as promising materials for bioimaging, sensing, optoelectronics, and photocatalysis (Zhu et al. 2015). These nanoparticles exhibit strong and tunable fluorescence, high water solubility, and low cytotoxicity, making them highly suitable for applications in biological imaging, environmental monitoring, drug delivery, and advanced optical devices (Lim et al. 2015; Zhu et al. 2015).

5.3.1.4 Other Carbon Nanomaterials

Other types of carbon nanomaterials are fullerenes, carbon nanohorns, carbon nanofibers, and activated carbon nanostructures, which have distinct structural and property features that enhance the use of carbon nanotechnology. Fullerenes, consisting of closed cage-like carbon structures, have high electron affinity and redox activity, which have been applied in photovoltaic cells, drug delivery, and molecular electronics (Kroto et al. 1985). Carbon nanohorns, having cone-shaped graphene structures, have high surface area and porosity, which have been applied in catalyst supports and drug delivery carriers (Iijima 1991). Carbon nanofibers, consisting of graphitic layers with high mechanical strength, have high electrical conductivity and mechanical strength, which have been applied in energy storage devices, composite materials, and electrocatalysis. Likewise, activated carbon nanostructures produced from biomass and organic waste have been found to have highly porous structures, large specific surface area, and high chemical stability, which make them highly useful in hydrogen storage, environmental remediation, water purification, and catalysis (Sevilla and Fuertes 2009). Taken together, these various carbon nanomaterials have a significant role in the development of sustainable technologies for efficient energy storage, pollution remediation, and catalysis.

5.3.2 Preparation Techniques of Carbon Nanomaterials

The synthesis route plays a decisive role in governing the structural, physicochemical, and functional properties of carbon nanomaterials, including their morphology, crystallinity, surface chemistry, porosity, and electronic characteristics. Precise control over synthesis parameters such as temperature, precursor composition, reaction time, and catalytic environment directly influences the size, shape, defect density, and surface functionalization of the resulting nanomaterials, thereby determining their performance in applications such as energy storage, catalysis, sensing, and environmental remediation. Among the various synthesis approaches, thermal methods such as pyrolysis enable efficient conversion of carbon-rich precursors into graphitic nanostructures through controlled thermal decomposition, while hydrothermal techniques offer a low-temperature, environmentally friendly route for producing highly functionalized and porous carbon nanomaterials with tunable surface properties (Sevilla and Fuertes 2009; Wang et al. 2023). Catalytic methods, particularly catalytic chemical vapor deposition (CVD), are widely employed for the controlled growth of high-quality carbon nanotubes and graphene with well-defined crystallinity and structural uniformity, owing to the role of metal catalysts in directing carbon atom assembly. Collectively, these synthesis strategies provide versatile and scalable pathways for tailoring carbon nanomaterials with desired structural and functional characteristics, enabling their widespread application in advanced technological and sustainable material systems.

5.3.2.1 Pyrolysis Method

Pyrolysis involves the thermal decomposition of carbon-rich precursors under an inert atmosphere at temperatures ranging from 400 to 1000 °C. Common precursors include biomass residues, polymers, and organic waste materials. This method is particularly attractive for sustainable and large-scale production of carbon nanomaterials (Sevilla and Fuertes 2009). It offers a simple and scalable process, enables the utilization of renewable or waste feedstocks, and is especially suitable for the synthesis of carbon dots and porous carbon materials, making it an efficient and environmentally sustainable approach for converting waste-derived carbon sources into high-value nanomaterials.

5.3.2.2 Hydrothermal and Solvothermal Methods

Hydrothermal synthesis involves the treatment of organic precursors in sealed reactors under high temperature and pressure in aqueous media. This approach allows precise control over particle size, surface functionality, and crystallinity, making it especially suitable for the synthesis of carbon dots and graphene oxide (Sevilla and Fuertes 2009). It offers advantages such as low processing temperature, an environmentally friendly ("green") process, and excellent control over morphology, enabling the production of carbon nanomaterials with tailored structural and surface properties.

5.3.2.3 Catalytic Chemical Vapor Deposition (CVD)

Catalytic chemical vapor deposition (CVD) is one of the most widely used techniques for synthesizing high-quality CNTs and graphene. In this method, hydrocarbon gases are decomposed over metal catalysts such as iron (Fe), nickel (Ni), or cobalt (Co) at elevated temperatures, resulting in the growth of highly ordered carbon nanostructures (Iijima 1991; Geim and Novoselov 2007). It offers advantages including high purity and crystallinity, controlled growth and alignment, and industrial scalability and relevance, making it a preferred method for the large-scale production of structurally uniform carbon nanomaterials.

5.4 Artificial Intelligence and Machine Learning in Waste Recycling

With swift rise in urbanization, there is an even higher rise of the amount of waste generated on a daily basis. However, converting these wastes into sustainable nanomaterials is an effective utilization of these wastes. Conversion of waste to carbon nanomaterials are garnering lot of attention owing to the unique features of carbon nanomaterials in terms of physical and chemical properties, thereby making these materials useful for different applications. The traditional conversion methodology needs a lot of straining procedure; however, with the use of AI and ML, it becomes easier and way more feasible for waste recycling. The use of AI and ML have not only revolutionized the process of conversion of wastes to CNM but also has improved other parameters in terms of accuracy and precision and reduction of operational cost (Bircanoğlu et al. 2018). Technological advancement that allowed deep learning models for waste classification, recycling, predictive analysis and working on data driven decision-making process, has helped a lot in facilitating automated sorting, real-time monitoring, predictive analysis, and also in terms of quality control.

5.4.1 *Deep Learning for Automated Sorting (CNNs, Transfer Learning)*

Deep learning is a widely acknowledged and widely recognized technology in terms of its's efficacy for waste classification, waste recycling, waste management. Deep learning is a comprehensive form of a subset of machine learning comprising various mathematical and statistical tools and methodologies used in the deployment of neural networks for classifying items. Deep learning is a form of widely recognized methodology in terms of its's efficacy for modern waste management technology. Deep learning allows computational models that are composed of multiple processing layers in order to learn representations of data with multiple levels of abstraction. Deep learning is a comprehensive form of a subset of machine learning comprising various mathematical and statistical tools and methodologies used in the deployment of neural networks for classifying items. Deep learning-based models can achieve successful classification even when traditional methods fail, particularly for waste categories with high visual similarity and intra-class variation. Deep learning, in terms of convolutional neural networks (CNNs) and transfer learning, has emerged as a powerful tool for automated sorting. Automated waste sorting—once completely reliant on manual labor, optical sensors, and rudimentary mechanical systems—now benefits hugely from CNN and transfer learning techniques, thus enabling accurate and precise classification of heterogeneous, mixed, and contaminated streams of wastes.

5.4.1.1 Role of CNN in Automated Waste Sorting

CNNs plays significant role in terms of modern automated waste sorting by providing the ability to automatically identify, classify, and separate various waste streams with high precision. Unlike traditional sorting approaches—such as manual picking, mechanical filters, or optical sensors—CNNs learn hierarchical visual features from large image datasets, making them ideal in terms of recognizing complex patterns in real-world waste materials. In particular, convolutional neural networks (CNN) have proven to be effective in discriminating between different forms of garbage due to their high accuracy and strong learning capabilities in image classification tasks (Bobulski and Kubanek 2021; Li and Chen 2023). CNN can learn highly abstract features and can identify objects efficiently. The key parameter for consideration of CNN includes it's reduced number of parameters that needs to be trained. This helps in a way that CNN can be trained smoothly and does not go through overfittings.

Their architecture consists of multiple layers that are hierarchically arranged; the network thereby automatically extracts from images spatial patterns, from simple edges to complex object structures. A broad overview of the architecture of a CNN is divided into feature extraction layers and classification layers. The architecture of CNN for waste sorting typically includes four layers: (1) convolution layer, (2) pooling layer, (3) activation function, and (4) fully connected layer. The hierarchical architecture of CNN enables it to directly learn discriminative visual features from the raw images themselves, without requiring any descriptors. This is an important ability when trying to sort real-world waste that is often deformed, contaminated, or partially occluded. It is effective at extracting low-level patterns like edges, textures, intermediate features of surface structures, and high-level material signatures that distinguish plastics, metals, paper, glass, and the components of electronic waste. Waste detection-oriented models, like YOLO, SSD, and R-CNN, enhance the capability of waste detection by detecting and localizing multiple waste products running at a high speed on a conveyor belt to further enable automated sorting process with precision. Considering this aspect of different detection models, CNN is considered to be invariant to waste orientation and waste object deformation thereby improving the efficiency, reliability, and throughput of automated sorting system.

Among different CNN models, fine-tuned CNNs are becoming important for achieving high accuracy in automated waste-sorting applications, as most of the datasets are either limited or are highly variable. Pretrained architectures, like ResNet-50, MobileNetV2, EfficientNet-B0/B4, DenseNet121, InceptionV3, and VGG-16, can be fine-tuned further ahead without having to train complete network from the very beginning, thereby allowing these models to recognize complex categories of wastes. The solution to detection of complex waste is pre-training the models on large amount of datasets, thereby enabling the models to train on learn rich generic visual features. During the process of fine-tuning these models, deeper layers are only retained on certain domain-specific waste images, thereby enabling these models to also specialize in subtle material characteristic recognition. EfficientNet models show very good performance because they can compound scale, which means they can reach high accuracy while still being computationally efficient. This makes them great for real-time sorting on fast-moving conveyor belts. Fine-tuned models like MobileNet and ShuffleNet also offer a lightweight solution that works well for edge devices and embedded systems used in industrial waste sorting. At least 93–97% classification accuracy has been consistently documented in existing studies employing fine-tuned CNNs for the multi-class recognition of waste, encompassing plastic type differentiation, metal versus nonmetal separation, paper grade identification, and the detection of e-waste components. The ability of fine-tuned CNN models to generalize visually ambiguous items of wastes makes the fine-tuned CNNs a critical enabler of a high-performance, industry-ready intelligent automated waste-sorting solution.

A model can be trained and improvised for one activity and then used in another activity that is closely related to it with the aid of transfer learning. This can be seen, for instance, when knowledge gained in one context is applied to improve performance and efficiency in the considered activity. Datasets smaller than the real training datasets were used to test pretrained models. Transfer learning exploits knowledge learned from a previously trained model on a larger general-purpose dataset and adapts to a new domain-specific task. Such an approach allows the system to share some feature representations learned, including edges, textures, shapes, and object structures, while fine-tuning only the upper layers to fit better in the target application. Pretrained models already possess rich and diverse visual features on millions of images, which contributes to the reduction of large annotated datasets required, shortening the required training time, and overall performance improvement. This is one of the most practical and powerful techniques for the implementation of deep learning-based automated waste-sorting systems.

5.4.2 ML-Based Process Optimization (Neural Networks, SVM, and Random Forests)

As a subset of artificial intelligence (AI), machine learning (ML) allows systems to learn from data and enhance performance of the system without the need for explicit programming. Using statistical models and algorithms, ML makes predictions, looks for patterns, and automates decision-making. ML is an important part of automation today because it makes businesses more productive, accurate, and flexible (Angelopoulos et al. 2019). Training data, models, algorithms, and performance evaluation are the key features of ML techniques. Training the data is the first step in learning, which later helps the models to identify shapes, patterns, relationships, and perform specified tasks. Machine learning algorithms like neural networks and support vector machines use this trained data to make predictions or perform specified tasks. Mean squared error, accuracy, and precision are few parameters of performance metrics that helps figure out how well an ML model is trained and if it works well for automation. Machine learning methods are now widely used to make many different kinds of operations better, especially in the areas of modern waste management and recycling. Unlike manually tuned processes, ML models can learn from past data and make smart guesses considering how to improvise efficiency, reliability, and resource use even more. ML algorithms like artificial neural network (ANN), support vector machine (SVM), and random forest (RF) are among the most renowned ML algorithms for performing specified tasks based on trained data and this feature of ML algorithms can be used for the task of process optimization. Each of these algorithms has its own characteristics when it comes to handling different types of data and making better form of decisions. These models can be used at different points in the waste process optimization process, starting from prediction of how waste materials will flow, optimising the process of waste processing, and also making sure to keep the quality of the output consistent. ML-based optimization process makes it possible for the systems to work much better by providing real-time operational insights on waste recycling process. Machine learning (ML)-based optimization enables waste recycling systems to operate more efficiently by delivering real-time operational insights. This, in turn, facilitates the development of highly automated, intelligent, and sustainable waste management solutions

More and more people are using artificial neural networks to model complicated, nonlinear processes. An ANN model uses the concept of human brain and consists of layers of interconnected processing units or neurons. In ANN architecture, there is an input layer, one or more hidden layers, and an output layer. The input layer contains the raw features, and each hidden layer can use weighted connections, summation, and activation functions to make nonlinear changes. In ANN, information only moves forward from one layer to the next. A neuron in a layer computes an activation over its inputs. The output layer gives the final prediction or classification. The network changes the weights on its connections during training using a method called backpropagation algorithm, which uses gradient-based optimization to reduce the prediction error. This architecture, which has layers and can change, gives ANNs the ability to learn about complex relationships in data. Artificial neural networks are now one of the most popular and widely used machine learning models for process optimization. This is because they can approximate complex nonlinear relationships that may exist between input variables and output responses. ANN architecture learns patterns by changing the weights of the connections between artificial neurons. Once the model is trained, the structure can be modelled for complex dependencies, subtle variations can be derived, and the model can do well on real-time data analysis. That means they can be used for optimizing industrial processes, predicting system performance, and finding the most important factors that affect efficiency. In process optimization applications, ANNs can work with large datasets and measurements that are not very accurate. They have been shown to be useful for modelling chemical reactions, energy systems, manufacturing processes, and even turning waste into resources. Overall, it can be mentioned that ANNs are a powerful tool for making better decisions based on complex form of data. ANNs are suitable for making accurate predictions, monitor things in real time, and improve performance in terms of waste management.

SVM is a supervised machine learning algorithm that is capable of classification, regressions, and outlier decision-making tasks. Essentially, the idea of SVM is to select the optimum hyperplane which best separates data points belonging to two different classes with the largest margin. SVM achieves very strong and general decision boundaries even in the case of overlapping datasets by concentrating only on the most significant data points, which are called support vectors. If the issue is nonlinear, the SVM algorithm employs kernel functions such as radial basis function (RBF), polynomial, or sigmoid kernels to convert input data to a higher-dimensional feature space where it can be separated. Firstly, SVM is excellent for process optimization when there is a scarcity of data, but the decision boundaries are complicated. It is extensively utilized in categorizing operational states, detecting faults, and modelling quality parameters in manufacturing, energy systems, and chemical processes. SVM is a trustworthy method of system performance forecasting as it has the ability to generalize well and not overfit. Such capability is in line with data-driven optimization strategies.

Random forest (RF) is one of the most popular ensemble learning methods that work by creating a set of decision trees with the aim of improving prediction stability, accuracy, and the ability to generalize. Different subsets of instances of the dataset are used to train each decision tree through bootstrap sampling, where at each split, a random subset of features is chosen. By introducing this controlled randomness, the problem of overfitting, which a single decision tree is prone to, is reduced; thus, better generalization is achieved. The output of the model, in fact, is the aggregation of the results of all the trees: majority voting for classification and averaging for regression. As a matter of fact, this is quite significant in process optimization, as RFs are capable of dealing with high-dimensional, heterogeneous, and noisy data without any problem. They are able to learn intricate patterns of the interactions among the variables, as well as, give the feature-importance rankings, which makes it easier for practitioners to identify the operational parameters that are of key relevance to the process outcomes. These RF model features could be very useful in waste processing for figuring out how accurate sorting will be, how to improve operating settings, and what affects system performance. RF is an excellent choice for data-driven optimization in industrial settings since it is strong and easy to understand when working with real-time data.

5.4.3 Real-Time Monitoring, Predictive Maintenance, and Quality Control

AI and ML are leading to a major leap by bringing real-time monitoring, predictive maintenance, and quality control to waste management as well as reusability in the case of carbon nanomaterials. To achieve the goals of the circular economy, very significant changes in the methods of processing and treating waste are necessary. Modern waste management facilities are designed around a closely integrated software and hardware architecture for real-time monitoring, where on-board AI processors, multimodal sensor arrays, and high-speed data acquisition units work together to ensure continuous situational awareness of the recycling line. The transformation of waste materials into carbon nanomaterials by the use of AI and ML is opening completely new frontiers in the areas of accuracy, precision, and process scale-up. Just as for the case of pyrolysis, catalytic CVD which are tightly coupled to real-time monitoring, AI-powered sensor elements, and data analytics systems have made the real-time analysis possible for the operations along CNM synthesis methods from wastes (Butler et al. 2018).

One of the significant ways of reducing carbon footprints is the implementation of AI-driven predictive maintenance techniques which provide an important aspect for enhancing the circular economy. Artificial intelligence (AI) is very instrumental in predictive maintenance through the usage of advanced algorithms and machine learning techniques to analyze huge volumes of data from machinery and equipment. AI can predict a breakdown of a machine by following real-time data such as vibration, temperature, and operating habits. Hence, maintenance can be scheduled before any major problems arise. This proactive approach by cutting on unplanned downtime and also by extending the life of the equipment reduces the rate of frequent replacements and therefore is in line with the circular economy's principles of resource reuse and waste minimization. The conversion process from waste to CNM largely includes the use of high-temperature reactors, catalytic chambers, gas flow systems, and thermal treatment units, which are, however, susceptible to wear, fouling, and mechanical degradation with time. In such cases, predictive maintenance systems based on AI and ML take care of the continuous streams of vibration, temperature, pressure, acoustic signature, and rotational speed data generated by sensors that monitor the equipment to identify the abnormal patterns or the changes in the normal operating behavior. Machine learning models like neural networks, SVM, and random forest can issue a prediction of the next faults way ahead of the time by analyzing the previous failure data and the real-time conditions. This notion of AI-powered predictive analysis facilitates the operators in executing the maintenance at the right time, and hence the reduction of unplanned downtime, the prevention of production losses, and the avoidance of expensive equipment damage become a reality.

Real-time quality control is essential for the efficient functioning of present-day waste management systems as it helps the system to sort, process, and recycle the different waste materials in an effective manner. The use of sensors, computer vision, and AI-based analytics in waste management operations makes it possible to have a continuous monitoring of the waste that it is moving from the stages of collection, segregation, and treatment. Immediate detection of contamination, misclassification, and process deviations is, therefore, possible which leads to an increase in the purity of recycling and the efficiency of the operations. ML algorithms work on real-time data analysis in order to determine the material composition, moisture content, and physical properties of the waste. Real-time quality control using ML algorithms contributes to an increase in resource recovery rates, a decrease in landfill disposal, and the promotion of environmentally friendly waste management that is in line with the principles of the circular economy (Chen et al. 2020).

5.4.4 Environmental Data Integration

Innovative AI and machine learning (ML) techniques are becoming more powerful thanks in large part to the integration of environmental data which is now recognized as the main driver behind these techniques in waste recycling sectors. The main focus is on systems aimed at transforming heterogeneous wastes into high-value products such as carbon nanomaterials. The integration of multisource environmental datasets that include parameters like moisture content, bulk density, and elemental composition real-time sensor outputs comprising hyperspectral images, thermal profiles, and acoustic-vibration signatures and operational parameters such as energy consumption, throughput variability, and equipment degradation gives an entire picture of the process conditions which is necessary for accurate modelling and decision-making (Liu et al. 2025). Modern data fusion infrastructures such as IoT-based networks, GIS-supported systems, and centralized data collected and processed enable high-frequency data synchronization and multivariate correlation analysis, thus reducing domain shifts and environmental uncertainties that usually lower model performances. Along with modern data fusion infrastructures, AI and ML algorithms are used for integrated environmental databases to take accuracy and performance of the waste recycling process to the next level. Real-time integrated environmental datasets also serve as a basis for life cycle-oriented assessments and sustainability-driven optimization, thereby providing recycling facilities the opportunity to minimize emissions, lower their energy footprints, and increase material recovery efficiency. From this point of view, environmental data integration is the core element that makes AI-based waste management systems reliable and scalable; at the same time, it paves the way for the development of adaptive and sustainable recycling routes that are in line with the principles of the circular economy.

5.5 Integration of AI, ML, and Nanotechnology in Circular Economy

The goal of the circular economy is to decrease the amount of waste and make the best use of resources by reusing, recycling, and managing the life cycle of waste in a way that is good for the environment. Using cutting edge technologies of AI and ML in terms of waste management has become a very effective way to make circular economy measures better. AI and ML techniques help use real-world data work on it specifically in terms of waste management and waste recycling process. Nanotechnology, on the other hand, is an advanced form of science and engineering focused on improvising the material's performance, recovery process, and thereby focusing on environment friendly ecosystem on a nanoscale level. The concept of AI, ML, and nanotechnology can be brought together to aid with smart waste management system, better recycling processes, and also extraction of new materials which are harmless to the environment. This integration and working up on waste management speeds up the shift to the concept of circular economy that is strong and makes appropriate usage of waste materials (Liu et al. 2025).

5.5.1 Material Recovery and Life Cycle Analysis

Due to the large global issues of resource depletion, environmental degradation, and the requirement of a shift toward sustainable and circular industrial processes, the researchers are looking for new methods to recover waste materials and produce chemicals. Although a range of innovative ideas are available in principle, they have to solve considerable problems and hurdles first before these new waste recovery processes can be implemented successfully. The variety and the complexity of waste materials is one of the biggest challenges in waste recovery. Methods for recovering valuable materials from waste in a single step are hardly ever found. Consequently, it is very important to have different waste recovery solutions designed and implemented, each of which can have their own technical and financial limitations. The process of material recovery is the main element of waste management or waste recycling, given the context of different heterogeneous materials present in the waste materials. Technological advancement in terms of AI and ML techniques have now been assigned a crucial role in the transformation of traditional recovery operations from wastes into more superior and highly automated, data-driven processes which are capable of achieving superior operational precision.

The life cycle assessment (LCA) of waste is a series of steps that happen from the time the trash is made until it is thrown away. In the context of circular economy, LCA provides a way to make decisions related to waste management. Municipal waste management is the best way to deal with waste because it includes different ways to collect, use, and get rid of it. These include collection (waste delivery system, collection at the point of generation, mobile collection points, post-component or fractional separate waste collection, or non-separated collection); material sorting and recycling (including the creation of refuse-derived fuel (RDF) briquettes); aerobic and anaerobic treatment of organic waste; thermal methods of utilization (incineration, pyrolysis, and gasification); and disposal (landfill disposal). Waste life cycle assessment is done in a number of steps: (1) defining the waste system's scope and boundaries; (2) inventorying the waste system's pertinent input and output flows; (3) evaluating any potential environmental effects; and (4) interpreting the findings of inventory analysis and the environmental effects evaluated during the life cycle impact assessment (LCIA) phase.

5.5.2 AI-Enhanced Pollution Reduction

Artificial intelligence is radically changing the way for dealing with waste by making pollution control more predictive, adaptive, and data-driven. In general, the traditional methods of treating and recycling waste limit the ability to optimize emissions, environmental changes, and process inefficiency in terms of real-time pollution reduction. To fill this gap, AI and ML algorithms implement sensor-driven data analytics, dynamic control strategies, and intelligent forecasting models that focus not only on maximizing environmental protection but also comply with the circular economy principles. Henceforth, AI is the main driver that radically reforms the waste management sector and leads to the substantial reduction of pollution not only by providing continuous monitoring, intelligent control, and predictive decision-making but also by enabling the capabilities that are impossible for even the most advanced conventional manual or sensor-driven systems. To this end, modern waste treatment facilities install sensor networks equipped with AI-enabled IoT gas analyzers, hyperspectral imaging, thermal cameras, and particulate counters to get the most accurate emissions, combustion efficiency, and environmental conditions data in the shortest time possible. Different ML algorithms process the data in real time to identify any deviations from normal operating conditions. They also detect any sudden increases in CO, NOx, SOx, VOCs, or fine particulates that may indicate periods of incomplete combustion or equipment malfunction (Chen et al. 2020). By incorporating past and present data, AI systems are capable of continuously adjusting operational parameters such as air-fuel ratio, chamber temperature, feedstock distribution, and catalyst performance toward achieving more complete oxidation and, therefore, less formation of toxic by-products, such as dioxins, furans, and tar residues, resulting from thermochemical processes like incineration, pyrolysis, and gasification. This is not all—AI has also contributed to the reduction of upstream material loss by making automated sorting more accurate through CNNs, which in turn drastically decrease cross-contamination of recyclable streams and thus lessen the pollution caused by landfill disposal and inefficient thermal treatment. On a larger scale, implementing AI in LCA frameworks offers the possibility of accurate prediction of greenhouse gas emissions, toxicity potentials, and energy footprints brought about by different routes of waste treatment, thus helping policymakers and engineers in making decisions that result in the least environmental impact. Altogether, the changes driven by AI in pollution control measures are a big step forward in controlling pollution, lessening the operational risks, and laying the foundation for the waste management systems that are cleaner, more sustainable, and environmentally friendly.

5.5.3 Smart Recycling and Circular Resource Management

Traditional CNM synthesis from virgin precursors is resource-intensive and environmentally taxing. Waste-derived feedstocks offer a sustainable alternative, and when integrated with AI and ML models, the transformation becomes more efficient, selective, and scalable. By combining predictive modeling, real-time optimization, and big-data analytics, AI enhances every stage of CNM production from feedstock sorting to industrial reactor management. The global escalation of municipal, electronic, and industrial waste has intensified interest in advanced valorization technologies that can convert these complex waste matrices into functional materials. Among such technologies, the synthesis of carbon nanomaterials including carbon nanotubes (CNTs), graphene derivatives, carbon dots (CDs), and nanofibers stands out due to their high surface area, tunable electronic properties, and diverse industrial applications. Recent breakthroughs show that integrating AI and ML into CNM production workflows enhances process efficiencies, improves material quality, and accelerates sustainable industrial deployment. This chapter presents a detailed examination of these technologies and their emerging applications. The convergence of artificial intelligence (AI), machine learning (ML), and nanotechnology is catalyzing a paradigm shift toward sustainable, circular economic systems. This triad empowers industries to reimagine waste not as refuse but as valuable feedstock, enabling intelligent recovery pathways, efficient resource management, and pollution mitigation. By embedding these technologies across material life cycles, we can extend product lifespans, reduce environmental footprints, and foster system-wide sustainability. This chapter explores three pivotal aspects where AI, ML, and nanotechnology synergize in circular economy frameworks: material recovery and life cycle analysis, AI-enhanced pollution reduction, and smart recycling with circular resource management (Geissdoerfer et al. 2017; Kirchherr et al. 2018).

Proper waste classification ensures efficient separation and purity of feedstocks destined for nanomaterial synthesis. AI-driven deep learning models trained on large image datasets enable high precision sorting of heterogeneous wastes distinguishing e-waste components, plastic polymers, biomass residues, and composite materials with remarkable accuracy and speed. Vision systems equipped with convolutional neural networks (CNNs) can process multi-sensor inputs including spectral imaging, near-infrared (NIR), Raman spectroscopy, and X-ray fluorescence (XRF) data to verify material identity and purity dynamically.

AI optimizes synthesis of these nanomaterials to maximize selectivity and capacity, refining surface functional groups through predictive modeling of nanostructure-property relationships. Such nano-enabled recovery promotes high-purity material recirculation and reduces reliance on environmentally intensive mining activities.

The integration of AI, ML, and nanotechnology is transforming circular economy systems by enabling intelligent material recovery, pollution reduction, and resource-efficient recycling. AI-driven classification tools, sensor fusion, and automated LCA accelerate accurate waste sorting and environmental impact assessment, while nano-enabled extraction techniques allow selective recovery of high-value metals and organics. Digital twins further enhance closed-loop material circulation by predicting system behavior and optimizing reintegration of recovered nanomaterials.

In pollution reduction, AI-powered predictive emission monitoring supports proactive control of industrial pollutants, while nano-engineered adsorbents and catalysts significantly improve remediation efficiency. ML-guided catalyst discovery enhances waste-to-carbon nanomaterial processes, lowering energy consumption and emissions. Environmental digital twins provide real-time simulations to design adaptive pollution control strategies.

Smart recycling is strengthened by AI-enabled robotic sorting, IoT sensor-based smart bins, and intelligent logistics optimization. Nanotechnology upgrades low-value wastes into high-performance carbon nanomaterials, with AI ensuring precise tuning of their properties. Circular manufacturing platforms use ML-driven quality control and demand forecasting to maintain consistent, closed-loop production. Together, these technologies enable high-purity feedstocks, efficient resource cycling, and sustainable material regeneration across the entire waste-to-nanomaterial value chain (Butler et al. 2018; Jensen et al. 2019).

5.6 Case Studies and Emerging Applications

The integration of artificial intelligence (AI), machine learning (ML), and nanotechnological innovations is reshaping how industries convert diverse waste streams into high-value carbon nanomaterials (CNMs). While traditional approaches to waste-to-nanomaterial synthesis often struggle with feedstock variability, catalyst inefficiency, and inconsistent product quality, AI-driven systems overcome these limitations through data-centric optimization, predictive modelling, and automated control mechanisms (Butler et al. 2018). The following case studies highlight real-world examples where AI and ML algorithms have accelerated catalyst discovery, enhanced reaction efficiency, reduced energy demands, and enabled scalable valorization of e-waste, organic residues, plastics, and industrial by-products. These applications illustrate not only the technological feasibility but also the economic and environmental potential of intelligent waste-to-CNM pathways (Goldsmith et al. 2018). By integrating deep learning, digital twins, sensor fusion, and generative modelling, researchers can now predict material behaviors, fine-tune reaction parameters, and maintain consistent nanomaterial quality across heterogeneous waste inputs. Collectively, these case studies demonstrate how AI-guided nanotechnology is ushering in a new era of circular manufacturing one where waste is continuously reimagined as a renewable, data-optimized resource for creating advanced carbon-based materials used in electronics, energy storage, environmental remediation, and smart manufacturing (Fadeel et al. 2018).

5.6.1 E-Waste to Carbon Nanotubes (AI-Optimized Catalysis)

Electronic waste (e-waste) is a rich secondary resource containing abundant transition metals such as nickel (Ni), iron (Fe), and cobalt (Co), which are widely used as catalysts in conventional CNT synthesis. These metals occur in printed circuit boards (PCBs), wires, batteries, capacitors, and magnetic components, making e-waste an ideal low-cost, circular feedstock for producing high-value carbon nanotubes (CNTs). However, the metal composition in e-waste is highly heterogeneous, and traditional catalyst preparation requires extensive purification, precise thermal activation, and tight control of process parameters. This is where artificial intelligence (AI) significantly enhances process efficiency and reproducibility.

AI and ML models particularly Bayesian optimization, artificial neural networks (ANNs), and genetic algorithms analyze the elemental profiles and oxidation states of metal residues from shredded e-waste. These models rapidly identify optimal catalyst formulations by predicting how different Ni:Fe:Co ratios, particle sizes, and surface morphologies influence CNT nucleation, growth kinetics, and defect formation. Instead of relying on trial-and-error catalyst tuning, AI quickly converges on high-performance catalytic fingerprints tailored to the specific e-waste batch.

AI-driven optimization also extends to catalyst activation parameters such as reduction temperature, dwell time, gas composition (H_2, CH_4, CO_2), and heating rate. By modeling thermal decomposition pathways, AI tools can predict the precise temperature windows that activate catalytic sites while preventing undesirable sintering or oxidation. This leads to more uniform nanoparticle dispersion and enhanced catalytic activity.

When e-waste is pyrolyzed or subjected to chemical vapor deposition (CVD), AI-optimized parameter sets significantly improve CNT yield, crystallinity, and diameter uniformity. Studies report reductions in energy consumption often by 20–40% because AI can identify the lowest effective reaction temperatures needed for graphitization. Additionally, ML-guided control systems continuously analyze in-line spectroscopic and thermal data to maintain stability and correct deviations during growth (Upadhyayula et al. 2018).

5.6.2 Organic Waste to Carbon Dots (ML-Guided Synthesis)

Organic waste including agro-residues, municipal food waste, lignocellulosic biomass, fruit peels, paper sludge, and wastewater-derived organics offers a highly abundant, renewable feedstock for synthesizing carbon dots (CDs). CDs derived from waste are particularly attractive because their optical properties (photoluminescence, quantum yield, emission tunability) can be modulated through precursor composition, reaction temperature, dopant integration, and surface chemistry. However, the inherent heterogeneity of organic waste streams often leads to inconsistent CD quality. Machine learning (ML) overcomes these limitations by predicting synthesis outcomes and optimizing parameters across complex, multidimensional datasets (Lim et al. 2015).

Organic waste typically contains variable proportions of cellulose, hemicellulose, lignin, lipids, proteins, sugars, and trace minerals. ML models such as random forest (RF), support vector regression (SVR), and gradient boosting machines (GBM) analyze the chemical fingerprints of different waste types to predict how variations in composition influence nucleation, carbonization efficiency, and fluorescence behavior. These models help select the most suitable waste streams or waste mixtures to achieve targeted CD properties. CD synthesis from organic waste typically involves hydrothermal carbonization, solvothermal routes, or microwave-assisted pyrolysis. ML algorithms trained on experimental datasets can predict optimum temperature-time combinations, determine ideal pH and precursor concentration, optimize the carbonization window to avoid quenching, and minimize energy consumption while preserving luminescent properties.

Stochastic optimization models, including Bayesian optimization and genetic algorithms, navigate the parameter space more efficiently than human-guided experimentation, reducing trial-and-error cycles by nearly 70–85%. Doping with nitrogen, sulfur, phosphorus, or metals is widely used to enhance CD brightness, shift emission wavelengths, or improve sensing capabilities. ML models correlate dopant ratios and precursors (e.g., protein-rich waste for N-doping and sulfur-containing biomass for S-doping) with photoluminescence outputs. Deep neural networks (DNNs) predict how functional groups and heteroatom configurations influence bandgap states, enabling precision tuning of multicolor emissions from blue to near-infrared. ML also accelerates the mapping of structure-property-application relationships. By analyzing CDs' optical signatures, surface chemistry, and defect profiles, ML models classify waste-derived CDs into application domains such as: bioimaging and fluorescence labeling, metal-ion sensing and environmental monitoring, photocatalysis and antimicrobial activities, energy storage, and opto-electronics. This enables a streamlined pipeline from waste selection to tailored CD production. Integrated ML models linked to real-time spectroscopy (UV-Vis, PL, Raman) allow dynamic feedback control during synthesis. These systems automatically adjust parameters to maintain consistent CD size distribution and fluorescence intensity, even when precursor compositions vary.

5.6.3 *Plastics Upcycling (AI for Catalyst Discovery)*

Plastic waste including polyethylene (PE), polypropylene (PP), polyethylene terephthalate (PET), polystyrene (PS), and mixed municipal plastic streams represents one of the most persistent global pollution challenges. Converting these carbon-rich polymers into high-value carbon nanomaterials (CNMs) such as carbon nanotubes (CNTs), graphene, carbon nanofibers (CNFs), and porous carbons presents a powerful route for circular resource recovery. However, effective upcycling requires high-performance catalysts capable of breaking strong carbon:carbon bonds, minimizing toxic by-products, and promoting selective graphitization. Traditional catalyst discovery is slow, expensive, and experimentally intensive. Artificial intelligence (AI) accelerates this process by predicting catalytic behavior, screening vast chemical spaces, and optimizing reaction conditions for selective CNM production. AI models including graph neural networks (GNNs), support vector machines (SVMs), random forests (RF), and deep learning (DL) architectures are trained on datasets of catalyst compositions, polymer degradation behaviors, and nanomaterial properties. These models learn structure-activity relationships that determine: catalytic cracking efficiency, graphitization potential, coke deposition tendency, activation energy profiles, carbon yield, and morphology of resulting CNMs (Miandad et al. 2016).

GNNs, in particular, model atomic interactions between polymers and catalyst surfaces, enabling highly accurate prediction of catalyst performance before experimental trials.

Generative AI algorithms such as variational autoencoders (VAEs), generative adversarial networks (GANs), and generative transformers explore enormous compositional spaces to recommend novel catalyst formulations. These systems can design bimetallic catalysts (e.g., Ni:Mo, Fe:Co, Co:Cu) with enhanced polymer-cracking selectivity, metal-oxide interfaces that promote graphitic carbon formation, and heteroatom-doped nanocatalysts that lower activation energy barriers and catalyst supports (Al_2O_3, MgO, SiO_2, zeolites) with optimized porosity and acidity. This accelerates catalyst discovery from years to weeks, reducing reliance on costly experimentation.

Reinforcement learning (RL) algorithms simulate reaction pathways under different catalysts and process variables, learning strategies that maximize CNM yield while minimizing emissions. RL agents iteratively adjust pyrolysis temperature profiles, heating rates, catalyst loading, gas atmospheres (H_2, N_2, CO_2), and residence times. These models identify energy-efficient, low-emission pathways that traditional methods often overlook. AI tools integrate experimental data with thermokinetic models to predict how catalysts influence polymer dehydrogenation, aromatization pathways, carbon ring fusion, and formation of tubular or sheet-like carbon structures. With AI guidance, researchers can tailor upcycling routes: Ni:Mo catalysts promote multi-walled CNT formation from PET. Fe:Co systems yield graphene-like nanosheets at lower temperatures (Goldsmith et al. 2018). Zeolite-supported metals enable ordered porous carbon formation for supercapacitors. The result is precise control over CNM morphology, purity, and functional properties. Plastics (especially PVC) can produce hazardous by-products such as HCl, dioxins, PAHs, and tar. AI models predict emission profiles under different catalytic conditions and identify catalyst compositions that suppress toxic intermediates. This contributes to cleaner, safer upcycling processes compliant with environmental regulations. AI-driven high-throughput computational screening evaluates thousands of catalyst:polymer combinations in silico, drastically reducing laboratory workload. These virtual experiments simulate nanoparticle-polymer interaction energies, degradation kinetics, carbonization pathways, and CNM growth mechanisms. AI-enabled catalyst discovery transforms plastic waste from an environmental burden into a resource for advanced materials manufacturing. By enabling predictable, selective, and energy-efficient upcycling pathways, AI supports reduction of plastic pollution; development of value-added CNMs for energy storage, sensing, and electronics; and establishment of circular manufacturing ecosystems that valorize waste streams (Geyer et al. 2017).

5.6.4 Big Data in Industrial Scale-Up

Industrial-scale transformation of heterogeneous waste streams into carbon nanomaterials (CNMs) is inherently constrained by feedstock variability, nonlinear thermochemical kinetics, catalyst deactivation, and difficulty in maintaining morphological uniformity (e.g., CNT diameter distribution, graphene layer number, and porosity) across large batch and continuous-flow reactors. Big Data analytics provides a robust framework to overcome these scale-up barriers by integrating high-volume, high-velocity, and high-variety datasets derived from distributed sensor networks, historical production logs, inline spectroscopic systems (Raman, NIR, FTIR), off-gas analyzers, catalyst health diagnostics, and feedstock compositional mapping. Through advanced data fusion architectures, machine learning (ML) algorithms detect subtle correlations between process variables and CNM quality attributes, enabling early anomaly detection, drift compensation, and predictive performance forecasting long before deviations compromise yield or graphitization degree. Digital twin frameworks dynamic, data-synchronized virtual replicas of industrial reactors continuously reconcile real-time plant data with thermokinetic and transport models, allowing autonomous adjustment of temperature gradients, gas flow rates, catalyst dispersion, and residence times to stabilize morphology and reduce batch-to-batch variability. High-throughput cloud computation further accelerates scale-up by simulating thousands of parameter combinations per hour, identifying bottlenecks such as flow maldistribution, coke-induced pore blockage, or catalyst poisoning, thereby shortening the transition from laboratory optimization to commercial deployment (Butler et al. 2018). Predictive maintenance models trained on historical vibration, pressure-drop, and thermal datasets anticipate reactor fouling, thermal runaway risk, and catalyst bed degradation, minimizing downtime and extending asset lifespan critical for economic feasibility of waste-derived CNM production. Importantly, Big Data platforms increasingly integrate life-cycle metrics including carbon footprint, energy intensity, emission profiles, and waste diversion rates into operational dashboards, aligning industrial CNM manufacturing with circular economy and ESG objectives (Geissdoerfer et al. 2017). Collectively, the convergence of sensor-rich infrastructure, data fusion, ML-driven optimization, digital twins, and sustainability analytics transforms waste-to-CNM scale-up from a trial-and-error endeavor into a predictive, self-optimizing manufacturing ecosystem capable of delivering consistent quality, energy efficiency, and regulatory compliance at industrial throughput.

5.7 Challenges and Limitations

AI- and ML-enabled waste-to-carbon nanomaterial systems face several challenges, including heterogeneous and noisy waste data, limited and biased datasets, and poor model generalization from lab to industry. Sensor drift and the black-box nature of deep learning reduce reliability and interpretability. Environmental concerns such as nanomaterial toxicity, secondary emissions, and unclear regulatory frameworks further complicate deployment. Economically, high capital costs, scale-up difficulties, and uncertain market demand hinder commercialization. Social and ethical issues including workforce displacement, data privacy risks, and uneven access also affect adoption. Addressing these barriers is essential for safe, scalable, and equitable circular nanotechnology.

5.7.1 Technical, Data, and Model Biases

The integration of AI and ML into waste-to-carbon nanomaterial (CNM) systems offers unprecedented advantages, but it also introduces several technical and data-centric challenges that can limit scalability, reliability, and real-world performance. These limitations arise from constraints in data quality, model design, sensor reliability, and the inherent variability of waste feedstocks. Waste streams such as plastics, e-waste, and biomass are highly variable in chemical composition, contamination levels, and morphology. AI models trained on limited or non-representative datasets may misinterpret sensor signals (e.g., NIR, Raman, and hyperspectral), leading to inaccurate classification or catalyst recommendations. Noise from industrial environments further complicates real-time analytics. High-quality, labeled datasets for waste-derived CNM synthesis are still rare. Many studies rely on small lab-scale datasets that cannot capture industrial-scale variability. Imbalanced data such as many examples of "successful" reactions but few of failed attempts can bias ML models, making them overly optimistic and less robust in real-world conditions. ML algorithms often overfit to controlled laboratory environments, failing when exposed to slightly different feedstocks or operating conditions. Bias can emerge from overrepresentation of certain waste types, limited catalyst diversity in training datasets, and experimental biases favoring certain reaction conditions.

This leads to suboptimal decision-making when transitioning from lab to commercial scale. Deep neural networks (CNNs, transformers, and generative models) can predict catalyst behavior or synthesis outcomes with high accuracy, but they lack interpretability. Poor transparency makes it difficult for researchers to understand why a model selects a specific catalyst composition or predicts a particular carbon dot morphology, hindering scientific validation and regulatory acceptance.

AI-driven synthesis and sorting pipelines depend on real-time sensor data. Instrument drift, calibration errors, and environmental variations (humidity, temperature) may introduce bias into model predictions. Without rigorous recalibration and error-correction, downstream processes (e.g., pyrolysis control and catalyst dosing) become unreliable. Models developed using one waste type (e.g., PET plastics) often perform poorly when applied to similar but chemically distinct wastes (e.g., PVC and multilayer plastics). Similarly, catalysts optimized for one feedstock may not generalize to others. This lack of cross-domain transferability limits broad deployment in mixed-waste recycling facilities. Industrial-scale AI systems integrate data from sensors, supply chains, reactors, and QC labs. However, inconsistencies, missing values, and batch-level variations generate uncertainty in ML predictions. Without uncertainty quantification (UQ), operators may rely on misleading outputs, leading to safety risks or failed batches. Training high-fidelity ML models (e.g., digital twins and generative catalyst design networks) demands large computational resources, high-speed data transfer, and robust cloud/edge integration. Many recycling industries and waste management facilities lack this infrastructure, restricting adoption of advanced AI frameworks. Biases in training data can propagate unfair allocation of resources or skewed environmental impact predictions. For example, if AI models undervalue waste generated in less-monitored regions, it may lead to inequitable distribution of recycling services or inaccurate LCA outcomes.

5.7.2 *Economic, Environmental, and Regulatory Hurdles*

The deployment of AI-enabled waste-to-carbon nanomaterial (CNM) technologies faces significant economic, environmental, and regulatory constraints that limit large-scale adoption and long-term sustainability. High capital expenditure is required for AI-integrated sorting systems, spectral imaging units, robotics, and digital twin infrastructure. Operational costs including energy-intensive pyrolysis or hydrothermal processes and maintenance of sensors and computational systems remain substantial. Market acceptance is another barrier: waste-derived CNMs often struggle to compete with commercially available high-purity CNTs and graphene, especially in electronics and aerospace applications. Supply chain inconsistencies, feedstock variability, and technology-transfer gaps further complicate industrial scalability. Although waste valorization reduces landfill and pollution burdens, it may generate secondary emissions such as VOCs, PAHs, and metal-laden fumes during thermal conversion. The environmental fate and potential toxicity of nanomaterials especially CNTs, graphene oxide, and small carbon dots remain insufficiently understood, raising concerns about inhalation, aquatic accumulation, and long-term ecosystem impacts. Additionally, improper disposal of catalyst residues or process effluents can create new environmental hazards, necessitating stringent life cycle monitoring and end-of-life management plans. Regulatory frameworks governing AI-controlled manufacturing, nanomaterial production, and automated waste processing are still evolving. Most countries lack standardized guidelines for environmental release, occupational exposure limits, nano-safety assessments, or certification of AI-assisted reactors and robotic systems. The absence of harmonized global standards complicates international trade, quality assurance, and compliance. Furthermore, regulatory agencies may be reluctant to approve black-box AI systems without transparent, interpretable mechanisms for decision-making, especially when linked to potential environmental or health risks.

5.7.3 Ethics and Social Acceptance

The integration of AI, ML, and nanotechnology into waste-to-carbon nanomaterial (CNM) systems introduces complex ethical concerns and societal acceptance challenges that influence deployment, governance, and long-term sustainability. AI-driven waste processing relies on vast datasets collected from industries, municipalities, and consumers. This raises issues of data ownership, surveillance, and privacy, especially when smart bins, IoT sensors, and geolocation systems track waste behaviors. Algorithmic decisions such as waste classification or emissions control often involve black-box models, creating transparency and accountability gaps. Ethical risks also emerge when automated systems inadvertently reinforce biases in data, influencing resource allocation or environmental burden across communities. Worker displacement due to robotics and automation intensifies concerns over livelihood security, demanding fair labor transitions and reskilling programs. Public perception of AI and nanotechnology strongly affects adoption. Communities may fear the toxicity of nanomaterials, potential nanoparticle release, or risks associated with large-scale thermal reactors operating near residential areas. Misconceptions about AI replacing human decision-making can lead to distrust, while low awareness of circular nanotechnology undermines stakeholder participation. Unequal access to advanced technologies can widen social and digital divides, with underserved regions potentially excluded from the benefits of sustainable nanomanufacturing. Meaningful community engagement, transparent communication, and participatory governance are essential to build trust and foster acceptance. Ethical deployment requires inclusive policymaking, involving scientists, local communities, policymakers, and industry. Frameworks based on responsible research and innovation (RRI), open-access reporting, risk-benefit communication, and continuous environmental monitoring can promote both safety and acceptance. Embedding fairness, explainability, and human oversight into AI systems ensure ethical integrity. Additionally, establishing codes of conduct for nanomaterial handling, long-term ecosystem risk research, and social impact assessments reinforces societal confidence (Jensen et al. 2019).

5.8 Future Prospects

The rate of innovation is increasingly being driven by advances in artificial intelligence (AI), computer architecture, and system integration approaches, which are transforming research, industry, and policy paradigms. The convergence of next-generation AI, edge computing, multimodal data fusion, and sustainable system design is expected to shape the next phase of the digital transformation era. These emerging technologies enable faster data processing, real-time decision-making, and intelligent automation while reducing dependence on centralized computing infrastructure. As a result, computing systems are becoming not only faster and smarter but also more efficient and environmentally responsible, with a growing emphasis on energy efficiency, resource optimization, and sustainability. The rising trend of the next-generation AI models is the application of decentralized intelligence, which involves the use of edge computing that relies on performing computations closer to the source of the data. Unlike the previous architecture that relied on the cloud, edge computing supports real-time processing of the data, faster speeds, increased privacy of the processed data, and optimized use of bandwidth. This is especially essential for use in autonomous systems, healthcare, automation, and IoT networks. In recent times, developments in light-weight deep learning models, neuromorphic computing, federated learning, etc. have rather added impetus to the feasibility of incorporating AI in edge technology. With edge AI, intelligent systems have the capability of learning from different sources of information. Energy efficiency in edge devices with AI in edge technology will be critical in ensuring sustainability in intelligent systems. A combination of different kinds of data, such as text data, image data, audio signals, sensor information, and/or biological/eco-data, is used in an effort to improve machine learning models of complex systems with a technique called multimodal data fusion. A combination of multiple kinds of data sources has been proven to improve accuracy and generality of machine learning models constructed using neural networks (Baltrušaitis et al. 2019).

Transformer models have brought a dramatic breakthrough in multimodal learning, which allowed for creating a single platform where multiple sources of information can be processed effectively. With regard to healthcare diagnostics, climate simulation, agriculture, and materials research, a dramatic change has occurred in making decisions with the help of multimodal information fusion. Future studies in this area will most likely address explainable AI and uncertainty quantification. Increased usage of AI and digital infrastructure has led to growing worries about energy consumption, greenhouse gas emissions, and resource requirements. As such, sustainability in design methodology, in such forms as energy-efficient architecture, green data centers, and green hardware and systems, among others, is greatly advocated for in AI innovation, bringing forth sustainability metrics.

Incentives and regulatory support have a major role to improve the pace of adoption of green digital technology. Governments have initiated support for green AI, carbon neutral computation, and circular economy paradigms in accordance with financial aid and tax benefits and regulatory requirements (Vinuesa et al. 2020). Amidst these, integration of technological innovation and sustainability aims to deliver far-reaching benefits. Integration of circular economy concepts with highly innovative artificial intelligence, machine learning, Big Data analysis, and nanotechnology technology solutions can be considered a paradigm shift in waste analysis, management, and utilization in most advanced industrial systems. This chapter will show that carbon nanomaterials derived from waste can represent a class of critically important sustainable materials, which address two major challenges simultaneously: a rapidly increasing waste crisis and an increasing need for high-performance materials with special functions.

From agricultural waste and biomass to plastics, organics, and e-waste, carbon nanotubes, graphene, carbon dots, and porous carbons with physicochemical properties matching and even surpassing those obtained through traditional methods have been fabricated. The high surface area, pore accessibility, electrical conductivities, chemical inertness, etc. of these materials unlock their applications in energy storage, catalysis, environmental remediation, sensing, and biomedical fields. Renewable waste carbon sources also cut the reliance on fossil carbon, thus making the waste-to-CNM routes even more financially attractive.

The convergence of AI and ML across the entire waste-to-CNM value chain plays the role of a key enabler. Intelligent waste sorting, using deep learning-based computer vision, sensor fusion, and transfer learning, improves the quality of feedstock. Besides preprocessing, ML-driven process optimization, predictive maintenance, and quality assurance provide for precise synthesis control, thereby reducing energy costs and variability while improving product quality. Big Data analytics and digital twin platforms take it further, with features such as real-time monitoring, adaptive scale-up, predictive failure detection, and life cycle sustainability analysis being crucial to industrial scalability.

The case studies described in this chapter make it abundantly clear that AI-assisted nanotechnology has been able to transform waste valorization from a trial-and-error, empiric way of working to a predictable, data-driven, and controllable way of working. The use of AI to discover catalytic systems, process controls, and structure-property-application relationships helps increase overall yields, uniformity, and functionalities of heterogeneous waste substrates. This is especially the case when it comes to more complicated wastes such as organic waste, plastics, etc.

Despite such breakthroughs, challenges that need to overcome for scaling are also brought into focus by this chapter. The challenges range from technical issues concerning the lack of data, generalization, drift, interpretability of the black box of AI models, economic factors, secondary emissions, and toxicity of nanotechnology, to societal challenges such as job replacement, privacy, bias, and acceptance of AI systems and nanotechnology.

Going forward, the future of the production of waste-based CNM, therefore, is going to be more influenced by the following areas, namely, the use of next-generation AI, edge computing, multimodal learning, and sustainability by design. The development of decentralized intelligence, edge computing, and edge-based AI, for instance, is going to bring greater speed, energy efficiency, and privacy-preserving mechanisms for making decisions, while multimodal learning developments are going to make discoveries more robust, interpretable, and aware of uncertainties.

In conclusion, the synergistic convergence of AI, ML, Big Data analytics, and nanotechnology has created a strong platform that is highly scalable, with vast potential for the conversion of waste into high-value carbon nanomaterials, within a circular economy perspective. The convergence of innovations with environmental consideration, viability, and responsibility has the potential to redefine waste as a strategic resource, with a significant impact on the development of a sustainable industry.

5.9 Conclusion

Increased usage of AI and digital infrastructure has led to growing worries about energy consumption, greenhouse gas emissions, and resource requirements. As such, sustainability in design methodology, in such forms as energy-efficient architecture, green data centers, and green hardware and systems, among others, is greatly advocated for in AI innovation, bringing forth sustainability metrics.

Incentives and regulatory support have a major role to improve the pace of adoption of green digital technology. Governments have initiated support for green AI, carbon neutral computation, and circular economy paradigms in accordance with financial aid and tax benefits and regulatory requirements (Vinuesa et al. 2020). Amidst these, integration of technological innovation and sustainability aims to deliver far-reaching benefits.

Integration of circular economy concepts with highly innovative artificial intelligence, machine learning, Big Data analysis, and nanotechnology technology solutions can be considered a paradigm shift in waste analysis, management, and utilization in most advanced industrial systems. This chapter will show that carbon nanomaterials derived from waste can represent a class of critically important sustainable materials, which address two major challenges simultaneously: a rapidly increasing waste crisis and an increasing need for high-performance materials with special functions.

From agricultural waste and biomass to plastics, organics, and e-waste, carbon nanotubes, graphene, carbon dots, and porous carbons with physicochemical properties matching and even surpassing those obtained through traditional methods have been fabricated. The high surface area, pore accessibility, electrical conductivities, chemical inertness, etc. of these materials unlock their applications in energy storage, catalysis, environmental remediation, sensing, and biomedical fields. Renewable waste carbon sources also cut the reliance on fossil carbon, thus making the waste-to-CNM routes even more financially attractive.

The convergence of AI and ML across the entire waste-to-CNM value chain plays the role of a key enabler. Intelligent waste sorting, using deep learning-based computer vision, sensor fusion, and transfer learning, improves the quality of feedstock. Besides preprocessing, ML-driven process optimization, predictive maintenance, and quality assurance provide for precise synthesis control, thereby reducing energy costs and variability while improving product quality. Big Data analytics and digital twin platforms take it further, with features such as real-time monitoring, adaptive scale-up, predictive failure detection, and life-cycle sustainability analysis being crucial to industrial scalability.

The case studies described in this chapter make it abundantly clear that AI-assisted nanotechnology has been able to transform waste valorization from a trial-and-error, empiric way of working to a predictable, data-driven, and controllable way of working. The use of AI to discover catalytic systems, process controls, and structure-property-application relationships helps increase overall yields, uniformity, and functionalities of heterogeneous waste substrates. This is especially the case when it comes to more complicated wastes such as organic waste, plastics, etc.

Despite such breakthroughs, challenges that need to overcome for scaling are also brought into focus by this chapter. The challenges range from technical issues concerning the lack of data, generalization, drift, interpretability of the black box of AI models, economic factors, secondary emissions, toxicity of nanotechnology, to societal challenges such as job replacement, privacy, bias, and acceptance of AI systems and nanotechnology.

Going forward, the future of the production of waste-based CNM, therefore, is going to be more influenced by the following areas, namely, the use of next-generation AI, edge computing, multimodal learning, and sustainability by design. The development of decentralized intelligence, edge computing, and edge-based AI, for instance, is going to bring greater speed, energy efficiency, and privacy-preserving mechanisms for making decisions, while multimodal learning developments are going to make discoveries more robust, interpretable, and aware of uncertainties.

In conclusion, the synergistic convergence of AI, ML, Big Data analytics, and nanotechnology has created a strong platform that is highly scalable, with vast potential for the conversion of waste into high-value carbon nanomaterials, within a circular economy perspective. The convergence of innovations with environmental consideration, viability, and responsibility has the potential to redefine waste as a strategic resource, with a significant impact on the development of a sustainable industry.

References

Angelopoulos A, Michailidis ET, Nomikos N, Trakadas P, Hatziefremidis A, Voliotis S, Zahariadis T (2019) Tackling faults in the industry 4.0 era—a survey of machine-learning solutions and key aspects. Sensors 20(1):109. https://doi.org/10.3390/s20010109

Avouris P, Chen Z, Perebeinos V (2007) Carbon nanotube electronics and optoelectronics. Nat Nanotechnol 2:605–615. https://doi.org/10.1038/nnano.2007.300

Baltrušaitis T, Ahuja C, Morency LP (2019) Multimodal machine learning: a survey and taxonomy. IEEE Trans Pattern Anal Mach Intell 41(2):423–443. https://doi.org/10.1109/TPAMI.2018.2798607

Bircanoğlu C, Atay M, Beşer F, Gençoğlu MT (2018) RecycleNet: intelligent waste sorting using deep neural networks. In: 2018 IEEE international conference on innovations in intelligent systems and applications (INISTA). IEEE, pp 1–7. https://doi.org/10.1109/INISTA.2018.8466276

Bobulski J, Kubanek M (2021) Deep learning for plastic waste classification system. Appl Comput Intellig Soft Comput 2021:1–7. https://doi.org/10.1155/2021/6626948

Butler KT, Davies DW, Cartwright H, Isayev O, Walsh A (2018) Machine learning for molecular and materials science. Nature 559:547–555. https://doi.org/10.1038/s41586-018-0337-2

Chen DMC, Bodirsky BL, Krueger T, Mishra A, Popp A (2020) The environmental sustainability of food waste. Nat Food 1:262–267. https://doi.org/10.1038/s43016-020-0074-0

Deng D et al (2022) Catalysis with carbon nanomaterials. Nat Nanotechnol 17:123–135. https://doi.org/10.1038/s41565-021-01043-1

Fadeel B, Farcal L, Hardy B et al (2018) Advanced tools for the safety assessment of nanomaterials. Nat Nanotechnol 13:537–543. https://doi.org/10.1038/s41565-018-0185-0

Food and Agriculture Organization of the United Nations (FAO) (2019) In The State of Food and Agriculture 2019: Moving forward on food loss and waste reduction (pp. 22–74). https://doi.org/10.4060/ca6030en

Galanakis CM (2012) Recovery of high added-value components from food wastes: Conventional, emerging technologies and commercialized applications, Trends in Food Science & Technology 26(2):68–87. https://doi.org/10.1016/j.tifs.2012.03.003. (https://www.sciencedirect.com/science/article/pii/S0924224412000532)

Geim AK, Novoselov KS (2007) The rise of graphene. Nat Mater 6:183–191. https://doi.org/10.1038/nmat1849

Geissdoerfer M, Savaget P, Bocken NMP, Hultink EJ (2017) The circular economy—a new sustainability paradigm? J Clean Prod 143:757–768. https://doi.org/10.1016/j.jclepro.2016.12.048

Geyer R, Jambeck JR, Law KL (2017) Production, use, and fate of all plastics ever made. Sci Adv 3(7):e1700782. https://doi.org/10.1126/sciadv.1700782

Goldsmith BR, Esterhuizen J, Liu J, Bartel CJ, Sutton C (2018) Machine learning for heterogeneous catalyst design and discovery. AICHE J 64(7):2311–2323. https://doi.org/10.1002/aic.16075

Iijima S (1991) Helical microtubules of graphitic carbon. Nature 354:56–58. https://doi.org/10.1038/354056a0

Jensen PB, Bialy A, Blanchard JW, Wilts BD (2019) Accelerated materials discovery using artificial intelligence. Adv Mater 31(13):1901111. https://doi.org/10.1002/adma.201901111

Kirchherr J, Reike D, Hekkert M (2018) Conceptualizing the circular economy: an analysis of 114 definitions. Resour Conserv Recycl 127:221–232. https://doi.org/10.1016/j.resconrec.2017.09.005

Kroto HW, Heath JR, O'Brien SC, Curl RF, Smalley RE (1985) C60: Buckminsterfullerene. Nature 318:162–163. https://doi.org/10.1038/318162a0

Lähde A, Haluska O, Alatalo S-M, Sippula O, Meščeriakovas A, Lappalainen R, Nissinen T, Riikonen J, Lehto V-P (2020) Synthesis of graphene-like carbon from agricultural side stream with magnesiothermic reduction coupled with atmospheric pressure induction annealing. Nano Express 1(1). https://doi.org/10.1088/2632-959X/ab82e5

Li N, Chen Y (2023) Municipal solid waste classification and real-time detection using deep learning methods. Urban Clim 49:101462. https://doi.org/10.1016/j.uclim.2023.101462

Lim SY, Shen W, Gao Z (2015) Carbon quantum dots and their applications. Chem Soc Rev 44: 362–381. https://doi.org/10.1039/C4CS00269E

Liu H, Chen D, Li Y, Zhao X (2025) Carbon quantum dots derived from biomass wastes: synthesis, properties, and agri-environmental applications. J Clean Prod 418:139822. https://doi.org/10.1016/j.jclepro.2024.139822

Miandad R, Barakat MA, Aburiazaiza AS, Rehan M, Nizami AS (2016) Catalytic pyrolysis of plastic waste: a review. Process Saf Environ Prot 102:822–838. https://doi.org/10.1016/j.psep.2016.06.022

Novoselov KS et al (2004) Electric field effect in atomically thin carbon films. Science 306:666–669. https://doi.org/10.1126/science.1102896

Sevilla M, Fuertes AB (2009) The production of carbon materials by hydrothermal carbonization of cellulose. Carbon 47:2281–2289. https://doi.org/10.1016/j.carbon.2009.04.026

Upadhyayula VKK, Meyer DE, Curran MA, Gonzalez MA (2018) Life cycle assessment as a tool to enhance the environmental performance of carbon nanotube products. J Clean Prod 172:419–432. https://doi.org/10.1016/j.jclepro.2017.10.131

Vinuesa R et al (2020) The role of artificial intelligence in achieving the sustainable development goals. Nat Commun 11:233. https://doi.org/10.1038/s41467-019-14108-y

Wang Y, Li X, Chen H (2023) Waste-derived carbon nanomaterials for sustainable applications. Carbon 205:123–145. https://doi.org/10.1016/j.carbon.2023.01.012

Yan J, Wang Q, Liu X, Zhang Y (2025) Biomass-derived carbon nanomaterials: structure–property relationships and applications in energy and environment. Renew Sustain Energy Rev 186: 114012. https://doi.org/10.1016/j.rser.2024.114012

Zhu S, Song Y, Zhao X, Shao J, Zhang J, Yang B (2015) The photoluminescence mechanism in carbon dots: current state and future perspective. Nano Res 8:355–381. https://doi.org/10.1007/s12274-014-0644-3

Chapter 6
Applications of Biosynthesized Carbon Nanomaterials in Agricultural Pesticide Synthesis and Crop Spoilage Sensing

Amuthalinkam T. Asaithambi, Lakshminarayanan Sivakumar, Arunagiri Santhosh Kumar, Punniyakotti Elumalai, Durairaj Thirumurugan, and Punniyakotti Parthipan

6.1 Introduction

Novel technologies are required for coping with major global problems such as environmental pollution, energy production, agriculture and food production. Sustainable development while reducing the pollutants produced, advanced technologies are required to enhance and automate the process that can effortlessly observe the environment and its effectiveness (Speranza 2021). Carbon exhibits high flexibility owing to its potential to exist in allotropic forms. They can form various chemical bonds with several orientations due to their diverse orbital hybridization (Hegde et al. 2024). Along with the advent of nanomaterials and nanotechnology, carbon nanomaterials (CNMs) have gained more popularity as a significant research domain due to their exceptional properties such as light weight, strength, and superior conductivity. In the past few decades, nanotechnology has advanced swiftly, and CNMs have been extensively utilized in several areas. The rapidly growing research and application of CNMs have drawn attention to their effects and behavior in the environment (Peng et al. 2020). CNMs include carbon nanotubes, fullerenes, graphene oxide, graphene, activated carbon, carbon-based quantum dots, and nanodiamonds. They have distinct optical properties such as

A. T. Asaithambi · L. Sivakumar · A. S. Kumar · D. Thirumurugan · P. Parthipan (✉)
Department of Biotechnology, Faculty of Science and Humanities, SRM Institute of Science and Technology, Kattankulathur, Chengalpattu, Tamil Nadu, India
e-mail: parthipp@srmist.edu.in

P. Elumalai
Department of Biotechnology, Karpagam Academy of Higher Education, Coimbatore, Tamil Nadu, India

Centre for Energy and Environment, Karpagam Academy of Higher Education, Coimbatore, Tamil Nadu, India

Y. K. Mohanta et al. (eds.), *Bioresource Waste to Carbon Nanomaterial*, Nanotechnology in the Life Sciences, https://doi.org/10.1007/978-981-92-2405-0_6

tunable narrow emission spectrum, intrinsic fluorescence, and high photostability. Moreover, surface modifications with functional groups such as carboxylic acid, hydroxyl, and epoxy provide a way to maximize their properties (Patel et al. 2019). The current state of agriculture is confronted with numerous challenges, including soil degradation, pest resistance, overuse of agrochemical products, and post-harvest losses, necessitating advanced and sustainable technological interventions (Brower-Toland et al. 2024). In the past, CNMs were produced by chemical and physical methods such as strong acid oxidation, laser ablation, and arc discharge which are high-energy dependent and rely on hazardous substances, leading to toxic intermediates and environmental problems. As a result, the need for sustainable, green, eco-friendly synthesis of CNMs gained interest for the utilization of renewable or low-cost precursors like biomass, biowaste, benign reagents, milder conditions, and synthesis of materials with enhanced biocompatibility for biomedical and environmental conditions (Iravani and Varma 2020; Goswami et al. 2021; Qasim et al. 2023). CNMs synthesized from biological sources like plants, microorganisms, and agricultural wastes are well reported to have huge functional surface groups and excellent chemical stability, which underpins their ability in agricultural systems (Jirimali et al. 2022). These properties make them a robust material for applications such as pesticide formulation, crop spoilage sensing, controlled agrochemical delivery, soil remediation, and disease diagnostics (Hegde et al. 2024; Carrillo-Lopez et al. 2024). Also, CNMs-based nano-sensors and fluorescent carbon quantum dots are used owing to its rapid, sensitive, and nondestructive tools for pesticide residue detection and microbial spoilage in food matrices and storage and supply chains (Khan et al. 2023; Tian et al. 2023). Thus, the integration of biosynthesized CNMs supports sustainable development by enabling safer crop protection, improved post-harvest handling, and enhanced farming; also they reduce environmental burden when compared with conventional methods (Jayasekhar Babu et al. 2022; Alam et al. 2024). This chapter focuses on biosynthesis of CNMs, the mechanisms of action of CNMs, emerging applications in pesticide synthesis, crop spoilage detection, and toxicological and environmental implications within sustainable agriculture.

6.2 Biosynthesis of Carbon Nanomaterials

6.2.1 Green Synthesis Methods of CNMs

Green synthesis of CNMs is known for its environmentally friendly alternatives when compared to traditional synthesis of CNMs, which requires excessive energy, high-cost instruments, and toxic intermediates (Makgabutlane et al. 2021). This method often uses renewable precursors such as leaf, stem, root, and crop residues and maintains mild reaction conditions while utilizing natural stabilizing agents derived from biological sources, adherence to the tenets of green chemistry and circular economics (Olaniyan et al. 2025). The fundamental mechanism of CNMs synthesis includes thermal or chemically induced carbonization of carbon-rich biomolecules, which includes condensation, aromatization, dehydration, and nucleation of carbonaceous clusters, which transform into nanostructures with large surface functional groups without relying on synthetic surfactants (Jing et al. 2023). There are several green synthesis methods for CNMs, but hydrothermal is the most preferred one. This method is widely used because it is easy to use, can be scaled up or down, and can make carbon dots, graphene-derived structures, and porous carbons at moderate temperatures, usually between 120 and 240 °C, under self-pressure conditions (Bressi et al. 2023). The hydrothermal process comprises four major steps: polymerization, carbonization and passivation, and hydrolysis. High-molecular-weight organic molecules are first hydrolyzed into small fragments in an aqueous condition, where chemical bonds are broken by high pressure and temperature. Following dehydration, which causes them to lose water molecules and become more reactive, these fragments go through polymerization. In this process they cluster together to form the intricate, sizable polymeric structure. The resultant polymers undergo carbonization, in which condensed carbon clusters are formed. They further nucleate and coalesce into the individual nanoscopic crystalline structures known as carbon quantum dots (Wang et al. 2025). The process, temperature, and pH vary for the synthesis of other CNMs. Microwave and ultrasound-assisted methods are well-known as quick, energy-efficient, and environmentally friendly ways to make CNMs, especially carbon dots for use in sensors. Microwave synthesis allows for quick volumetric energy transfer and even heating, which means that reaction times are short, particle size and fluorescence can be controlled well, and luminescent carbon dots can be made on a larger scale for use in chemical and food safety sensing. Sonochemical synthesis relies on sonic cavitation, which induces localized carbonization, precursor fragmentation, and in situ stabilization to produce small, homogeneous nanoparticles with highly reactive surfaces (Adeola et al. 2023). Bio-templating and enzymatic or bioengineered synthesis are regarded as significant contemporary green methods to produce nanomaterials from biological sources, including cellulose fibers, fungal mycelial networks, and bacterial cells. They serve as natural scaffolding that govern the development of CNMs, facilitating regulated structures with minimum reagents (Manan et al. 2021; Srujana et al. 2025).

6.2.2 Sources and Biological Precursors

Plants are recognized as the primary green precursors for CNMs synthesis due to their abundance, cost-effectiveness, and exceptionally high levels of metabolites such as sugars, organic acids, phenolics, flavonoids, and terpenoids. These metabolites serve as carbon sources, reducing agents, and stabilizers during the formation of nanomaterials. A wide variety of plant-based matrices, such as whole extracts, fruits, peels, leaves, seeds, and roots, were utilized to produce fluorescent carbon dots and graphene-like carbons through hydrothermal, pyrolytic, and microwave-assisted techniques (Jing et al. 2023; Andleeb et al. 2025; Liu et al. 2025). For instance, lemon peel, orange peel, and other citrus peel fruits are often used as they contain a high level of citric acid, which assists in producing highly photoluminescent carbon dots (Prasannan and Imae 2013; Kundu et al. 2023). Analogously extracts from tea leaves, neem leaves, onion peels, aloe vera, and ginger have been used to formulate carbon dots that fluorescent strongly in blue or green, offer greater stability, and have many surfaces functional groups that make them useful for sensing and detecting pesticides (Yadav et al. 2019; Paveethra et al. 2024; Sinha and Ray 2024). Through controlled pyrolysis, plant biomass like rice husk, coconut shell, bamboo, sugarcane bagasse, and peanut shells have also been transformed into porous activated carbon or graphene-like structures. These structures have configurable porosity and adsorption capacities that make them useful for delivering agrochemicals and cleaning up soil (Kumar Mishra et al. 2024). The structural diversity of plant biomolecules allows for precise changes in size, shape, and photoluminescent properties of CNMs, making it possible to customize them for specific uses. Figure 6.1a shows the various synthesis mechanisms of CNMs and their precursors. Bacteria are another promising biological source because they have unique metabolic pathways and can make nanostructures in mild, aqueous conditions. Some species of bacteria, like *Escherichia coli*, *Bacillus subtilis*, *Shewanella oneidensis*, and *Klebsiella pneumoniae*, can be used to synthesize CNMs by enzyme catalysis, bio-reduction, and biomineralization (Ghosh et al. 2021; Koul et al. 2021). Enzymes like oxidoreductases, dehydrogenases, and peroxidases are often involved in the biosynthesis of microbial nanoparticles. These enzymes cause redox changes in organic or inorganic substrates, which facilitate the formation and growth of nanoscale domains (Yadav et al. 2023). Studies highlight that controlled biological growth conditions can support improved reproducibility in the production of biogenic nanomaterials. Bacteria-based routes have been reported to produce small, quasi-spherical carbon dots with narrow/defined average sizes in aqueous dispersions (Lin et al. 2018). Bacterial synthesis has potential for future industrial-level CNMs manufacturing as it enables continuous production using bioreactors.

6.3 Mechanisms of Action of Carbon Nanomaterials

6.3.1 Surface Functionalization and Reactivity

The choice of the adsorbent material has a critical role in ascertaining the efficiency, sustainability, and stability of the material used in diverse research applications. Among these, CNMs have attracted increasing attention due to their distinctive physicochemical properties (Sharma and Soni 2024). Carbon-based materials possess a high surface area, good dispersibility, thermal stability, and the potential for surface modification, which makes them versatile for many environmental applications (Hoque and Saikia 2025). In addition, their ability to be reused, strong adsorption performance, and low cost have led many studies to focus on carbon materials for pesticide degradation. Research that evaluates different adsorbents

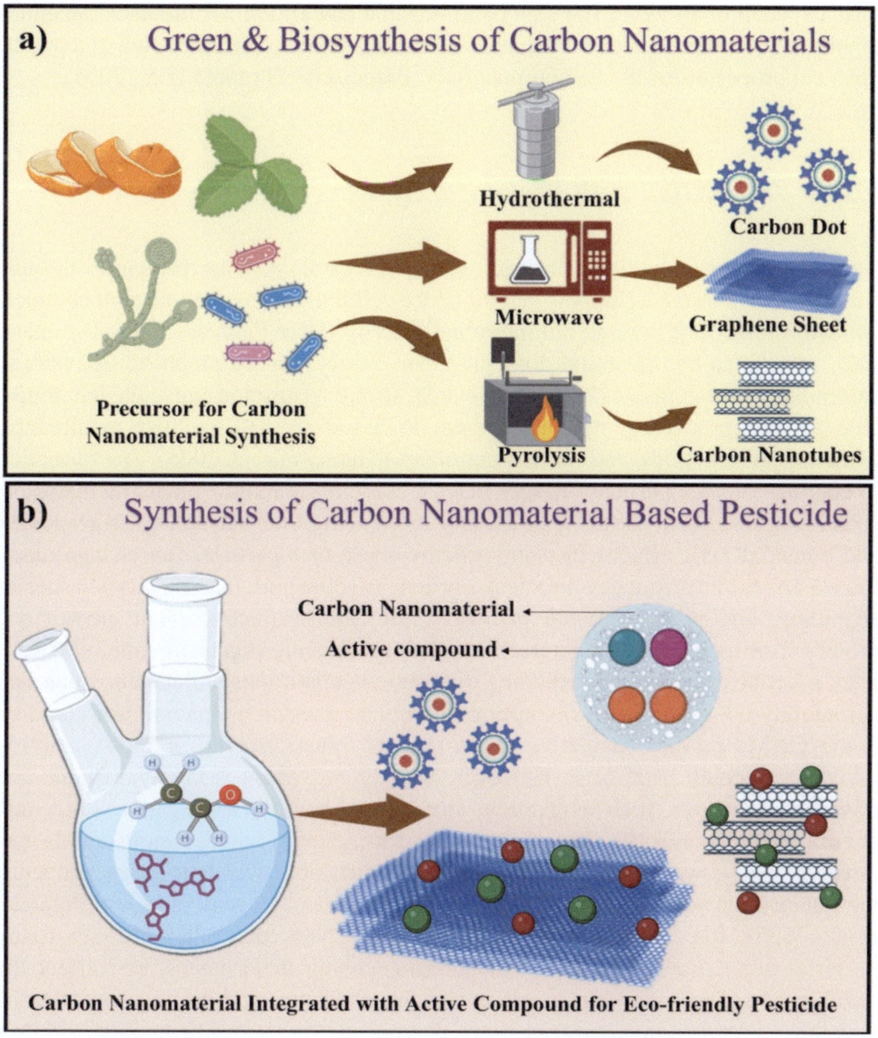

Fig. 6.1 (**a**) Biosynthetic pathway of carbon nanomaterials; (**b**) carbon nanomaterials integrated with active compounds for eco-friendly pesticide synthesis

shows that carbon-based materials often perform better than other nanomaterials, especially with regard to adsorption efficiency and their capacity for frequent use. A study investigated activated carbon for pesticide adsorption in aqueous systems, reporting high removal efficiency under acidic conditions within 50–60 minutes and confirmed that the nanoscale pore structure improves pesticide uptake, demonstrating its suitability for water treatment applications (Zieliński et al. 2022). A study showed that carbon-based materials, particularly carbon nanotubes, achieved exceptionally high adsorption efficiencies for atrazine, simazine, and diuron, with removal rates exceeding 96–99%. The study showed that adsorption performance depended mainly on chemical interactions and treatment-stage application rather than the physical properties of the carbon materials (Francisco Fukumoto et al. 2024).

6.3.2 Interaction with Biological Systems and Pathogens

Carbon-based nanoparticles interact with biological systems primarily through surface-adsorption mechanisms that influence cellular uptake, modulate biochemical pathways, and contribute to antimicrobial activity (Díez-Pascual 2021). The materials, such as carbon quantum dots, graphene oxide, reduced graphene derivatives, carbon nanotubes, possess high surface area, strong adsorption capacity, and adjustable surface properties that allow them to associate with cellular membranes, extracellular structures, and soil environments (Thamarai et al. 2025). The formation of coatings on nanoparticle surfaces plays a crucial role in how plant and microbial cells recognize these particles, ultimately influencing their physiological responses and potential toxic effects. In plant systems, these nanoparticles can change membrane permeability, root development, nutrient ion transport, reactive oxygen species signalling, and phytohormonal pathways, ultimately affecting plant growth and protection responses. Their interactions with pathogenic organisms may also alter plant microbe dynamics by reshaping rhizospheric conditions, influencing microbial community composition, or disrupting microbial adhesion and membrane stability. Some CNMs produce oxidative stress through photocatalytic or redox activity, inhibiting specific pathogens. However, their effectiveness is highly variable and rely on parameters such as particle size, concentration, aggregation state, and environmental condition. In soil environments, carbon-based nanomaterials can affect porosity, water retention, and nutrient diffusion, which together influence the balance between beneficial microorganisms and pathogenic species (Prajapati et al. 2025). Recent studies have shown that CNMs may alter nitrogen-fixing bacterial populations, interfere with quorum-sensing mechanisms, and affect the stability of soil organic matter, all of which contribute to shifts in plant-microbe dynamics. Previous investigations have demonstrated that graphene oxide exhibits potent antibacterial activity against Gram-negative bacteria, including *Pseudomonas aeruginosa*, through oxidative stress induction, membrane disruption, and nanosheet insertion into lipid layers (Gurunathan et al. 2012). Reduced graphene oxide has similarly been reported to induce intracellular ROS accumulation, mitochondrial dysfunction in microbial cells, and fragmentation of genomic DNA, confirming its role as a strong bactericidal agent (Sengupta et al. 2019). Additional research on carbon quantum dots indicates that these materials may interfere with microbial electron-transport chains and enzyme activity, further contributing to pathogen suppression (Xue et al. 2026). Although CNMs may improve plant stress tolerance and affect pathogen interactions, their use in agriculture requires careful evaluation of their environmental fate, bioaccumulation, ecotoxicity, and long-term impacts on soil ecosystems. Significant research gaps persist concerning nanoparticle transformation in soil, trophic transfer, and potential impacts on beneficial organisms such as mycorrhizal fungi and rhizobia (Mukherjee et al. 2016; Alagendran et al. 2024).

6.4 Role of Biosynthesized Carbon Nanomaterials in Pesticide Synthesis

6.4.1 Nanocatalytic Applications in Eco-Friendly Pesticide Formation

Nanobiopesticides are increasingly recognized as a promising strategy for developing more sustainable and efficient methods of pesticide synthesis, particularly as the agricultural sector adopts alternatives to conventional pathways characterized by high energy or harsh chemical conditions (Harun-Ur-Rashid et al. 2025). Among these, biosynthesized CNMs have gained considerable attention due to their sustainable production pathways and multifunctional physicochemical properties (Ayilara et al. 2023). These nanomaterials, typically derived through microbial, plant, or enzymatic processes, possess intrinsic surface functionalities and high structural stability that make them effective platforms for catalytic transformation, adsorption, and controlled delivery of pesticidal agents (Jasim et al. 2024). During pesticide synthesis, biosynthesized CNMs can act as heterogeneous catalysts that facilitate bond formation, enhance reaction kinetics, and improve the efficiency of active ingredient incorporation. Their large surface area and tunable porosity further support high loading capacities and enable the design of formulations with enhanced solubility, reduced volatility, and greater resistance to premature degradation. In addition to improving the physicochemical performance of pesticides, the biological source of these nanomaterials offers potential advantages in terms of reduced toxicity, lower energy requirements for production, and enhanced compatibility within agroecosystems. As a result, biosynthesized CNMs improve reaction efficiency and limit the formation of unwanted side products, supporting more sustainable synthesis by reducing temperature requirements and minimizing solvent use (Fig. 6.1b). Composite systems that combine CNMs with small amounts of metals or metal oxides have also attracted significant interest, as these structures offer additional catalytic sites and improved reaction efficiency while requiring substantially less metal than traditional catalysts (Liang et al. 2021). This reduced metal loading lowers production costs and mitigates environmental impacts associated with catalyst disposal. Moreover, many carbon-based catalytic materials can be reused multiple times with minimal loss of activity, increasing their practicality for large-scale industrial applications. Enhanced catalytic performance also supports the synthesis of cleaner and more stable active ingredients for modern pesticide formulations. As efforts continue to reduce the environmental footprint of agricultural chemicals, biosynthesized CNMs are developing as an optimistic route for advancing sustainable pesticide production (Tok et al. 2025). Further research is needed to clarify their reaction mechanisms, optimize synthesis conditions, and evaluate their behavior to ascertain safety and potential use in industrial applications. Future studies should also integrate life-cycle assessment frameworks to properly understand long time sustainability impact (Upadhyayula et al. 2012). However, further research is required to elucidate their reaction mechanisms, optimize synthesis conditions, and evaluate their behavior to ensure safe and effective industrial application. Future studies can incorporate life cycle assessment frameworks to evaluate the long-term sustainability of CNMs enabled pesticide synthesis (Ke et al. 2025).

6.4.2 Encapsulation and Controlled Release of Biopesticides

Encapsulation has become an important method for improving the stability, persistence, and effectiveness of biopesticides under field conditions (do Nascimento Junior et al. 2021). Biologically derived active ingredients, including microbial metabolites, plant-derived compounds, and peptide-based substances, are susceptible to environmental degradation and frequently show limited residual activity (Singh et al. 2023). The encapsulation provides a protective barrier that shields sensitive bioactive compounds from degradative factors, thereby maintaining their activity over more extended periods and enhancing their field performance (Yu et al. 2025). A wide range of materials has been investigated as encapsulation matrices, including natural and synthetic polymers, lipid-based systems, inorganic nanocarriers, and carbon-based nanomaterials. These formulations isolate the active ingredient within a physical or chemical barrier that regulates its degradation and release behavior during pesticide application (Pavithran et al. 2024). The effectiveness of an encapsulation system is governed by parameters such as carrier porosity, surface properties, and strength of interactions between the encapsulating material and the biopesticide. Controlled-release behavior is usually obtained through mechanisms such as diffusion, breakdown of the encapsulating material, or stimulus-responsive release initiated by changes in pH, moisture, or enzymatic activity (Fig. 6.2). The implications of nanomaterials have improved controlled-release performance, due to its high surface area and adaptable chemical properties allow more precise control over the release of biopesticide agents (Singh et al. 2021). Carbon-based nano-systems have strong adsorption capacity and good stability, enabling a slower and more sustained delivery of biopesticides while reducing the risk of premature loss through dissolution or evaporation. Such formulations can also enhance the solubility and mobility of poorly water-soluble biopesticides, thereby increasing their bioavailability at the target site (Uddin 2021). Additionally, encapsulation can protect living microbial biopesticides such as entomopathogenic fungi and bacteria from desiccation and ultraviolet stress, significantly improving their survival and infectivity under field conditions. Encapsulation and controlled-release approaches contribute to reduced application frequency, lower environmental loading, and improved compatibility of biopesticides with integrated pest management programs.

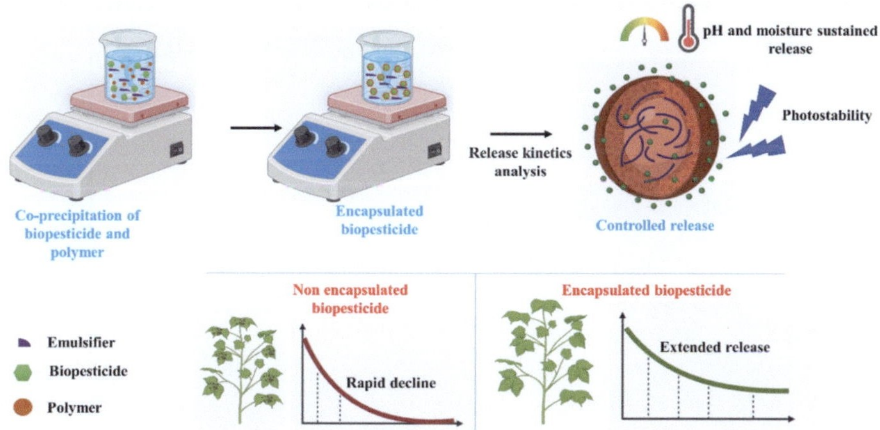

Fig. 6.2 Encapsulation-based protection and controlled release of biopesticide

6.4.3 Enhancing Pesticidal Efficiency and Target Specificity

Improving the efficiency and target specificity of biopesticides has become an important focus in sustainable crop protection research. Although biopesticides provide advantages over synthetic chemicals, including lower toxicity, biodegradability, and reduced environmental persistence, their field performance is frequently limited by rapid degradation, restricted mobility, and variable contact with the target pest (Fenibo et al. 2022). Recent advances in formulation methods and application approaches have improved the performance of these bioactive compounds once they are introduced into the crop environment (Kyriakoudi et al. 2021). Improved pesticidal efficiency is commonly achieved through enhanced stability and availability of bioactive compounds during application and subsequent exposure to target organisms (Fig. 6.1b). Many natural compounds undergo rapid degradation when exposed to sunlight, temperature changes, and microbial activity. Modern encapsulation methods and nanoscale carriers provide physical protection and modulate the release behavior of these compounds over time. This controlled release helps maintain effective biopesticide levels for longer durations, thereby reducing the frequency of application. CNMs can improve adhesion to plant surfaces or insect cuticles, helping the biopesticide remain in areas where target organisms are most active. Figure 6.3a depicts that improving target specificity is important for reducing effects on nontarget insects, soil microorganisms, and the surrounding environment (Kumar et al. 2021). One approach is to develop carrier systems that function in response to biological or environmental conditions associated with the specific organisms. For example, pH-responsive carriers can release the potential compound in the high pH gut of certain insect larvae, whereas enzyme-responsive carriers can initiate release in the presence of digestive enzymes specific to the target species.

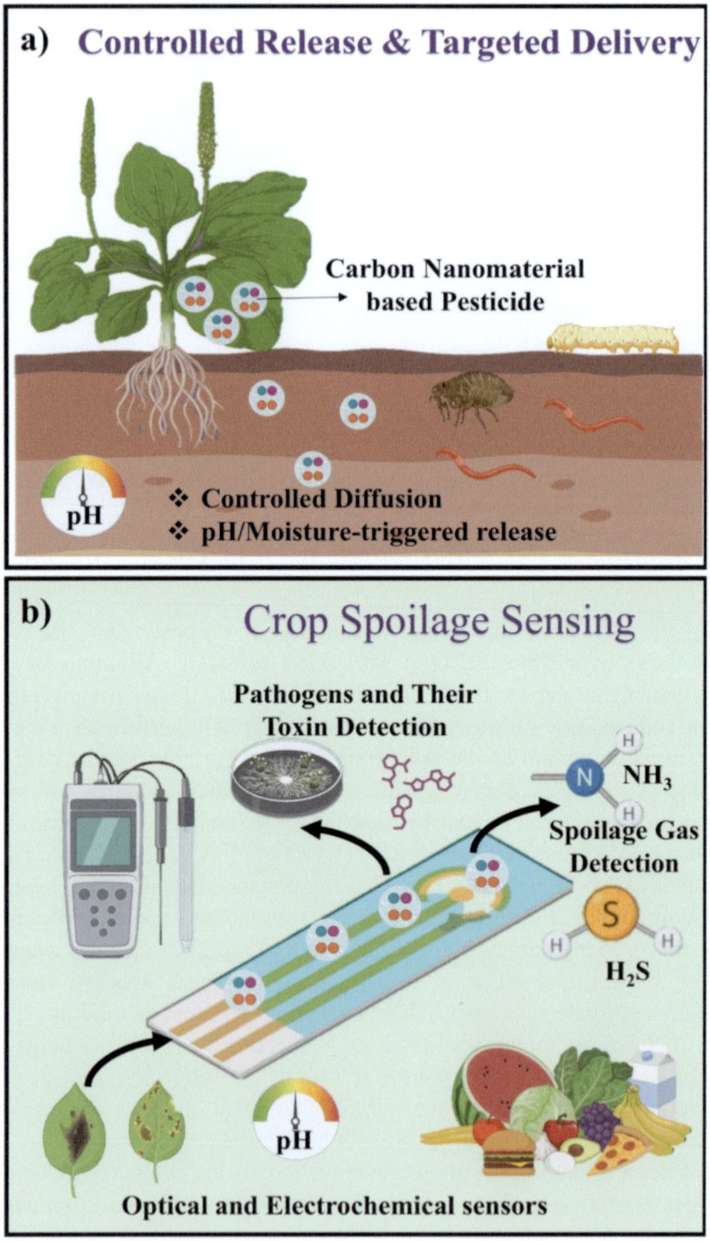

Fig. 6.3 (**a**) Carbon nanomaterial-based pesticides for improved target delivery; (**b**) carbon nanomaterials as a crop spoilage-sensing device

6.5 Crop Spoilage Detection and Sensing Applications

6.5.1 Nano-Biosensors for Pathogen and Toxin Detection

Nanotechnology has a pivotal role in agriculture, particularly in crop spoilage detection, by enabling rapid, sensitive, and real-time detection of plant health, pathogens, and storage condition. Pathogens such as *Aspergillus*, *Penicillium*, *Acidovorax* sp., *Pseudomonas syringae*, *Ralstonia* sp., and *Leptosphaeria* sp., along with toxins like aflatoxins, ochratoxins, and fumonisins, pose severe threats to agricultural products. Nanomaterial-based sensors have emerged as powerful tools for the rapid and highly sensitive detection of plant viruses, including tomato ring spot virus, bean pod mottle virus, and arabis mosaic virus, offering promising solutions for early disease management in agriculture. Nanoparticles functionalized with antibodies, DNA probes, or enzymes bind specifically to pathogen biomarkers. Rispail et al. (2014) biosynthesized CNMs provide a versatile platform due to their large surface area and ability to immobilize biomolecules without denaturation. Plant-extract-derived carbon dots offer excellent photostability, biocompatibility, and excitation-dependent fluorescence. They can be functionalized with aptamers targeting specific mycotoxins (aflatoxin B1) (Rispail et al. 2014). A recent study by Rahmani et al. (2023) developed an imaging system combining a DNA-single-walled carbon nanotube biosensor and a Wasserstein distance-based convolutional neural network, achieving 98% accuracy in classifying tea leaf. A magnetic paper biosensor integrating $Fe_3O_4/SiO_2\text{-}NH_2$ immunomagnetic microspheres and MWCNTs was developed for rapid pseudorabies virus detection. Binding-induced fluorescence quenching or enhancement forms the basis of ultrasensitive optical detection (Guo et al. 2021). CDs also assist in rapid bacterial identification through their ability to preferentially bind to Gram-positive or Gram-negative cell walls (Fig. 6.3b). GO and rGO sheets serve as outstanding transducer surfaces. Their π–π interactions allow efficient immobilization of DNA probes and antibodies, making them suitable for detecting *Fusarium* DNA sequences, *Salmonella* antigens, or toxin biomarkers (Zhang et al. 2024). GO-based FRET (fluorescence resonance energy transfer) biosensors have demonstrated extremely low detection limits (picomolar to nanomolar range) for pathogens and toxins. GO-peptide FRET biosensor was developed for detecting botulinum neurotoxins, using a SNAP-25-green fluorescence protein substrate immobilized on GO with BSA passivation (Shi et al. 2015). Sensor showed high specificity and ultrasensitivity, offering a 1-fg/mL detection limit and a linear range from 1 fg/mL to 1 pg/mL. CNTs enhance signal transduction due to 1D electron transport and high conductivity. CNT-based immunosensors have been used to detect *E. coli* O157:H7 in fresh produce, aflatoxin M1 in milk, and ochratoxin A in cereals (Viswanathan et al. 2012; Joksović et al. 2024). Activated biochar-based immunosensor to quickly detect *E. coli* O157:H7 without incubation. Steam-activated biochar enhanced impedance from 3.5 to 8.5 kΩ upon antibody cell binding, achieving a detection limit of 4 log CFU/mL, confirmed by SEM. Functionalization with chitosan, metal oxide nanocomposite, or aptamers further improves sensitivity (Sobhan et al. 2022). Combining biosynthesized CNMs with metallic nanoparticles (AuNPs, AgNPs) or magnetic nanoparticles results in multimodal detection platforms (Li et al. 2025). Wang et al. (2020) investigated aptamer-functionalized chitosan magnetic nanoparticles (Fe_3O_4-chitosan-aptamer) were synthesized and characterized for selective ochratoxin A (OTA) extraction. In food samples spiked with 5–10 μg/kg OTA, the method achieved 91–99%, comparable to immunoaffinity chromatography, enabling rapid, efficient OTA detection. Immunomagnetic sensors has gained more attention for mycotoxin capture and pre-concentration using colorimetric sensors and AuNPs aggregation controlled by carbon nanomaterial scaffolds, Surface Enhanced Raman Scattering nanoprobes can be utilized for detecting traces of pesticides and toxins (Terry et al. 2022).

6.5.2 Optical and Electrochemical Sensing Platforms

Optical and electrochemical detection systems constitute two major technological platforms for crop spoilage and pathogen sensing. Biosynthesized carbon nanomaterials (CNMs) significantly enhance the analytical performance of these systems owing to their distinctive optical signatures, excellent electrical conductivity, large surface area, and tunable surface chemistry. Early detection of postharvest diseases is critical to minimizing economic losses and maintaining produce quality. For example, Pieczywek et al. (2018) demonstrated that optical sensing techniques enabled earlier detection of bull's eye rot in apples artificially inoculated with *Pezicula malicorticis* compared to conventional visual inspection. Carbon dots (CDs), in particular, exhibit strong and tunable fluorescence, making them highly suitable for detecting spoilage-associated molecules such as ammonia (released during meat and fish deterioration), hydrogen sulfide, pH fluctuations, and organic acids produced during fruit and grain spoilage. Green-synthesized dual-emissive phosphorus-doped carbon dots have been employed for sensitive tetracycline detection, achieving limits of detection (LODs) of 5.78 and 6.18 μM, while simultaneously enabling visual pH monitoring through distinct color transitions (Kayani et al. 2024). Fluorescence-based detection commonly relies on mechanisms such as photoinduced electron transfer (PET) and the inner filter effect (IFE), which facilitate selective analyte recognition. Wang et al. (2024) developed a dual halochromy-fluorescence sensor incorporating quantum dot (QD)-silane coatings and blueberry anthocyanin for noncontact freshness monitoring of protein-rich foods. The system responded within 1 minute to volatile nitrogen compounds, exhibiting an ammonia LOD of 0.01 ppm over a dynamic range of 0.001–50 ppm. Smart packaging systems integrating graphene oxide (GO) or CDs provide real-time visual spoilage indicators via observable color changes. Incorporation of gold nanoparticles (AuNPs) into carbon nanomaterial matrices further amplifies optical signals, enabling rapid naked-eye detection of toxins and pathogenic compounds (Xu et al. 2023). Moreover, graphene-coated plasmonic substrates enhance surface plasmon resonance (SPR) and surface-enhanced Raman scattering signals, allowing ultrasensitive detection of trace-level mycotoxins, spoilage-related volatile organic compounds (VOCs), and pesticide residues (Akkaş et al. 2025; El Hosry and Bou-Maroun 2025). Electrochemical biosensors represent another powerful detection platform, measuring changes in current, voltage, or impedance upon analyte interaction. These systems offer high sensitivity, portability, rapid response, and cost-effectiveness (Grieshaber et al. 2008; Wei et al. 2009; Abdul Ghani et al. 2022). Reduced graphene oxide (rGO) and carbon nanotube (CNT)-modified electrodes significantly improve electron transfer kinetics and signal amplification and have been successfully applied for fungal metabolite detection, bacterial pathogen identification, and enzymatic sensing of spoilage biomarkers (Chethan et al. 2024). Label-free electrochemical impedance spectroscopy (EIS) sensors detect changes in interfacial impedance when pathogens bind to carbon nanomaterial-coated electrodes, enabling real-time monitoring in complex matrices. For instance, a reduced graphene oxide/molybdenum disulfide/polyaniline-AuNP electrochemical aptasensor demonstrated ultra-trace detection of aflatoxin B_1 with an exceptionally low detection limit of 0.002 fg mL^{-1}, along with a wide linear detection range, high selectivity, and excellent reproducibility in food samples (Geleta et al. 2018). In another study, a multiplexed Förster resonance energy transfer (FRET)-based aptasensor employing multicolor upconversion nanoparticles as donors and GO as a universal acceptor enabled simultaneous detection of ochratoxin A and fumonisin B_1 in maize samples (Wu et al. 2012). Foodborne microorganisms often produce biogenic amines such as putrescine and histamine during spoilage. Portable opto-electrochemical biosensors integrating impedance and luminescence detection have demonstrated sensitive quantification of these markers in food systems (Zamfir et al. 2025). Collectively, optical and electrochemical sensing strategies incorporating biosynthesized carbon nanomaterials provide enhanced specificity, sensitivity, and operational simplicity in complex agricultural and food matrices. These advancements are facilitating the transition from laboratory-scale prototypes to practical, field-deployable pathogen monitoring systems aimed at reducing both preharvest and postharvest crop spoilage.

6.5.3 Real-Time Monitoring Devices for Postharvest Loss Management

Real-time monitoring devices play a critical role in maintaining the integrity of stored crops by continuously detecting environmental variables and spoilage markers (Selvam and Al-Humairi 2023). Biodegradable films embedded with carbon dots or GO can respond to spoilage gases or humidity changes. For instance, petunia dye enables sensitive SO_2 detection for food packaging, showing an adsorption energy of 74.8- and a 1.52-ppm detection limit (Yuan et al. 2023). These serve as low-cost tools for farmers to monitor grains, pulses, fruits and vegetables without requiring laboratory analyses. Carbon nanomaterial-based sensors can be integrated with radio-frequency identification (RFID) tags, Bluetooth modules, near field communication (NFC) systems, and cloud-based data analytics. An ultralow power RFID/NFC wireless sensor tag capable of autonomously logging potentiometric signals from pH and ion-selective electrodes was demonstrated by Kassal et al. (2013). It enables mobile-based chemical sensing for IoT applications, showing pH measurement performance comparable to a standard laboratory pH meter. These smart systems allow remote monitoring of storage silos, cold storage units, and transport containers. Real-time monitoring helps prevent mold outbreaks, moisture accumulation, and temperature fluctuations that trigger spoilage. Nano-enabled gas sensors detect ethylene (fruit ripening), ammonia (protein degradation), CO_2 buildup, and fungal VOCs. Yin et al. (2023) developed a gas sensor-based prototype to monitor apple spoilage by detecting VOCs (CO_2, O_2, C_2H_4) with temperature-humidity data and chemometric modelling. The optimal simulated annealing-partial least squares model enabled real-time spoilage warnings and remote warehouse visualization. CNT and rGO gas sensors demonstrate a rapid response, low detection limits, and strong selectivity. Time-resolved reflectance spectroscopy (TRS) effectively tracks mango ripening by measuring chlorophyll loss and carotenoid rise in intact fruit. Absorption at 540 and 670 nm, combined into a biological age index, accurately modelled ethylene production and firmness decline (Eccher Zerbini et al. 2015). Miniaturized electrochemical analyzers with carbon nanomaterial-modified electrodes allow farmers to measure spoilage indicators, toxin levels, and pathogen load directly at the field or warehouse. Such devices reduce dependency on centralized labs and support early intervention. A miniature poly (styrene-co-maleic anhydride) wireless sensor was developed for mobile-based detection of spoilage via volatile biogenic amines. Integrated into packaged chicken and beef, the sensor showed a 700% signal increase by day 3 at room temperature, while frozen samples showed minimal change (Jabbar and East 2016). Data from real-time sensors using AI can feed into predictive models for estimating spoilage progression, optimizing fumigation schedules, and determining safe storage periods. Guo et al. (2025) fabricated a multi-sensor, deep-learning spoilage warning system that monitored apple VOCs from fungal infections. Variable selection, optimized features, and synergy interval-gated recurrent unit achieved the maximum correlation efficiency >0.98 in multifactor prediction. A vehicle-mounted MSFT platform enabled real-time data transmission, visualization, and graded alerts. This integration of AI and carbon nanomaterial-based sensors significantly reduces economic losses and enhances food security.

6.6 Toxicological and Environmental Implications

6.6.1 Safety Assessment of Nanomaterials in Agriculture

Nanotechnology's rapid growth raises concerns due to nanoparticles' small size and high reactivity, enabling easy body entry and complex toxicity (Kaur et al. 2014). Ensuring the safe use of biosynthesized carbon nanomaterials in agriculture requires a comprehensive and systematic safety assessment framework that evaluates their physicochemical properties, biological effects, and potential human exposure pathways. The behavior of CNMs in living systems depends heavily on factors such as particle size, surface charge, chemical composition, aggregation tendency, and functional groups. These characteristics determine how nanomaterials interact with biological membranes, proteins, and nucleic acids, ultimately influencing their toxicity (Wang et al. 2022). Safety assessment typically begins with rigorous in vitro studies, where cell-based assays are used to evaluate cytotoxicity, oxidative stress generation, membrane integrity disruption, and genotoxic potential (Yuan et al. 2019). For agricultural applications, it is also critical to assess ecotoxicity using soil organisms like earthworms, beneficial insects, and aquatic species that may be exposed through environmental dispersion. According to Xu et al. (2021), carbon-based nanomaterials showed morphology-dependent toxicity in earthworms, with high-aspect-ratio SWCNTs and rGO causing greater metabolic, antioxidant, and cellular stress. Biomarkers such as cell viability and amino-acid ratios revealed membrane disruption, highlighting particle shape as a key determinant of soil environmental risk. Additionally, when CNMs are incorporated into food packaging or postharvest treatments, potential migration into edible products must be evaluated (Lalpekhlua et al. 2024). To complement toxicological testing, a risk-benefit analysis is essential to determine whether the agricultural advantages, such as reduced pesticide usage, improved crop protection, or early spoilage detection, and justify the introduction of CNMs into the environment. According to Rana et al. (2024), nanotechnology enhances agriculture through precision farming, improved production, smart packaging, nano-sensors, controlled agrochemical delivery, and bio-based innovations.

6.6.2 Biodegradability and Regulatory Aspects

The biodegradability of biosynthesized CNMs is a critical factor in assessing their environmental sustainability. In contrast to conventional nanomaterials that remain persistent for extended periods, CNMs, especially carbon dots and graphene oxide, undergo degradation more readily. This enhanced degradability is attributed to the presence of oxygen-containing surface functional groups, reduced particle size, and their compatibility with naturally occurring biochemical pathways (Dutta et al. 2022; Ayanda et al. 2024). Microorganisms present in soil and aquatic systems can enzymatically oxidize or metabolize these materials, converting them into smaller, less reactive fragments or mineralizing them completely into carbon dioxide and water. Environmental factors such as pH, temperature, the presence of organic matter, and sunlight further influence degradation rates. As reported by Johansen et al. (2008), soil exposure to C60 fullerenes reduced fast-growing bacteria and slightly affected protozoans, though overall respiration and biomass remained unchanged. However, not all CNMs degrade uniformly; for instance, CNTs remain relatively more persistent and may require specialized microbial consortia or catalytic processes for breakdown (Zhang et al. 2013). According to Dhiman et al. (2025), indigenous *Bacillus*, *Comamonas*, and *Lysinibacillus* strains survived and showed structural damage in MWCNTs, with a microbial consortium showing the highest efficiency of degradation of CNTs. As agricultural nanotechnology expands, regulatory frameworks must evolve to ensure responsible development and deployment. Globally, agencies such as the US Environmental Protection Agency, European Food Safety Authority, and Food Safety and Standards Authority of India are gradually integrating specific guidelines for the usage of nanomaterials into existing regulatory policies. These guidelines mandate detailed characterization, toxicological profiling, environmental fate studies, and labelling requirements for nano-enabled agricultural products (Chowdhury et al. 2023; Djebaili et al. 2025). However, regulatory challenges persist due to the lack of standardized testing methods, variability in nanomaterial synthesis, and difficulty in detecting CNMs in complex matrices such as soil or food. Developing harmonized regulations and clear definitions for nanoforms is essential to facilitate safe commercialization while fostering innovation in sustainable agricultural technologies.

6.6.3 Addressing Knowledge Gaps and Ethical Considerations

Despite promising advancements in biosynthesized carbon nanomaterials for agricultural applications, significant knowledge gaps remain that must be addressed to ensure safe and ethical implementation (Khan et al. 2025; Raveena et al. 2026). One major gap lies in understanding the long-term environmental fate and transformations of CNMs under realistic field conditions, where interactions with soil minerals, organic matter, microorganisms and co-contaminants can alter their structure and toxicity (Zhang et al. 2020; Pikula et al. 2024). Limited information is available on the potential for trophic transfer, bioaccumulation or biomagnification through the food chain. For instance, Shi et al. (2020) found ^{13}C-labelled fullerenols bioaccumulated in algae, transferred to *Daphnia*, and entered fish tissues, concentrating mainly in the intestine and liver. Strong biomagnification occurred from algae to *Daphnia* but not to fish, indicating reduced risk for higher-trophic aquatic organisms despite tissue uptake. Ethical considerations also come to the forefront, particularly regarding equitable access to nanotechnology-based solutions (Verma et al. 2024; Thakur et al. 2024). Smallholder farmers may struggle to adopt these technologies due to cost or lack of technical training, potentially widening socio-economic disparities. Transparency in risk communication, proper labelling of nano-treated produce, and active involvement of stakeholders, including farmers, consumers, scientists, and policymakers, are essential to maintain public trust (Capangpangan et al. 2025). Ethical deployment also requires preventing misuse, ensuring environmental justice, and adopting "safe-by-design" principles, where nanomaterials are engineered to minimize toxicity, enhance biodegradability, and reduce ecological impact. Addressing these knowledge and ethical gaps through multidisciplinary research, public engagement, and robust policy development will guide the responsible integration of nanotechnology into future agricultural systems.

6.7 Future Directions

Future research should focus on optimization and standardization of plant-based synthesis to control the size of CNMs, their surface modifications and functional group distribution as these factors influence the catalytic, sensing, and biological interactions. Plant and microbe-based CNMs can be beneficial for the environment, but to make them on a large scale and in a way that can be repeated, bioprocess optimization, and automation need to be done continuously. CNMs could change farming, especially when used with new technologies like machine learning, IoT-based monitoring systems, and smart packaging. Integrating these systems might help farmers keep an eye on rotting gases, follow the development of diseases, and find signs of plant stress in real time. Future research should also work on making multimodal sensors that use optical, electrochemical, and plasmonic technologies to provide very sensitive, quick, and ready-to-use diagnostics. It is also vital to create strict rules, standardized procedures for measuring ecotoxicity, and recommendations that are the same across the world to make sure that commercialization is safe.

6.8 Conclusion

The biosynthesized CNMs have immense potential in pesticide synthesis, crop spoilage sensing, pesticidal activity, and crop monitoring. Integrating CNMs with biosensors, smart packaging, and monitoring systems enables enhanced agricultural and supply chain management. Despite these benefits, CNMs can react with biological and environmental systems encompassing their assessment for safe dosage, toxicity evaluation, and bioaccumulation. Altogether, biosynthesized CNMs act as a connecting bridge between nanotechnology, biotechnology, and agriculture. Developing CNMs with proper regulation, ethical deployment will pave the path to novel bio and environmentally compatible CNMs useful for agrochemical products, microbe detection, and crop spoilage sensing. Thus, CNMs have the potential to enhance agricultural productivity significantly while maintaining ecosystem health through the integration of scientific innovation and sustainability principles.

References

Abdul Ghani MA, Nordin AN, Zulhairee M et al (2022) Portable electrochemical biosensors based on microcontrollers for detection of viruses: a review. Biosensors (Basel) 12:666. https://doi.org/10.3390/bios12080666

Adeola AO, Duarte MP, Naccache R (2023) Microwave-assisted synthesis of carbon-based nanomaterials from biobased resources for water treatment applications: emerging trends and prospects. Front Carbon 2. https://doi.org/10.3389/frcrb.2023.1220021

Akkaş T, Reshadsedghi M, Şen M et al (2025) The role of artificial intelligence in advancing biosensor technology: past, present, and future perspectives. Adv Mater 37:2504796. https://doi.org/10.1002/adma.202504796

Alagendran S, Mounika C, Ashoka P et al (2024) Carbon-based nanomaterials for soil amendment: a paradigm shift in agriculture practices. J Exp Agric Int 46:833–849. https://doi.org/10.9734/jeai/2024/v46i123192

Alam MW, Junaid PM, Gulzar Y et al (2024) Advancing agriculture with functional NM: "pathways to sustainable and smart farming technologies". Discov Nano 19:197. https://doi.org/10.1186/s11671-024-04144-z

Andleeb N, Zafar S, Rahim Z et al (2025) Carbon quantum dots as versatile nanomaterials for improving soil health and plant stress tolerance: a comprehensive review. Planta 262:44. https://doi.org/10.1007/s00425-025-04758-2

Ayanda OS, Mmuoegbulam AO, Okezie O et al (2024) Recent progress in carbon-based nanomaterials: critical review. J Nanopart Res 26:106. https://doi.org/10.1007/s11051-024-06006-2

Ayilara MS, Adeleke BS, Akinola SA et al (2023) Biopesticides as a promising alternative to synthetic pesticides: a case for microbial pesticides, phytopesticides, and nanobiopesticides. Front Microbiol 14. https://doi.org/10.3389/fmicb.2023.1040901

Bressi V, Balu AM, Iannazzo D, Espro C (2023) Recent advances in the synthesis of carbon dots from renewable biomass by high-efficient hydrothermal and microwave green approaches. Curr Opin Green Sustain Chem 40:100742. https://doi.org/10.1016/j.cogsc.2022.100742

Brower-Toland B, Stevens JL, Ralston L et al (2024) A crucial role for technology in sustainable agriculture. ACS Agric Sci Technol 4:283–291. https://doi.org/10.1021/acsagscitech.3c00426

Capangpangan RY, Alguno AC, Halog A (2025) Nanotextiles for sustainability and circular economy. In: Muthu SS (ed) Sustainable nano textile technology: case studies. Springer Nature Switzerland, Cham, pp 43–63

Carrillo-Lopez LM, Villanueva-Verduzco C, Villanueva-Sánchez E et al (2024) Nanomaterials for plant disease diagnosis and treatment: a review. Plants 13:2634. https://doi.org/10.3390/plants13182634

Chethan B, Sunilkumar A, Prasad V, Thomas S (2024) Surfactant electrochemical sensor based on carbon nanotubes for the analysis of fungicides. In: Manjunatha JG (ed) Advances in surfactant biosensor and sensor technologies. Springer Nature Switzerland, Cham, pp 217–228

Chowdhury AR, Sarkar DJ, Singh BK, Sarkar B (2023) Nanotoxicological issues in agriculture and related regulatory framework. In: Biological applications of nanoparticles. Springer, Singapore, pp 217–231

Dhiman L, Anand S, Singh SK (2025) Bioremediation potential of the consortium of indigenous microbial isolates in degrading multiwall carbon nanotube (MWCNT) present in contaminated water. Arch Microbiol 207:49. https://doi.org/10.1007/s00203-025-04250-1

Díez-Pascual AM (2021) State of the art in the antibacterial and antiviral applications of carbon-based polymeric nanocomposites. Int J Mol Sci 22. https://doi.org/10.3390/ijms221910511

Djebaili R, Farda B, Fares R, Pellegrini M (2025) Regulatory and safety concerns related to the use of nanomaterials. In: Sayyed R, Ilyas N, Fernando D (eds) Nanobiostimulants in innovative agriculture. Springer Nature, Singapore, pp 261–281 do Nascimento Junior DR, Tabernero A, Cabral Albuquerque EC et al (2021) Biopesticide encapsulation using supercritical CO2: a comprehensive review and potential applications. Molecules 26. https://doi.org/10.3390/molecules26134003

Dutta V, Verma R, Gopalkrishnan C et al (2022) Bio-inspired synthesis of carbon-based nanomaterials and their potential environmental applications: a state-of-the-art review. Inorganics (Basel) 10. https://doi.org/10.3390/inorganics10100169

Eccher Zerbini P, Vanoli M, Rizzolo A et al (2015) Optical properties, ethylene production and softening in mango fruit. Postharvest Biol Technol 101:58–65. https://doi.org/10.1016/j.postharvbio.2014.11.008

El Hosry L, Bou-Maroun E (2025) Molecularly imprinted polymer-based sensors in food contaminants analysis: advances, applications, and future trends. Chemosensors 13:420. https://doi.org/10.3390/chemosensors13120420

Fenibo EO, Ijoma GN, Matambo T (2022) Biopesticides in sustainable agriculture: current status and future prospects. In: De Mandal S, Ramkumar G, Karthi S, Jin F (eds) New and future development in biopesticide research: biotechnological exploration. Springer Nature, Singapore, pp 1–53

Francisco Fukumoto AA, Alves Pimenta JA, Hirooka EY, Kuroda EK (2024) Pesticides removal from water using activated carbons and carbon nanotubes. Environ Technol 45:431–453. https://doi.org/10.1080/09593330.2022.2112979

Geleta GS, Zhao Z, Wang Z (2018) A novel reduced graphene oxide/molybdenum disulfide/polyaniline nanocomposite-based electrochemical aptasensor for detection of aflatoxin B 1. Analyst 143:1644–1649. https://doi.org/10.1039/C7AN02050C

Ghosh S, Ahmad R, Zeyaullah M, Khare SK (2021) Microbial nano-factories: synthesis and biomedical applications. Front Chem 9. https://doi.org/10.3389/fchem.2021.626834

Goswami AD, Trivedi DH, Jadhav NL, Pinjari DV (2021) Sustainable and green synthesis of carbon nanomaterials: a review. J Environ Chem Eng 9:106118. https://doi.org/10.1016/j.jece.2021.106118

Grieshaber D, MacKenzie R, Vörös J, Reimhult E (2008) Electrochemical biosensors – sensor principles and architectures. Sensors 8:1400–1458. https://doi.org/10.3390/s80314000

Guo X, Hou J, Yuan Z et al (2021) A novel paper biosensor based on $Fe_3O_4@SiO_2$ –NH_2 and MWCNTs for rapid detection of pseudorabies virus. Nanotechnology 32:355102. https://doi.org/10.1088/1361-6528/ac002a

Guo Z, Zhang Y, Xiao H et al (2025) Multi-sensor fusion and deep learning for batch monitoring and real-time warning of apple spoilage. Food Control 172:111174. https://doi.org/10.1016/j.foodcont.2025.111174

Gurunathan S, Han JW, Dayem AA et al (2012) Oxidative stress-mediated antibacterial activity of graphene oxide and reduced graphene oxide in Pseudomonas aeruginosa. Int J Nanomedicine 7: 5901–5914. https://doi.org/10.2147/IJN.S37397

Harun-Ur-Rashid M, Jahan I, Imran AB (2025) Nanocatalysts and environmental safety: a modern approach to pollution control. In: Singh AK, Singh AK, Susan MDABH (eds) Advances in nanomaterials for detection, control, and removal of environmental pollutants. Springer Nature Switzerland, Cham, pp 87–121

Hegde V, Bhat MP, Lee J-H et al (2024) Carbon-based nanomaterials: multifaceted role in agrochemical recognition, remediation, and release. Nano Today 57:102388. https://doi.org/10.1016/j.nantod.2024.102388

Hoque N, Saikia BK (2025) Coal-derived carbon materials: pathways to graphene, carbon nanotubes, carbon quantum dots, and nanodiamond for energy and environmental solutions. ACS Omega 10:53673–53701. https://doi.org/10.1021/acsomega.5c06313

Iravani S, Varma RS (2020) Green synthesis, biomedical and biotechnological applications of carbon and graphene quantum dots. A review. Environ Chem Lett 18:703–727. https://doi.org/10.1007/s10311-020-00984-0

Jabbar A, East AR (2016) Quantifying the ethylene induced softening and low temperature breakdown of 'Hayward' kiwifruit in storage. Postharvest Biol Technol 113:87–94. https://doi.org/10.1016/j.postharvbio.2015.11.002

Jasim SA, Rachchh N, Pallathadka H et al (2024) Recent advances in carbon-based materials derived from diverse green biowaste for sensing applications: a comprehensive overview from the perspective of synthesis method and application. RSC Adv 14:39787–39803. https://doi.org/10.1039/D4RA07693A

Jayasekhar Babu P, Saranya S, Longchar B, Rajasekhar A (2022) Nanobiotechnology-mediated sustainable agriculture and post-harvest management. Curr Res Biotechnol 4:326–336. https://doi.org/10.1016/j.crbiot.2022.07.004

Jing HH, Bardakci F, Akgöl S et al (2023) Green carbon dots: synthesis, characterization, properties and biomedical applications. J Funct Biomater 14:27. https://doi.org/10.3390/jfb14010027

Jirimali H, Singh J, Boddula R et al (2022) Nano-structured carbon: its synthesis from renewable agricultural sources and important applications. Materials 15:3969. https://doi.org/10.3390/ma15113969

Johansen A, Pedersen AL, Jensen KA et al (2008) Effects of C60 fullerene nanoparticles on soil bacteria and protozoans. Environ Toxicol Chem 27:1895–1903. https://doi.org/10.1897/07-375.1

Joksović S, Kundačina I, Milošević I et al (2024) Single-walled carbon nanotube-modified gold leaf immunosensor for *Escherichia coli* detection. ACS Omega 9:22277–22284. https://doi.org/10.1021/acsomega.4c01599

Kassal P, Steinberg IM, Steinberg MD (2013) Wireless smart tag with potentiometric input for ultra low-power chemical sensing. Sens Actuators B Chem 184:254–259. https://doi.org/10.1016/j.snb.2013.04.049

Kaur IP, Kakkar V, Deol PK et al (2014) Issues and concerns in nanotech product development and its commercialization. J Control Release 193:51–62. https://doi.org/10.1016/j.jconrel.2014.06.005

Kayani KF, Mohammed SJ, Mohammad NN et al (2024) Ratiometric fluorescence detection of tetracycline in milk and tap water with smartphone assistance for visual pH sensing using innovative dual-emissive phosphorus-doped carbon dots. Food Control 164:110611. https://doi.org/10.1016/j.foodcont.2024.110611

Ke M, Zhang K, Hicks AL et al (2025) A life cycle risk assessment of nanopesticides in freshwater. Environ Sci Ecotechnol 25:100565. https://doi.org/10.1016/j.ese.2025.100565

Khan A, Ezati P, Kim J-T, Rhim J-W (2023) Biocompatible carbon quantum dots for intelligent sensing in food safety applications: opportunities and sustainability. Mater Today Sustain 21:100306. https://doi.org/10.1016/j.mtsust.2022.100306

Khan A, Ashiq B, Qadri TA et al (2025) Nanotechnology solutions for advancing sustainable development goals (SDGs): integrating nanomaterials for global sustainability. BioNanoScience 15:590. https://doi.org/10.1007/s12668-025-02214-9

Koul B, Poonia AK, Yadav D, Jin J-O (2021) Microbe-mediated biosynthesis of nanoparticles: applications and future prospects. Biomolecules 11:886. https://doi.org/10.3390/biom11060886

Kumar Mishra R, Singh B, Acharya B (2024) A comprehensive review on activated carbon from pyrolysis of lignocellulosic biomass: an application for energy and the environment. Carbon Resour Convers 7:100228. https://doi.org/10.1016/j.crcon.2024.100228

Kumar J, Ramlal A, Mallick D et al (2021) An overview of some biopesticides and their importance in plant protection for commercial acceptance. Plants 10. https://doi.org/10.3390/plants10061185

Kundu A, Basu S, Maity B (2023) Upcycling waste: Citrus limon peel-derived carbon quantum dots for sensitive detection of tetracycline in the nanomolar range. ACS Omega 8:36449–36459. https://doi.org/10.1021/acsomega.3c05424

Kyriakoudi A, Spanidi E, Mourtzinos I et al (2021) Innovative delivery systems loaded with plant bioactive ingredients: formulation approaches and applications. Plants 10. https://doi.org/10.3390/plants10061238

Lalpekhlua K, Tirkey A, Saranya S, Babu PJ (2024) Post-harvest management strategies for quality preservation in crops. Int J Veg Sci 30:587–635. https://doi.org/10.1080/19315260.2024.2392551

Li H, Murugesan A, Shoaib M et al (2025) Functionalized metal-organic frameworks with biomolecules for sensing and detection applications of food contaminants. Crit Rev Food Sci Nutr 65:5218–5250. https://doi.org/10.1080/10408398.2024.2406482

Liang W, Wang B, Cheng J et al (2021) 3D, eco-friendly metal-organic frameworks@carbon nanotube aerogels composite materials for removal of pesticides in water. J Hazard Mater 401:123718. https://doi.org/10.1016/j.jhazmat.2020.123718

Lin F, Li C, Chen Z (2018) Bacteria-derived carbon dots inhibit biofilm formation of Escherichia coli without affecting cell growth. Front Microbiol 9. https://doi.org/10.3389/fmicb.2018.00259

Liu Q, Chen H, Mi R et al (2025) Biomass-derived carbon dots: preparation, properties, and applications. Nanomaterials 15:1279. https://doi.org/10.3390/nano15161279

Makgabutlane B, Nthunya LN, Maubane-Nkadimeng MS, Mhlanga SD (2021) Green synthesis of carbon nanotubes to address the water-energy-food nexus: a critical review. J Environ Chem Eng 9:104736. https://doi.org/10.1016/j.jece.2020.104736

Manan S, Ullah MW, Ul-Islam M et al (2021) Synthesis and applications of fungal mycelium-based advanced functional materials. J Bioresour Bioprod 6:1–10. https://doi.org/10.1016/j.jobab.2021.01.001

Mukherjee A, Majumdar S, Servin AD et al (2016) Carbon nanomaterials in agriculture: a critical review. Front Plant Sci 7:172. https://doi.org/10.3389/fpls.2016.00172

Olaniyan OF, Ariwaodo CA, Ibrahim SO et al (2025) Advances in green synthesis and application of nanoparticles from crop residues: a comprehensive review. Sci Afr 28:e02654. https://doi.org/10.1016/j.sciaf.2025.e02654

Patel KD, Singh RK, Kim H-W (2019) Carbon-based nanomaterials as an emerging platform for theranostics. Mater Horiz 6:434–469. https://doi.org/10.1039/C8MH00966J

Paveethra S, Manisekaran H, Sasidharan S (2024) Medicinal plants derived green carbon dots: synthesis, characterization and their potential applications in cancer therapy. Asian Pac J Cancer Prev 25:3393–3411. https://doi.org/10.31557/APJCP.2024.25.10.3393

Pavithran RK, Reddy SG, Kumar BS, Kugabalasooriar S (2024) Enhancing sustainability in agriculture: natural polymer-based controlled release systems for effective pest management and environmental protection. ES Food Agrofor 18:1276

Peng Z, Liu X, Zhang W et al (2020) Advances in the application, toxicity and degradation of carbon nanomaterials in environment: a review. Environ Int 134:105298. https://doi.org/10.1016/j.envint.2019.105298

Pieczywek PM, Cybulska J, Szymańska-Chargot M et al (2018) Early detection of fungal infection of stored apple fruit with optical sensors – comparison of biospeckle, hyperspectral imaging and chlorophyll fluorescence. Food Control 85:327–338. https://doi.org/10.1016/j.foodcont.2017.10.013

Pikula K, Johari SA, Santos-Oliveira R et al (2024) Joint toxicity and interaction of carbon-based nanomaterials with co-existing pollutants in aquatic environments: a review. Int J Mol Sci 25. https://doi.org/10.3390/ijms252111798

Prajapati D, Prajapati N, Chaudhary A et al (2025) Impact of carbon-based nanomaterials on soil microbiomes and associated activities. In: Emerging carbon nanomaterials for sustainable agricultural practices: synthesis, plant growth, performance, production and protection. Springer Nature Singapore, Singapore, pp 333–354

Prasannan A, Imae T (2013) One-pot synthesis of fluorescent carbon dots from orange waste peels. Ind Eng Chem Res 52:15673–15678. https://doi.org/10.1021/ie402421s

Qasim M, Clarkson AN, Hinkley SFR (2023) Green synthesis of carbon nanoparticles (CNPs) from biomass for biomedical applications. Int J Mol Sci 24:1023. https://doi.org/10.3390/ijms24021023

Rahmani MKI, Ghanimi HMA, Jilani SF et al (2023) Early pathogen prediction in crops using Nano biosensors and neural network-based feature extraction and classification. Big Data Res 34: 100412. https://doi.org/10.1016/j.bdr.2023.100412

Rana L, Kumar M, Rajput J et al (2024) Nexus between nanotechnology and agricultural production systems: challenges and future prospects. Discov Appl Sci 6:555. https://doi.org/10.1007/s42452-024-06265-7

Raveena R, Rajkishore SK, Maheswari M et al (2026) Mechanistic insights into carbon nanomaterials as potential plant biostimulants: enhancing photosynthesis and stress tolerance for climate-resilient agriculture. Environ Sci Nano. https://doi.org/10.1039/D5EN00814J

Rispail N, De Matteis L, Santos R et al (2014) Quantum dot and superparamagnetic nanoparticle interaction with pathogenic fungi: internalization and toxicity profile. ACS Appl Mater Interfaces 6:9100–9110. https://doi.org/10.1021/am501029g

Selvam AP, Al-Humairi SNS (2023) The impact of IoT and sensor integration on real-time weather monitoring systems: a systematic review. Springer Science and Business Media LLC

Sengupta I, Bhattacharya P, Talukdar M et al (2019) Bactericidal effect of graphene oxide and reduced graphene oxide: influence of shape of bacteria. Colloid Interface Sci Commun 28:60–68. https://doi.org/10.1016/j.colcom.2018.12.001

Sharma V, Soni V (2024) Advancement in carbon nanoparticle synthesis and their application: a comprehensive review. Am J Nano Res Appl 12:1–14. https://doi.org/10.11648/j.nano.20241201.11

Shi J, Tian F, Lyu J, Yang M (2015) Nanoparticle based fluorescence resonance energy transfer (FRET) for biosensing applications. J Mater Chem B 3:6989–7005. https://doi.org/10.1039/C5TB00885A

Shi Q, Wang CL, Zhang H et al (2020) Trophic transfer and biomagnification of fullerenol nanoparticles in an aquatic food chain. Environ Sci Nano 7:1240–1251. https://doi.org/10.1039/C9EN01277J

Singh H, Sharma A, Bhardwaj SK et al (2021) Recent advances in the applications of nano-agrochemicals for sustainable agricultural development. Environ Sci Process Impacts 23:213–239. https://doi.org/10.1039/D0EM00404A

Singh S, Prasad SM, Bashri G (2023) Fate and toxicity of nanoparticles in aquatic systems. Acta Geochim 42:63–76. https://doi.org/10.1007/s11631-022-00572-9

Sinha N, Ray S (2024) Application of carbon quantum dots derived from waste tea for the detection of pesticides in tea: a novel biosensor approach. ACS Omega 9:50201–50213. https://doi.org/10.1021/acsomega.4c04449

Sobhan A, Jia F, Kelso LC et al (2022) A novel activated biochar-based immunosensor for rapid detection of E. coli O157:H7. Biosensors (Basel) 12:908. https://doi.org/10.3390/bios12100908

Speranza G (2021) Carbon nanomaterials: synthesis, functionalization and sensing applications. Nanomaterials 11:967. https://doi.org/10.3390/nano11040967

Srujana TL, Rao KJ, Korumilli T (2025) Natural biogenic templates for nanomaterial synthesis: advances, applications, and environmental perspectives. ACS Biomater Sci Eng 11:1291–1316. https://doi.org/10.1021/acsbiomaterials.4c02075

Terry LR, Sanders S, Potoff RH et al (2022) Applications of surface-enhanced Raman spectroscopy in environmental detection. Anal Sci Adv 3:113–145. https://doi.org/10.1002/ansa.202200003

Thakur A, Thakur P, Suhag D, Khurana SMP (2024) Green nanobiotechnology. CRC Press, Boca Raton

Thamarai P, Deivayanai VC, Ragini YP et al (2025) Recent advances in carbon nanomaterials: pioneering sustainable environmental contamination remediation and advancing circular economy practices. Rev Environ Contam Toxicol 263:19. https://doi.org/10.1007/s44169-025-00094-4

Tian J, An M, Zhao X et al (2023) Advances in fluorescent sensing carbon dots: an account of food analysis. ACS Omega 8:9031–9039. https://doi.org/10.1021/acsomega.2c07986

Tok K, Oztekin T, Yilmaz S et al (2025) Safe-by-design green nanopesticides as a sustainable approach for pest management, plant health, and environmental safety. Sci Total Environ 994: 180077. https://doi.org/10.1016/j.scitotenv.2025.180077

Uddin S (2021) Removal of pesticides using carbon-based nanocomposite materials. In: Jawaid M, Ahmad A, Ismail N, Rafatullah M (eds) Environmental remediation through carbon based nano composites. Springer Nature, Singapore, pp 365–385

Upadhyayula VKK, Meyer DE, Curran MA, Gonzalez MA (2012) Life cycle assessment as a tool to enhance the environmental performance of carbon nanotube products: a review. J Clean Prod 26:37–47. https://doi.org/10.1016/j.jclepro.2011.12.018

Verma V, Gupta P, Singh P, Pandey NK (2024) Considerations in the development and deployment of nanotechnology. In: Nanotechnology in societal development. Springer Nature Singapore, Singapore, pp 505–540

Viswanathan S, Rani C, Ho JA (2012) Electrochemical immunosensor for multiplexed detection of food-borne pathogens using nanocrystal bioconjugates and MWCNT screen-printed electrode. Talanta 94:315–319. https://doi.org/10.1016/j.talanta.2012.03.049

Wang S, Niu R, Yang Y et al (2020) Aptamer-functionalized chitosan magnetic nanoparticles as a novel adsorbent for selective extraction of ochratoxin A. Int J Biol Macromol 153:583–590. https://doi.org/10.1016/j.ijbiomac.2020.03.035

Wang X, Liang D, Wang Y et al (2022) A critical review on the biological impact of natural organic matter on nanomaterials in the aquatic environment. Carbon Res 1:13. https://doi.org/10.1007/s44246-022-00013-5

Wang S, Zhang L, Wang H et al (2024) Halochromy incorporated with inner filter effect-based fluorescence quenching: a dual-response strategy for spoilage sensing of proteinous foods with rapid and irreversible readout. Sens Actuators B Chem 409:135631. https://doi.org/10.1016/j.snb.2024.135631

Wang Z, Changotra R, Dasog M et al (2025) Carbon quantum dots: synthesis via hydrothermal processing, doping strategies, integration with photocatalysts, and their application in photocatalytic hydrogen production. Sustain Mater Technol 44:e01386. https://doi.org/10.1016/j.susmat.2025.e01386

Wei D, Bailey MJA, Andrew P, Ryhänen T (2009) Electrochemical biosensors at the nanoscale. Lab Chip 9:2123. https://doi.org/10.1039/b903118a

Wu S, Duan N, Ma X et al (2012) Multiplexed fluorescence resonance energy transfer aptasensor between upconversion nanoparticles and graphene oxide for the simultaneous determination of mycotoxins. Anal Chem 84:6263–6270. https://doi.org/10.1021/ac301534w

Xu K, Wang X, Lu C et al (2021) Toxicity of three carbon-based nanomaterials to earthworms: effect of morphology on biomarkers, cytotoxicity, and metabolomics. Sci Total Environ 777: 146224. https://doi.org/10.1016/j.scitotenv.2021.146224

Xu Y, Hassan MM, Sharma AS et al (2023) Recent advancement in nano-optical strategies for detection of pathogenic bacteria and their metabolites in food safety. Crit Rev Food Sci Nutr 63: 486–504. https://doi.org/10.1080/10408398.2021.1950117

Xue H, Yang J, Yu H et al (2026) The carbon quantum dots: preparation, antibacterial mechanisms, and application in food packaging. Curr Res Food Sci 12:101316. https://doi.org/10.1016/j.crfs.2026.101316

Yadav PK, Singh VK, Chandra S et al (2019) Green synthesis of fluorescent carbon quantum dots from Azadirachta indica leaves and their peroxidase-mimetic activity for the detection of H2O2 and ascorbic acid in common fresh fruits. ACS Biomater Sci Eng 5:623–632. https://doi.org/10.1021/acsbiomaterials.8b01528

Yadav R, Kumar M, Tomar RS (2023) Revisiting the microbial biosynthesis of metal nanoparticles and their applications. J Appl Pharm Sci. https://doi.org/10.7324/JAPS.2023.89282

Yin L, Jayan H, Cai J et al (2023) Spoilage monitoring and early warning for apples in storage using gas sensors and chemometrics. Foods 12:2968. https://doi.org/10.3390/foods12152968

Yu X, Liu J, Wang W et al (2025) Plant-derived peptides: from identification to agronomic applications. Mol Plant 18:1963–1982. https://doi.org/10.1016/j.molp.2025.11.004

Yuan X, Zhang X, Sun L et al (2019) Cellular toxicity and immunological effects of carbon-based nanomaterials. Part Fibre Toxicol 16:18. https://doi.org/10.1186/s12989-019-0299-z

Yuan L, Gao M, Xiang H et al (2023) A biomass-based colorimetric sulfur dioxide gas sensor for smart packaging. ACS Nano 17:6849–6856. https://doi.org/10.1021/acsnano.3c00530

Zamfir L-G, Răut I, Constantin M et al (2025) Assessment of biogenic amines produced by microorganisms as food spoilage indicators by sensitive detection using portable opto-electrochemical tools based on biosensors. Food Control 172:111161. https://doi.org/10.1016/j.foodcont.2025.111161

Zhang L, Petersen EJ, Habteselassie MY et al (2013) Degradation of multiwall carbon nanotubes by bacteria. Environ Pollut 181:335–339. https://doi.org/10.1016/j.envpol.2013.05.058

Zhang M, Tao S, Wang X (2020) Interactions between organic pollutants and carbon nanomaterials and the associated impact on microbial availability and degradation in soil: a review. Environ Sci Nano 7:2486–2508. https://doi.org/10.1039/D0EN00515K

Zhang Z, Ou X, Ma L et al (2024) A double methylene blue labeled single-stranded DNA and hairpin DNA coupling biosensor for the detection of Fusarium oxysporum f. sp. cubense race 4. Bioelectrochemistry 156:108612. https://doi.org/10.1016/j.bioelechem.2023.108612

Zieliński B, Miądlicki P, Przepiórski J (2022) Development of activated carbon for removal of pesticides from water: case study. Sci Rep 12:20869. https://doi.org/10.1038/s41598-022-25247-6

Chapter 7
Rice Waste-Derived Carbon Nanomaterials: A Sustainable Approach to Nanotechnology

Jyotirmayee Giri, Laxmidhar Mallick, Sambit Sahu, Lopamudra Subudhi, Shibani Mohapatra, and Alok Kumar Panda

7.1 Introduction

One of the main producers of extensive volumes of waste is the agricultural sector, which defined as the science and art of cultivating plants and livestock and has raised concerns to everyone and threatened the viability of agricultural systems (Ali et al. 2022; Ufitikirezi et al. 2024). Agricultural waste is described as an extensive collection of various materials produced from a number of different agricultural operations in large volumes. These include crop residues, animal byproducts, agricultural processing wastes, and home/business wastes from agriculture (Fig. 7.1) (Bhardwaj et al. 2025). Field-drained salt and silt; pesticides that contaminate water, air, or soil; fertilizer; runoff from fields; manure wastes; and many more are the main components of agricultural waste. Because they contain numerous contaminants, hazardous metals, toxic gases, and other substances, these wastes are extremely dangerous to both the environment and humanity. Additionally, these waste

J. Giri · L. Mallick · S. Sahu
Environmental Science Laboratory, School of Applied Sciences, Kalinga Institute of Industrial Technology (Deemed to Be University), Bhubaneswar, Odisha, India

L. Subudhi · S. Mohapatra (✉)
Centre for Biotechnology, Siksha 'O' Anusandhan (Deemed to Be University), Bhubaneswar, Odisha, India
e-mail: shibanimohapatra@soa.ac.in

A. K. Panda (✉)
Environmental Science Laboratory, School of Applied Sciences, Kalinga Institute of Industrial Technology (Deemed to Be University), Bhubaneswar, Odisha, India

Centre for Water Research and Climate Change, Kalinga Institute of Industrial Technology (Deemed to Be University), Bhubaneswar, Odisha, India
e-mail: alok.pandafch@kiit.ac.in

Y. K. Mohanta et al. (eds.), *Bioresource Waste to Carbon Nanomaterial*, Nanotechnology in the Life Sciences, https://doi.org/10.1007/978-981-92-2405-0_7

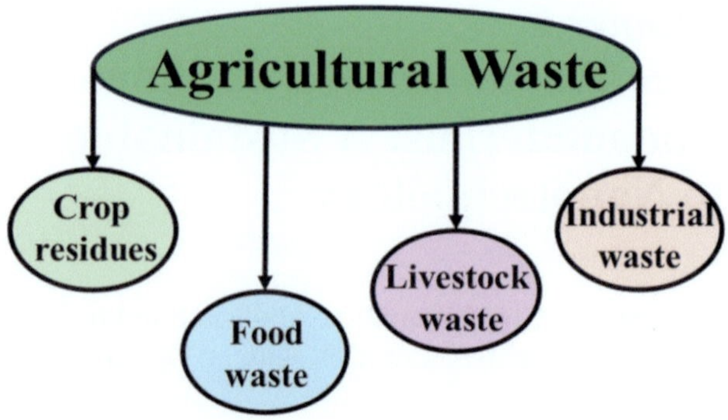

Fig. 7.1 Classification of agricultural wastes

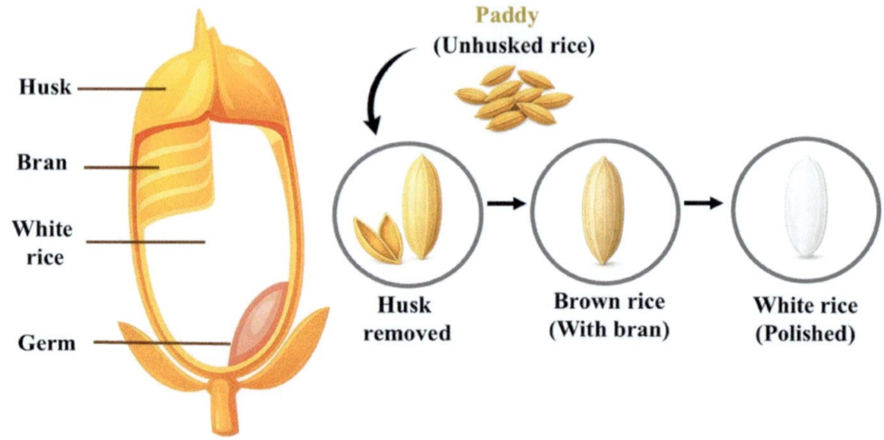

Fig. 7.2 Paddy grain (left) and its products after husking

products contribute in some way to the cause of global warming, which is currently the world's biggest concern (Ali et al. 2022).

Among all the agricultural waste, rice is one of the oldest crops in the world, having been cultivated for centuries. It is now grown in more than 100 countries, and nearly half of the world's population consumes it as a staple food. When rice-based foods are produced, a significant amount of waste is generated. The inability to properly dispose of this waste has harmful effects on both human health and the environment. Rice husk (RH), rice straw (RS), rice husk ash (RHA), and broken rice (Fig. 7.2) are all found in rice waste which historically were burned or discarded (Chakroborty et al. 2023). Paddy rice is typically composed of 69% starchy endosperm, 20% rice husk or hull, and 11% grain/bran layers. After burning, about 18–20% of the RH is converted to rice husk ash (RHA). Rice husk, which is composed of 15–20% inorganic materials like silica SiO_2 and 65–75% organic materials like cellulose, hemicellulose, and lignin, is used as a precursor for the production of highly valued nanomaterials. Consequently, it is important to convert rice husk into different carbon and silica nanomaterials (Mubarik et al. 2021).

The main aim of this review is to explore and emphasize sustainable, green synthesis approaches for producing advanced carbon-based nanomaterials—such as graphene, carbon nanotubes, and carbon dots—using rice-derived agricultural waste as renewable precursors. Moreover, it converts a low-value waste into high value while using environmentally friendly, low-cost, and energy-efficient processing techniques to make nanostructure materials from rice byproducts which creates an opportunity for farmers to benefit from the rice they harvest. It will also examine the many multifunctional uses of these nanostructured materials for various disciplines (e.g., medical imaging, environmental remediation, and energy conversion and storage) and connect the food/agriculture industry's waste stream to global objectives of environmental sustainability and circular economies.

7.2 Composition and Properties of Rice Waste

The basic composition of the rice dust and the elemental composition of different portions of rice dust. It was noted that silica, cellulose, and lignin are the main components of the rice dust which is shown in Table 7.1 (Korotkova et al. 2016). Rice residues such as husk and straw constitute a silica-rich lignocellulosic biomass that can be converted into biochar and other carbon materials with distinctive 3D carbon silica architectures and strong potential for environmental remediation. These residues are available at very large scales in major rice-producing regions, making rice waste-derived biochar an attractive, regionally abundant platform for water, soil, and emerging contaminant treatment within circular bioeconomy frameworks (Mubarik et al. 2021).

7.2.1 Composition of Rice Waste and Resulting Biochar

Paddy processing generates several byproducts, notably rice husk and rice straw, which together reached about 137 and 685 million tonnes globally in 2012, respectively, highlighting the magnitude of this resource. Rice husk typically contains about 65–75% organic constituents and 15–20% inorganic silica (SiO_2), while

Table 7.1 The elemental composition of different parts of the grain dust (Korotkova et al. 2016)

Elemental composition	Outer surface of husk (wt%)	Inside the husk (wt %)	Inner surface of husk (wt%)
C	6.91	62.54	30.20
O	47.93	35.19	42.53
Si	45.16	2.27	27.27

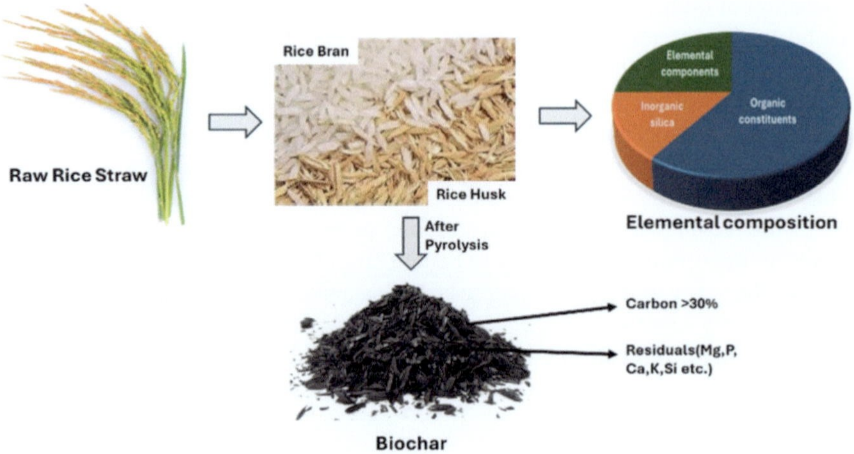

Fig. 7.3 Composition of rice waste and resulting biochar

moisture and minor mineral oxides account for the balance. Elemental analysis of rice husk biochar produced at 500 °C shows carbon contents above 30% (9) and residual macronutrients (K_2O, CaO, MgO, P), indicating that the original lignocellulosic mineral matrix is partially preserved in the char (Fig. 7.3) (Barzallo et al. 2025).

During pyrolysis (300–700 °C), progressive devolatilization of cellulose, hemicellulose, and lignin reduces biochar yield but increases aromatic carbon content, surface area, and pore volume, with rice husk biochar BET area rising from 17.54 to 88.21 $m^2\ g^{-1}$ between 300 and 500 °C. Chemical activation, for example, with H_3PO_4, further enhances these textural properties in rice husk biochar, giving surface areas around 207 $m^2\ g^{-1}$ and more oxygenated surface groups, which improve sorption capacity toward aqueous pollutants (Barzallo et al. 2025).

7.2.2 *Availability and Geographical Distribution of Rice Waste*

Rice is among the most widely cultivated cereals, and its processing generates large byproduct streams that are unevenly distributed but heavily concentrated in Asia, with additional contributions from Africa and Latin America. Estimates cited for 2012 indicate worldwide production of about 137 million tonnes of rice husk and 685 million tonnes of rice straw, much of which has historically been burned in open fields or dumped, causing air pollution and greenhouse gas emissions (Mubarik et al. 2021).

Because rice husk comprises around 20% of paddy mass and is produced continuously wherever rice is milled, it provides a regionally reliable feedstock for

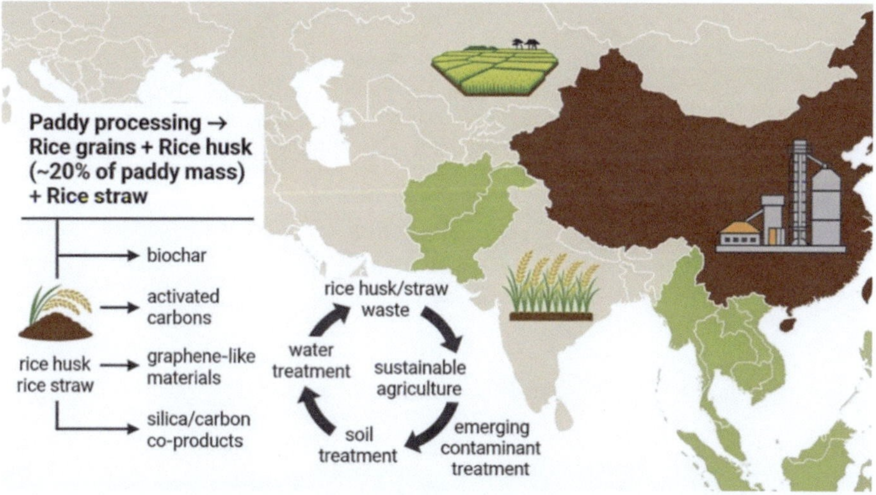

Fig. 7.4 Availability and geographical distribution of rice waste

biochar, activated carbons, graphene-like materials, carbon nanotubes, and silica content also enables integrated routes where silica and carbonaceous adsorbents are co-produced by maximizing material recovery from a single agricultural residue stream (Fig. 7.4) (Barzallo et al. 2025).

7.3 Synthesis Techniques for Carbon Nanomaterials from Rice Waste

Various fabrication methods have been developed to convert rice waste (primarily husk and straw) into carbon-based nanomaterials. Each technique has its advantages, limitations, and yields. The major synthesis routes include thermal decomposition (pyrolysis), hydrothermal carbonization, chemical activation, and catalytic CVD, among others. A comparative evaluation of these methods is summarized at the end of this section and Fig. 7.5.

7.3.1 Pyrolysis

Pyrolysis is a thermochemical process that produces solid carbon-rich biochar, bio-oil, and non-condensable gases by heating lignocellulosic biomass, such as rice straw or rice husk, between 300 and 10,000 °C in an oxygen-free or oxygen-limited atmosphere. The distribution of the products is primarily influenced by temperature, heating rate, and vapor residence time (Bushra and Remya 2024;

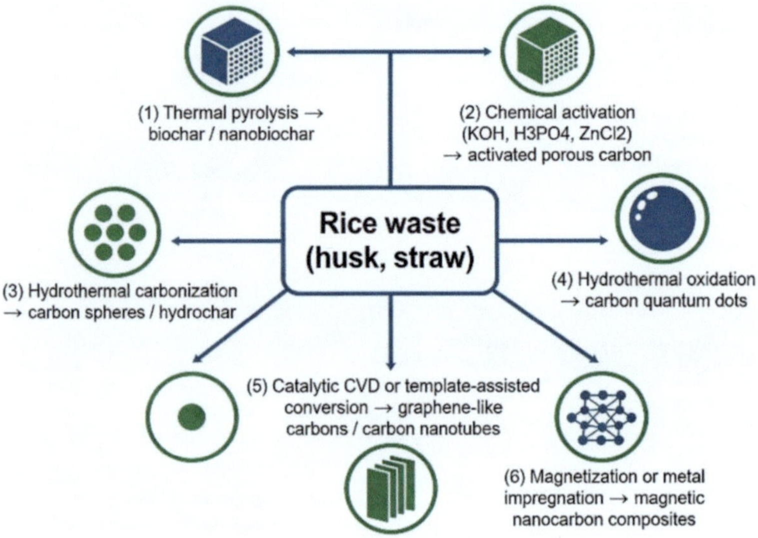

Fig. 7.5 Synthesis techniques for carbon nanomaterials from rice waste

Choudhary et al. 2025; Pandey et al. 2025; Zou et al. 2024). Biochar produced from slow pyrolysis (at low heating rates) and high residence times and moderate temperatures (400–600 °C) has been demonstrated to have significantly increased fixed carbon; lower H/C and O/C ratios; higher levels of aromatic compounds; and greater stability, making it an effective amendment to improve the quality of soils, capture carbon, and adsorb pollutants (Chen et al. 2025; Choudhary et al. 2025; Pandey et al. 2025; Sahoo and Remya 2022). In general, fast and flash pyrolysis at high temperatures to create bio oil and gas by using high heating rate and high temperatures (500–800 °C), this method produces both; however, increasing temperature regimes that exceed approximately 800–900 °C, while reducing the amount of char yield, also significantly promote the rate of graphitization, the development of microporous structures, and the retention of oxygen in the form of stable carbonyl and carboxyl groups which make biochar an advanced material for catalytic use, energy storage, and metallurgical applications (Choudhary et al. 2025; Herrera et al. 2022; Mubarik et al. 2021; Zou et al. 2024).

There are many different methods to tweak the yield and structure of biochar from different types of feedstocks including wet pyrolysis, hydrothermal carbonization, torrefaction, microwave-assisted pyrolysis, and co-pyrolysis of mixed agricultural residues by using different compositions of lignin, ash, and alkali/alkaline earth metals found in the various feed stocks used; by changing how the heating is applied to the biomass being processed; and also by adjusting the amount of ash produced during/after biochar production as well as surface functionality, porosity, and nutrient concentrations present in the final product which all have an impact on successfully using straw derived biochar as an environmentally sound agricultural practice.

7.3.2 Hydrothermal Carbonization

Hydrothermal carbonization (HTC) is a water-based thermochemical process conducted typically at 150–3000c and autogenous pressures in which wet biomass is converted into hydrochar, an aqueous phase rich in organics and nutrients and a minor gas phase, avoiding the energy-intensive drying required for conventional pyrolysis (Abe et al. 2023; González-Arias et al. 2022).

In HTC, hydrolysis of cellulose, hemicellulose, and lignin is followed by dehydration, decarboxylation, and condensation or polymerization, progressively lowering the H/C and O/C ratios of the solid and yielding a carbon-dense product suitable as a substitute for low-rank coal or a precursor for advanced carbons (Abe et al. 2023; González-Arias et al. 2022). Rice husk and rice straw have been extensively treated via HTC, giving high solid yields often 50–80 wt% with upgraded higher heating values; improved energy density; and, depending on conditions, enhanced mineral and nutrient content that support applications in wastewater treatment, energy, and soil management. Across the cited works, HTC of rice husk at 1800 °C and 70n bar yields about 58 wt% biochar with increased bulk density, surface area, and porosity, improved fuel indices, and reduced SiO_2 content, enabling simultaneous use as a solid fuel and as an adsorbent for metals and organics in water (Hossain et al. 2020). Review studies show that HTC of diverse wet wastes is thermodynamically and economically attractive compared with pyrolysis, particularly when integrated with anaerobic digestion, gas upgrading, or membrane concentration to valorize liquid and gas fractions (González-Arias et al. 2022; Zhang et al. 2024). For environmental materials, HTC of rice husk under tailored conditions can generate hydrochars and derived activated carbons or mesoporous carbon spheres with hierarchical porosity and graphitic domains, suitable for supercapacitor electrodes, hydrogen storage media, and catalyst or photocatalyst supports (Kızılduman et al. 2021; Mubarik et al. 2021).

7.3.3 Chemical Activation

Chemical activation of rice waste biochar has emerged as a central strategy to transform low-cost residues such as rice husk and rice straw into high-performance sorbents and activated carbons for environmental applications and resource recovery. Rice husk biochar obtained by slow or fast pyrolysis typically exhibits modest surface area and partially blocked pores and is therefore subjected to posttreatments including physical activation with steam or CO_2, alkali, or acid activation and metal impregnation to tailor porosity and surface chemistry. Steam and CO_2 activation partially gasify amorphous carbon, widening existing pores and generating a more developed micropore network, while retaining aromatic carbon domains important for structural stability (Emenike et al. 2025; Li et al. 2023; Sahoo and Remya 2022).

Alkali activation using NaOH or KOH is particularly effective in removing ash and unpyrolyzed organics from rice-derived biochar, increasing surface area and enhancing micropore volume and basic surface sites that promote adsorption of nonpolar organic pollutants and some metal ions. KOH activation of lignocellulosic precursors analogous to rice residues can raise BET surface areas into the 500–800 m^2/g range with pore development driven by dehydration, decarboxylation, and redox reactions between hydroxides, and similar reaction schemes are applied in feasibility studies on algal and other biochar. Acid activation with mineral acids such as H_3PO_4, H_2SO_4, or HNO_3 increases hydrophilicity and oxygen-containing groups (carboxyl, lactone, phenolic), improving affinity for polar contaminants and nutrient species but may partially block pores through hydration of surface functionalities (Bushra and Remya 2024; Li et al. 2023; Premchand et al. 2024; Tsarpali et al. 2024).

Metal and metal-oxide functionalization of rice waste biochar and fabrication of magnetic or composite materials further diversify functionality, coupling ion exchange, surface complexation, and redox with the underlying porous carbon matrix. Such engineered sorbents show markedly enhanced uptake of heavy metals, oxyanions, and dyes while magnetic phases facilitate easy recovery and regeneration. Reviews on rice husk biochar and rice husk-based biochar consistently emphasize that combining appropriate pyrolysis or HTC conditions with targeted chemical activation is crucial to optimize surface area, pore architecture, functional groups, and stability for applications ranging from water and wastewater treatment to soil remediation and soil release fertilization.

7.3.4 Catalytic Chemical Vapor Deposition

Catalytic chemical vapor deposition (CCVD) is the most widely used and industrially scalable route for carbon nanotube (CNT) synthesis. Process variants include thermal CVD, plasma-enhanced CVD, hot or cold wall reactors, floating catalyst, and fluidized bed CVD, each offering additional control over alignment, diameter distribution, and purity, but at the cost of higher complexity and energy demand. Recent work has extended CCVD to renewable or waste feedstocks such as rice straw and rice husk hydrochars, vegetable and essential oils, turpentine, and other biomass-derived volatiles using in situ Fe or Ni catalysts supported on hydrothermally treated lignocellulosic residues, enabling CNTs with diameters typically in the 20–70 nm range and bundles anchored on porous carbon frameworks. These biomass-based CCVD routes reduce reliance on fossil hydrocarbons and exploit inherent mineral or metal content as sustainable catalyst sources (Baghel et al. 2022; Fathy 2017; Licht and Licht 2025; Mostafavi et al. 2022; Mubarik et al. 2021). Microwave-assisted and microwave plasma CVD constitute a complementary CCVD family in which selective volumetric heating and plasma activation promote ultrafast CNT growth at somewhat lower bulk temperatures and can be combined with biochar or waste polymers as solid precursors. However, CCVD and its derivatives remain energy-intensive and carbon positive; life cycle analyses indicate significantly higher energy requirements and greenhouse gas footprints per gram of CNT than for bulk metals, motivating comparisons with emerging electrochemical CO_2 to CNT routes and driving research toward greener catalysts, lower temperature operation, and biomass-based carbon sources.

7.3.5 *Comparative Evaluation of Methods*

Pyrolysis, hydrothermal carbonization, chemical activation, and catalytic chemical vapor deposition form a complementary toolbox for upgrading rice husk and straw into high-value carbon nanomaterials instead of open-field burning or landfilling.

Pyrolysis is the simplest route, thermally decomposing rice residues in limited oxygen to give biochar, gases, and oils. Biochar from rice waste can be further converted into graphene oxide, CNT, or carbon dots and is already used as a low-cost adsorbent for water purification and gas storage, thereby diverting large rice waste streams from uncontrolled combustion. However, conventional pyrolysis alone offers limited control over nanostructure and often requires post-processing (Mubarik et al. 2021). Hydrothermal carbonization (HTC) treats rice residues or pulps in hot compressed water (180–2500 °C) to produce oxygen-rich hydrochars with abundant surface functional groups. These hydrochars act as excellent supports or in situ precursors for CNT growth and other nanocarbons, as shown for hydrothermally treated rice straw used as a substrate for CNT bundles in subsequent CVD or pyrolysis. HTC therefore enables wet, low-temperature valorization of high-moisture rice wastes close to the point of generation (Fathy 2017; Mubarik et al. 2021). Chemical activation, particularly KOH activation of rice husk or rice husk ash, generates highly porous carbons and graphene-like materials with very high specific surface areas and well-developed micro- or meso-porosity. These carbons are applied in supercapacitors, methane or CO2 adsorption, and heavy metal removal, turning problematic husk accumulations into sorbents and electrodes (Mubarik et al. 2021). CCVD uses rice-derived chars or pretreated straw (often after HTC) as both carbon source and catalyst support, feeding camphor vapor or other botanical hydrocarbons over Fe or Fe-Ni nanoparticles at 700–9000c to grow CNT with controlled diameters and graphitization. Such biomass-assisted CCVD significantly reduces reliance on fossil feed stocks and expensive substrates while generating high-value CNT for composites, adsorbents, and energy devices (Fig. 7.6).

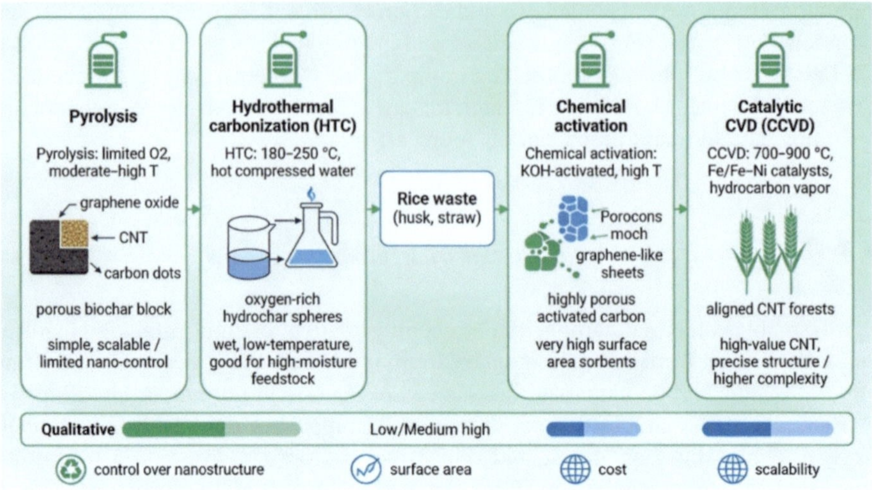

Fig. 7.6 Comparative evaluation of synthetic technique methods

7.4 Types of Carbon Nanomaterials Obtained from Rice Waste

Carbon nanomaterials produced from biomass are eco-friendly, affordable, and possess unique characteristics like a large surface area for functionalization with different functional groups and minerals (Wang et al. 2018b). Many methods are used to produce carbon nanomaterials from biomass, including physical activation, chemical activation, and hydrothermal carbonization. Rice husk, rice straw, and related residues are one of the most common and plentiful sources for creating viable carbon nanomaterials such as carbon nanotubes, carbon dots, graphene, graphene quantum dots, and carbon quantum dots (Mubarik et al. 2021) (Table 7.2).

7.4.1 Carbon Nanotubes (CNTs)

Carbon nanotubes (CNT) were first detected by Iijima in 1991 (Chakroborty et al. 2023). CNTs are regarded as an important category of carbon nanomaterial based on their one-dimensional character—their outstanding properties and performance characteristics as an electrical conductor, chemical reaction product, thermal conductivity, mechanical support structure, optical devices, and electronic circuits (Chakroborty et al. 2023). Rice waste is a suitable material for creating CNTs (carbon nanotubes). Rice husk contains approximately 45–55% carbon, which is a significantly greater percentage than most biomass materials (Tiwari et al. 2016). Because of its low cost, rice husk is also a good choice as a starting material for creating CNTs. Scientists have already successfully produced CNTs using methods

Table 7.2 Various versatile carbon-based nanomaterials from rice residue, i.e., rice husk and rice straw, their synthesis methods, and applications

Rice biomass waste	Method of synthesis	Carbon nanostructure	Application	References
Rice husk	Combustion/ chemical activation with KOH	Graphene sheets	Additives in polymer composites, electrochemical performance	Wang et al. (2013)
Rice husk	Microwave CVD	Thin Graphene (2–6 layers) MWCNTs	–	Debalina et al. (2017)
Rice husk	Pyrolysis	C-dots	Bio-imaging	Wang et al. (2016)
Rice husk	Hydrothermal, pyrolysis	GQDs	Fluorescent sensor	Wang et al. (2016)
Rice husk	KOH activation	Graphene oxide	Desalination membrane	Mansurov (2020) and Seitzhanova et al. (2019)
Rice husk	Pyrolysis	Graphene oxide	–	Naik et al. (2020)
Rice straw	Conventional CVD	CNTs	–	Fathy (2017)
Rice straw	Microwave-assisted	Graphene oxide	Ni(II) adsorption	Tohamy et al. (2020)
Rice grains	Pyrolysis	GQDs	–	Kalita et al. (2016) and Wang et al. (2018a, b)
Rice residue	Hydrothermal	C-dots	Fe^{3+} ions and tetracycline detection	Qi et al. (2019)

that include catalytic CVD, pyrolysis (typically following some form of pretreatment), and then also microwave-assisted plasma techniques. In one example, bundles of well-graphitized multi-walled carbon nanotubes (CNTs) with outer diameters between ~22 and 60 nm were created from the use of a catalyst-impregnated hydrochar made from rice straw in a chemical vapor deposition (CVD) process (Mubarik et al. 2021). This process involves the use of camphor vapor.

7.4.2 *Carbon Quantum Dots (CQDs)*

Carbon quantum dots (CQDs), which are semiconductor nanomaterials, are a byproduct of the purification process of single-walled carbon nanotubes, which occurred in 2004. Carbon quantum dots are spherical and have sizes of less than 10 nanometers. Because they are composed of carbon and their size is in the nanometer region, carbon quantum dots can have crystalline and amorphous structures (including a nanocrystalline phase) (de Oliveira and da Silva Abreu 2021). Both structures exhibit some unique properties, including strong emission of photoluminescence, wavelength-dependent photoluminescence, excellent solubility, low toxicity, ease of functionalizing, and biocompatibility (Kim et al. 2015). Carbon quantum dots derived from green biomass waste have many benefits over those made from traditional chemical carbon sources including plenty of feedstock, cheaper production costs, and higher levels of biocompatibility and water solubility (Wu et al. 2023). The method of preparing CQDs from rice straw is often adopted by different researchers using varying methods. Kumari et al. (2024) reported that the storage time of rice straw affected the carbonization and following ultrasonication process. CQDs synthesized using this method had a quantum yield (QY) value of 64% and exhibited excitation-dependent emissions which were used to fluorosensitively monitor the ellagic acid (EA) present in cosmetic products. CQDs have been proven to be biocompatible for use with breast cancer cell lines (MCF-7). This means that CQDs can be used for sensing and bioimaging purposes, while the rice husk CQDs mentioned above were prepared in such a way as to increase the durability of an epoxy-based coating. The CQDs produced were found to be 8.3 nm in size and had a QY of 56.2% along with high photostability and excitation-dependent fluorescence. The CQDs produced were used to selectively detect Fe^{3+} ions via fluorescence quenching with a measured detection limit of 149 nM. Fluorescence detection was carried out through charge transfer and hydrogen-bonding mechanisms created by the fluoroquinolone compounds. CQDs produced from the residue rice (2.7 nm with QY of 2.48%) were used to sense Fe^{3+} ions, detect tetracycline, and bioimage HepG2 cells (Tripathi et al. 2025; Kundu et al. 2022).

7.4.3 Graphene-Like Nanosheets

Graphene has been labelled as a wonder material of the twenty-first century, just as plastic was labelled as a wonder material of the twentieth century (Raghavan et al. 2017). Graphene is composed of a single atomic layer of carbon atoms bonded using sp^2 hybridization and arranged in a hexagonal lattice, which creates an incredibly large surface area, excellent electrical conductivity, excellent chemical stability, and a tremendous amount of mechanical strength (Avouris and Dimitrakopoulos 2012), and can also be derived from rice waste, often referred to as "graphene-like" nanosheets because of their morphology and structure. The method used to create graphene using agricultural waste material such as rice husk is by using potassium hydroxide (KOH), because the properties of KOH have been shown to aid in the production of enhanced pore structures for carbon-based materials (carbon nanotube, graphene, and carbon fiber) which will therefore improve their electrochemical properties. KOH also produces synthetic, high-purity, stable, and clean-edge graphene materials (Arifin et al. 2020). In summary, rice waste has the potential to be converted into the type of graphene nanosheets used in many applications by applying the activation process and/or oxidation process for creating this "wonder material" (Raghavan et al. 2017).

7.4.4 *Activated Porous Carbon*

Activated carbon (AC) has many applications due to its porous nature, i.e., its ability to absorb contaminants from the environment and to prepare electrodes for supercapacitors and other applications (Yan et al. 2015). Also AC can provide support for the catalytic activity of several organic compounds and can be used as a platform for developing new ways to store energy (e.g., batteries, fuel cells, etc.). In addition, AC can withstand very high temperatures and has an abundance of chemical functional groups, which create the potential for multiple uses in development and application (Tsai et al. 2019). Activated carbons of micro- to mesoporous structure can be produced from rice waste using carbonization (pyrolysis, HTC) and activation (physical using CO_2/steam, chemical using KOH, etc.) (Rybarczyk et al. 2016). Activated carbon produced from rice husk (RH) offers many advantages, including being both plentiful and inexpensive and has been used to remove many types of pollutants, including textile dyes, organic compounds, inorganic ions, pesticides, and heavy metals. Activated carbon produced from rice husk is also used in supercapacitors, lithium-ion battery electrodes, catalytic supports, energy storage devices, etc.; thus, the cost-effective synthesis of activated carbon presents potential benefits to AC production (Alam et al. 2020).

7.5 Physicochemical Properties of Rice Waste-Derived Carbon Nanomaterials

Rice waste-derived carbon nanomaterials have excellent physicochemical properties that are used for a wide range of applications. Owing to their physicochemical properties, i.e., great chemical stability, increased electrical conductance, manageable porosity, increased specific area, and impenetrable electroactive sites, these promising nanomaterials have acquired a lot of interest. Carbon nanomaterials made from biomass offer more favorable future application opportunities than many other types of carbon nanomaterials due to their affordability, environmentally descent manufacturing features, and extraordinary characteristics, including many functionalized surface areas for many functionalized groups and minerals to be applied. Rice from biomass can be used to create multiple different types of carbon nanomaterials (like carbon quantum dots, graphene, carbon nanotubes, graphene quantum dots), and it is used throughout the world for carbon-based nanomaterials. The production of bio-carbon-based nanomaterials varies in its manufacturing

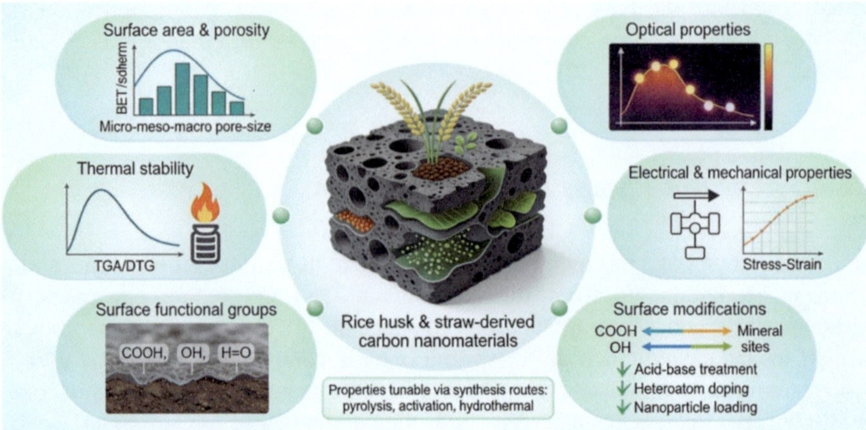

Fig. 7.7 Physicochemical properties of rice waste-derived carbon nanomaterials

techniques with three main methods of biomass-derived carbon nanomaterials: physical activation, chemical activation, and hydrothermal carbonization (Fig. 7.7) (Mubarik et al. 2021). Below is an overview of each aspect:

7.5.1 Surface Area and Porosity

Surface area and porosity are critical physical properties of biochar, influencing its performance in various applications, including wastewater treatment and soil remediation. Pore size distribution is a key factor in characterizing the structural heterogeneity of biochar. Notably, the surface area is closely linked to porosity, with micropores contributing most significantly (Leng et al. 2021). The ratio of the total surface area to the total mass of the biochar particles is known as the specific surface area (SSA), and it is expressed in m^2/g. Usually, Brunauer-Emmett-Teller (BET) analysis is used to determine it (Mobarak et al. 2024). Surface area and porosity of rice waste-derived carbon nanomaterials vary widely depending on feedstock, activation method, and target application, ranging from very low for raw biochar to ultrahigh for chemically activated carbons and graphene-like structures. Rice husk and straw can yield porous carbons, activated carbons, carbon dots, graphene-like carbons and CNT-based materials with specific surface areas from $<5\ m^2/g$ up to $>2000\ m^2/g$, and pore volumes typically between about 0.01 and 3 cm^3/g, dominated by micro-mesoporous networks tuned by activation conditions (Shen 2018). Simple pyrolysis of rice straw or husk without activation often produces low surface area biochar ($\sim$1–5 m^2/g) with small pore volumes (~0.008–0.01 cm^3/g), mainly meso-macroporous structures suitable for bulk adsorption or soil amendment. Despite low BET values, these materials can still show good removal of metals such as Pb^{2+} due to surface functional groups and accessible macro/mesopores (Singh et al. 2025).

Table 7.3 Textural properties of activated carbons and biochars derived from rice waste feedstocks prepared using different chemical activation agents

Feedstock	Activation	S_BET (m^2/g)	V_total (cm^3/g)	Micropore vol. (cm^3/g)	Avg. pore size (nm)	References
Rice husk	KOH	755	0.393	0.243	0.3–0.6	Nandi et al. (2023)
Rice husk	KOH	3145	–	–	–	Gao et al. (2015)
Rice straw	K_2CO_3	>2000	>1.6		–	Motlagh et al. (2021)
Rice husk biochar, raw	–	228	Not specified		Micro-meso	Longo et al. (2025)
Rice straw	K_2FeO_4	440.5	0.274		Meso-microporous	Bao et al. (2024)

KOH-activated rice husk-derived porous carbon can reach surface areas around 755–1007 m^2/g with total pore volume ~0.39–1.5 cm^3/g, predominantly microporous with some mesoporosity, favorable for CO_2 capture, and adsorption of small organic molecules (Nandi et al. 2023; Sudhan et al. 2017). Other activation routes (e.g., optimised RHAC-90 M) report BET surface area ~ 588 m^2/g and pore volume ~0.37 cm^3/g, confirming that chemical activation greatly enhances textural properties compared to the parent biochar (Yadav and Kumar 2025). Consequently, rice straw-derived biochar and carbon dots prepared in a combined route show relatively low BET surface area for the biochar (~1.71 m^2/g) but still effective Pb^{2+} removal, while the nanodots contribute high surface reactivity at the nanoscale (Table 7.3) (Singh et al. 2025).

7.5.2 *Thermal Stability*

Thermal stability of carbon nanomaterials is due, in large part, to the strength of covalent bonds among carbon atoms, specifically the sp^2 hybridization of carbon-carbon bonds forming graphene-like hexagonal lattices (Painuli et al. 2022). In the case of graphene and carbon nanotubes, each carbon atom is connected to three adjacent carbon atoms by σ-bonding; also, they are associated with π-electrons that remain collectively in an area above and below the plane of carbon atoms making up the hexagonal lattice (Tian et al. 2025). Bond dissociation energies of these bonds are extremely high, so a significant amount of thermal energy must be applied to disassociate them. In addition to bond strength, another mechanism contributing to the thermal stability of carbon nanomaterials is the high degree of crystallinity and graphitization in many carbon nanomaterials. Very high degrees of graphitization create ordered domains that facilitate the transport of phonons through heat dissipation, which reduces localized thermal stresses that could otherwise cause structural failure. Thus, the ability of CNM's to transport heat quickly versus that of conventional materials results in less structural damage resulting from heat exposure. A third factor influencing the high thermal stability of carbon-based nanomaterials is the very low density of defects in the structure. Defects such as voids, grain boundaries, or oxygen-containing functional groups act as reactive sites for the initial oxidation or thermal decomposition of CNMs. Generally, the fewer the defects present within a CNM, the greater will be its thermal stability due simply to the ability of an ordered defect-free lattice to prevent premature failure of the material (Mishra and Verma 2024).

7.5.3 Optical Properties

Optical properties of the carbon dots were established using ultraviolet/visible light (UV-Vis) and fluorescence spectroscopy analysis techniques (Mubarik et al. 2021). Rice husk-derived CQDs show dual-emissive fluorescence with peaks around 430–450 nm (blue emission at ~350 nm excitation), excitation-dependent red shifts (220–380 nm excitation wavelengths), and high quantum yields enhanced by N/S-doping or metal doping (Singh et al. 2025). N/Bi-doped rice husk CQDs exhibit tunable PL for photocatalysis, with emission intensity modulated by surface passivation and doping for methylene blue degradation (~70% efficiency) (Hui et al. 2021). Rice straw-derived graphene shows broad XRD-confirmed amorphous carbon peaks and flake morphology, with optical bandgap tunability from sp^2-hybridized domains, suitable for photothermal or sensing applications (Uda et al. 2021). Porous carbons from rice husk/straw display weak visible fluorescence under UV but strong near-IR/electroluminescence potential post-activation, with absorption edges indicating bandgaps 3.2 eV influenced by graphitization degree (Mercy et al. 2023). Doping (N, S, Bi) and hydrothermal conditions enhance PL quantum yield, shift emission wavelengths, and improve photostability against bleaching, while surface functional groups prevent aggregation-induced quenching (Hui et al. 2021; Yang et al. 2024; Zhou et al. 2020). These characteristics facilitate applications in photocatalysis, antibiotic detection (such as ofloxacin/ciprofloxacin), heavy metal sensing (Pb^{2+} quenching via complexation), and bioimaging (Kundu et al. 2022; Singh et al. 2025). The optical characteristics of carbon nano-materials made from rice are predominantly evident in the nanoscale region and have an extremely high degree of fluorescence. This allows them to be employed as fluorescent sensors or for biological imaging. Modification of these carbon nano-dot materials provides the opportunity for the creation of highly photostable materials that emit multiple colors when stimulated by light. The optical structures of larger carbon structures (graphene and carbon nanotubes) provide optical properties for applications mainly through the absorption of light or by being part of optical devices (e.g., conductive transparent films). It should be pointed out that waste-derived precursors to make the nanostructures do not adversely affect the optical performance of the carbon nanostructures, such as rice straw carbon quantum dots (CQDs), are able to function as highly sensitive fluorescent sensors (Mubarik et al. 2021).

7.5.4 Surface Functional Groups and Modifications

Rice waste-derived carbon nanoparticles from rice husk exhibit oxygen-rich surface functional groups such as hydroxyl (–OH), epoxy (C–O–C), carbonyl (C=O), and carboxyl (–COOH), identified via FTIR and XPS, which enhance adsorption and catalytic properties. Modifications like heteroatom doping (N, P, S) or acid activation introduce tailored groups for improved dispersibility and reactivity. These features make them ideal for environmental remediation and energy applications (Angellinnov et al. 2025; Longo et al. 2025; Mesa et al. 2025; Mubarik et al. 2021; Negm et al. 2024). The research completed by Thongsai et al. has shown that carbon dots can be synthesized easily using a hydrothermal synthesis technique. The carbon dots generated through this technique are brown solids with approximately 3% quantum yield, have diameters of only 4–5 nm, and exhibit extreme photostability when illuminated under UV-Vis light. Using X-ray photoelectron spectroscopy and Fourier transform infrared spectroscopy (FT-IR), it was demonstrated that the external surface of the carbon dots contained hydrophilic functional groups. Due to their hydrophilic properties, these functional groups have a high affinity toward small molecules, which makes them ideal chemical probes. Additionally, the strong interaction between the hydrophilic groups of the carbon dots and small molecules is what produces the emission characteristics of the carbon dots (Mubarik et al. 2021; Pal et al. 2018; Thongsai et al. 2019; Wu et al. 2013). Carbon nano-sorbents like graphene, carbon nanotubes, graphene oxide, and carbon dots are extensively used for purifying water due to their high surface area. From these carbon nanomaterials, GO is emerging as a promising candidate for use as an adsorbent and has many uses. Because GO has a unique structure that includes several types of reactive oxygen-containing functional groups, including carbonyl (–C=O), phenol, and carboxyl (–COOH) at the edge of each sheet and epoxy (C–O–C) and hydroxyl (–OH) on each basal plane, it is possible to modify the surface of GO to make it hydrophilic. Graphene oxide (GO) is very effective in dispersing water, and it can be easily modified in numerous ways to create a broad range of possible uses (Chigo-Anota et al. 2011; Mubarik et al. 2021). These functional groups can be beneficial: they provide active sites for adsorption of pollutants (via electrostatic attraction or covalent bonding) and enable further chemical functionalization (e.g., attaching polymers, nanoparticles, or doping with heteroatoms) (Rahdar et al. 2025). In the case of activated carbon, surface functional groups (often acidic or basic groups depending on activation and posttreatment) influence adsorption properties and catalysis (Yang et al. 2019). Functionalization of CNTs (f-CNTs) from rice waste—e.g., by acid treatment to attach carboxyl groups—greatly improves their dispersion in solvents and interaction with contaminants or matrix materials. In summary, rice waste-derived carbon nanomaterials (CNMs) or rice-straw-derived graphene oxide (GO) have many functional groups available for chemical modification or treatment and chemical/functional group addition and can be integrated into creating composite films. These materials exhibit a wide array of reactivity and versatility due to their abundance of surface functional groups; therefore, they are ideal for many applications (Akram et al. 2025; Mubarik et al. 2021). For example, rice straw GO (RSGO) can exhibit reactivity/affinity with various metal ions (through the oxygen-containing groups on the GO molecule) or be crosslinked into composite films, which makes them extremely useful materials for the development of the present invention (Mubarik et al. 2021).

7.6 Applications of Rice Waste-Derived Carbon Nanomaterials

Beyond classical biochar, rice husk and straw serve as precursors for graphene, graphene oxide, carbon nanotubes, and carbon dots, which provide higher surface areas, diverse functionalities, and enhanced optical or electrochemical properties that can be exploited in advanced water treatment, sensing, and energy storage applications. Collectively, these studies position rice materials that address environmental remediation while simultaneously mitigating residue disposal problems in rice-based agroecosystems (Barzallo et al. 2025).

7.6.1 Environmental Remediation

Carbon nanomaterials derived from rice wastes are being increasingly recognized as low-cost environmental remediation techniques owing to their large surface area, high porosity, and rich surface chemistry. This ability has allowed them to adsorb metals, dyes, and microbial agents, making them an excellent candidate for environmental remediation processes (Baby et al. 2019; Regkouzas et al. 2023). Activated carbon materials derived from rice wastes have well-developed mesoporous structures and oxygenated functional groups, which result in their high adsorption capacities for heavy metals.

Carbon nanotubes (CNTs) are carbon-based nanostructures that have specific properties including an incredible high tensile strength, excellent thermal and electrical conduction capabilities, and great chemical stability. CNTs are classified as either single-walled (SWCNT) or multi-walled (MWCNT). MWCNTs are made up of bundled SWCNTs. CNTs are a potentially effective means of environmental remediation, particularly for removing pollutants like heavy metals, pesticides, and estrogen from aquatic environments (Chung et al. 2022). Due to their increased adsorption sites and highly porous structure, the CNT and CNT-based membranes have high rejection rates for heavy metal ions like Cr, As, and Pb (Baby et al. 2019). These capabilities make them a good candidate for water pollutant adsorption and filtration system. When these CNTs are incorporated into membranes, they yield porous and high-flux filters (Khan et al. 2021). Despite some health and environmental concerns, their applications play a crucial role in sustainable pollution management. CNTs serve as contaminants adsorbents (membrane filtration, catalysis, disinfection, pollution detection systems) in addition to being used to produce green alternatives to traditional environmental protection methods and to enhance the quality of drinking water. Research on the environment has indicated that the use of CNT products or technologies could provide the opportunity to improve the quality of the air, water, soil, and other natural resources while also helping to reduce the number of air pollutants released into the atmosphere. CNT technologies provide several solutions for both remediation of various types of contaminants and their associated toxins. Carbon nanotubes (CNTs) provide an effective means of removing contaminants from the environment and treating water using CNTs as an adsorbent. Adsorption provides one method for treating pollutants, but it also allows the removal of gases, liquids, and organic pollutants. Additionally, CNTs can be used for membrane filtration, disinfection, catalytic, and monitoring and sensor applications. Heavy metals are considered a pollution source (toxic substance) regardless of how low they are in concentration because the biological effects and/or physiological impact of the excess metal in the environment are harmful to living organisms (Ali et al. 2021; Oladimeji et al. 2024). The biological impact and/or prolonged exposure of an organism will cause an accumulation of specific heavy metals in the organism's body and will result in a toxic accumulation within the tissues. Accumulated toxic metals will inhibit and/or cause the dysfunction of specific body systems and ultimately lead to disease. In summary, the conclusion drawn regarding improper management of the industrial waste from mining and metals is the major cause of environmental and biological problems related to heavy metals (Briffa et al. 2020; Singh et al. 2011).

CDs, a distinct fluorescent nanomaterial, are promising for colorimetric and fluorescent detection and can also be used to remove contaminants from the environment. The functions of the CD surfaces are also extremely important for environmental pollution control to the extent that CD surface functional groups (e.g., carboxyl, hydroxyl, and amine) are capable of interacting with pollutants via different routes (e.g., adsorption, surface modification, and chemical reaction) (Madima et al. 2020). CDs play an essential role in the wastewater treatment industry through different approaches like photocatalysis, adsorption, and/or advanced composites. A straightforward and adaptable method was created by Issaka and co-workers for fabricating a composite film of polyvinyl alcohol (PVA)-nanosilica carbon dots (NCDs) for capturing cadmium ions from industrial wastewater, utilizing carboxymethyl cellulose (CMC) derived from palm oil palm empty fruit bunches as a binder for the composite dot film. After elution with 0.1 molar hydrochloric acid, the CD film attained a maximum capacity of 113.6 milligrams per gram (mg/g) for adsorbing cadmium ions from a pH-neutral (7.0) aqueous solution. The CD film removed 95.34% of the cadmium ions from tap water, while 90.9% were removed from drinking water (Issaka et al. 2022). Another study by Abidin and coworkers showed that functionalized QCDs derived from rice straw can help improve Cd binding and quenching, thus removing Cd from water (Abidin et al. 2020). Rice waste-derived biochar effectively binds pesticides and toxins, thereby reducing leaching and contamination risk into the soil (Baby et al. 2019). Nanomaterials derived from rice wastes can induce oxidative stress, therefore damaging the bacterial cell membranes. Rice husk ash impregnated with silver nanoparticles can help in removing contaminants; hence, it is being used in filtration units to treat floodwater to achieve safe drinking water (Dao et al. 2021). Similarly, activated carbon exhibit strong CO_2 and VOCs uptake, thus being used in air purification systems. This overall property not only reduces pollution but also adds value to agricultural waste, thus setting up sustainability and following green environmental approaches in remediation methods (Regkouzas et al. 2023).

7.6.2 Sensors and Biosensors

Carbon quantum dots (CQDs) are considered excellent candidates for use in biosensing, bioimaging, and drug/gene therapies because they are small, less than 10 nanometers in size, and they possess all of the qualities that make them attractive for fluorescence emission, photo and chemical stability, ability to dissolve in water, and be found in biological systems (Azam et al. 2021; Tavan et al. 2025). The properties of CQDs allow for many different types of chemistry that can also modify surface properties or introduce additional atoms into the carbon lattice. This paper provides a thorough review of recent advances in CQDs with respect to the development of sensitive detectors for target analytes. In particular, we highlight CQDs derived from waste materials (biomass) as a source of low-cost, renewable carbon to produce CQDs that are not only very efficient for biosensing but also help to create valuable products and minimize environmental impacts. Low-cost carbon sources can reduce sensor/biosensor costs by replacing expensive and sometimes toxic metal precursors with CQDs. CQDs are an area of great interest in the areas of theranostics and biosensing (Meng et al. 2019; Šafranko et al. 2021).

7.6.3 Catalysis

Rice processing residues such as husk and straw have emerged as versatile precursors for engineered carbon nanomaterials offering simultaneous solutions for waste management, pollution control, and low carbon energy systems. Thermochemical routes, particularly pyrolysis and hydrothermal carbonization, convert these silica and carbon-rich feedstocks into (Chung et al. 2022) biochar or hydrochar with internal porosity, surface functionality, and mineral composition, which can be further upgraded to nanostructured carbons and catalytic composites (Bushra and Remya 2024; Sharma et al. 2024).

Recent work on rice straw-derived carbons highlights the role of activation chemistry in generating high-performance nanocarbons for electrochemical energy applications. Zinc halide activation of hydrothermally pre-carbonized straw, followed by high-temperature pyrolysis, produces N-doped carbons with surface areas up to 1400 $m^2\ g^{-1}$, abundant micro- or meso-porosity and enriched graphitic or pyrrolic nitrogen delivering oxygen-reduction activity and durability comparable to

Pt/C in alkaline media. This demonstrates hoe intrinsic N in biomass that can be exploited for in situ heteroatom doping, while avoiding complex synthetic precursors. Parallel developments in engineered biochar show that posttreatments such as gas-steam activation, ball milling, heteroatom doping, and nanoparticle impregnation can raise surface into the 1000–2000 m^2 g^{-1} range, create tailored functional groups, and introduce redox-active sites, enabling applications from supercapacitor electrodes to adsorbents and catalysts (Ou et al. 2024; Sharma et al. 2024).

7.7 Challenges and Future Directions

Large volumes of rice husks, agricultural waste from milling rice, are produced annually worldwide, and a large portion of this biomass is burned, leading to pollution of the environment. Because of their carbon content and silica levels, rice husks have started to gain popularity as a sustainable, inexpensive starting material for making carbon nanomaterials (CNMs) out of rice husks. However, creating quality CNMs from rice husks is challenging. One of the main issues is the pretreatment required to properly prepare rice husk as an efficient feedstock prior to chemical conversion into nanomaterials. The pretreatments that need to occur (washing, chemical activation, acid/alkaline treatment, and thermal processing) are required for the cleaning, lowering silica, and modification of the biomass prior to being converted to CNMs. The pretreatment of the rice husks improves the quality and yield of the CNMs but also increases the cost of producing CNMs, increases the energy consumption of producing CNMs, and increases the time to produce CNMs. Excessive pretreatment may lower the amount of biomass available and may limit the ability to produce in large quantities or scale. Increasing interest in new methods for producing carbon nanomaterials from rice husks is driving innovation since these methods typically require very little or no pretreatment. Researchers are looking for ways to create simpler processes of producing CNMs, lowering the operational costs compared to current methods, and using more sustainable biomass-derived raw materials. Examples of these process improvements include direct pyrolysis, hydrothermal carbonization, and microwave-assisted synthesis techniques to convert raw rice husks into different types of CNMs including carbon nanotubes, graphene-like materials, carbon quantum dots, and activated carbon. Rice husk-derived CNMs have been found to possess a variety of desirable properties, including high surface area, good electrical conductivity, good chemical stability, and tunable porosity. Due to their properties, rice husk-derived carbon nanomaterials may be suitable for a range of advanced applications. Examples of potential applications are as components in energy-storage devices (e.g., supercapacitors and lithium-ion batteries), as filtration materials for water purification and environmental remediation, as heterogeneous catalysts, and for gas-sorption technologies. Because of their biocompatibility and functional surface chemistries, rice husk-derived CNMs may also be suitable for various biomedical applications such as biosensor and drug-delivery systems and bioimaging. Overall, using rice husks to produce carbon nanomaterials represents a sustainable method for transforming a waste material into a high-value product while also addressing environmental and waste management issues.

7.8 Conclusion

Carbon nanomaterials produced from rice waste are a unique and sustainable resource that can be exploited in the production of useful functional nanomaterials from the enormous amount of agricultural waste. Rice straw and husks contain a large amount of silica and carbon and can therefore be used as low-cost, clean, and renewable precursors for the manufacture of graphene, carbon nanotubes, carbon dots, and activated carbon. There are several environmentally friendly synthetic methods for producing carbon from rice husks and straw, including pyrolysis, hydrothermal carbonization, chemical activation, and biomass-assisted chemical vapor deposition. Using one of these methods allows for precise control of the carbon nanomaterials' structure, surface chemistry, and physicochemical characteristics. The resulting carbon nanomaterials are characterized by having high specific surface area, tailored porosity, good thermal stability, excellent electrical conductivity, and abundant functional groups. This means that carbon nanomaterials made from rice waste will be of great interest in environmental remediation, energy storage and conversion, sensing, catalysis, and biomedical applications; in fact, rice waste carbon nanomaterials can help reduce greenhouse gas emissions, prevent pollution, and comply with the principles of a circular economy. Despite this great opportunity presented by rice waste-derived carbon nanomaterials, there are still some barriers to be overcome in terms of industrial-scale production, energy efficiency, standardization of manufacturing processes, and providing life cycle assessment. Future studies should focus on the development of low-pretreatment methods for producing rice waste-derived carbon nanomaterials and incorporating them into scalable green manufacturing technologies and biorefinery concepts. Thus, rice waste-derived carbon nanomaterials represent a possible link between sustainable agricultural practices and advanced nanotechnology, while also addressing global environmental issues.

Declaration Artificial intelligence (AI) tools were used only for images in the preparation of this manuscript. No AI tools were used to generate any kind of content, including text and tables.

References

Abe H, Nakayasu Y, Haga K, Watanabe M (2023) Progress on separation and hydrothermal carbonization of rice husk toward environmental applications. Global Chall 7(8):2300112

Abidin NHZ, Wongso V, Hui KC, Cho K, Sambudi NS, Ang WL, Saad B (2020) The effect of functionalization on rice-husks derived carbon quantum dots properties and cadmium removal. J Water Process Eng 38:101634

Akram R, Anwar N, Arshad A, Khan MM (2025) Rice waste-derived carbon nanomaterials for environmental applications. In: Waste-derived carbon nanostructures: synthesis and applications. Springer, pp 177–192

Alam MM, Hossain MA, Hossain MD, Johir M, Hossen J, Rahman MS, Zhou JL, Hasan AK, Karmakar AK, Ahmed MB (2020) The potentiality of rice husk-derived activated carbon: from synthesis to application. Processes 8(2):203

Ali MM, Hossain D, Khan MS, Begum M, Osman MH (2021) Environmental pollution with heavy metals: a public health concern. In: Heavy metals-their environmental impacts and mitigation. IntechOpen

Ali SW, Bairagi S, Bhattacharyya D (2022) Valorization of agricultural wastes: an approach to impart environmental friendliness. In: Handbook of biomass valorization for industrial applications, pp 369–393

Angellinnov F, Subhan A, Rengga WDP, Jumari A, Syahrial AZ (2025) Effect of rice husk derived activated carbon surface coating on NMC 811 characteristics and performance as lithium-ion battery cathode. Results Surf Interfaces 19:100567

Arifin NFT, Yusof N, Ismail AF, Jaafar J, Aziz F, Salleh WWJ, Sciences A (2020) Graphene from waste and bioprecursors synthesis method and its application: a review. Malay J Fundam Appl Sci 16(3):342–350

Avouris P, Dimitrakopoulos C (2012) Graphene: synthesis and applications. Mater Today 15(3): 86–97

Azam N, Najabat Ali M, Javaid Khan T (2021) Carbon quantum dots for biomedical applications: review and analysis. Front Mater 8:700403

Baby R, Saifullah B, Hussein MZ (2019) Carbon nanomaterials for the treatment of heavy metal-contaminated water and environmental remediation. Nanoscale Res Lett 14(1):341

Baghel P, Sakhiya AK, Kaushal P (2022) Ultrafast growth of carbon nanotubes using microwave irradiation: characterization and its potential applications. Heliyon 8(10):e10943

Bao Z, Lotfy VF, Zhou X, Fu S, Basta AH (2024) Assessment of porous carbon from rice straw residues with potassium ferrate-assisted activation as cationic and anionic dye adsorbents. Ind Crop Prod 212:118298

Barzallo D, Carrasquero E, Andrade M, Heredia Jara DA, Palmay P (2025) Preparation and characterization of unactivated, activated, and γ-Fe2O3 nanoparticle-functionalized biochar from rice husk via pyrolysis for dyes removal in aqueous samples: comparison, performance, and mechanism. ChemEngineering 9(2):30

Bhardwaj AK, Thakur B, Tripathi SK, Turabayev ME, Aljulaih AA, Kharlamova TS, Kulinich SA (2025) Recent advances in valorizing agricultural waste: a sustainable approach. Waste Biomass Valoriz:1–28

Briffa J, Sinagra E, Blundell R (2020) Heavy metal pollution in the environment and their toxicological effects on humans. Heliyon 6(9):e04691

Bushra B, Remya N (2024) Biochar from pyrolysis of rice husk biomass – characteristics, modification and environmental application. Biomass Convers Biorefinery 14(5):5759–5770

Chakroborty S, Pal K, Nath N, Singh V, Barik A, Soren S, Panda P, Asthana N, Kyzas GZ (2023) Sustainable synthesis of multifunctional nanomaterials from rice wastes: a comprehensive review. Environ Sci Pollut Res Int 30(42):95039–95053

Chen L, Al-Dhabi NA, Ai W, Sun S, Peng Y, Gao C, Zhuo Y, Tang W, Zhou Y, Sun C (2025) Impact of low-temperature pyrolysis on ash and physicochemical properties of straw biochar: multivariate analysis and implications for agricultural and environmental use. Ind Crop Prod 224:120431

Chigo-Anota E, Salazar-Villanueva M, Hernández-Cocoletzi H (2011) Electronic properties of boron nitride oxide nanoclusters. J Nanosci Nanotechnol 11(6):5515–5518

Choudhary T, Singh P, Bhunia H (2025) Pyrolysis of rice straw: a critical review. Biomass Convers Biorefinery:1–12

Chung JH, Hasyimah N, Hussein N (2022) Application of carbon nanotubes (CNTs) for remediation of emerging pollutants – a review. Trop Aquat Soil Pollut 2(1):13–26

Dao MU, Hoang HY, Tran AK, Cong HH (2021) Assessment and treatment of floodwater in the Vietnamese Mekong Delta using a simple filter system based on silver nanoparticles coated onto activated carbon derived from rice husk. RSC Adv 11(63):39838–39847 de Oliveira BP, da Silva Abreu FOM (2021) Carbon quantum dots synthesis from waste and by-products: perspectives and challenges. Mater Lett 282:128764

Debalina B, Reddy RB, Vinu R (2017) Production of carbon nanostructures in biochar, bio-oil and gases from bagasse via microwave assisted pyrolysis using Fe and Co as susceptors. J Anal Appl Pyrol 124:310–318

Emenike EC, Adeleke JA, Iwuozor KO, Egbemhenghe A, Bello-Hassan MT, Adu AO, Okoro HK, Eletta OA, Adeniyi AG (2025) Application of rice husks-based biochar for removal of pollutants from water and wastewater. Sep Purif Rev:1–22

Fathy NA (2017) Carbon nanotubes synthesis using carbonization of pretreated rice straw through chemical vapor deposition of camphor. RSC Adv 7(45):28535–28541

Gao Y, Li L, Jin Y, Wang Y, Yuan C, Wei Y, Chen G, Ge J, Lu H (2015) Porous carbon made from rice husk as electrode material for electrochemical double layer capacitor. Appl Energy 153:41–47

González-Arias J, Sánchez ME, Cara-Jiménez J, Baena-Moreno FM, Zhang Z (2022) Hydrothermal carbonization of biomass and waste: a review. Environ Chem Lett 20(1):211–221

Herrera K, Morales LF, Tarazona NA, Aguado R, Saldarriaga JF (2022) Use of biochar from rice husk pyrolysis: Part A: Recovery as an adsorbent in the removal of emerging compounds. ACS Omega 7(9):7625–7637

Hossain N, Nizamuddin S, Griffin G, Selvakannan P, Mubarak NM, Mahlia TMI (2020) Synthesis and characterization of rice husk biochar via hydrothermal carbonization for wastewater treatment and biofuel production. Sci Rep 10(1):18851

Hui KC, Ang WL, Sambudi NS (2021) Nitrogen and bismuth-doped rice husk-derived carbon quantum dots for dye degradation and heavy metal removal. J Photochem Photobiol A Chem 418:113411

Issaka E, Amu-Darko JN-O, Yakubu S, Fapohunda FO, Ali N, Bilal M (2022) Advanced catalytic ozonation for degradation of pharmaceutical pollutants – a review. Chemosphere 289:133208

Kalita H, Mohapatra J, Pradhan L, Mitra A, Bahadur D, Aslam M (2016) Efficient synthesis of rice based graphene quantum dots and their fluorescent properties. RSC Adv 6(28):23518–23524

Khan FSA, Mubarak NM, Tan YH, Khalid M, Karri RR, Walvekar R, Abdullah EC, Nizamuddin S, Mazari SA (2021) A comprehensive review on magnetic carbon nanotubes and carbon nanotube-based buckypaper for removal of heavy metals and dyes. J Hazard Mater 413:125375

Kim Y, Jang G, Lee TS (2015) New fluorescent metal-ion detection using a paper-based sensor strip containing tethered rhodamine carbon nanodots. ACS Appl Mater Interfaces 7(28): 15649–15657

Kızılduman BK, Turhan Y, Doğan M (2021) Mesoporous carbon spheres produced by hydrothermal carbonization from rice husk: optimization, characterization and hydrogen storage. Adv Powder Technol 32(11):4222–4234

Korotkova TG, Ksandopulo SJ, Donenko AP, Bushumov SA, Danilchenko AS (2016) Physical properties and chemical composition of the rice husk and dust. Orient J Chem 32(6):3213–3219

Kumari M, Chaudhary GR, Chaudhary S, Huang M, Guo Z (2024) Transformation of waste rice straw to carbon quantum dots and their potential chemical sensing application: green and sustainable approach to overcome stubble burning issues. Biomass Convers Biorefinery 14(6): 7507–7518

Kundu A, Maity B, Basu S (2022) Rice husk-derived carbon quantum dots-based dual-mode nanoprobe for selective and sensitive detection of Fe3+ and fluoroquinolones. ACS Biomater Sci Eng 8(11):4764–4776

Leng L, Xiong Q, Yang L, Li H, Zhou Y, Zhang W, Jiang S, Li H, Huang H (2021) An overview on engineering the surface area and porosity of biochar. Sci Total Environ 763:144204

Li Z, Zheng Z, Li H, Xu D, Li X, Xiang L, Tu S (2023) Review on rice husk biochar as an adsorbent for soil and water remediation. Plants 12(7):1524

Licht G, Licht S (2025) Carbon nanotube production pathways: a review of chemical vapor deposition and electrochemical CO2 conversion, such as C2CNT. Crystals 15(10):887

Longo L, Baldassin D, Di Michele A, Bittencourt C, Menegazzo F, Signoretto M (2025) Activation strategies for rice husk biochar: enhancing porosity and performance as a support for Pd catalysts in hydrogenation reactions. Cat Sci Technol 15(17):5101–5115

Madima N, Mishra S, Inamuddin I, Mishra A (2020) Carbon-based nanomaterials for remediation of organic and inorganic pollutants from wastewater. A review. Environ Chem Lett 18(4): 1169–1191

Mansurov Z (2020) Seitzhanova Makpal Azizovna. University of Naples Federico II, Naples

Meng W, Bai X, Wang B, Liu Z, Lu S, Yang B (2019) Biomass-derived carbon dots and their applications. Energy Environ Mater 2(3):172–192

Mercy DJ, Kiran V, Thirumalai A, Harini K, Girigoswami K, Girigoswami A (2023) Rice husk assisted carbon quantum dots synthesis for amoxicillin sensing. Results Chem 6:101219

Mesa S, Castro-Ladino JR, Amaya SL, Manrique C, Echavarría A, Hoyos-Ayala DA, Uran L (2025) Comprehensive characterization of carbonaceous material derived from rice husk pyrolysis and its potential for CO2 adsorption. Materials 18(22):5151

Mishra RK, Verma K (2024) Defect engineering in nanomaterials: impact, challenges, and applications. Smart Mater Manuf 2:100052

Mobarak MB, Pinky NS, Mustafi S, Chowdhury F, Nahar A, Akhtar US, Quddus MS, Yasmin S, Alam MA (2024) Unveiling the reactor effect: a comprehensive characterization of biochar derived from rubber seed shell via pyrolysis and in-house reactor. RSC Adv 14(41): 29848–29859

Mostafavi E, Iravani S, Varma RS, Khatami M, Rahbarizadeh F (2022) Eco-friendly synthesis of carbon nanotubes and their cancer theranostic applications. Mater Adv 3(12):4765–4782

Motlagh EK, Asasian-Kolur N, Sharifian S, Pirbazari AE (2021) Sustainable rice straw conversion into activated carbon and nano-silica using carbonization-extraction process. Biomass Bioenergy 144:105917

Mubarik S, Qureshi N, Sattar Z, Shaheen A, Kalsoom A, Imran M, Hanif F (2021) Synthetic approach to rice waste-derived carbon-based nanomaterials and their applications. Nano 1(3): 109–159

Naik M, Debbarma J, Saha M, Bhargava A (2020) Graphene oxide nanoflakes from various agrowastes. Materialwissenschaft und Werkstofftechnik 51(3):368–374

Nandi R, Jha MK, Guchhait SK, Sutradhar D, Yadav S (2023) Impact of KOH activation on rice husk derived porous activated carbon for carbon capture at flue gas alike temperatures with high CO2/N2 selectivity. ACS Omega 8(5):4802–4812

Negm NA, Altalhi AA, Ahmed HM, Mohamed EA (2024) Synergistic effect of rice husk-derived activated carbon modified by Ni/Al-layered double hydroxides for lead removal from industrial wastewater. Sci Rep 14(1):28411

Oladimeji T, Oyedemi M, Emetere M, Agboola O, Adeoye J, Odunlami O (2024) Review on the impact of heavy metals from industrial wastewater effluent and removal technologies. Heliyon 10(23)

Ou D, Zheng S, Zhu M, Chen R, Gao T, Guo H, Guo J, Tang J (2024) Excellent electrocatalytic performance of rice straw-derived carbon catalysts activated by zinc halides for oxygen reduction reaction. Int J Hydrog Energy 50:1373–1380

Painuli R, Kumar C, Kumar D (2022) Carbon nanotubes and graphene nanomaterials for biomedical applications. In: Design, principle and application of self-assembled Nanobiomaterials in biology and medicine. Elsevier, pp 215–226

Pal T, Mohiyuddin S, Packirisamy G (2018) Facile and green synthesis of multicolor fluorescence carbon dots from curcumin: in vitro and in vivo bioimaging and other applications. ACS Omega 3(1):831–843

Pandey H, Pandey A, Sinha A, Agrahari GK, Pyrolysis GRG (2025) Conversion of rice straw and sugarcane bagasse waste to biochar through pyrolysis: influence of feedstock and process conditions. J Mater Cycles Waste Manag 27(5):3603–3618

Premchand P, Demichelis F, Galletti C, Chiaramonti D, Bensaid S, Antunes E, Fino D (2024) Enhancing biochar production: a technical analysis of the combined influence of chemical activation (KOH and NaOH) and pyrolysis atmospheres (N2/CO2) on yields and properties of rice husk-derived biochar. J Environ Manag 370:123034

Qi H, Teng M, Liu M, Liu S, Li J, Yu H, Teng C, Huang Z, Liu H, Shao Q (2019) Biomass-derived nitrogen-doped carbon quantum dots: highly selective fluorescent probe for detecting Fe3+ ions and tetracyclines. J Colloid Interface Sci 539:332–341

Raghavan N, Thangavel S, Venugopal G (2017) A short review on preparation of graphene from waste and bioprecursors. Appl Mater Today 7:246–254

Rahdar S, Dehghan A, Davoudi M, Shams M (2025) Functionalized graphene oxide for Cr (VI) removal: a systematic review of functional groups, mechanisms, and environmental implications. Results Eng 28:107385

Regkouzas P, Sygellou L, Diamadopoulos E (2023) Production and characterization of graphene oxide-engineered biochars and application for organic micro-pollutant adsorption from aqueous solutions. Environ Sci Pollut Res 30(37):87810–87829

Rybarczyk MK, Peng H-J, Tang C, Lieder M, Zhang Q, Titirici M (2016) Porous carbon derived from rice husks as sustainable bioresources: insights into the role of micro−/mesoporous hierarchy in hosting active species for lithium–sulphur batteries. Green Chem 18(19): 5169–5179

Šafranko S, Goman D, Stanković A, Medvidović-Kosanović M, Moslavac T, Jerković I, Jokić S (2021) An overview of the recent developments in carbon quantum dots – promising nanomaterials for metal ion detection and (bio) molecule sensing. Chemosensors 9(6):138

Sahoo D, Remya N (2022) Influence of operating parameters on the microwave pyrolysis of rice husk: biochar yield, energy yield, and property of biochar. Biomass Convers Biorefinery 12(8):3447–3456

Seitzhanova M, Mansurov Z, Yeleuov M, Roviello V, Di Capua R (2019) The characteristics of graphene obtained from rice husk and graphite. Eurasian Chem Technol J 21(2):149–156

Sharma HB, Vanapalli KR, Bhatia D, Singh S, Arora G, Panigrahi S, Dubey BK, Ramamurthy PC, Mohanty B (2024) Engineered biochar/hydrochar derived from organic wastes for energy, environmental, and agricultural applications. Clean Techn Environ Policy 26(12):4059–4093

Shen Y (2018) Rice husk-derived activated carbons for adsorption of phenolic compounds in water. Global Chall 2(12):1800043

Singh R, Gautam N, Mishra A, Gupta R (2011) Heavy metals and living systems: an overview. Indian J Pharmacol 43(3):246–253

Singh J, Bhattu M, Verma M, Bechelany M, Brar SK, Jadeja R (2025) Sustainable valorization of rice straw into biochar and carbon dots using a novel one-pot approach for dual applications in detection and removal of lead ions. Nano 15(1):66

Sudhan N, Subramani K, Karnan M, Ilayaraja N, Sathish M (2017) Biomass-derived activated porous carbon from rice straw for a high-energy symmetric supercapacitor in aqueous and non-aqueous electrolytes. Energy Fuel 31(1):977–985

Tavan M, Yousefian Z, Bakhtiar Z, Rahmandoust M, Mirjalili MH (2025) Carbon quantum dots: multifunctional fluorescent nanomaterials for sustainable advances in biomedicine and agriculture. Ind Crop Prod 231:121207

Thongsai N, Tanawannapong N, Praneerad J, Kladsomboon S, Jaiyong P, Paoprasert P (2019) Real-time detection of alcohol vapors and volatile organic compounds via optical electronic nose using carbon dots prepared from rice husk and density functional theory calculation. Colloids Surf A Physicochem Eng Asp 560:278–287

Tian MA, Yang Q, Jin Y, Hu B (2025) 3d graphene/carbon nanotube composites: synthesis, properties, and applications. ChemNanoMat 11(6):e202400675

Tiwari SK, Kumar V, Huczko A, Oraon R, Adhikari AD, Nayak GC (2016) Magical allotropes of carbon: prospects and applications. Crit Rev Solid State Mater Sci 41(4):257–317

Tohamy H-AS, Anis B, Youssef MA, Abdallah AE, El-Sakhawy M, Kamel S (2020) Preparation of eco-friendly graphene oxide from agricultural wastes for water treatment. Desalin Water Treat 191:250–262

Tripathi M, Bansal S, Tripathi SC, Singh R, Singh T, Rab SO, Srivastava N, Gupta VK (2025) Lignocellulosic biomass inspired fabrication of carbon quantum dots for biomedical and environmental applications: a review. Waste Biomass Valoriz:1–24

Tsai W-T, Huang P-C, Lin Y-Q (2019) Reusing cow manure for the production of activated carbon using potassium hydroxide (KOH) activation process and its liquid-phase adsorption performance. Processes 7(10):737

Tsarpali M, Kuhn JN, Philippidis GP (2024) Activated carbon production from algal biochar: chemical activation and feasibility analysis. Fuel Commun 19:100115

Uda MA, Gopinath SC, Hashim U, Halim N, Parmin N, Uda MA, Anbu P (2021) Production and characterization of graphene from carbonaceous rice straw by cost-effect extraction. 3 Biotech 11(5):205

Ufitikirezi J d DM, Filip M, Ghorbani M, Zoubek T, Olšan P, Bumbálek R, Strob M, Bartoš P, Umurungi SN, Murindangabo YT (2024) Agricultural waste valorization: exploring environmentally friendly approaches to bioenergy conversion. Sustainability 16(9):3617

Wang H, Xu Z, Kohandehghan A, Li Z, Cui K, Tan X, Stephenson TJ, King'Ondu CK, Holt CM, Olsen BC (2013) Interconnected carbon nanosheets derived from hemp for ultrafast supercapacitors with high energy. ACS Nano 7(6):5131–5141

Wang Z, Yu J, Zhang X, Li N, Liu B, Li Y, Wang Y, Wang W, Li Y, Zhang L (2016) Large-scale and controllable synthesis of graphene quantum dots from rice husk biomass: a comprehensive utilization strategy. ACS Appl Mater Interfaces 8(2):1434–1439

Wang W, Wang Z, Liu J, Peng Y, Yu X, Wang W, Zhang Z, Sun L (2018a) One-pot facile synthesis of graphene quantum dots from rice husks for Fe3+ sensing. Ind Eng Chem Res 57(28): 9144–9150

Wang Z, Shen D, Wu C, Gu S (2018b) State-of-the-art on the production and application of carbon nanomaterials from biomass. Green Chem 20(22):5031–5057

Wu ZL, Zhang P, Gao MX, Liu CF, Wang W, Leng F, Huang CZ (2013) One-pot hydrothermal synthesis of highly luminescent nitrogen-doped amphoteric carbon dots for bioimaging from Bombyx mori silk–natural proteins. J Mater Chem B 1(22):2868–2873

Wu J, Chen T, Ge S, Fan W, Wang H, Zhang Z, Lichtfouse E, Van Tran T, Liew RK, Rezakazemi M (2023) Synthesis and applications of carbon quantum dots derived from biomass waste: a review. Environ Chem Lett 21(6):3393–3424

Yadav BK, Kumar A (2025) Sustainable preparation of porous activated carbons with graphitic-like domains from waste rice husk via two-step thermochemical conversion. J Indian Chem Soc 102: 102045

Yan Y, Miao J, Yang Z, Xiao F-X, Yang HB, Liu B, Yang Y (2015) Carbon nanotube catalysts: recent advances in synthesis, characterization and applications. Chem Soc Rev 44(10): 3295–3346

Yang X, Wan Y, Zheng Y, He F, Yu Z, Huang J, Wang H, Ok YS, Jiang Y, Gao B (2019) Surface functional groups of carbon-based adsorbents and their roles in the removal of heavy metals from aqueous solutions: a critical review. Chem Eng J 366:608–621

Yang X, Lotfy VF, Basta AH, Liu H, Fu S (2024) Carbon quantum dots derived from rice straw doped with N and S and its nanocomposites with hydroxypropyl cellulose nanocomposite. Int J Biol Macromol 278:134925

Zhang L, Sun Y, Li C, Zhang S, Gholizadeh M, Hu X (2024) Hydrothermal carbonization pretreatment makes a remarkable difference in activation of rice and lettuce in food waste. Green Chem Eng 5(3):348–363

Zhou J, Ge M, Han Y, Ni J, Huang X, Han S, Peng Z, Li Y, Li S (2020) Preparation of biomass-based carbon dots with aggregation luminescence enhancement from hydrogenated rosin for biological imaging and detection of Fe3+. ACS Omega 5(20):11842–11848

Zou X, Debiagi P, Amjed MA, Zhai M, Faravelli T (2024) Impact of high-temperature biomass pyrolysis on biochar formation and composition. J Anal Appl Pyrolysis 179:106463

Chapter 8
Biomedical Applications of Waste-Recycled CNMs: Exploring Antioxidant and Anti-cancerous Efficacy

Biswabhusan Dash, Himadri Tanaya Behera, Ambika Sahoo, Sanjeeb Mohapatra, and Santosh Mahanty

8.1 Introduction

8.1.1 Significance of CNMs in Contemporary Biomedical Applications

Carbon nanomaterials (CNMs) are structures composed of carbon atoms organized at the nanometer scale. Typically, they possess a minimum of one dimension ranging from 1 to 100 nanometers. These materials are distinctive because to their robust mechanical properties, excellent electrical and thermal conductivity, chemical stability, optical characteristics, and extensive surface area (Speranza 2021). Their unusual behavior comes from how carbon atoms are organized and linked at such a small scale and provide the functionality that ordinary carbon materials such as graphite or diamond cannot provide. Because of this, they have become strong candidates for work in tissue engineering, drug delivery, biosensing, imaging, and cancer therapy (Parveen et al. 2025; Jayaprakash et al. 2024). Their high surface area and good biocompatibility allow them to carry therapeutic molecules with control and precision (Zhang et al. 2011). This has made them promising tools for targeted drug delivery (Debnath and Srivastava 2021), where the goal is to release drugs at

B. Dash (✉) · A. Sahoo
Centre for Biotechnology, Siksha 'O' Anusandhan (Deemed to be University), Bhubaneswar, Odisha, India
e-mail: biswabdash@soa.ac.in

H. T. Behera · S. Mahanty
Imgenex India Pvt. Ltd., Bhubaneswar, Odisha, India

S. Mohapatra
Department of Water Management, Delft University of Technology, Delft, The Netherlands

Y. K. Mohanta et al. (eds.), *Bioresource Waste to Carbon Nanomaterial*, Nanotechnology in the Life Sciences, https://doi.org/10.1007/978-981-92-2405-0_8

the right place while reducing side effects. Many of these materials also show special optical behavior (Parveen et al. 2025), which supports their use in imaging techniques such as fluorescence and photoacoustic imaging.

8.1.1.1 Main Types of CNMs

Graphene is a 1-atom-thick sheet of carbon arranged in a flat hexagonal pattern, known for its remarkable strength, conductivity, and transparency (Rasyotra et al. 2024). Carbon nanotubes are hollow cylinders made from rolled graphene layers. They can be single-walled or multi-walled and are valued for their mechanical strength and excellent electrical and thermal transport (Vidyalakshmi et al. 2022). Fullerenes, such as C60, are closed carbon cages shaped like spheres, ellipsoids, or tubes. They show unusual chemical behavior and have notable antioxidant activity (Rasyotra et al. 2024; Choudhary et al. 2014). Carbon nanofibers are carbon-based filaments with nanoscale diameters. They resemble nanotubes but usually have a less uniform structure (Kim et al. 2013; Choudhary et al. 2014). Nanodiamonds are tiny diamond particles that combine high hardness with good biocompatibility and a reactive surface (Choudhary et al. 2014). Carbon quantum dots are small, nearly spherical carbon nanoparticles that emit light and are useful in imaging and sensing (Choudhary et al. 2014) (Table 8.1).

Table 8.1 Key properties and biomedical applications of various carbon nanomaterials

Carbon nanomaterial	Key properties	Biomedical applications
Graphene	High electrical conductivity, mechanical strength, flexibility, chemical stability, large surface area (Huang et al. 2011)	Biosensors, drug delivery, tissue engineering, bioimaging (Lee et al. 2019)
Carbon nanotubes (CNTs)	Excellent tensile strength, high conductivity, chemical versatility, hollow structure (Popov 2004)	Targeted drug delivery, imaging contrast agents, cancer therapy, gene delivery (Ibrahim 2013)
Fullerenes	Antioxidant activity, ability to trap radicals, spherical structure, chemical reactivity (Khalil et al. 2019)	Antioxidant therapy, photodynamic therapy, MRI contrast agents (Chen et al. 2012)
Carbon nanofibers	High strength, electrical conductivity, fibrous morphology (Feng et al. 2014)	Wound healing scaffolds, biosensors, tissue engineering (Feng et al. 2014)
Nanodiamonds	Biocompatibility, surface reactivity, mechanical hardness, fluorescence (Rehor et al. 2014)	Drug delivery, cancer therapy, bioimaging, dental applications (Jung and Neuman 2021)
Carbon quantum dots	Photoluminescence, biocompatibility, small size, chemical reactivity (Kamal et al. 2025)	Bioimaging, biosensors, gene delivery, antibacterial agents

8.1.2 Environmental Rationale and Sustainability Imperatives for Waste Recycling into CNMs

Turning waste into CNMs helps tackle two major problems at once: it cuts down on pollution from long-lasting waste and offers a more sustainable, affordable way to produce advanced nanomaterials (Mohamed et al. 2025). From an environmental standpoint, huge amounts of plastics, crop residues, and municipal waste are produced every year. These materials often end up in landfills, contribute to microplastic contamination, and release greenhouse gases as they break down. Converting them into CNMs captures and reuses their carbon content, limits their environmental impact, and slows the buildup of nonbiodegradable waste (Mohamed et al. 2025; Tatrari et al. 2021).

8.1.2.1 Antioxidant and Anti-cancerous Therapeutic Potential of CNMs

CNMs are promising candidates as antioxidants and anticancer molecules due to their unique physicochemical properties and their highly dynamic interactions with biological systems (Famutimi et al. 2025).

Antioxidant Potential

CNMs such as fullerenes, carbon quantum dots, and graphene derivatives display distinct free radical scavenging potential due to an electron-rich surface and distinctive structural characteristics. These materials are beneficial in scavenging reactive oxygen species (ROS) and reactive nitrogen species (RNS) to prevent cellular injury due to oxidative stress (important for aging, inflammation, neurodegenerative diseases, and the development of cancer) (Algarra et al. 2025). Fullerenes act as "radical sponges" that directly quench oxidative species, and graphene oxide and reduced graphene oxide can restore redox homeostasis through cellular and tissue pathways. Likewise, carbon quantum dots exhibit potent antioxidant activity, highlighting their utility for antioxidant therapeutics against oxidative-stress-related pathologies (Algarra et al. 2025).

Anti-cancerous Therapeutic Potential

CNMs possess biocompatibility, a considerable surface area for functionalization, and their capacity for tailored anticancer therapy delivery to tumor sites (Mohan et al. 2023), making them promising candidates for anticancer and cancer therapy. For instance, engineered CNMs such as carbon nanotubes, graphene, and nanodiamonds can deliver anticancer chemotherapeutics, nucleic acids, or targeted therapies directly onto malignant cells. That means the treatments are more effective and less toxic to the entire body. In photothermal and photodynamic therapy, the use of light energy to create localized heat or cytotoxic species enables precise tumor ablation with little damage to healthy adjacent tissue. In addition, functionalized carbon nanomaterials (CNMs) could also be exploited to modulate immune responses, by increasing antitumor immunity or changing the tumor microenvironment, which could make them versatile platforms in oncological (Mohan et al. 2023).

8.1.3 Objectives of Chapter and Research Questions

This chapter endeavors to analyze the conversion of various waste sources into useful CNMs using sustainable and green synthesis protocols and to better understand the biochemical basis of antioxidant and anticancer activities (Famutimi et al. 2025; Omar et al. 2024). In particular, it investigates how CNMs attenuate oxidative stress by scavenging reactive oxygen and nitrogen species and how they participate in cancer therapy through the induction of apoptosis, tumor growth inhibition, restricting metastasis, and targeted drug and gene delivery through efficient surface approaches. The chapter further integrates experimental evidence for therapeutic potential from both in vitro and in vivo research into this with regard to cytotoxicity concerns about particle size, functionalization, and dosage (Algarra et al. 2025).

In this regard, the chapter tackles the following goals: How can waste be efficiently converted into high-quality nanomaterials through green synthesis? Which of the following physicochemical properties has the highest impact on antioxidant activity? How can CNMs exert anticancer activity and enable the targeted administration? How do the present experimental studies suggest their safety and potency? How can these materials be implemented in therapeutic and diagnostic platforms? And what challenges, whether clinical or regulatory, must be overcome in order to enable its translation to real-world biomedical applications?

8.2 Waste-Derived CNMs: Sources and Green Synthesis Technologies

8.2.1 Comprehensive Overview of Waste Sources

Waste resources from *agricultural biomass*, *plastic residues*, and *industrial by-products* are important feedstocks for sustainable material recovery and energy production (Udourioh et al. 2025; Pandey et al. 2026).

Waste products of the industrial process—which include spent catalysts, sludge, used batteries, and electronic waste—can also be used as a substantial secondary resource. With the rich carbon content of most of them, they are precursors to carbon nanomaterial (CNM) synthesis (Mehta et al. 2025). These wastes can be converted to very advanced carbon forms from carbon nanotubes, graphene, and carbon dots using hydrothermal, solvothermal, and catalytic processes which can be used in sensors, energy storage, and biomedical technologies (Aslam et al. 2024).

All together, these waste streams become low-cost sustainable feedstocks for CNM production as well as bioenergy generation, which in turn helps to waste valorization and environmental sustainability.

8.2.2 Green Synthesis Methodology

8.2.2.1 Hydrothermal and Solvothermal Carbonization

Hydrothermal carbonization (HTC) and solvothermal carbonization are green synthesis methods for converting biomass and waste into carbon-rich materials under mild and environmentally friendly conditions (González-Arias et al. 2022). HTC generally works at 160–300 °C with higher temperatures (240–300 °C) resulting in higher carbonization, enhanced carbon content, superior energy density, and formation of porous hydrochar. Solvothermal carbonization typically occurs under subcritical conditions below 200 °C, depending on the method (Sivaprasad et al. 2021).

HTC works at lower temperatures than pyrolysis, and it is capable of treating wet biomass without precondition drying, resulting in lower energy use and lower production costs. The resulting hydrochar is high in carbon, has good fuel properties, and is easy to tune both surface chemistry and porosity; thus, it can be used for soil amendment, energy, and precursor for carbon nanomaterials (CNMs). In addition, HTC mimics natural coal formation on an accelerated timescale, producing fewer toxic by-products, and employing different feedstocks such as agricultural residues, food waste, and industrial biomass wastes (Hejna et al. 2022; Ndlwana et al. 2021).

8.2.2.2 Pyrolysis and Controlled Carbonization

Pyrolysis is a common green technology for carbon nanomaterial (CNM) production through thermal decomposition of biomass and waste in an oxygen-limited or inert atmosphere (usually 400–1000 °C) (Devi et al. 2021).

The method is accomplished by bond cleavage which generates reactive radicals, followed by molecular rearrangement, cyclization, and aromatic condensation to produce both polycyclic aromatic hydrocarbons (PAHs) and graphitic carbon structures. Volatile compounds are eliminated through the pores in pyrolysis and typically adopt hollow or cage-like structures. Catalysts can also affect the morphology, yield, and crystallinity of the material by regulating the growth of nanotubes or carbon films. This method can be used to generate multiple CNMs, such as carbon nanotubes, graphene, carbon quantum dots, and porous carbons, and morphology is strongly influenced by temperature: at lower temperatures (~800 °C), hollow structure is preferred, while at high temperatures (~900 °C), nanotube formation is favored (Duraia et al. 2024).

The prepared CNMs generally provide low toxicity, high surface area, and rich functional groups that improve performances for fluorescence and adsorption applications. In addition, the application of renewable biomass as a feedstock decreases environmental deterioration and promotes sustainability, although catalyst residues could require post-treatment purification for more critical reactions. The advantages of pyrolysis are scalability, economy versus the conventional alternatives, flexibility in utilizing alternative feedstocks (from lignin to agricultural residues), and tailoring the nanostructured carbon morphology (Venkataraman and Sundramoorthy 2026). Crucially, pyrolysis also supports the co-production of value-added gases and char to enhance the cost of the process. Finally, pyrolysis is an efficient and sustainable pathway of converting biomass and waste via controlled thermal decomposition to the highest quality CNMs (Aparicio et al. 2025).

8.2.2.3 Microwave-Assisted and Ultrasonic Techniques

In a practical environmental aspect, microwave-assisted synthesis has been presented as successful green processing for synthesizing CNMs through microwave-assisted preparation, which effectively heats the reaction medium in a rapid and uniform manner under microwave irradiation conditions. This approach allows effective volumetric heating and rapid reaction times up to seconds or minutes and precise control of particle size, morphology, and surface chemistry by adjusting parameters such as power, duration, or solvent selection (Qureshi et al. 2022).

8.2.3 Surface Modification Strategies to Enhance Biocompatibility and Bioactivity

Improving the biocompatibility and biological performance of CNMs usually means changing the surface properties of their surfaces. This can either be chemical functionalization or the application or gradual incorporation of mild physical coatings. Chemical functionalization, or the application or gradual incorporation of mild physical coatings, are methods of doing so. This adaptation is thought to increase the dispersion in living creatures of the materials and decrease their toxic tendency, so predictable responses of the compounds can be produced effectively and proactively inside cells/tissues in this phase. The strategies are touched upon in short below:

8.2.3.1 Covalent Functionalization (Madkour 2024)

Introduce hydrophilic groups (COOH, OH, NH_2) for increased water solubility and a decrease in aggregation. It improves interactions with biomolecules and cells, allowing for directed drug delivery. Methods employed are acid oxidation, click chemistry, and polymer grafting (e.g., PEG and gelatin). This helps alleviate cytotoxicity, reduce hydrophobicity, and prevent nonspecific protein adsorption.

8.2.3.2 Non-covalent Functionalization (Karakoti et al. 2018; Behnam et al. 2013)

Biomolecular adsorption (proteins, peptides, DNA, and surfactants) preserves innate electronic properties. Additionally, biocompatible interfaces designed to mimic the extracellular matrix (ECM) structure facilitate cellular uptake and reduce immune responses.

8.2.3.3 Doping and Heteroatom Incorporation (Jayaprakash et al. 2024)

Addition of nitrogen, sulfur, or phosphorus creates active sites that increase bioactivity. It promotes cell adhesion and antimicrobial activity.

8.3 Antioxidant Properties of Waste-Derived CNMs

8.3.1 Molecular Mechanisms of ROS Scavenging and Redox Modulation

As such, waste-derived CNMs, especially carbon dots generated from biomass, graphene-like sheets, and porous carbons, can be used as efficient redox modulators by scavenging reactive oxygen species (ROS) directly as well as mimicking the activity of antioxidant enzymes and regulating cellular redox signaling pathways (Das et al. 2018). It is mainly ascribed to the sp^2-rich carbon domains, which are also decorated with abundant edge and defect functional groups including hydroxyl (OH), carboxyl (COOH), amine (NH_2), and quinone moieties (Dong et al. 2022; Famutimi et al. 2025). These functional groups are frequently extracted from the primary waste materials and are able to be further modified by green synthesis methods and are proposed to improve the redox properties and biological activity of the developed nanostructures (Wu et al. 2023).

8.3.1.1 Direct ROS Scavenging

The antioxidant benefit of waste-based CNMs includes the active antioxidant potential of several synergistic mechanisms, acting synergistically to counteract oxidative stress. An example of an important mechanism involves electron and proton donation where surface phenolic, carboxyl, and amino groups donate electrons or protons to reactive oxygen species (ROS) such as hydroxyl radicals, superoxide anions, and peroxyl radicals (Famutimi et al. 2025). This transformation of these species to less reactive ones results in the cessation of radical chain reactions. Finally, the π conjugated sp^2 carbon domains and fullerene-like cages are capable to foster radical adduct formation by capturing reactive species and delocalizing unpaired electrons throughout the extended carbon network to stabilize them as "radical sponges" (Akhtar et al. 2017).

8.3.1.2 Enzyme-Mimetic Mechanisms

Waste-derived CNMs, especially carbon dots and doped carbons, have tremendous enzyme-derived functionalities which could be the major sources of their antioxidant activity. Waste-derived carbon dots exhibit superoxide dismutase (SOD)-like activities, catalyzing the dismutational degradation of superoxide anions via redox-acting quinone and phenolic motifs or heteroatom dopants, cycling between oxidation and reduction states (Dong et al. 2024). Moreover, some biomass carbon dots and Fe- or N-doped carbons exhibit catalase (CAT) and peroxidase-like activities that decompose hydrogen peroxide or use it to oxidize co-substrates through surface redox centers that transit electrons between H_2O_2 and associated biomolecules (Wareing et al. 2021). More importantly, CNMs can exhibit these various enzyme-like activities and work as a series of synergistic reactions in which superoxide is first converted to hydrogen peroxide and then hydrolyzed to water. By mimicking multiple enzymes, it lowers the average oxidative load in cells, indicating the pharmaceutical potential of waste-derived CNMs that can serve as sustainable nanomaterials for regulation of redox pathways and in the formulation of biomedical materials (Xu et al. 2025).

8.3.1.3 Redox Signalling and Cellular Pathways

Waste-derived CNMs have biological actions and bioactive activity other than direct ROS scavenging, and in addition, they can also influence vital cellular signaling pathways. The moderate impact of the ROS buffer, together with the electrophilic surface groups, can trigger the Nrf2/ARE pathway and result in the upregulation of endogenous antioxidants such as SOD, CAT, GPx, HO 1, and glutathione-related enzymes in cells, enhancing the endogenous cellular defense mechanisms (Takhar and Singh 2025). In addition, CNMs reduce ROS in mitochondria and cytosolic

ROS, decrease redox-sensitive inflammatory signaling by means of ROS, and inhibit the activity of inflammatory signals by decreasing the mitochondrial and cytosolic ROS level that downregulates NF κB and NLRP3 inflammasome activity and therefore decreases the production of pro-inflammatory cytokines such as TNF α, IL 1β, and IL 6 secretion in inflammatory and wound healing models (Dong et al. 2022). Moreover, CNMs protect mitochondrial integrity by maintaining membrane potential, reducing lipid peroxidation levels as reflected in malondialdehyde (MDA) levels, and maintaining ATP production. What is more, together with anti-apoptotic and anti-necrotic actions against apoptosis and necrosis under oxidative stress, such protective effects contribute to the therapeutic potential of CNMs as multifunctional modulators of the cellular redox balance and inflammation that affect mitochondrial health (Famutimi et al. 2025) (Fig. 8.1).

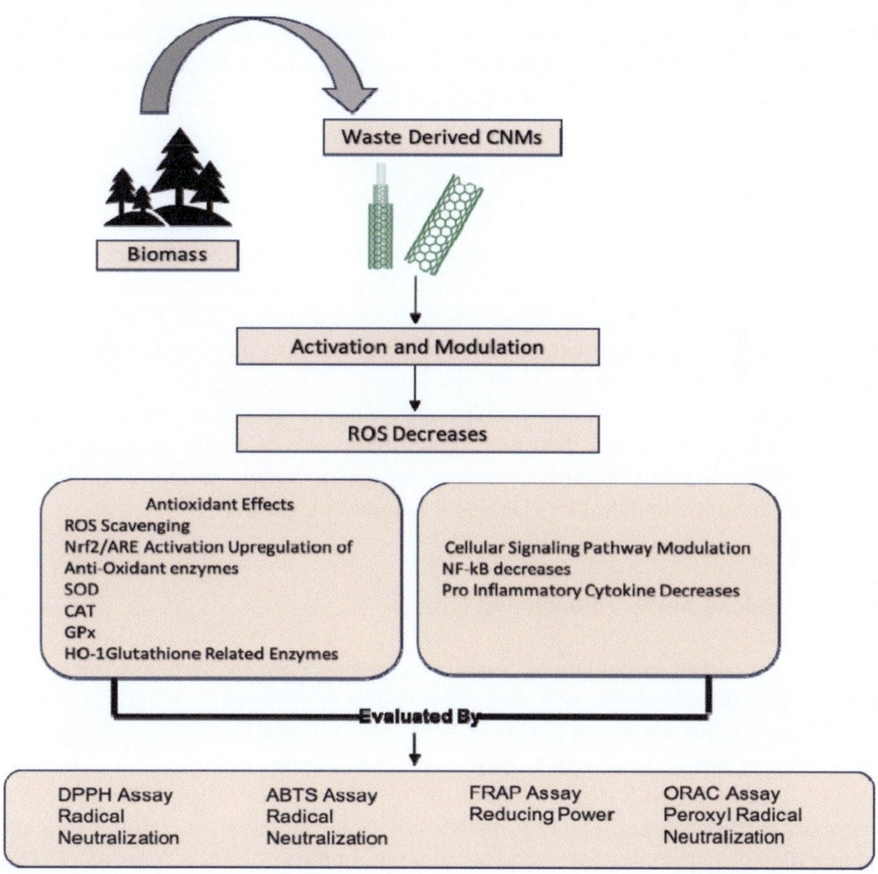

Fig. 8.1 Schematic diagram showing mechanistic overview of antioxidant and anti-inflammatory actions of waste-derived CNMs and their in vitro evaluation

8.3.2 *Influence of Size, Surface Functional Groups, and Structural Defects on Antioxidant Activity*

Size, surface functional groups, and structural defects serve as tightly interconnected determinants of the antioxidant activity of CNMs, particularly carbon dots, and graphene based systems (Innocenzi and Stagi 2023; Zhao et al. 2019). Smaller particle sizes generally enhance surface-to-volume ratios, increasing the availability of reactive sites for radical scavenging. Meanwhile, functional groups such as hydroxyl, carboxyl, and amino moieties contribute to electron and proton donation, thereby facilitating redox reactions and neutralizing reactive oxygen species. Structural defects, including edge sites and vacancies, further augment electron transfer capacity by providing localized states that interact with radicals or redox-active biomolecules (Innocenzi and Stagi 2023). By strategically tuning these parameters through synthesis and post modification, it becomes possible to modulate radical scavenging efficiency, electron transfer dynamics, and downstream impacts on cellular redox signaling (Qiu et al. 2014). This tunability underscores the versatility of CNMs as sustainable nanomaterials for biomedical and chemical applications where precise control of oxidative balance is critical (Qiu et al. 2014).

8.3.2.1 Effect of Size

Smaller CNMs, such as sub 10-nm carbon dots, exhibit a markedly higher surface to volume ratio, thereby increasing the number of accessible reactive sites for interaction with reactive oxygen species (ROS) and radical probes in antioxidant assays (Cui et al. 2021). However, extremely small sizes can also promote higher levels of cellular uptake and, in certain contexts, induce prooxidant effects. As a result, optimal antioxidant performance is typically achieved within an intermediate size window, where the balance between high surface area, reduced aggregation, and minimized prooxidant behavior ensures effective redox modulation without compromising biocompatibility (Daré and Lautenschlager 2025).

8.3.2.2 Role of Surface Functional Groups

Oxygen- and nitrogen-containing functional groups such as hydroxyl (OH), carboxyl (COOH), carbonyl (C=O), and amino (NH_2) moieties play a central role in the radical scavenging activity of CNMs (Ji et al. 2019). These groups act as electron or proton donors and serve as redox active sites, enabling the neutralization of reactive oxygen species through direct electron or hydrogen atom transfer. In particular, phenolic, carboxylic, and carbonyl motifs contribute to the stabilization of radical intermediates via resonance and conjugation within the sp^2 carbon framework. This delocalization of unpaired electrons underlies the efficiency of H atom transfer and single electron transfer mechanisms, thereby enhancing the antioxidant capacity of CNMs and reinforcing their potential as sustainable nanomaterials for biomedical and environmental applications (Qiu et al. 2014).

8.3.3 *Comparative Analysis with Conventional Antioxidants and Synthetic Nanomaterials*

Waste-derived CNMs offer several advantages over conventional antioxidants and synthetic nanomaterials in terms of their antioxidant activity and application (Famutimi et al. 2025; Sager et al. 2014). Waste-derived CNMs are not consumed in the process of their application, unlike conventional antioxidants such as vitamin C, vitamin E, and polyphenolic compounds, which are used in the direct scavenging of free radicals and are consumed in the process (Liu et al. 2025; Flores 2023). The activity of waste-derived CNMs can be justified by their integration with multiple mechanisms of electron/proton donation, adduct formation, and enzyme-like catalytic activities (Bedhiafi et al. 2023). Waste-derived CNMs are not as toxic and less biocompatible compared to synthetic materials such as metal oxide nanoparticles and semiconductor quantum dots, which are not stable in physiological conditions (Kashtiban et al. 2024). The application of waste-derived CNMs is also in compliance with the concept of a circular economy, which makes it an eco-friendly and cost-effective approach to the synthesis of multifunctional materials (d'Amora et al. 2016). The attributes of waste-derived CNMs make them promising materials with great application in various fields compared to conventional antioxidants and synthetic materials (Flores 2023).

8.3.4 *Applications in Mitigating Oxidative Stress-Related Disorders and Chronic Inflammation*

Waste-derived CNMs such as carbon dots, graphene quantum dots, and carbon nanotubes have major potential as therapeutic agents in the context of oxidative stress and chronic inflammation. They exert benefit superior to classical antioxidants as they can actively scavenge harmful reactive oxygen and nitrogen species, simulate natural antioxidant enzymes, and are biocompatible when functionalized appropriately (Nie et al. 2024). In contrast to vitamins or plant-based polyphenols that indiscriminately kill radicals, these nanomaterials are able to selectively attack the true harmful species, such as hydroxyl radicals and peroxynitrite, and do not destroy signaling molecules like superoxide and hydrogen peroxide. This selective manipulation of oxidative balance addresses the core mechanisms that have a major impact on diseases such as neurodegeneration, diabetes, cardiovascular disease, and persistent inflammatory conditions, making CNMs future stewards of cellular health (Tang et al. 2023).

8.3.4.1 Neurodegenerative Diseases and Brain Injury

CNMs have promising potential to safeguard the CNS from oxidative insults where high oxygen requirements and low natural antioxidant defense mechanisms make the neurons highly susceptible to ROS-induced death (Nilewski et al. 2015). In experimental models of stroke, traumatic brain injury, and Parkinson's disease, antioxidant carbon dots and functionalized fullerenes have been demonstrated to quickly reduce harmful species like hydroxyl radicals (OH) and peroxynitrite (ONOO). It acts to enhance mitochondrial activity, decrease lipid peroxidation, and safeguard neuronal survival. Phenylenediamine-based carbon dots, which ameliorate neuroinflammation related to acute kidney injury by hindering NF κB and NLRP3 inflammasome signaling, and selenium-doped graphene quantum dots that prevent dopaminergic neurons from aggregation of α-synuclein through Nrf2 activation and glutathione replenishment in particular are illustrative examples of these factors (Huq et al. 2016). Importantly, these materials also show translational promise as when surface-modified with PEG or transferrin, they may cross the blood-brain barrier (BBB) to facilitate targeted neuroprotection without inducing immunosuppressant effects (Tang et al. 2023).

8.3.4.2 Metabolic and Cardiovascular Diseases

In light of diabetes and disease pathogenesis, CNMs could contribute to counter hyperglycemia induced oxidative stress in pancreatic β cells and vascular endothelium. Carbon nanotube nanoparticles have also been demonstrated to help restore insulin secretion in the β cells by scavenging ROS-induced β cell dysfunction and reverse the oxidative stress state in diabetes (Liu et al. 2022; Jiang et al. 2023; Yousaf et al. 2025). Furthermore, the citric acid/glutathione-derived carbon dots are found to scavenge intracellular ROS in macrophages, limiting advanced glycation end products (AGEs) and increasing effective glycemic control in diabetic wound models. In addition to diabetes, graphene oxide derivatives offer protective actions in the cardiovascular system by chelating transition metals Fe^{2+} and Cu^{+}, suppressing Fenton chemistry and stabilizing atherosclerotic plaques via modulation of endothelial inflammation, reflected in decreased IL-6 and TNF-α levels (Liu et al. 2022). The plant-derived carbon dots (e.g., cultivated from honeysuckle and dandelion) also showed dynamic ROS modulation in diabetic wounds, increasing the rate of repair by selective removal of hydroxyl radicals and peroxynitrite and promoting photodynamic bacterial clearance. Altogether, these results collectively reflect the versatile pharmaceutical opportunities of CNMs in oxidative stress related sequelae of metabolic and cardiovascular diseases (Nie et al. 2024).

8.3.4.3 Chronic Inflammatory Conditions

CNMs hold tremendous potential in long-term management of inflammation by interrupting ROS induced cytokine storms and suppressing the over-activation of immune cells. These T-cell targeted carbon nanoparticles selectively scavenged superoxide radicals in the activated lymphocytes of chronic diseases such as rheumatoid arthritis and inflammatory bowel disease (IBD), promoting suppression of autoimmune responses without induction of widespread immunosuppression (Yu et al. 2008; Huq et al. 2016). Likewise, polydopamine-like carbon dots and Mn_3O_4-integrated carbon composites create multistep ROS clearance in a manner matching the sequential activities of superoxide dismutase, catalase, and peroxidase which aid in restoring gut barrier integrity and inhibit NLRP3-mediated pyroptosis in models of ear inflammation and colitis. For example, in periodontal disease and chronic wounds, herbal carbon dots obtained from taxus leaves (TACD) work together to scavenge anti-inflammatory ROS and generate red light-activated superoxide, enhancing sterilization (Nie et al. 2024). This dual action provides over 95% bacterial clearance and twofold acceleration of tissue repair, highlighting the multifaceted therapeutic relevance of CNMs through inflammation- and infection-related diseases.

8.4 Anticancer Efficacy and Mechanisms of Waste-Recycled CNMs

8.4.1 Cellular and Molecular Mechanisms: Apoptosis Induction and Cell Cycle Arrest

8.4.1.1 Inhibition of Migration and Angiogenesis

Waste-recycled carbon nanomaterials (WR CNMs) are a new frontier in sustainable biomedical applications with potential uses utilizing waste streams for therapeutic purposes (Rabiee et al. 2022; Abd Elkodous et al. 2022). Though WR CNMs are not completely understood, and although mechanisms of action are still a work in progress, most biologically similar features and properties to conventionally made CNMs, such as carbon nanotubes, graphene, fullerenes, carbon dots, etc. (Kharlamova et al. 2023). These materials have shown promise in the modulation of key cellular processes (specifically cancer), through apoptosis, cell cycle arrest, and by inhibition of cellular migration and angiogenesis.

Apoptosis induction: Like other CNMs, W-R CNMs could induce apoptosis (programmed cell death) in target cells. This is frequently accompanied by production of reactive oxygen species (ROS) with potential to produce oxidative stress (Murr and Garza 2009). CNMs have been demonstrated to induce elevated levels of ROS with oxidative modification of proteins which can then activate signaling pathways such as JNK/p38 pathways leading to hepatocyte apoptosis (Gao et al. 2024). Some CNMs have also been shown to lead to an increase in apoptotic cell percentages mediated by modified regulation of some important protein genes including BAX and BCL2, constitutive of mitochondrial apoptotic cascading pathways (Yousefi et al. 2022). The extent of apoptosis induction was previously investigated, and it was reported that it is often determined by specific nanomaterial properties, including surface functionalization and the applied dose (Havrdova et al. 2016; Kyriakidou et al. 2020).

8.4.1.2 Cell Cycle Arrest

The uncontrolled proliferation features of cancer can be addressed by addition of cell cycle arrest. W-R CNMs can disrupt the normal progression of the cell cycle, thus halting cell division. This may be achieved by directly perturbing master cell cycle regulators. These nanomaterials can suppress the tumorigenesis of malignant cells by arresting proliferation at defined checkpoints (Elemike et al. 2025).

8.4.1.3 Inhibition of Migration and Angiogenesis

Cancer metastasis and tumor growth depend heavily on cell migration and formation of new blood vessels (angiogenesis). CNMs have been reported to be able to effectively inhibit these processes. Studies have demonstrated that functionalized single-walled carbon nanotubes inhibit the migration of breast cancer cells through the regulation of signaling pathways, especially the PI3K/Akt/mTOR pathway (Badea et al. 2022). These effects are attributed to their capacity to impair endothelial cell functions, interfere with growth factor signaling, and create an environment less conductive to new blood vessel formation.

8.4.2 Modulation of Cancer-Related Signalling Pathways (e.g., p53, Bcl-2, and ROS-Mediated Cytotoxicity)

The potential of waste-recycled carbon nanomaterials has been proposed as a sustainable and promising platform for cancer therapy, and also, they exhibit anticancer activity by influencing key cellular signaling pathways. Typically based on biowaste materials, this has been suggested through their unique physicochemical properties to evoke reactive oxygen species-mediated cytotoxicity, which plays a role in the regulation of important key proteins such as p53 and BCL2 proteins (Rabiee et al. 2022).

8.4.2.1 ROS-Mediated Cytotoxicity

The principal mechanism of W-R CNMs' anticancer effects is reactive oxygen species (ROS) generation, resulting in oxidative damage and cytotoxicity to cancer cells. Cancer cells often operate under elevated basal ROS levels, making them particularly vulnerable to further oxidative insults and subsequent cell death (Getachew et al. 2021; Lin et al. 2023). Nanoparticles of carbon functionalized, even when drug carriers, can induce apoptosis by producing ROS (Liu et al. 2024). The oxidative stress resulting is to trigger downstream signaling pathways playing an important role in the anticancer effect (Wang et al. 2021).

8.4.2.2 Modulation of the p53 Pathway

The p53 tumor suppressor protein is central to cellular responses to stress, regulating cell cycle arrest and apoptosis (Krishnamurthy et al. 2020). W-R CNMs show a capability to control the p53 pathway. Some nanomaterials, for instance, titania nanosheets, can induce p53; this property can mediate apoptosis producing ability (Soonnarong et al. 2021). Moreover, functionalized multiwalled carbon nanotubes have been shown to produce the anticancer action in particular via apoptosis and G0/G1 cell cycle arrest, partially through ROS overproduction that induces DNA damage-mediated p53 phosphorylation (Wan et al. 2015).

8.4.2.3 Regulation of Bcl-2 Family Proteins

The pro-apoptotic (e.g., Bax) and anti-apoptotic (e.g., Bcl-2) proteins of the Bcl-2 group each act together to determine cell fate. W-R CNMs are able to affect this balance and induce apoptosis. Organic carbon dots can interact preferentially with overexpressed proteins in cancer cells, which eventually induce apoptosis usually via the mitochondrial pathway (Ahmad et al. 2023).

8.4.3 Integration into Synergistic Therapies

For instance, waste-recycled CNMs represent sustainable platforms for advanced biomedical applications including cancer therapy (Chary et al. 2023; Elahi and Zeinalipour-Yazdi 2025). This versatility has been proven in terms of special properties that allow incorporating the compounds into combinations of therapies to potentially improve therapeutic effectiveness as well as to compensate single approach limitations (Gao et al. 2024; Mousavi-Kiasary et al. 2025).

8.4.3.1 Targeted Drug Delivery Systems Leveraging High Surface Area and Functionalization

The large surface area to volume ratio of nanoparticles makes possible tremendous drug loading and improved interaction with biological systems (Hami 2021; Unnikrishnan et al. 2023). Functionalization, by attaching certain ligands or functional groups, is central for directing these carriers to diseased cells and reducing off target effects by making them more biocompatible (Alavi et al. 2024; Thakur et al. 2024). Photothermal and Photodynamic Therapy Using Waste-Recycled Carbon Nanomaterials.

8.4.3.2 Photothermal and Photodynamic Therapeutic (PPT) Modalities

Due to their high photothermal and photodynamic potential, carbon nanomaterials have received increasing interest as noninvasive and highly selective approaches in disease treatment, including cancer. Carbon nanomaterials of photothermal and photodynamic therapeutics transform absorbed light, often near-infrared, into heat, prompting local hyperthermia and killing specific cells (Wang et al. 2022; Ji et al. 2025; Dhall 2023).

8.4.3.3 Gene and RNA Interference Delivery Platforms

Among other nonviral vectors for nucleic acids, carbon nanomaterials such as carbon nanotubes, grapheme, and carbon quantum dots have been shown to be strong for delivering gene and RNA interference (NAP) (Yazdani et al. 2024). Its large surface area and relatively simple surface functionalization provide a high loading capacity and allow for the rapid delivery of heterogeneous genetic material including plasmid DNA (pDNA), small interfering RNA (siRNA), microRNA (miRNA), and messenger RNA (mRNA) (Yazdani et al. 2024). Because of their needle-like morphologies, CNTs can cross cell membranes to carry nucleic acids. Grapheme-based platforms (Liu et al. 2021) are often functionalized by biomolecules.

8.4.4 Preclinical Studies: In Vitro Cancer Cell Line Efficacy and In Vivo Tumor Models

Carbon nanomaterials, irrespective of their origin, possess unique physicochemical properties that make them highly attractive for various oncological applications, including drug delivery, imaging, and direct therapeutic interventions.

Waste-recycled CNMs leverage these fundamental characteristics while also offering environmental and economic benefits.

8.4.4.1 In Vitro Cancer Cell Line Efficacy

In vitro studies are crucial for understanding the direct interaction of CNMs with cancer cells and elucidating their mechanisms of action. For waste-recycled CNMs, the core principles observed with conventionally synthesized CNMs largely apply.

8.4.4.2 Direct Cytotoxicity and Apoptosis Induction

Waste-recycled nanomaterials have been demonstrated to exhibit anticancer properties. Similar to other CNMs, they can induce apoptosis and inhibit proliferation in various cancer cell lines. This direct cytotoxic effect is often mediated through mechanisms such as oxidative stress generation, disruption of cellular organelles, and interference with key cellular pathways involved in cancer cell survival and growth. The functional groups present on the surface of CNMs, which can vary depending on the waste source and recycling process, are critical in determining their interaction with cell membranes and intracellular components, thus influencing their anticancer and antibacterial properties (Benko et al. 2023).

8.4.4.3 Enhanced Drug Delivery

CNMs (waste-derived CNMs) have been promising nanocarriers for chemotherapeutic drugs, which are excellent candidates in these situations. They can carry anticancer drugs, including doxorubicin, with both physical adsorption and chemical conjugation. Such drug-carrying ability implies a tailored payload to cancer cells, resulting in better accumulation of drugs in tumors, as well as potentially reducing systemic toxicity to healthy tissues (Gayathri and Vidya 2024).

8.4.4.4 In Vivo Tumor Models

Beyond these promising in vitro investigations, waste-recycled CNMs have also been studied in vivo in models of tumors, typically involving xenograft or syngeneic mice.Tumor growth inhibition: Some CNMs, including carbon dots, in vivo models showed effective antitumor growth as monotherapies as well as when combined with other therapies (Xia et al. 2019). It has been shown that waste-recycled CNMs can concentrate in tumor tissues by the enhanced permeability and retention effect, or actively by their surface modification targeting, which account for their activities in vivo as antitumor agents.

8.4.4.5 Targeted Therapeutic Approaches

As such, CNMs can be targeted through the functionalization of CNMs based on the presence of targeting ligands and then targeted into selected tumor areas or internalized cells within specific cancers for enhanced efficacy and decrease of off-target effect. In the same way, carbon nanotubes with tumor microenvironmental targeting have indicated promising anti-metastasis action in preclinical melanoma models. While this specific study does not explicitly state "waste-recycled," the principle of targeted delivery applies directly to CNMs from sustainable sources. Antibody-functionalized carbon nanocapsules have also been studied as potential targets for cancer radiotherapy and show the capability to deliver therapeutic agents precisely into cancer cells in vivo (Wang et al. 2020).

8.4.4.6 Combination Therapies and Modalities

The use of waste-recycled CNMs can be combined with different therapeutic methods. For example, their optical characteristics make CNMs able to be used as part of phototherapy, which can absorb light energy quickly and convert it into heat as in photothermal and photodynamic therapy or by the emission of reactive oxygen species that help in selective destruction of tumor. Moreover, CNMs have now become valuable tools in immuno-oncology, where they can modulate the immune system to combat cancer thus providing a multipronged strategy for in vivo tumor treatment (Wang et al. 2022).

8.5 Safety, Toxicological Challenges, and Translational Barriers

8.5.1 Toxicological Profiles: Dose Dependence, Functional Group Effects, and Size Variability

The toxicity profile of carbon nanomaterials is complex and depends primarily on several properties: dose dependence, functional group effects, and size variability. However, earlier studies constantly highlights that the molecular toxicity of CNMs is dose-dependent so adverse effects generally increase as exposure concentrations increase (Jiang et al. 2021). And apart from the amount of drug introduced, the specific functional group and general surface chemistry are a factor that defines their interaction to biological systems and thus toxicity (Jiang et al. 2020). Earlier study (Jiang et al. 2020) documented that the functional groups and electronic structures of single-walled carbon nanotubes directly shape their toxicological potential. In addition, size variability is among the key determinant in CNM safety assessment. The structure, size, and length of the entire nanomaterial are critically important in determining their biological behavior and toxicological effect. Thus, thorough elucidation of these factors is key for safe design and use of all carbon nanomaterials, even the more and more green alternatives derived from waste-recycled sources, to gain their therapeutic advantage and minimize risk.

8.5.2 *Immunogenicity and Biodistribution Concerns*

CNM immunogenicity and biodistribution are pivotal for clinical translation. Their interaction with the immune system—potentially triggering innate or adaptive responses—is dictated by physicochemical properties such as size, shape, and surface functionalization. While non-functionalized nanotubes may lack inherent immunogenicity, systemic administration often leads to accumulation in the liver, spleen, and lungs. Small, well-functionalized CNMs are typically cleared rapidly via renal pathways (Bolshakova et al. 2022); however, larger components may persist in the liver for months before biliary excretion. Consequently, precise engineering is essential to mitigate immunotoxicity and ensure long-term safety and optimal biodistribution (Xia et al. 2019) (Fig. 8.2).

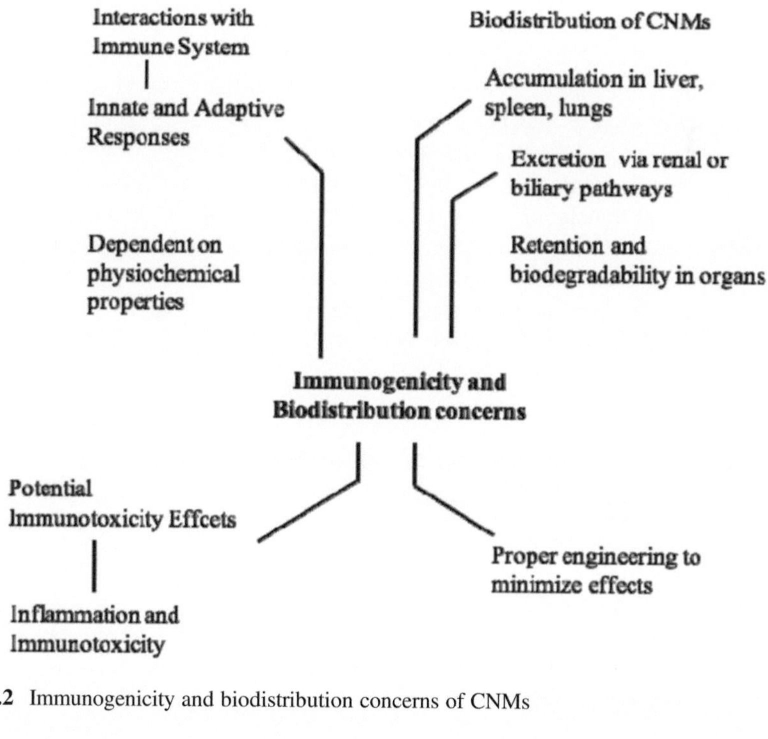

Fig. 8.2 Immunogenicity and biodistribution concerns of CNMs

8.5.3 Batch-to-Batch Variability and Quality Control Challenges from Waste Precursors

Waste obtained carbon nanomaterials provide a cost-effective and green solution but present many obstacles. High batch-to-batch variability and poor-quality control hinder large-scale production, particularly for graphene-type materials (Balqis et al. 2023). The variations in carbon content, contaminants, and synthesis conditions have a great impact on the quality and properties of CNMs (Sharaf Addin et al. 2025). Amorphous carbon and metallic nanoparticles and other contaminants need further purification which leads to cost and process complexity (Venkataraman and Sundramoorthy 2026). Reproducibility, homogeneity, and consistent product features are the remaining challenge. Nowadays, research is aimed to formulate green and cost-effective approaches that will produce high-grade, greener, and sustainable CNMs that can be manufactured.

8.5.4 Regulatory Frameworks and Ethical Considerations in Biomedical Applications

The biomedical application of CNMs faces critical challenges concerning regulatory frameworks and ethical considerations (Parveen et al. 2025; Jayaprakash et al. 2024). It is necessary to establish protocols for nanomedicine using specific guidelines and standards currently unmoored from current guidelines regarding how drugs can be evaluated the standardization is often based on the same as with traditional drugs. Agencies such as the FDA and EMA are establishing specific preclinical standards for testing, toxicity screening, and manufacturing; however, there is insufficient policy, regulatory, or technical framework to guarantee both patient safety and clinical translation (Gujar et al. 2025). At an ethical level, the advent of nanomedicine has also drawn attention to the issues of informed consent because of missing knowledge regarding long-term outcomes, and the challenges in communicating risk during clinical trials due to the various biological interactions of nanomaterials.

8.5.5 Environmental Impact of Nano-Bio Interactions and Life Cycle Assessment

Carbon nanomaterials (CNMs), despite their advantages, pose potential environmental risks throughout their life cycle. Increased production and use raise the likelihood of their release during manufacturing, usage, and disposal. Once in the environment, CNMs can exhibit toxicity toward aquatic and terrestrial organisms, including bacteria and microalgae (Pikula et al. 2022). Carbon nanotubes, in particular, show ecotoxicological effects in aquatic systems and are highly persistent. Therefore, understanding their colloidal behavior, transformation, and biodegradation is essential for designing biodegradable-by-design nanomaterials and reducing ecological risks. Standardized assessment protocols and comprehensive life cycle studies are necessary to ensure safer CNM development.

8.6 Emerging Trends and Future Directions

8.6.1 Engineering Multifunctional Waste-Recycled CNMs for Personalized Medicine and Theranostics

The advancement of carbon nanomaterials in personalized medicine and theranostics will increasingly focus on their engineering. Multifunctional variants of carbon nanomaterials, derived from recycled waste, represent an essential aspect of the "waste-to-wealth" concept. This sustainable technique makes use of large volumes of biomass and industrial waste in order to create sophisticated systems that can diagnose and treat diseases (Rabiee et al. 2022). Ultimately, we aim to realize very specific and new therapies to patients, where treatment advancement and efficacy can be monitored in real time and the patient can receive feedback and be personalized as the individual responded according to their particular treatment needs (Mauro et al. 2021) (Fig. 8.3).

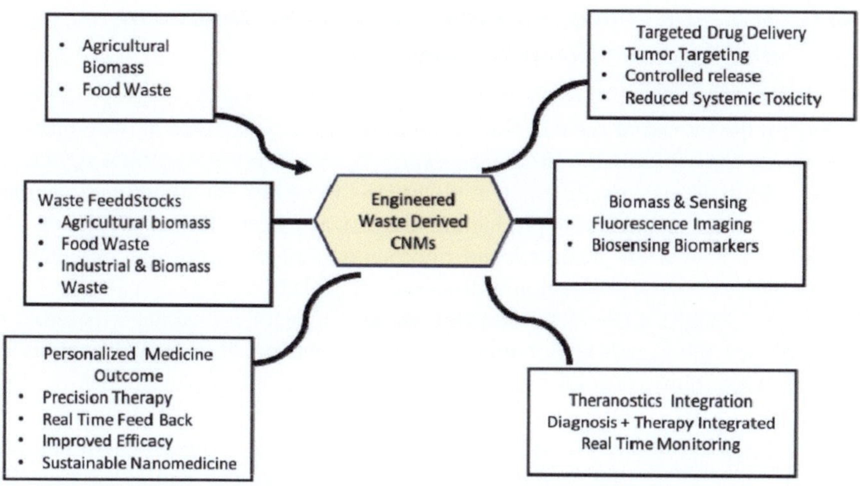

Fig. 8.3 Schematic representation of the development and biomedical application of engineered carbon nanomaterials derived from waste resources

8.6.2 Integration with Biosensors and Diagnostic Devices for Early Cancer Detection

The utilization of carbon nanomaterials as biosensors and diagnostic materials could transform the detection of cancer early, with the promising benefits of unmatched sensitivity, selectivity, and speed. It is in the carbon nanotube (CNT) and graphene-based transistor (GBT) biosensor that it has now been the first to be harnessed, exploiting its unique electrical and mechanical properties to identify cancer biomarkers accurately at low concentrations and rapidly. These benefits are essential since traditional cancer screening methods are time-consuming and labor-intensive and offer the demand for streamlined screening techniques (Sengupta and Hussain 2023). High-performance biosensors relying on both electrochemical and optical transduction foundations are also based on carbon dots and glassy carbon electrodes for rapid diagnosis of early cancers. The application of these nanomaterials in a miniature point-of-care device has enabled the increased overall sensing performance and enables noninvasive and quick cancer diagnosis, which has great implications for patient results and survival rates. Artificial intelligence in addition to a better biosensor efficiency will also be promoted by improvements in device architectures, signal processing and in the integration with clinical routines. Especially, graphene and its derivatives exhibit exceptional sensitivity, selectivity, stability, and broad detection range across diverse biosensing platforms, including field-effect transistors, fluorescence sensors, and electrochemical biosensors, thereby positioning them as promising candidates for future cancer screening and diagnostic applications (Pourmadadi et al. 2022).

8.6.3 Scale-Up Processes, Industrial Commercialization, and Circular Economy Strategies

Scaling up the industrial commercialization of waste-recycled carbon nanomaterials is necessary for a sustainable circular economy; however, it is not without significant hurdles. While it has a positive potential "waste-to-wealth" paradigm, the transition from laboratory to industrial production has always been obstructed due to the variability of raw waste materials and rigorous quality assurance requirements, so reliable batch-to-batch production of high-quality CNMs remains a major challenge (Balqis et al. 2023). Current frameworks are also inferior in making it possible to mass produce biomimetic compounds, such as graphene, from a wide variety of waste streams (biowastes, coal, industrial wastes), and for this to happen at scale (Balqis et al. 2023). However, the technology and synthesis of industrial-level functionalized carbon nanostructures is a burgeoning field with potential applications far reaching beyond biomedicine for sustainable energy and electronics. To resolve this problem, the focus on improving bulk synthesis and the commercialization of processing as well as manufacturing techniques is essential. This will require an overarching shift in thinking about the concept of waste, that is, that waste represents precious properties, for an industry, a process for sustainable, large-scale sustainable manufacturing in a circular economy (Liu et al. 2024).

8.7 Conclusion

This waste-recycled CNMs have great potential for promoting sustainable biomedical innovation in personalized medicine and theranostics. These CNMs, derived from biomass and industrial waste, can be cost-effective, biocompatible, and multifunctional platforms for drug delivery, bioimaging, and highly sensitive biosensing, especially for the early detection of cancer. Their unique physiochemical characteristics allow the realization of intelligent systems for accurate diagnosis and real-time therapeutic monitoring. In the meantime, large-scale application is limited by lack of harmonized guidelines, difficulties to maintain consistent quality from variable waste feed stocks, and insufficient understanding of long-term environmental and biological effects. These challenges need a systematic safety evaluation, appropriate quality control, and clearly established translational pathways from lab studies to clinical application. Advancing waste-recycled CNMs as a sustainable nanomedicine strategy, especially for antioxidant and anticancer therapies, require an emphasis on interdisciplinary cooperation across materials science, molecular biology, and clinical oncology. By combining circular economy principles with novel recycling strategies, CNM-based therapeutics will be developed in a targeted, environmentally responsible manner with better patient outcomes.

References

Ahmad MA, Aung Y-y, Widati AA, Sakti SCW, Sumarsih S, Irzaman, Yuliarto B, Chang J-y, Fahmi MZ (2023) A perspective on using organic molecules composing carbon dots for cancer treatment. Nanotheranostics 7:187

Akhtar MJ, Ahamed M, Alhadlaq HA, Alshamsan A (2017) Mechanism of ROS scavenging and antioxidant signalling by redox metallic and fullerene nanomaterials: potential implications in ROS associated degenerative disorders. Biochim Biophys Acta Gen Subj 1861:802–813

Alavi SE, Alavi SF, Koohi M, Raza A, Shahmabadi HE (2024) Nanoparticle-integrated metal–organic frameworks: a revolution in next-generation drug delivery systems. J Pharm Investig 54: 751–783

Algarra M, Vinacua S, Gil-Korilis A, Gil A (2025) Recent developments in the use of carbon-based nanomaterials in cancer therapy. J Control Release 386:114100

Aparicio RR, de Lima GG, Perissutti GE, Bassi M d J, Jacumazo J, Schiavon MA, Roman LS, de Muniz GIB, Magalhães WLE, de Cademartori PHG (2025) Pyrolytic lignin: a promising precursor for the green synthesis of fluorescent carbon nanoparticles. ACS Omega 10:11054–11062

Aslam J, Aslam R, Hussain CM (2024) Waste-derived nanoparticles: synthesis, applications and sustainability. Elsevier

Badea MA, Balas M, Prodana M, Cojocaru FG, Ionita D, Dinischiotu A (2022) Carboxyl-functionalized carbon nanotubes loaded with cisplatin promote the inhibition of PI3K/Akt pathway and suppress the migration of breast cancer cells. Pharmaceutics 14:469

Balqis N, Jan BM, Metselaar HSC, Sidek A, Kenanakis G, Ikram R (2023) An overview of recycling wastes into graphene derivatives using microwave synthesis; trends and prospects. Materials 16:3726

Bedhiafi T, Idoudi S, Fernandes Q, Al-Zaidan L, Uddin S, Dermime S, Billa N, Merhi M (2023) Nano-vitamin C: a promising candidate for therapeutic applications. Biomed Pharmacother 158: 114093

Behnam B, Shier WT, Nia AH, Abnous K, Ramezani M (2013) Non-covalent functionalization of single-walled carbon nanotubes with modified polyethyleneimines for efficient gene delivery. Int J Pharm 454:204–215

Benko A, Medina-Cruz D, Wilk S, Ziąbka M, Zagrajczuk B, Menaszek E, Barczyk-Woźnicka O, Guisbiers G, Webster TJ (2023) Anticancer and antibacterial properties of carbon nanotubes are governed by their functional groups. Nanoscale 15:18265–18282

Bolshakova OI, Slobodina AD, Sarantseva SV (2022) Carbon nanoparticles as promising neuroprotectors: pro et contra. II. Application of carbon nanoparticles in neurobiology and neurology. Nanobiotechnology Rep 17:141–154

Chary KJS, Sharma A, Singh A (2023) Carbon quantum dots in healthcare: a promising solution for sustainable healthcare and biomedical practices. E3S Web Conf 453:01017. EDP Sciences

Chen Z, Ma L, Liu Y, Chen C (2012) Applications of functionalized fullerenes in tumor theranostics. Theranostics 2:238

Choudhary N, Hwang S, Choi W (2014) Carbon nanomaterials: a review. In: Handbook of nanomaterials properties. Springer Science & Business Media, pp 709–769

Cui L, Ren X, Sun M, Liu H, Xia L (2021) Carbon dots: synthesis, properties and applications. Nanomaterials 11:3419

d'Amora M, Rodio M, Bartelmess J, Sancataldo G, Brescia R, Zanacchi FC, Diaspro A, Giordani S (2016) Biocompatibility and biodistribution of functionalized carbon nano-onions (f-CNOs) in a vertebrate model. Sci Rep 6:33923

Daré RG, Lautenschlager SOS (2025) Nanoparticles with antioxidant activity. MDPI, p 221

Das B, Pal P, Dadhich P, Dutta J, Dhara S (2018) In vivo cell tracking, reactive oxygen species scavenging, and antioxidative gene down regulation by long-term exposure of biomass-derived carbon dots. ACS Biomater Sci Eng 5:346–356

Debnath SK, Srivastava R (2021) Drug delivery with carbon-based nanomaterials as versatile nanocarriers: progress and prospects. Front Nanotechnol 3:644564

Devi M, Rawat S, Sharma S (2021) A comprehensive review of the pyrolysis process: from carbon nanomaterial synthesis to waste treatment. Oxf Open Mater Sci 1:itab014

Dhall S (2023) Carbon nanomaterials and their nanocomposite-based chemiresistive gas sensors: applications, fabrication and commercialization. Elsevier

Dong C, Ma X, Huang Y, Zhang Y, Gao X (2022) Carbon dots nanozyme for anti-inflammatory therapy via scavenging intracellular reactive oxygen species. Front Bioeng Biotechnol 10: 943399

Dong C, Wang Y, Chen T, Ren W, Gao C, Ma X, Gao X, Wu A (2024) Carbon dots in the pathological microenvironment: ROS producers or scavengers? Adv Healthc Mater 13:2402108

Duraia E-s M, Opoku M, Beall GW (2024) Efficient eco-friendly synthesis of carbon nanotubes over graphite nanosheets from yellow corn: a one-step green approach. Sci Rep 14:16405

Elahi N, Zeinalipour-Yazdi CD (2025) Advances in the synthesis of carbon nanomaterials towards their application in biomedical engineering and medicine. C 11:35

Elemike EE, Onunkwo IC, Ughumiakpor O, Alawuru F, Mukoro A, Ishom P, Obarakpor F, Hossain I, Aziza AE (2025) Bio-nanomaterials: promising anticancer properties and treatment strategies. Nano TransMed 4:100076

Elkodous A, Mohamad HA, Hamad MIA, Maksoud A, Ali GAM, El Abboubi M, Bedir AG, Eldeeb AA, Ayed AA, Gargar Z, Zaki FS (2022) Cutting-edge development in waste-recycled nanomaterials for energy storage and conversion applications. Nanotechnol Rev 11:2215–2294
Famutimi OG, Masha S, Maluleke R, Ncapayi V, Lebepe TC, Mgedle N, Kungwa CM, Fanoro OT, Adewale IO, Oluwafemi OS (2025) Assessment of antioxidant potential of carbon-based nanomaterials from different sources. Antioxidants 14:1227
Feng L, Xie N, Zhong J (2014) Carbon nanofibers and their composites: a review of synthesizing, properties and applications. Materials 7:3919–3945
Flores DG, Espino J, Pariente JA (2023) Antioxidant potential of nanomaterials. Turk J Biol 47: 218–235
Gao X, Fang W, Li W, Wang P, Khan K, Tang Y, Wang T (2024) Effects of multidimensional carbon-based nanomaterials on the low-carbon and high-performance cementitious composites: a critical review. Materials 17:2196
Gayathri K, Vidya R (2024) Carbon nanomaterials as carriers for the anti-cancer drug doxorubicin: a review on theoretical and experimental studies. Nanoscale Adv 6:3992–4014
Getachew G, Korupalli C, Rasal AS, Chang J-Y (2021) ROS generation/scavenging modulation of carbon dots as phototherapeutic candidates and peroxidase mimetics to integrate with polydopamine nanoparticles/GOx towards cooperative cancer therapy. Compos Part B 226: 109364
González-Arias J, Sánchez ME, Cara-Jiménez J, Baena-Moreno FM, Zhang Z (2022) Hydrothermal carbonization of biomass and waste: a review. Environ Chem Lett 20:211–221
Gujar V, Li H, Paull TT, Neumann CA, Weyemi U (2025) Unraveling the nexus: genomic instability and metabolism in cancer. Cell Rep 44:115540
Hami Z (2021) A brief review on advantages of nano-based drug delivery systems. Ann Mil Health Sci Res 19(1):1–6
Havrdova M, Hola K, Skopalik J, Tomankova K, Petr M, Cepe K, Polakova K, Tucek J, Bourlinos AB, Zboril R (2016) Toxicity of carbon dots–effect of surface functionalization on the cell viability, reactive oxygen species generation and cell cycle. Carbon 99:238–248
Hejna M, Świechowski K, Rasaq WA, Białowiec A (2022) Study on the effect of hydrothermal carbonization parameters on fuel properties of chicken manure hydrochar. Materials 15:5564
Huang X, Yin Z, Wu S, Qi X, He Q, Zhang Q, Yan Q, Boey F, Zhang H (2011) Graphene-based materials: synthesis, characterization, properties, and applications. Small 7:1876–1902
Huq R, Samuel ELG, Sikkema WKA, Nilewski LG, Lee T, Tanner MR, Khan FS, Porter PC, Tajhya RB, Patel RS (2016) Preferential uptake of antioxidant carbon nanoparticles by T lymphocytes for immunomodulation. Sci Rep 6:33808
Ibrahim KS (2013) Carbon nanotubes properties and applications: a review. Carbon Lett 14:131–144
Innocenzi P, Stagi L (2023) Carbon dots as oxidant-antioxidant nanomaterials, understanding the structure-properties relationship. A critical review. Nano Today 50:101837
Jayaprakash N, Elumalai K, Manickam S, Bakthavatchalam G, Tamilselvan P (2024) Carbon nanomaterials: revolutionizing biomedical applications with promising potential. Nano Mater Sci. https://doi.org/10.1016/j.nanoms.2024.11.004
Ji Z, Sheardy A, Zeng Z, Zhang W, Chevva H, Allado K, Yin Z, Wei J (2019) Tuning the functional groups on carbon nanodots and antioxidant studies. Molecules 24:152
Ji Y, Yan S, Xu W, Yang M, Chen A, Hou S, Du J (2025) Nanocarbon dots enhancing photothermal therapy for cancer: mechanisms, synergistic strategies, and frontier advances. J Mater Chem B 13:14194
Jiang T, Amadei CA, Gou N, Lin Y, Lan J, Vecitis CD, Gu AZ (2020) Toxicity of single-walled carbon nanotubes (SWCNTs): effect of lengths, functional groups and electronic structures revealed by a quantitative toxicogenomics assay. Environ Sci Nano 7:1348–1364
Jiang T, Lin Y, Amadei CA, Gou N, Rahman SM, Lan J, Vecitis CD, Gu AZ (2021) Comparative and mechanistic toxicity assessment of structure-dependent toxicity of carbon-based nanomaterials. J Hazard Mater 418:126282

Jiang H, Xia C, Lin J, Garalleh HAL, Alalawi A, Pugazhendhi A (2023) Carbon nanomaterials: a growing tool for the diagnosis and treatment of diabetes mellitus. Environ Res 221:115250
Jung H-S, Neuman KC (2021) Surface modification of fluorescent nanodiamonds for biological applications. Nanomaterials 11:153
Kamal A, Hong S, Ju H (2025) Carbon quantum dots: synthesis, characteristics, and quenching as biocompatible fluorescent probes. Biosensors 15:99
Karakoti M, Dhali S, Rana S, Prasanna SRVS, Mehta SPS, Sahoo NG (2018) Surface modification of carbon-based nanomaterials for polymer nanocomposites. In: Carbon-based polymer nanocomposites for environmental and energy applications. Elsevier
Kashtiban AE, Okpala COR, Karimidastjerd A, Zahedinia S (2024) Recent advances in nano-related natural antioxidants, their extraction methods and applications in the food industry. Explor Foods Foodomics 2:125–154
Khalil I, Yehye WA, Etxeberria AE, Alhadi AA, Dezfooli SM, Julkapli NB, Basirun WJ, Seyfoddin A (2019) Nanoantioxidants: Recent trends in antioxidant delivery applications. Antioxidants 9(1):24
Kharlamova MV, Kramberger C, Chernov AI (2023) Progress in carbon nanostructures: from synthesis to applications, vol 2181. MDPI
Kim YA, Hayashi T, Endo M, Dresselhaus MS (2013) Carbon nanofibers. In: Springer handbook of nanomaterials. Springer
Krishnamurthy K, Lindsey AM, Estrada C-A, Martinez CC, Cusnir M, Schwartz M, Sriganeshan V, Poppiti R (2020) Genomic landscape of squamous cell carcinoma-different genetic pathways culminating in a common phenotype. Cancer Treat Res Commun 25:100238
Kyriakidou K, Brasinika D, Trompeta AFA, Bergamaschi E, Karoussis IK, Charitidis CA (2020) In vitro cytotoxicity assessment of pristine and carboxyl-functionalized MWCNTs. Food Chem Toxicol 141:111374
Lee J-H, Park S-J, Choi J-W (2019) Electrical property of graphene and its application to electrochemical biosensing. Nanomaterials 9:297
Lin L, Zheng Z, Li X, Park S, Zhang W, Diao G, Piao Y (2023) Design strategy for porous carbon nanomaterials from rational utilization of natural rubber latex foam scraps. Ind Crop Prod 192: 116036
Liu H, Chen J, Qiao S, Zhang W (2021) Carbon-based nanomaterials for bone and cartilage regeneration: a review. ACS Biomater Sci Eng 7:4718–4735
Liu Y, Zeng S, Ji W, Yao H, Lin L, Cui H, Santos HA, Pan G (2022) Emerging theranostic nanomaterials in diabetes and its complications. Adv Sci 9:2102466
Liu Y, Zhang J, Chunying W, Lai Y, Fan H, Wang Q, Lin Z, Chen J, Zhao X, Jiang X (2024) Nanoplatform based on carbon nanoparticles loaded with doxorubicin enhances apoptosis by generating reactive oxygen species for effective cancer therapy. Oncol Lett 27:288
Liu ZJ, Roy A, Zheng Y, Annabi N (2025) Harnessing carbon nanomaterials for reactive oxygen species regulation: insights into generation, scavenging, and sensing. Adv Drug Deliv Rev 224: 115651
Madkour LH (2024) Carbon-based nanomaterials for sustainable and technological applications. CRC Press
Mauro N, Utzeri MA, Varvarà P, Cavallaro G (2021) Functionalization of metal and carbon nanoparticles with potential in cancer theranostics. Molecules 26:3085
Mehta J, Dilbaghi N, Deep A, Hai FI, Hassan AA, Kaushik A, Kumar S (2025) Plastic waste upcycling into carbon nanomaterials in circular economy: synthesis, applications, and environmental aspects. Carbon 234:119969
Mohamed MG, Mansour NA, Mousa AA, Mazrouaa AM (2025) Applications of solid waste-derived carbon nanomaterials in packaging material. In: Waste derived carbon nanomaterials, vol 2. ACS Publications
Mohan H, Fagan A, Giordani S (2023) Carbon nanomaterials (CNMs) in cancer therapy: a database of cnm-based nanocarrier systems. Pharmaceutics 15:1545

Mousavi-Kiasary SMS, Ahmood Senabreh, Ashkan Zandi, Rogelio Pena, Frances Cruz, Ali Adibi, Hooshmand N (2025) Synergistic cancer therapies enhanced by nanoparticles: advancing Nanomedicine through multimodal strategies. Pharmaceutics 17:682

Murr LE, Garza KM (2009) Natural and anthropogenic environmental nanoparticulates: their microstructural characterization and respiratory health implications. Atmos Environ 43:2683–2692

Ndlwana L, Raleie N, Dimpe KM, Ogutu HF, Oseghe EO, Motsa MM, Msagati TAM, Mamba BB (2021) Sustainable hydrothermal and solvothermal synthesis of advanced carbon materials in multidimensional applications: a review. Materials 14:5094

Nie R, Zhang J, Jia Q, Li Y, Tao W, Qin G, Liu X, Tao Y, Zhang Y, Li P (2024) Structurally oriented carbon dots as ROS nanomodulators for dynamic chronic inflammation and infection elimination. ACS Nano 18:22055–22070

Nilewski LG, Sikkema WKA, Kent TA, Tour JM (2015) Carbon nanoparticles and oxidative stress: could an injection stop brain damage in minutes? Taylor & Francis, pp 1677–1679

Omar RA, Talreja N, Chuhan D, Ashfaq M (2024) Waste-derived carbon nanostructures (WD-CNs): an innovative step toward waste to treasury. Environ Res 246:118096

Pandey K, Saharan BS, Singh Y, Sadh PK, Duhan JS, Jabborova D (2026) Biodegradable innovations: harnessing agriculture for eco-friendly plastics. J Xenobioth 16:8

Parveen A, Chatterjee A, Karak P (2025) Biomedical applications of carbon-based nanomaterials: exploring recent advances in therapeutics, diagnostics, and tissue engineering. Adv Pharm Bull 15:232

Pikula K, Johari SA, Golokhvast K (2022) Colloidal behavior and biodegradation of engineered carbon-based nanomaterials in aquatic environment. Nanomaterials 12:4149

Popov VN (2004) Carbon nanotubes: properties and application. Mater Sci Eng R Rep 43:61–102

Pourmadadi M, Dinani HS, Tabar FS, Khassi K, Janfaza S, Tasnim N, Hoorfar M (2022) Properties and applications of graphene and its derivatives in biosensors for cancer detection: a comprehensive review. Biosensors 12:269

Qiu Y, Wang Z, Owens ACE, Kulaots I, Chen Y, Kane AB, Hurt RH (2014) Antioxidant chemistry of graphene-based materials and its role in oxidation protection technology. Nanoscale 6:11744–11755

Qureshi SS, Shah V, Nizamuddin S, Mubarak NM, Karri RR, Dehghani MH, Ramesh S, Khalid M, Rahman ME (2022) Microwave-assisted synthesis of carbon nanotubes for the removal of toxic cationic dyes from textile wastewater. J Mol Liq 356:119045

Rabiee N, Ahmadi S, Rabiee M, Bagherzadeh M, Vahabi H, Jouyandeh M, Saeb MR (2022) Green carbon-based nanocomposite biomaterials through the lens of microscopes. Emerg Mater 5:665–671

Rasyotra A, Thakur A, Gaykwad B, Jani V, Jasuja K (2024) Carbon nanomaterials: novel structures for biological applications. In: Nanobiotechnology. CRC Press

Rehor I, Slegerova J, Kucka J, Proks V, Petrakova V, Adam M-P, Treussart F, Turner S, Bals S, Sacha P (2014) Fluorescent nanodiamonds embedded in biocompatible translucent shells. Small 10:1106–1115

Sager TM, Wolfarth MW, Andrew M, Hubbs A, Friend S, Chen T-h, Porter DW, Wu N, Yang F, Hamilton RF (2014) Effect of multi-walled carbon nanotube surface modification on bioactivity in the C57BL/6 mouse model. Nanotoxicology 8:317–327

Sengupta J, Hussain CM (2023) The emergence of carbon nanomaterials as effective nano-avenues to fight against COVID-19. Materials 16:1068

Sharaf Addin AH, Ramli RM, Sufian S, Al-Mahbashi N, Idris SS, Farooqi AS, Shamsu Bakri SH (2025) Sustainable synthesis of carbon nanomaterials from waste tire-derived char using microwave pyrolysis. J Ecol Eng 26:189

Sivaprasad S, Manandhar A, Shah A (2021) Hydrothermal carbonization: upgrading waste biomass to char. The Ohio State University Extension, Columbus

Soonnarong R, Tungsukruthai S, Nutho B, Rungrotmongkol T, Vinayanuwattikun C, Maluangnont T, Chanvorachote P (2021) Titania nanosheet generates peroxynitrite-dependent S-nitrosylation and enhances p53 function in lung cancer cells. Pharmaceutics 13:1233

Speranza G (2021) Carbon nanomaterials: synthesis, functionalization and sensing applications. Nanomaterials 11:967

Takhar V, Singh S (2025) Nanomaterials ROS: a comprehensive review for environmental applications. Environ Sci Nano 12:2516–2550

Tang N, Ding Z, Zhang J, Cai Y, Bao X (2023) Recent advances of antioxidant low-dimensional carbon materials for biomedical applications. Front Bioeng Biotechnol 11:1121477

Tatrari G, Karakoti M, Tewari C, Pandey S, Bohra BS, Dandapat A, Sahoo NG (2021) Solid waste-derived carbon nanomaterials for supercapacitor applications: a recent overview. Mater Adv 2: 1454–1484

Thakur A, Kumar A, Berdimurodov E (2024) Biomass wastes: fundamentals, classification and properties. In: Biomass wastes for sustainable industrial applications. CRC Press

Tsachouridis S, Papaioannidou P (2010) Fullerenes: chemical structure and properties. Front Pharmacol

Udourioh GA, Solomon MM, Okolie JA (2025) A review of the valorization of dairy industry wastes through thermochemical, biological, and integrated processes for value-added products. Food Sci Anim Resour 45:375

Unnikrishnan G, Anjumol Joy M, Megha EK, Senthilkumar M (2023) Exploration of inorganic nanoparticles for revolutionary drug delivery applications: a critical review. Discov Nano 18: 157

Venkataraman S, Sundramoorthy AK (2026) Waste-derived carbon nanomaterials for electrochemical applications: toward a circular and sustainable future. RSC Adv 16:733–747

Vidyalakshmi IS, Raj A, Neelima S, Vidya L, Thomas RK (2022) Environmental applications of carbon nanotubes. In: Carbon nanotubes. Apple Academic Press

Wan D, Jiang C, Hua X, Wang T, Chai Y (2015) Cell cycle arrest and apoptosis induced by aspidin PB through the p53/p21 and mitochondria-dependent pathways in human osteosarcoma cells. Anti-Cancer Drugs 26:931–941

Wang L, Liu F, Ning Y, Bradley R, Yang C, Yong K-T, Zhao B, Wu W (2020) Biocompatible mesoporous hollow carbon nanocapsules for high performance supercapacitors. Sci Rep 10: 4306

Wang Z-G, Liu S-L, Pang D-W (2021) Quantum dots: a promising fluorescent label for probing virus trafficking. Acc Chem Res 54:2991–3002

Wang Z, Li L, Ren G, Duan X, Guo J, Liu W, Ang Y, Zhu L, Ren X (2022) A comprehensive review on stability of therapeutic proteins treated by freeze-drying: induced stresses and stabilization mechanisms involved in processing. Dry Technol 40:3373–3388

Wareing TC, Gentile P, Phan AN (2021) Biomass-based carbon dots: current development and future perspectives. ACS Nano 15:15471–15501

Wu T, Li M, Li T, Zhao Y, Yuan J, Zhao Y, Tian X, Kong R, Zhao Y, Kong H (2023) Natural biomass-derived carbon dots as a potent solubilizer with high biocompatibility and enhanced antioxidant activity. Front Mol Biosci 10:1284599

Xia C, Zhu S, Feng T, Yang M, Yang B (2019) Evolution and synthesis of carbon dots: from carbon dots to carbonized polymer dots. Adv Sci 6:1901316

Xu G, Tang Y, Xiong D, Zhang W, Liu Z, Chu PK, Wang G (2025) Carbon dots for reactive oxygen species modulation. Mater Sci Eng R Rep 166:101024

Yazdani S, Mozaffarian M, Pazuki G, Hadidi N (2024) Carbon nanostructure: cutting-edge platforms for advanced genetic transport. Cell Signal 2:148–154

Yousaf T, Ahmad I, Younas Z, Hamdard MH, Mashwani Z-u-R (2025) Emerging plant-based nanotechnological advances and molecular insights for type-2 diabetes, diagnosis and treatments-recent trends and future prospects. ACS Omega 10:35310–35326

Yousefi A-M, Pourbagheri-Sigaroodi A, Fakhroueian Z, Salari S, Fateh K, Momeny M, Bashash D (2022) Anticancer effects of ZnO/CNT@ Fe3O4 in AML-derived KG1 cells: shedding light on promising potential of metal nanoparticles in acute leukemia. Int J Hematol Oncol Stem Cell Res 16:140

Yu Y, Zhang Q, Mu Q, Zhang B, Yan B (2008) Exploring the immunotoxicity of carbon nanotubes. Nanoscale Res Lett 3:271

Zhang W, Zhang Z, Zhang Y (2011) The application of carbon nanotubes in target drug delivery systems for cancer therapies. Nanoscale Res Lett 6:555

Zhao L, Wang Y, Li Y (2019) Antioxidant activity of graphene quantum dots prepared in different electrolyte environments. Nanomaterials 9:1708

Chapter 9
Food-Waste Bio-refinery Implications of Carbon Nanostructure Synthesis and Fabrication

Ramya Sree, Bishwambhar Mishra, Sanjeeb Kumar Mandal, Ashoutosh Panday, Dinesh Chand Agrawal, and Rajeev Kumar Jha

9.1 Introduction

Inefficiencies regarding the production, distribution and consumption of food underline the global issue of food waste. It represents the loss of vital resources and has detrimental effects economically, environmentally and socially. Estimates show that almost 17% of food produced globally is wasted every year, resulting in lose-lose situations for both consumers and producers (Alexander et al. 2017). More dangerously, food waste is responsible for nearly 8% of the total global greenhouse gas emissions. This worsens climate change and the degradation of the environment. In the last few years, various research and studies have shown promising results for the conversion of food waste into different forms of carbon allotropes (Hossain et al. 2024). These carbon allotropes hold great importance for the improvement of performance in energy generation and storage devices due to their surface areas. High surface area carbon materials have been synthesised from food waste. Considering these concerns, new approaches for the valorisation of food waste have been developed as a means to promote circular economy initiatives and sustainable waste management (Davidraj et al. 2024). Figure 9.1 illustrates the key theme of food

R. Sree · B. Mishra (✉) · S. K. Mandal · A. Panday
Department of Biotechnology, Chaitanya Bharathi Institute of Technology, Hyderabad, Telangana, India
e-mail: bishwambhar_biotech@cbit.ac.in

D. C. Agrawal
Department of Biotechnology, School of Life and Basic Science, Jaipur National University, Jaipur, Rajasthan, India

R. K. Jha
School of Health Sciences, The Neotia University, Kolkata, West Bengal, India

Y. K. Mohanta et al. (eds.), *Bioresource Waste to Carbon Nanomaterial*, Nanotechnology in the Life Sciences, https://doi.org/10.1007/978-981-92-2405-0_9

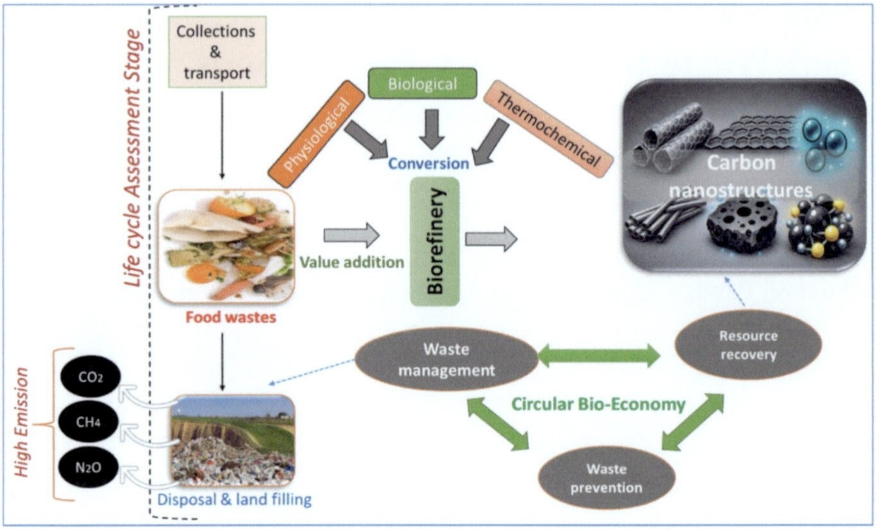

Fig. 9.1 Overview of the biorefinery process for food waste and the production of carbon nanostructures within the circular bioeconomy framework

waste biorefinery under the boundary of circular economy for the production of carbon nanomaterials.

During the collection and transport of food waste, conventional disposal methods, such as landfilling, can occur, leading to greenhouse gas (CO_2, CH_4 and N_2O) emissions. Alternatively, the food waste can be valued through biorefinery-based conversion methods, which include one or more of the biological, physiological and thermochemical pathways. This approach fosters the transformation and valorisation of food waste into high-value carbon nanostructures, thus supporting the sustainable resource recovery, waste management and waste prevention principles within a circular bioeconomy framework. One innovative solution includes the conversion of food waste into carbon-based nanomaterials (CNMs). Organic carbon is abundant in food residues such as the by-products of plants (fruit peels, husks, shells and stalks), the remains of animals (bones, fat and proteins) and remnants from food processing (spent grains, coffee grounds and pulp). This makes them excellent starting substances for the development of carbon nanostructures like carbon nanotubes (CNTs), graphene, carbon nanofibres, activated carbon and quantum dots (Chand et al. 2025). CNMs from food waste address the problem of waste in food while also producing high-valued functional materials for energy storage, catalysis, the purification of water, environmental remediation, sensors and biomedicine. Therefore, the conversion of food waste into nanostructured carbon materials is a solution for waste management and advanced material science. It offers a pathway to sustainability and technological advancement (François et al. 2024).

Recent advancements in the use of carbon nanomaterials (CNMs) made from agro-industrial waste show promise in new application areas such as energy storage and conversion, environmental remediation, catalysis, biomedicine, agriculture and materials science. Organisations involved in carbon nanomaterials (CNMs) and their conversion from agro-industrial waste utilise their high surface area and chemical stability, along with high mechanical strength and electrical conductivity, to develop nanomaterials that can be used in innovative ways. For example, CNMs produced from food waste can be employed in energy storage devices, in the treatment of wastewater, as delivery systems for target drugs and in enhancing soil for agriculture (Chand et al. 2025; Venkataraman and Sundramoorthy 2026). Thus, the conversion of agro-industrial waste to CNMs provides a sustainable solution to waste management as well as the creation of new functional materials. CNMs derived from food waste demonstrate the environmental and economic potential of providing sustainable solutions, extending the lifespan of a product's usefulness or moving towards a zero waste model of linear consumption and disposal. Integrating biorefinery technology within the circular bioeconomy framework illustrates a practical approach to (i) closing the carbon and nutrient loops, (ii) reducing greenhouse gas emissions and (iii) lessening the adverse effects of pollution. In the circular bioeconomy systems, biomass and organic waste are transformed into renewable fuels, chemicals and advanced materials. With this, they promote the sustainable use of natural resources and the sustainable development of the economy (Zakaria et al. 2024; Fahad Halim et al. 2025).

This chapter focuses on the biorefineries of food waste and their potential in sustainably producing carbon nanostructures. The chapter explains the biorefinery concept and circular bioeconomy, the carbon-rich precursor, food waste and the primary methods of synthesis and fabrication of carbon nanomaterials. The applications, barriers and the prospects of the food-waste-derived carbon nanostructures that drive the development of sustainable materials are discussed in this chapter.

9.2 Food Waste as a Carbon Source

Food waste is increasingly recognized as a valuable renewable carbon-rich resource for the production of value-added materials, including carbon-based nanostructures. Food waste is generated in large amounts in the household, agricultural, food processing industry and food retail sectors. Up until now, food waste has been disposed of by means of landfilling or incineration, and these waste disposal methods pose a serious environmental pollution problem and also result in greenhouse gases (Ahmed et al. 2025; Vatieri et al. 2025). From a scientific research viewpoint, the high level of organic and carbon content in food waste makes the disposal of food waste as a precursor for producing a wide range of carbon materials such as activated carbon, carbon nanotubes, graphene and carbon quantum dots, a very environmentally damaging practice. From this perspective, food waste as a precursor for producing carbon-based materials is a means of achieving sustainable waste management and also solidly supports the principles of resource recovery and the circular bioeconomy (Dhandapani et al. 2022; Giwa et al. 2026).

9.2.1 *Major Types of Food Waste*

Food waste generated across the food supply chain can be broadly classified into four major categories based on its source and characteristics:

9.2.1.1 Household Food Waste

Food waste from households consists of food materials discarded during the food preparation and consumption processes within a household. Typical examples include leftover cooked food, spoiled fruits and vegetables, food peels and eggshells (Todd and Faour-Klingbeil 2024). Packaged food that is no longer within the sell-by date also counts here. Food waste from households is a large component of the municipal organic waste and abundant in carbohydrate, protein and lipid content. This means that food waste from households can also be a suitable feedstock for the production of buries (Mohamed Yusoff et al. 2024). Figure 9.2 describes sources of food waste and various biorefinery technology.

9.2.1.2 Agricultural and Post-harvest Waste

This category includes the wastes obtained during the harvesting of crops and the subsequent handling of the crops. Examples of these wastes include peels of fruits, stems of vegetables, cobs of corn, husks of rice, bran of wheat, bagasse of sugarcane

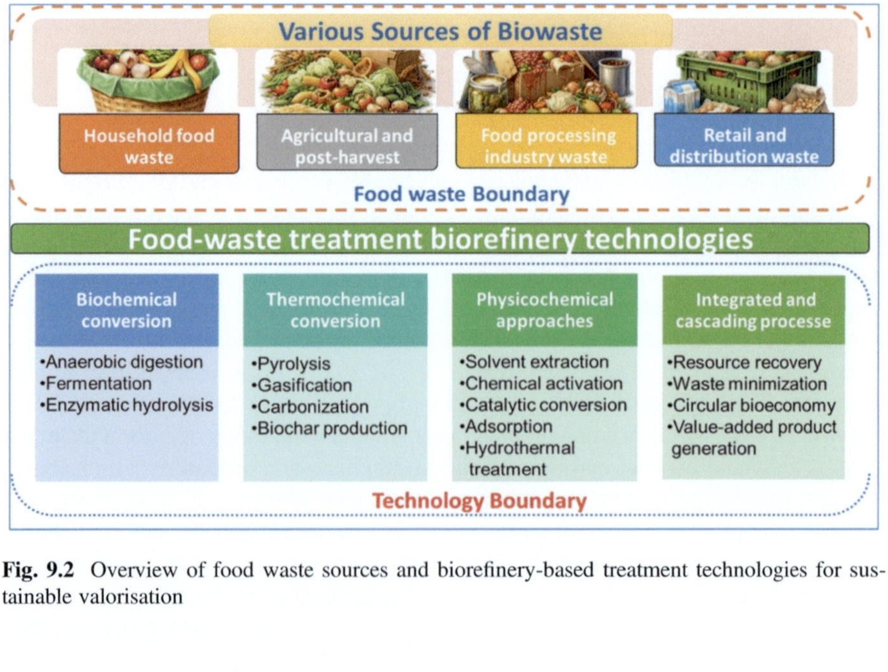

Fig. 9.2 Overview of food waste sources and biorefinery-based treatment technologies for sustainable valorisation

and other wastes of plant origin. The main composition of these materials is a type of organic polymer known as lignocellulose, which is made up of cellulose, hemicellulose and lignin. Because of their high carbon content and their stable structure, agricultural wastes are abundantly used as raw materials for the production of porous carbon materials and carbon nanostructures (Yafetto et al. 2023; Wagh et al. 2024).

9.2.1.3 Food Processing Industry Waste

While processing food products, food processing industries generate large quantities of organic by-products such as fruit pomaces, pulp, dairy, whey, oilseed cakes, brewery residues, fish scales and shellfish shells (Mishra et al. 2023). Processing industry waste has a high concentration of organic processing waste, which consists of carbon, proteins, and lipids, and therefore can be processed into carbon materials by thermochemical or biochemical processing (Lanjekar et al. 2024).

9.2.1.4 Retail and Distribution Waste

Retail and distribution waste includes food items that are thrown away during storage, transport and sales because they are damaged, spoiled or are discarded due to poor aesthetic qualities. Examples of retail and distribution waste are expired packaged food items, damaged produce and food items that remain unsold in supermarkets and restaurants. Retail and distribution waste is a type of waste that is readily overlooked; however, it contains a high concentration of organic matter that can be utilised as a potential carbon source in a biorefinery (Todd and Faour-Klingbeil 2024).

9.2.2 Chemical Composition and Suitability for Carbon Materials

The potential for synthesising carbon materials from food waste depends on the specific food waste's chemical constituents, chiefly, carbohydrates, proteins, lipids, lignin and cellulose which are made mostly of carbon and can be converted into stable carbon structures via pyrolysis, hydrothermal carbonisation and chemical activation. In food waste, the most abundant constituents are likely to be some form of carbohydrate, especially starch, cellulose or hemicellulose (Liu et al. 2023). These materials are thermochemically converted via reactions of dehydration, polymerisation and carbonisation to yield formation of structures rich in carbon. Food waste rich in proteins supplies waste nitrogen which can be incorporated into carbon structures yielding the carbon materials, now doped with nitrogen, which have enhanced conductivity and catalytic activity. In addition to previously mentioned constituents, food waste rich in lipids (i.e. oils and fats) provides further carbon content and in the thermal conversion process, aids in the formation of porous structures in the carbon formed. The stable aromatic framework of lignin, cellulose and hemicellulose, abundant in agricultural residues, are excellent precursors for the preparation of activated carbon and other materials with porous carbons (Nicolae et al. 2020; Zhou et al. 2023).

Due to these compositional features, food waste is an inexpensive, plentiful and sustainable source of carbon feedstock for the production of new carbon nanomaterials. The process of transforming food waste into carbon materials, within the circular and bioeconomy and waste valorisation strategies, not only diminishes environmental pollution but also fosters the sustainable development of materials (Bhatia et al. 2023).

9.3 Biorefinery Approaches for Food Waste Valorisation

The biorefinery approach for food waste valorisation focuses on organic wastes with potential for developing profitable biobased products including biofuels, bioenergy, biochemicals and other biobased materials (Tsegaye et al. 2021). A biorefinery is defined as a facility that utilises several conversion techniques for the efficient and maximum recovery of the useful components from given biomass. Food wastes, due to their high carbohydrates, proteins and lipids, as well as lignocellulosic constituents, are an appropriate feedstock for a variety of biorefinery processes (Schmidt et al. 2018).

9.3.1 Biochemical Conversion

Biochemical conversion is the most common approach of food waste biorefineries, as it relies on microorganisms and enzymes to process organic substrates into useful, biobased products. The most common processes in food waste biorefineries are anaerobic digestion, fermentation and enzymatic hydrolysis (De Buck et al. 2020). Anaerobic digestion of organic waste leads to the formation of biogas (methane and carbon dioxide), which can be transformed into useful energy. Similarly, the fermentation of carbohydrate-rich food waste streams can yield bioethanol, organic acids and biopolymers, which are also of great value (Sevillano et al. 2021).

9.3.2 Thermochemical Conversion

Another technique of great importance is thermochemical conversion, which applies heat with a controlled degree of oxygen to convert biomass to energy and other carbon-rich products. The conversion of food waste to biochar, syngas and liquid biofuels through pyrolysis, gasification and hydrothermal carbonisation is especially applicable for the transformation of lignocellulosic food waste into carbonaceous materials, which can be used as precursors for sophisticated carbon nanostructures (Amalina et al. 2022; Wang and Wu 2023).

9.3.3 *Physicochemical Approaches*

Aside from biochemical and thermochemical techniques, the physicochemical approaches to food waste biorefineries include the use of some biopolymers such as solvent extraction, catalytic conversion and chemical activation, which are used in the recovery of valuable materials such as lipids, antioxidants and bioactive substances. The recovered products can be used in the food, pharmaceutical and materials sciences industries (Capanoglu et al. 2022).

9.3.4 *Integrated and Cascading Processes*

In modern food waste biorefineries, integrated cascade systems are highlighted in which several conversion routes are integrated to optimise the use of available resources. In such systems, a sequential approach is employed where high-value substances are recovered first, followed by the recovery of energy and the generation of carbon-rich residues, which can be used in the synthesis of materials. This cascading approach maximises the use of resources for multiple purposes and minimises unsought waste (Lin et al. 2014; Osman et al. 2021). Overall, the biorefinery process offers a sustainable method for converting food waste into useful products and mitigating the negative impact on the environment. With the combination of biochemical, thermochemical and physicochemical processes, food waste can be converted into fuels, chemicals and advanced carbon products. This process enables the circular bioeconomy and sustainable management of resources (García-Morales and Fernández-Morales 2024).

9.4 Carbon Nanostructures: Types and Properties

Carbon nanostructures are nanoscale materials made mostly of carbon atoms formed into specific architectures. Carbon's special versatility has to do with its ability to form very different types of chemical bonds, especially covalent bonds of the types called sp^2 and sp^3. These types of bonds result in the formation of many different nanostructures with many different properties and characteristics (Slepičková Kasálková et al. 2021). The last few decades have seen a very large amount of research and materials development of nanoscale carbon structures because of extraordinary characteristics such as great mechanical strength, large surface area and high electrical and thermal conductivities, as well as thermal and chemical stabilities. Because of these properties, carbon nanostructures are used in energy storage devices, environmental clean-up, catalysts, electronics, sensors and biotechnology. Recently, researchers have focused their attention on the sustainable development of carbon nanostructures from biomass materials like food waste. Carbon-rich compounds in food waste are especially suitable for producing advanced carbon nanomaterials (Speranza 2021).

9.4.1 Carbon Nanotubes

As cylindrical nanostructures created from rolled-up sheets of graphene, carbon nanotubes (CNTs) have tube-like structures that have diameters of 1 nanometre. Their discovery in 1991 by Iijima launched an explosion of interest and research due to their remarkable structural and functional attributes. CNTs can be divided into single-walled carbon nanotubes (SWCNTs) containing one graphene layer and multi-walled carbon nanotubes (MWCNTs) containing several layers of graphene. CNTs' excellent mechanical, thermal and electrical properties are attributable to their highly ordered graphene structures. CNTs are extremely strong, flexible and rugged, outperforming most available structural materials in strength (Saifuddin et al. 2013; Ali et al. 2024). Their unique structural attributes facilitate a highly ordered graphitic structure that offers superior electrical and thermal conductivity. Due to their extremely large surface area and their high aspect ratio, CNTs are excellent for use in adsorption, catalysis and energy storage. CNTs have become well established for use in a variety of applications including nanoelectronics, sensors, composite reinforcement materials and electrochemical devices (Hughes et al. 2024).

9.4.2 Graphene and Graphene Oxide

Graphene consists of an atomic layer of carbon arranged in a two-dimensional hexagonal pattern (honeycomb structure) with astounding characteristics. Its immense scientific value is the result of its chemical and physical properties. Notably, its high chemical and physical properties include superior electrical conductivity and mobilization of electrons, one of the most sought-after nanostructures in optoelectronic and electronic industries. Along with outstanding thermal conductivity, extraordinary tensile strength and an enormous high surface area, it is suitable for applications of nanocomposites, sensors, batteries and supercapacitors (Walimbe and Chaudhari 2019; Urade et al. 2023). Graphene oxide (GO) is a derivative of graphene which, through the oxidation process, has surface functional groups (fg) containing oxygen, such as hydroxyl, epoxy and carboxyl fg. The presence of these functional groups allows the graphene oxide to improve its chemical reactivity and its dispersion in water, allowing for easier functionalisation and modifications. These qualities apply graphene oxide and its reduced form to catalysis, environmental remediation, adsorption and biomedical applications (Shahriari et al. 2021).

9.4.3 Carbon Quantum Dots

Carbon quantum dots (CQDs; also known as carbon dots) are nanoscale molecules and typically measure less than 10 nm. CQDs are gaining popularity because they are capable of strong and adjustable photoluminescence. In addition to their photoluminescence, CQDs are highly soluble in water, chemically stable and less toxic than standard quantum dots (Cui et al. 2021). They also exhibit great biocompatibility and possess fluorescence properties that are useful in bioimaging, biosensing, drug delivery and photocatalysis applications. CQDs also exhibit good electron transfer, allowing them to be used in optoelectronic devices and catalytic systems. Recently, hydrothermal, microwave-assisted and pyrolytic methods have been used to create CQDs from biomass and food waste, making these methods environmentally friendly (Molaei 2019).

9.4.4 Carbon Nanofibres

One-dimensional filamentous carbon materials with diameters of a few to hundreds of nanometres and lengths of several microns are referred to as carbon nanofibres (CNFs). Carbon nanofibres are formed with multiple graphene layers, which are organised at different angles with respect to the axis of the fibre, providing them with unique structural properties. They exhibit exceptional mechanical strength and rigidity as well as remarkable electrical and thermal stability (Osorio-Aguilar et al. 2024). Their considerable surface area and porous structure also make them useful in a range of electrochemical and catalytic applications. Due to the network of CNFs, batteries and supercapacitors are able to utilise them to enhance conductivity and optimise electron transport. Furthermore, CNFs are employed in polymer composites to provide reinforcement and also to serve as supports for catalysts in various chemical processes (Zhou et al. 2020).

9.4.5 Activated Carbon and Porous Carbon Materials

One important class of carbon materials are activated carbons and porous carbons, which are manufactured from numerous carbon-rich materials, including biomass and agro-industrial wastes. These types of carbon materials are characterised by high surface area and porosity. Carbon-rich precursors are subjected to carbonisation and activation, which results in the development of a micro-, meso- and macropore structure. Activated carbon and porous carbon materials are used in a wide range of applications that include water and air treatment, pollutant removal and also in energy stores (batteries, supercapacitors and fuel cells) due to their high porosity and good electrical conductivity (Soffian et al. 2022; Kumar Mishra et al. 2024). In conclusion, the use of renewable biomass and food wastes as carbon sources for the production of structures consisting of carbon nanomaterials contributes to the sustainable development of materials and enhances the circular bioeconomy.

9.5 Conversion of Food Waste into Carbon Nanostructures

Food waste is increasingly being explored as an economical and sustainable precursor for the synthesis of carbon nanostructures because it contains abundant carbon-rich biomolecules such as carbohydrates, proteins, lipids, cellulose, hemicellulose and lignin. These organic components can undergo thermal or chemical transformations to produce graphitic carbon frameworks and nanostructured materials. Various studies have demonstrated that biomass-derived carbon materials can possess high surface areas, tunable pore structures and functional groups that make them suitable for applications in energy storage, catalysis, environmental remediation and sensing technologies (Chand et al. 2025). Several technological approaches have been developed to convert food waste into carbon nanostructures, including thermochemical conversion, catalytic synthesis, microwave-assisted processing and environmentally friendly green synthesis methods (Danial 2024).

9.5.1 Thermochemical Conversion Processes

Biomass thermochemical conversion techniques are methods that create biochar, hydrochar and other precursors through controlled heating methods to create other valuable carbon materials. The techniques involve heating raw, organic materials to create carbon-rich products.

9.5.1.1 Pyrolysis

Pyrolysis is one of the most studied conversion techniques and is valuable for the creation of carbon products from food waste. This technique is the heating of biomass waste in the absence of oxygen at temperatures of 400 to 900 °C. While the biomass is in this environment, the organic constituents of the food waste undergo dehydration, depolymerisation and aromatisation reactions to leave behind a carbon-rich residue, called biochar (Ungureanu et al. 2025). Research has demonstrated that under the right activation parameters, the pyrolysis of biomass materials can yield porous carbons with surface area greater than 1000 $m^2 g^{-1}$. The pyrolysis of some agricultural waste and fruit peel has been shown to produce porous carbon materials with high electrical conductivity and adsorption and hence a high value for supercapacitor and pollutant adsorption applications (El Ouahabi et al. 2026).

9.5.1.2 Hydrothermal Carbonisation

Another useful thermochemical method for transforming food waste into carbon products is hydrothermal carbonisation (HTC) which takes place at temperatures between 180 and 250 °C and under autogenous pressure in an aqueous medium. A major advantage of the HTC process is the ability to process wet biomass without any prior drying steps (Kassim et al. 2024). HTC results in a solid carbonaceous material called hydrochar, which is produced via hydrolysis, dehydration and polymerisation and which is often rich in numerous oxygenated functional groups that can be further activated or modified. Several studies have shown that the hydrothermal carbonisation of some food waste (fruits and starch-rich materials) can yield carbon spheres and some nanostructured carbon materials with enhanced adsorption and electrochemical properties (Masoumi et al. 2021; Khan et al. 2022).

9.5.1.3 Gasification

Gasification refers to another method of thermochemical conversion of biomass that produces energy-rich gaseous and solid carbonaceous products. In gasification, biomass is treated at high temperatures above 700 °C, and at a limited supply of oxygen, air or steam. This process transforms biomass into a mixture of gases, or syngas, consisting of hydrogen, carbon monoxide, methane and carbon dioxide (Maitlo et al. 2022). Though gasification is mostly conducted for energy purposes, the carbonaceous residuals produced during gasification can also be employed in the synthesis of advanced carbon nanostructures. The residuals produced during gasification, in general, have a porous and graphitic structure, which are advantageous in catalytic and electrochemical applications (Ng et al. 2024).

9.5.2 Chemical and Catalytic Synthesis Methods

Carbon nanostructures with specific, desired morphologies and structural characteristics can routinely be produced using chemical and catalytic synthesis methods. These methods use either chemical activating agents or metal catalysts to facilitate the formation of nanostructured carbon frameworks. The chemical activation of biomass-derived precursors using agents such as KOH, $ZnCl_2$ or H_3PO_4 is a common practice to develop porous carbons. KOH activation of biomass-derived carbons has produced porous carbons with surface areas viable for supercapacitor electrodes and adsorption applications of 1500–3000 m^2g^{-1} (Oubenali et al. 2024).

Iron, nickel and cobalt as metal catalysts promote the growth of carbon nanotubes and carbon nanofibres via catalytic decomposition of carbon sources. One of the most common methods of generating CNTs is called catalytic chemical vapour deposition (CVD) in which hydrocarbon gases are deposited onto metallic catalysts and decompose to generate tubular carbon structures. The use of biomass-derived carbon sources has in addition been studied to feed the catalytic synthesis of carbon nanomaterials (Acomb et al. 2016).

9.5.3 Synthesis with the Assistance of Microwaves

The synthesis of carbon nanomaterials from biomass and food waste using microwave assistance has proven to be a fast and efficient way to synthesise materials. Specifically, microwave heating allows the electromagnetic radiation to heat the precursor materials, and the reaction medium is heated evenly, allowing for increased overall heating compared to more of the conventional thermal heating methods (Adeola et al. 2023). Carbonisation using microwaves has been effective for the synthesis of carbon quantum dots, graphene-like materials and porous carbons from a variety of biomass sources. It was reported that microwave-assisted techniques produced fluorescent carbon quantum dots in minutes with strong photoluminescence and high quantum yields, making them suitable for use in sensing and bioimaging (Yu et al. 2018).

9.5.4 Approaches for Green Synthesis

The positive environmental impact of green synthesis methods for producing carbon nanostructures from renewable biomass sources and non-toxic precursors is crucial. Green synthesis using food waste is especially appealing because food waste can be used in synthesis due to the organic components present in food waste and does not require the use of any toxic chemicals. Several green synthesis methods for producing carbon nanomaterials using food waste, such as hydrothermal treatment, low-temperature carbonisation and solvent-less methods, have been developed to minimise environmental impact (Qasim et al. 2023). For example, the synthesis of carbon quantum dots and graphene-like materials from fruit and vegetable waste via hydrothermal treatment has been demonstrated to be viable without the use of toxic chemicals. These materials can be used in biomedicine and sensors because of their biocompatibility and fluorescence (Iravani and Varma 2020).

The use of thermochemical, chemical and green synthesis methods for the conversion of food waste to carbon nanostructures enables the organic waste to be converted into substantially more valuable functional materials. In addition to diminishing the environmental issues related to the disposal of food waste, these methods also conserve the environmentally friendly nanomaterials manufacturing processes in the context of a circular bioeconomy and sustainable nanomaterials manufacturing (Zhang et al. 2021).

9.6 Fabrication and Functionalization of Carbon Nanostructures

From the view of its raw material origin, carbon nanostructures can be produced using various biomasses and food wastes. These structures often require further processing and modification of their functionalised surfaces. Such modifications are aimed at bolstering the structures' physical and chemical characteristics, thereby widening the potential of the structures' use in various technological applications. Functionalisation, in this instance, implies adding, removing or substituting existing chemical groups or constituents on the surface of structural carbon nanomaterials, thereby improving the material's dispersion, catalytic and electrochemically active characteristics (Zhu et al. 2020). Functionalisation can also be implemented within the context of composite formulations, wherein carbon nanostructures, along with either polymers or metals or their oxides, are synthesised and transformed into composite formulations with enhanced functionalities. In the literature, numerous accounts of modification involving surface alterations, doping with heteroatoms, composite formation and the tuning of structures are prevalent and documented within research and review studies to modify the characteristics of carbon nanostructures for specific intended and desired uses (Farmand et al. 2022).

9.6.1 Techniques of Surface Modulation

Surface modulation is among the most implemented surface exercises aimed at modifying and improving the chemical reactivity, dispersibility and interactive ability of carbon nanostructures and their components. It is the structure of carbon nanomaterials, e.g. graphite, carbon nanotubes and porous carbons, that has been noted as one of the structures that is less susceptible to a wide range of chemical modifications when surfaces are either left untouched or not modified to be graphitised. The surface modifications are aimed at enhancing performance in a wide variety of applications (Alosime 2023; Bulla et al. 2024).

The chemical oxidation technique is commonly used to add certain functional groups to the surface of carbon nanostructures, such as the hydroxyl, carboxyl and epoxy groups. A case in point is the functionalisation of graphene oxide, which is the result of the oxidation of graphite. This process of oxidation produces an enormous

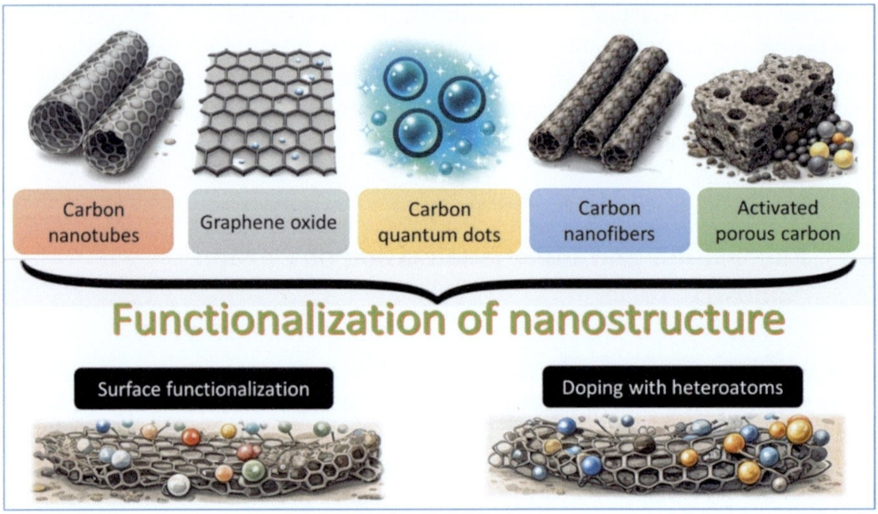

Fig. 9.3 Schematic representation of carbon nanostructure types and their functionalization strategies. Various carbon nanostructures—including carbon nanotubes, graphene oxide, carbon quantum dots, carbon nanofibers and activated porous carbon—are illustrated as major classes of nanomaterials derived from carbon

increase in the hydrophilicity and chemical reactivity of the oxide. Furthermore, these functional groups create additional active sites to undergo further chemical reactions or bond with other materials. Surface functionalisation of carbon nanotubes by acid treatment with nitric acid and/or sulphuric acid has been reported to increase the dispersion of the modified carbon nanotubes in aqueous and polymeric systems (Speranza 2019). The surface modification of carbon nanomaterials through physical modification techniques such as plasma treatment and thermal activation has been documented to change the surface characteristics of carbon nanomaterials. These methods alter the surface characteristics without significant harm to the structural framework of the nanomaterials. These improvements also increase the adsorption capacity, catalytic activity and overall compatibility with other elements of composite systems (Lim et al. 2021). Figure 9.3 illustrates types of carbon nanostructure and their functionalization process.

9.6.2 Doping with Heteroatoms

Another method to enhance the electronic and catalytic functions of carbon nanostructures is heteroatom doping. In this case, atoms such as nitrogen (N), sulphur (S), phosphorus (P) or boron (B) are synthesised into the carbon lattice. The introduction of heteroatoms disrupts the lattice structure of the carbon matrix, creating defects that can favourably alter the material's electronic structure and surface chemistry. Modification via amino doping is one of the most popular, as nitrogen's atomic dimensions are comparable to those of carbon, yet nitrogen also offers different levels of electronegativity. The resulting nitrogen-containing polymers reveal exemplary electrical performance and electrochemistry, enhancing catalysis, including but not limited to, the electrochemistry of oxygen reduction reactions in fuel cells. In like manner, the doping of sulphur and phosphorus leads to the creation of more active sites and changes in the electrochemistry of the carbon structure (Shinde et al. 2025). Also, the implantation of boron into the structure of carbon nanomaterials increases their electrical performance and electrochemistry. Numerous reports detail how the incorporation of heteroatoms in doped carbon materials leads to advancements in supercapacitors, batteries and catalytic systems. For example, the supercapacitor application of biomass-derived nitrogen-doped porous carbon materials results in high specific capacitance values of over 300 F g^{-1} due to the presence of more active sites and faster charge transport (Xu et al. 2023).

9.6.3 Composite Fabrication with Polymers and Metals

An additional significant area of carbon nanostructures fabrication is the development of nanocomposites with polymers, metals or metal oxides. In such hybrid materials, carbon nanostructures and another material are combined to improve mechanical, electrical or catalytic properties. In flexible electronics, sensors and structural materials, the use of polymer-carbon nanocomposites is widespread. The incorporation of carbon nanotubes and graphene into polymer matrices is a strategy to improve the mechanical stability, electrical conductivity and the thermal stability of the matrices. For instance, the incorporation of small quantities of graphene or CNTs has been shown to enhance the tensile strength and electrical conductivity of the composites significantly (Fritea et al. 2021).

Metal and metal oxide nanoparticles, like Fe_3O_4, TiO_2, ZnO and MnO_2, are frequently integrated with carbon nanostructures and used to create functional nanocomposites. These types of materials are essential for use in catalysis, environmental cleanup and energy storage. The role of the carbon nanostructures is to improve conductivity. They provide support in enhancing electron transport while offering added surface area for the dispersion of metal nanoparticles. These types of composites have shown increased catalytic activity and improved performance electrochemically in comparison to the use of the components separately (Al-Rawashdeh et al. 2020; Nagaraj 2025).

9.6.4 Enhancing Performance Through Structural Modifications

Another important aspect for optimising the performance of carbon nanostructures is structural tuning. Other parameters such as pore size distribution, surface area and structural imbalance enable the tuning of these parameters for specific applications. Porous carbon materials are favourable for inclusion in systems with variable sized porous structures at the micro-, meso- and macro-scales. These structures are important for applications in adsorptive and electrochemical energy storage materials. The micro-sized pores provide area for constructive use of adsorptive surfaces while, the meso- and macro-sized pores enhance the ease of transport for ions and molecules. There are studies that have shown that porous carbon structures that have been uniformly constructed have a surface area greater than 2000 $m^2\ g^{-1}$. This feature greatly improves the adsorptive and electrochemical storage performance (Barzegar et al. 2021).

Furthermore, controlling the morphology of carbon nanostructures, whether it be nanosphere, nanofibre or nanosheet structures, will be able to control electron transport pathways and the extent to which catalysts will be active. Hence, the structural engineering of carbon materials obtained from biomass or food waste is a relevant research focus to enhance the functionality of carbon-based nanomaterials across a wider range of modern technological applications. In summary, fabrication and functionalisation approaches are vital for the enhanced use of carbon nanostructures. Carbon nanomaterials may be used for energy storage, environmental remediation, catalysis and sensing technologies by way of surface modification, heteroatom doping, composite formation and structural modification (Barzegar et al. 2021; Maliutina et al. 2023).

9.7 Applications of Food-Waste-Derived Carbon Nanostructures

There are numerous advantages to the use of carbon nanostructures developed from food waste. Some of these advantages are high surface area, adjustable porosity, high electrical conductivity and modification of surface structures. The production of advanced carbon nanomaterials from organic waste not only offers economical resources to technology but also reinforces sustainable waste management. Studies done recently and reviews illustrate the hydrogen-rich carbon nanostructures derived from biomass and their great potential for use in environment, energy, bio-, and water technologies and in catalytic applications.

9.7.1 Environmental Remediation

Food waste carbon nanostructures are predominantly researched for environmental remediation, especially for the absorption and removal of heavy metals and organic contaminants in water and wastewater. The use of biomass-derived carbon nanomaterials such as activated carbon and graphene-like and porous carbons provides strong adsorption of contaminants due to their sufficient surface areas and favourable functional groups (Hamda et al. 2025).

Recent research has shown that porous carbon created from food and agricultural waste has surface areas of more than 1000–2000 $m^2\ g^{-1}$. This remarkable surface area produces a significant increase in adsorption. Furthermore, porous carbons from food and agricultural waste have also been shown to remove heavy metals from contaminated water, including lead (Pb^{2+}), cadmium (Cd^{2+}), mercury (Hg^{2+}) and chromium (Cr^{6+}). For example, activated carbon from biomass has been shown to have adsorption capacities of over 150–300 mg/g for Pb^{2+} ions, and this is influenced by functionalisation and pore structure. Furthermore, porous carbon from food and agricultural waste has also been shown to remove organic contaminants such as pesticides, dyes and drugs from wastewater because of their π–π interactions and surface charge (Mechnou et al. 2025). Figure 9.4 describes the major applications of carbon nanomaterials.

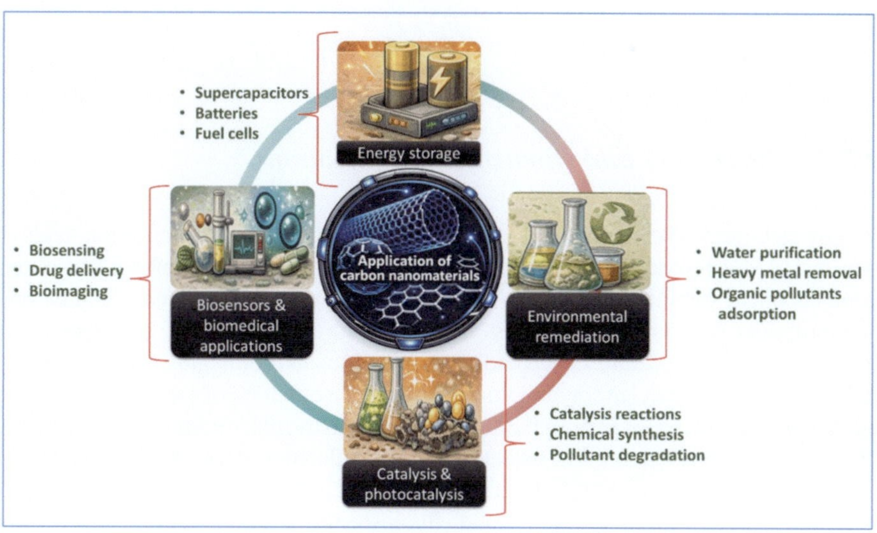

Fig. 9.4 Various emerging application of carbon nanomaterials

9.7.2 Energy Storage and Conversion

Food waste-derived carbon nanostructures have excellent potential for use in electrochemical energy storage (EES) and conversion (EEC) devices, particularly in batteries, fuel cells and supercapacitors. Additionally, carbon nanostructures have excellent conductivity, a high surface area and a hierarchically porous structure, which is ideal for the efficient transport of electrons and diffusion of ions and is critical for the functioning of energy storage devices (Davidraj et al. 2024). The carbon materials for electrodes made from biomass are being used in supercapacitors. These biomass-made carbon electrodes exhibit specific capacitance in the range of 200 to 400 F g^{-1}. This range is dependent on the construction of the pores as well as heteroatom-doping. The electrochemical performance of the material is enhanced when a material contains nitrogen or oxygen functional groups due to enhanced wettability, as well as the introduction of pseudocapacitive behaviour. Food-waste-derived carbon nanostructures have also been used in batteries, including lithium-ion and sodium-ion batteries, where they provide a conductive framework for the falls and rises in volume of the electrodes that occurs during the charge/discharge cycle. This also helps to provide the speedy transport of charge carriers. The porous carbons with high surface areas also assist in the storage of lithium ions and cycling stability. Heteroatom-doped carbon nanomaterials from biomass have been used in fuel cells as affordable catalysts for the oxygen reduction reaction (ORR). In particular, nitrogen-doped carbon materials have offered a promising solution as a catalyst with good activity and durability, with a motivation to eliminate the dependence on costly catalysts that are made from platinum (Yıldırım et al. 2025).

9.7.3 Biosensors and Biomedical Applications

Food-waste-derived carbon nanostructures also have a lot of promise for the fields of biosensing and biomedicine. This can be attributed to their good biocompatibility, good chemical stability, and unique, good electronic properties. Synthesis of carbon quantum dots (CQDs) using biomass as a precursor also possesses excellent water solubility and strong fluorescence, aiding applications in bioimaging and biosensing. In biosensing systems, carbon nanomaterials act as a conductive medium to increase the rate of electron transfer from biological species to the electrodes (Cui et al. 2021). They have been used for the detection of biomolecules such as glucose, DNA, proteins and neurotransmitters with remarkable sensitivity and selectivity. Moreover, the low cytotoxicity and good fluorescence of biomass-derived carbon nanodots also make them good candidates for use in cellular imaging and diagnostics. In addition, the use of functionalised carbon nanomaterials for drug delivery systems has also been studied. Their large surface area and surface functional groups facilitate the effective loading and controlled release of therapeutic agents (Aksu and Güzdemir 2025).

9.7.4 *Catalysis and Photocatalysis*

Another important application of carbon nanostructures from food waste is in catalysis and photocatalysis. Carbon nanomaterials can be used either as a catalyst themselves or as a support for metal or metal oxide catalysts. Due to their high surface area and good conductivity, a large extent of electron transfer is achieved, leading to enhanced catalytic performance (Ng et al. 2024). For instance, in the case of hydrogenation, oxidation and environmental catalytic processes, the use of porous carbon materials as catalyst supports for metal nanoparticles has been prominent, and as such, the combinations of carbon nanostructures with semiconductors like TiO_2 or ZnO facilitate improvements in photocatalytic charge carrier separation and reductions in recombination of the catalysts. Under the catalytic degradation of organic pollutants such as dyes and pharmaceuticals, the photocatalytic systems with carbon catalysts have been found to be effective. Utilising biomass-derived carbon materials as photocatalyst supports effectively integrates sustainability and efficiency into the environmental and energy aspects (Mohapatra et al. 2023). In general, the carbon nanostructures from food waste provide a diverse range of useful materials for a wide array of applications and processes, including the effective utilisation of food waste within the circle of a bioeconomy.

9.8 Challenges, Limitations, and Future Perspectives

Food waste is gaining attention for its ability to produce carbon nanostructures. However, various technical, economic and environmental issues remain for the large-scale production and commercialisation of carbon compounds derived from food waste. To create commercially viable food waste-derived carbon compounds, the issues outlined below must be solved. For food waste to be commercially viable, one of the most concerning issues is the diverse, inconsistent composition of food waste. Food waste from households, agriculture and the food processing industry varies in moisture, carbohydrate, protein, lipid and lignocellulosic content. All of these factors play a critical role in determining conversion process efficiency and the quality of the carbon nanomaterials produced (Oladzadabbasabadi et al. 2023). Variability of the feedstock composition can result in differing pore structures, surface areas and functional groups and ultimately affect the nanomaterials' performance. One of the significant challenges is the energy and processing demands to synthesise carbon nanostructures. Thermochemical processes, like pyrolysis and gasification, require high temperatures and the ability for operational processes to remain stable during the thermochemical process to decrease or minimise consumed energy. While processes such as hydrothermal carbonisation and microwave-assisted methods are considerably less energy consuming and operationally affordable, process control remains one of the primary challenges for the large-scale production of these methods (Davidraj et al. 2024).

Another concern is the control of structural properties and morphology of carbon nanomaterials originating from biomass. For instance, it is usually complex and challenging to achieve uniform control of size, shape and porosity of the carbon nanostructures from complex biomass such as food waste. Therefore, nanomaterials to use in more advanced applications must have the needed properties, and to achieve this, there should be certain levels of control over the synthesis conditions, activation methods and catalytic processes. Furthermore, the environment and safety aspects should be taken into consideration. In the course of synthesis, the use of chemical activating agents and metal catalysts may give rise to unwanted by-products and secondary waste streams that could be hazardous. To ensure environmentally friendly production, the hazardous chemicals should be treated and recovered. Additionally, the widespread use of carbon nanomaterials may be limited by the unknown long-term effects on health and the environment (Singh et al. 2024; Fahad Halim et al. 2025). The economic feasibility of producing nanomaterials from food waste is also worth considering. While food waste is readily available and low priced, the costs associated with the reverse logistics, preprocessing and advanced conversion technologies could all significantly increase the overall process costs. However, the development of integrated biorefinery systems that can produce fuels and chemicals along with the nanomaterials could enhance the economic feasibility and the efficient utilisation of resources (Oladzadabbasabadi et al. 2023).

The future of carbon nanostructures from food waste appears positive despite several obstacles. Production process efficiency and sustainability are expected to be improved with the ongoing development of green synthesis techniques, low-energy conversion methods and improved catalysts. Furthermore, the intersection of biotechnology, nanotechnology and material science will likely facilitate the emergence of other carbon nanostructures with advanced properties and increased functionality. Future studies are directed towards identifying scalable synthesis techniques and performing life cycle assessments and techno-economic analyses to ascertain the practicality of large-volume production. Furthermore, the advanced characterisation techniques coupled with process optimisation should be developed to acquire better control of the desired structural and functional properties of carbon nanostructures derived from biomass. Converting food waste into carbon nanostructures seeks to advance sustainable materials development and waste management, with increasing research and the refining of new technologies capable of providing materials to aid environmental conservation, renewable energy and circular bioeconomy systems.

9.9 Conclusion

The growing global generation of food waste poses serious challenges on social, environmental and economic fronts. Yet, food waste can be viewed as a potential source of renewable materials as it contains a number of useful, bio-nanostructure forming, natural polymers and carbon-based compounds, such as carbohydrates, proteins, lipids and lignocellulosic materials. Therefore, in combination with biorefinery methods and new conversion methods, such organic materials can be utilised to produce valuable biocarbons in the form of nanostructures. Hence, the development of advanced sustainable approaches to produce functional nanomaterials from organic food waste is a positive contributor to waste valorisation, recovery of resources, sustainable approaches and environmental protection. Thermochemical and chemical methods (pyrolysis, hydrothermal carbonisation, gasification, catalytic conversion and microwave-assisted techniques) can be successfully utilised to produce food waste-derived carbon-based nanomaterials. These processes allow for the production of a wide range of food waste-derived carbon nanomaterials, including carbon nanotubes, graphene, carbon quantum dots and porous carbon. In addition, fabrication and functionalisation methods—including surface modification, heteroatom doping and formation of composites—are of paramount importance for the enhancement of the physicochemical properties and the overall performance of these materials. Food waste-derived carbon nanostructures have shown potential use in various applications such as environmental clean-up, energy storage and conversion, biosensing, biomedicine and catalysis. Advanced engineering applications require materials that are highly porous have tunable pores, high conductivity and surface functionalisation. Carbon nanostructures derived from food waste possess these properties. In particular, food waste-derived carbon nanostructures are able to address some of the world's most pressing problems once applied to supercapacitors, batteries, adsorbent materials and in catalytic applications.

Though some progress has been made, there are still issues to address in terms of variability of feedstock, scalability and energy requirements of the processes, and assessing if the processes are economically viable. More work needs to be done in optimising processes and assessing integrated biorefinery systems if these processes are to be made commercially viable and sustainable.

References

Acomb JC, Wu C, Williams PT (2016) The use of different metal catalysts for the simultaneous production of carbon nanotubes and hydrogen from pyrolysis of plastic feedstocks. Appl Catal B 180:497–510. https://doi.org/10.1016/j.apcatb.2015.06.054

Adeola AO, Duarte MP, Naccache R (2023) Microwave-assisted synthesis of carbon-based nanomaterials from biobased resources for water treatment applications: emerging trends and prospects. Front Carbon 2. https://doi.org/10.3389/frcrb.2023.1220021

Ahmed MM, Nabi MHB, Mia MS, Ahmad I, Zzaman W (2025) Valorization of plant-based agro-waste into sustainable food packaging materials: current approaches and functional applications. Appl Food Res 5:101368. https://doi.org/10.1016/j.afres.2025.101368

Aksu M, Güzdemir Ö (2025) Food waste-derived carbon quantum dots and their applications in food technology: a critical review. Food Bioprocess Technol 18:6753–6778. https://doi.org/10.1007/s11947-025-03854-1

Alexander P, Brown C, Arneth A, Finnigan J, Moran D, Rounsevell MDA (2017) Losses, inefficiencies and waste in the global food system. Agric Syst 153:190–200. https://doi.org/10.1016/j.agsy.2017.01.014

Ali Z, Yaqoob S, Yu J, D'Amore A (2024) Critical review on the characterization, preparation, and enhanced mechanical, thermal, and electrical properties of carbon nanotubes and their hybrid filler polymer composites for various applications. Compos C Open Access 13:100434. https://doi.org/10.1016/j.jcomc.2024.100434

Alosime EM (2023) A review on surface functionalization of carbon nanotubes: methods and applications. Discov Nano 18:12. https://doi.org/10.1186/s11671-023-03789-6

Al-Rawashdeh NAF, Allabadi O, Aljarrah MT (2020) Photocatalytic activity of graphene oxide/zinc oxide nanocomposites with embedded metal nanoparticles for the degradation of organic dyes. ACS Omega 5:28046–28055. https://doi.org/10.1021/acsomega.0c03608

Amalina F, Razak ASA, Krishnan S, Sulaiman H, Zularisam AW, Nasrullah M (2022) Biochar production techniques utilizing biomass waste-derived materials and environmental applications—A review. J Hazard Mater Adv 7:100134. https://doi.org/10.1016/j.hazadv.2022.100134

Barzegar F, Pavlenko V, Zahid M, Bello A, Xia X, Manyala N et al (2021) Tuning the nanoporous structure of carbons derived from the composite of cross-linked polymers for charge storage applications. ACS Appl Energy Mater 4:1763–1773. https://doi.org/10.1021/acsaem.0c02908

Bhatia L, Jha H, Sarkar T, Sarangi PK (2023) Food waste utilization for reducing carbon footprints towards sustainable and cleaner environment: a review. Int J Environ Res Public Health 20:2318. https://doi.org/10.3390/ijerph20032318

Bulla M, Kumar V, Devi R, Kumar S, Dahiya R, Singh P et al (2024) Exploring the frontiers of carbon nanotube synthesis techniques and their potential applications in supercapacitors, gas sensing, and water purification. J Environ Chem Eng 12:114504. https://doi.org/10.1016/j.jece.2024.114504

Capanoglu E, Nemli E, Tomas-Barberan F (2022) Novel approaches in the valorization of agricultural wastes and their applications. J Agric Food Chem 70:6787–6804. https://doi.org/10.1021/acs.jafc.1c07104

Chand S, Kumar R, Thakur N, Kumar K, Umar A, Almas T et al (2025) Sustainable synthesis and multifunctional applications of biowaste-derived carbon nanomaterials and metal oxide composites: a review. Chemosphere 385:144540. https://doi.org/10.1016/j.chemosphere.2025.144540

Cui L, Ren X, Sun M, Liu H, Xia L (2021) Carbon dots: synthesis, properties and applications. Nanomaterials 11:3419. https://doi.org/10.3390/nano11123419

Danial WH (2024) Synthesis of carbon nanostructures from waste materials. In: Handbook of functionalized carbon nanostructures. Springer, Cham, pp 1–38. https://doi.org/10.1007/978-3-031-14955-9_25-1

Davidraj JM, Sathish CI, Benzigar MR, Li Z, Zhang X, Bahadur R et al (2024) Recent advances in food waste-derived nanoporous carbon for energy storage. Sci Technol Adv Mater 25. https://doi.org/10.1080/14686996.2024.2357062

De Buck V, Polanska M, Van Impe J (2020) Modeling biowaste biorefineries: a review. Front. Sustain. Food Syst. 4. https://doi.org/10.3389/fsufs.2020.00011

Dhandapani E, Duraisamy N, Mohan Raj R (2022) Green synthesis of carbon quantum dots from food waste. Mater Today Proc 51:1696–1700. https://doi.org/10.1016/j.matpr.2020.10.025

Fahad Halim AFM, Poinern GEJ, Fawcett D, Sharma R, Surendran S, R R (2025) Biomass-derived carbon nanomaterials: synthesis and applications in textile wastewater treatment, sensors, energy storage, and conversion technologies. CleanMat 2:4–58. https://doi.org/10.1002/clem.15

Farmand M, Jahanpeyma F, Gholaminejad A, Azimzadeh M, Malaei F, Shoaie N (2022) Carbon nanostructures: a comprehensive review of potential applications and toxic effects. 3 Biotech 12:159. https://doi.org/10.1007/s13205-022-03175-6

François M, Lin K-S, Rachmadona N, Khoo KS (2024) Utilization of carbon-based nanomaterials for wastewater treatment and biogas enhancement: a state-of-the-art review. Chemosphere 350: 141008. https://doi.org/10.1016/j.chemosphere.2023.141008

Fritea L, Banica F, Costea T, Moldovan L, Dobjanschi L, Muresan M et al (2021) Metal nanoparticles and carbon-based nanomaterials for improved performances of electrochemical (bio)sensors with biomedical applications. Materials 14:6319. https://doi.org/10.3390/ma14216319

García-Morales JL, Fernández-Morales FJ (2024) Sustainable development of food waste biorefineries. Fermentation 10:301. https://doi.org/10.3390/fermentation10060301

Giwa HS, Abd Hamid H, Zulkifli FH, Xiao J (2026) Green synthesis of carbon quantum dots from organic waste on environmental protection: a bibliometric analysis. Biomass Bioenergy 211: 109120. https://doi.org/10.1016/j.biombioe.2026.109120

Hamda AS, Areti HA, Gudeta RL, Abo LD, Jayakumar M (2025) Carbon-based nanomaterials for water treatment: a comprehensive review of recent advances and mechanisms. Chem Eng J Adv 23:100834. https://doi.org/10.1016/j.ceja.2025.100834

Hossain MA, Ferdous N, Ferdous E (2024) Crisis-driven disruptions in global waste management: impacts, challenges and policy responses amid COVID-19, Russia-Ukraine war, climate change, and colossal food waste. Environ Chall 14:100807. https://doi.org/10.1016/j.envc.2023.100807

Hughes KJ, Iyer KA, Bird RE, Ivanov J, Banerjee S, Georges G et al (2024) Review of carbon nanotube research and development: materials and emerging applications. ACS Appl Nano Mater 7:18695–18713. https://doi.org/10.1021/acsanm.4c02721

Iravani S, Varma RS (2020) Green synthesis, biomedical and biotechnological applications of carbon and graphene quantum dots. A review. Environ Chem Lett 18:703–727. https://doi.org/10.1007/s10311-020-00984-0

Kassim FO, Sohail M, Taylor B, Afolabi OOD (2024) Hydrothermal carbonisation of mixed agri-food waste: process optimisation and mechanistic evaluation of hydrochar inorganic chemistry. Biomass Bioenergy 180:107027. https://doi.org/10.1016/j.biombioe.2023.107027

Khan MA, Hameed BH, Siddiqui MR, Alothman ZA, Alsohaimi IH (2022) Hydrothermal conversion of food waste to carbonaceous solid fuel—A review of recent developments. Foods 11: 4036. https://doi.org/10.3390/foods11244036

Kumar Mishra R, Singh B, Acharya B (2024) A comprehensive review on activated carbon from pyrolysis of lignocellulosic biomass: an application for energy and the environment. Carbon Resour Convers 7:100228. https://doi.org/10.1016/j.crcon.2024.100228

Lanjekar PR, Panwar NL, Patel MR, Divyangkumar N (2024) Exploring sustainable energy: an overview of biochemical and thermochemical conversion of dairy and food waste. Environ Pollut Manag 1:152–166. https://doi.org/10.1016/j.epm.2024.09.003

Lim J-V, Bee S-T, Tin Sin L, Ratnam CT, Abdul Hamid ZA (2021) A review on the synthesis, properties, and utilities of functionalized carbon nanoparticles for polymer nanocomposites. Polymers (Basel) 13:3547. https://doi.org/10.3390/polym13203547

Lin CSK, Koutinas AA, Stamatelatou K, Mubofu EB, Matharu AS, Kopsahelis N et al (2014) Current and future trends in food waste valorization for the production of chemicals, materials and fuels: a global perspective. Biofuels Bioprod Biorefin 8:686–715. https://doi.org/10.1002/bbb.1506

Liu Z, Zhao L, Yao Z, Jia J, Wang Z, Liu Z (2023) Behaviors and interactions during hydrothermal carbonization of protein, cellulose and lignin. Chem Eng J 476:146373. https://doi.org/10.1016/j.cej.2023.146373

Maitlo G, Ali I, Mangi KH, Ali S, Maitlo HA, Unar IN et al (2022) Thermochemical conversion of biomass for syngas production: current status and future trends. Sustainability 14:2596. https://doi.org/10.3390/su14052596

Maliutina KM, Omoriyekomwan JE, He C, Fan L, Folli A (2023) Biomass-derived carbon nanostructures and their applications as electrocatalysts for hydrogen evolution and oxygen reduction/evolution. Front Environ Eng 2. https://doi.org/10.3389/fenve.2023.1228992

Masoumi S, Borugadda VB, Nanda S, Dalai AK (2021) Hydrochar: a review on its production technologies and applications. Catalysts 11:939. https://doi.org/10.3390/catal11080939

Mechnou I, Benabdallah A, Chham A-I, Rachdi Y, Hlaibi M, El kartouti A et al (2025) Activated carbons for effective pharmaceutical adsorption: impact of feedstock origin, activation agents, adsorption conditions, and cost analysis. Results Eng 27:105966. https://doi.org/10.1016/j.rineng.2025.105966

Mishra B, Mohanta YK, Reddy CN, Reddy SDM, Mandal SK, Yadavalli R et al (2023) Valorization of agro-industrial biowaste to biomaterials: an innovative circular bioeconomy approach. Circ Econ 2:100050. https://doi.org/10.1016/j.cec.2023.100050

Mohamed Yusoff NIB, Godsell J, Woolley E (2024) Towards zero waste: a comprehensive framework for categorizing household food waste. Sustain Prod Consum 48:1–13. https://doi.org/10.1016/j.spc.2024.05.002

Mohapatra L, Cheon D, Yoo SH (2023) Carbon-based nanomaterials for catalytic wastewater treatment: a review. Molecules 28:1805. https://doi.org/10.3390/molecules28041805

Molaei MJ (2019) Carbon quantum dots and their biomedical and therapeutic applications: a review. RSC Adv 9:6460–6481. https://doi.org/10.1039/C8RA08088G

Nagaraj K (2025) Metal oxides in biomedicine: advances in imaging, drug delivery, tissue engineering, and biosensing. Microchem J 215:114481. https://doi.org/10.1016/j.microc.2025.114481

Ng CH, Mistoh MA, Teo SH, Galassi A, Taufiq-Yap YH, Siambun NJ et al (2024) The roles of carbonaceous wastes for catalysis, energy, and environmental remediation. Catal Commun 187:106845. https://doi.org/10.1016/j.catcom.2024.106845

Nicolae SA, Au H, Modugno P, Luo H, Szego AE, Qiao M et al (2020) Recent advances in hydrothermal carbonisation: from tailored carbon materials and biochemicals to applications and bioenergy. Green Chem 22:4747–4800. https://doi.org/10.1039/D0GC00998A

Oladzadabbasabadi N, Dheyab MA, Nafchi AM, Ghasemlou M, Ivanova EP, Adhikari B (2023) Turning food waste into value-added carbon dots for sustainable food packaging application: a review. Adv Colloid Interf Sci 321:103020. https://doi.org/10.1016/j.cis.2023.103020

Osman AI, Mehta N, Elgarahy AM, Al-Hinai A, Al-Muhtaseb AH, Rooney DW (2021) Conversion of biomass to biofuels and life cycle assessment: a review. Environ Chem Lett 19:4075–4118. https://doi.org/10.1007/s10311-021-01273-0

Osorio-Aguilar D-M, Terán-Salgado E, García-Betancourt M-L (2024) Hierarchical porous hollow carbon nanofibers. In: Handbook of functionalized carbon nanostructures. Springer, Cham, pp 1–40. https://doi.org/10.1007/978-3-031-14955-9_9-1

El Ouahabi H, Bayessi O, Mouaky A, hirt A, Rachidi S (2026) Biochar revolution: harnessing pyrolysis for climate resilience and circular environmental solutions. Carbon Trends 23:100628. https://doi.org/10.1016/j.cartre.2026.100628

Oubenali M, Kasbaji M, Mennani M, Mbarki M, Moubarik A (2024) Introduction to carbon nanostructures: history, classifications, and recent advances. In: Handbook of functionalized carbon nanostructures. Springer, Cham, pp 3–56. https://doi.org/10.1007/978-3-031-32150-4_1

Qasim M, Clarkson AN, Hinkley SFR (2023) Green synthesis of carbon nanoparticles (CNPs) from biomass for biomedical applications. Int J Mol Sci 24:1023. https://doi.org/10.3390/ijms24021023

Saifuddin N, Raziah AZ, Junizah AR (2013) Carbon nanotubes: a review on structure and their interaction with proteins. J Chem 2013. https://doi.org/10.1155/2013/676815

Schmidt LM, Andersen LF, Dieckmann C, Lamp A, Kaltschmitt M (2018) The biorefinery approach. In: Encyclopedia of sustainability science and technology. Springer, New York, NY, pp 1–30. https://doi.org/10.1007/978-1-4939-2493-6_1050-1

Sevillano CA, Pesantes AA, Peña Carpio E, Martínez EJ, Gómez X (2021) Anaerobic digestion for producing renewable energy—the evolution of this technology in a new Uncertain Scenario. Entropy 23:145. https://doi.org/10.3390/e23020145

Shahriari S, Sastry M, Panjikar S, Singh Raman R (2021) Graphene and graphene oxide as a support for biomolecules in the development of biosensors. Nanotechnol Sci Appl 14:197–220. https://doi.org/10.2147/NSA.S334487

Shinde PA, Shrestha LK, Ariga K (2025) Heteroatom-functionalized carbon nanoarchitectonics: unlocking the doping effects for supercapacitor electrode design. Green Energy Environ 10: 1838–1862. https://doi.org/10.1016/j.gee.2025.02.007

Singh AK, Kiraula J, Jha D, Labhane PK, Sonawane SL (2024) Surface functionalization of bio-derived carbon nanostructures. In: Bio-derived carbon nanostructures. Elsevier, pp 195–229. https://doi.org/10.1016/B978-0-443-13579-8.00009-7

Slepičková Kasálková N, Slepička P, Švorčík V (2021) Carbon nanostructures, nanolayers, and their composites. Nano 11:2368. https://doi.org/10.3390/nano11092368

Soffian MS, Abdul Halim FZ, Aziz F, A. Rahman M, Mohamed Amin MA, Awang Chee DN (2022) Carbon-based material derived from biomass waste for wastewater treatment. Environ Adv 9:100259. https://doi.org/10.1016/j.envadv.2022.100259

Speranza G (2019) The role of functionalization in the applications of carbon materials: an overview. C (Basel) 5:84. https://doi.org/10.3390/c5040084

Speranza G (2021) Carbon nanomaterials: synthesis, Functionalization and Sensing Applications. Nanomaterials 11:967. https://doi.org/10.3390/nano11040967

Todd ECD, Faour-Klingbeil D (2024) Impact of food waste on society, specifically at retail and foodservice levels in developed and developing countries. Foods 13:2098. https://doi.org/10.3390/foods13132098

Tsegaye B, Jaiswal S, Jaiswal AK (2021) Food waste biorefinery: pathway towards circular bioeconomy. Foods 10:1174. https://doi.org/10.3390/foods10061174

Ungureanu N, Vlăduț N-V, Biriș S-Ș, Gheorghiță N-E, Ionescu M (2025) Biomass pyrolysis pathways for renewable energy and sustainable resource recovery: a critical review of processes, parameters, and product valorization. Sustainability 17:7806. https://doi.org/10.3390/su17177806

Urade AR, Lahiri I, Suresh KS (2023) Graphene properties, synthesis and applications: a review. JOM 75:614–630. https://doi.org/10.1007/s11837-022-05505-8

Vatieri C, Cirillo T, Esposito F (2025) Waste to worth: bioplastic synthesis from lignocellulosic food waste in the age of the circular bioeconomy. Front Sustain Food Syst 9. https://doi.org/10.3389/fsufs.2025.1698348

Venkataraman S, Sundramoorthy AK (2026) Waste-derived carbon nanomaterials for electrochemical applications: toward a circular and sustainable future. RSC Adv 16:733–747. https://doi.org/10.1039/D5RA08357E

Wagh MS, S S, Nath PC, Chakraborty A, Amrit R, Mishra B et al (2024) Valorisation of agro-industrial wastes: circular bioeconomy and biorefinery process—A sustainable symphony. Process Saf Environ Prot 183:708–725. https://doi.org/10.1016/j.psep.2024.01.055

Walimbe P, Chaudhari M (2019) State-of-the-art advancements in studies and applications of graphene: a comprehensive review. Mater Today Sustain 6:100026. https://doi.org/10.1016/j.mtsust.2019.100026

Wang Y, Wu JJ (2023) Thermochemical conversion of biomass: potential future prospects. Renew Sust Energ Rev 187:113754. https://doi.org/10.1016/j.rser.2023.113754

Xu Y, Xiao M, Liu T, Lin R, Meng Y, Zhang Y et al (2023) Nitrogen, boron and fluorine tri-doped carbon nanotubes on carbon cloth as the electrode materials for supercapacitors. J Electroanal Chem 948:117812. https://doi.org/10.1016/j.jelechem.2023.117812

Yafetto L, Odamtten GT, Wiafe-Kwagyan M (2023) Valorization of agro-industrial wastes into animal feed through microbial fermentation: a review of the global and Ghanaian case. Heliyon 9:e14814. https://doi.org/10.1016/j.heliyon.2023.e14814

Yıldırım ID, Bakhtiari B, Güngör A, Erdem E (2025) Supercapacitor electrodes based on biowaste. Discov Electrochem 2:30. https://doi.org/10.1007/s44373-025-00045-7

Yu T, Wang H, Guo C, Zhai Y, Yang J, Yuan J (2018) A rapid microwave synthesis of green-emissive carbon dots with solid-state fluorescence and pH-sensitive properties. R Soc Open Sci 5. https://doi.org/10.1098/rsos.180245

Zakaria NZJ, Rozali S, Mubarak NM, Ibrahim S (2024) A review of the recent trend in the synthesis of carbon nanomaterials derived from oil palm by-product materials. Biomass Convers Biorefinery 14:13–44. https://doi.org/10.1007/s13399-022-02430-3

Zhang YS, Zhu HL, Yao D, Williams PT, Wu C, Xu D et al (2021) Thermo-chemical conversion of carbonaceous wastes for CNT and hydrogen production: a review. Sustain Energy Fuels 5: 4173–4208. https://doi.org/10.1039/D1SE00619C

Zhou X, Liu B, Chen Y, Guo L, Wei G (2020) Carbon nanofiber-based three-dimensional nanomaterials for energy and environmental applications. Mater Adv 1:2163–2181. https://doi.org/10.1039/D0MA00492H

Zhou Y, Remón J, Pang X, Jiang Z, Liu H, Ding W (2023) Hydrothermal conversion of biomass to fuels, chemicals and materials: a review holistically connecting product properties and marketable applications. Sci Total Environ 886:163920. https://doi.org/10.1016/j.scitotenv.2023.163920

Zhu L, Shen D, Wu C, Gu S (2020) State-of-the-art on the preparation, modification, and application of biomass-derived carbon quantum dots. Ind Eng Chem Res 59:22017–22039. https://doi.org/10.1021/acs.iecr.0c04760

Chapter 10
Food Waste Valorization: Implication in Circular Economy Paradigm

Veerapandi Loganathan, Perraulta Lavanya, Aruna Selvaraj, Priyanga Govindarasu, and Radhika Theagarajan

10.1 Introduction

In low- and middle-income countries, the major cause of food loss is the inefficiency in the postharvest process, whereas in high-income countries, the major cause of food loss is the excessive production, improper storage, and consumer-driven waste generation. In addition, a large amount of agro-industrial by-products, i.e., fruit peels, pulps, shells, etc., is produced annually, which are usually disposed by open dumping and burning, resulting in significant air and soil pollution. If the same pattern of waste generation continues, municipal solid waste may rise to over 3.4 billion tonnes annually by 2050. The food production/distribution system alone is responsible for almost one-third of the overall food loss, which represents a major concern further necessitating the implementation of various waste valorization approaches. Therefore, reducing the waste as well as utilizing the same in the production of value-added products is highly encouraged (Davidraj et al. 2024).

V. Loganathan
Department of Food Technology, TKM Institute of Technology, Kollam, Kerala, India

Department of Food science and Technology, Hamelmalo Agricultural College, Keren, Eritrea

P. Lavanya · P. Govindarasu
Department of Food science and Technology, Pondicherry University, Puducherry, India

A. Selvaraj · R. Theagarajan (✉)
Department of Food Technology, Manakula Vinayagar Institute of Technology, Puducherry, India

Y. K. Mohanta et al. (eds.), *Bioresource Waste to Carbon Nanomaterial*, Nanotechnology in the Life Sciences, https://doi.org/10.1007/978-981-92-2405-0_10

10.1.1 Circular Bioeconomy

The focus on sustainable development has evolved the requirement for the development of alternative economic systems that reduce environmental impacts while enhancing the efficiency resource utilization. The circular bioeconomy is considered an alternative approach that assists in addressing the challenges associated with the depletion of resources, climate change, and the accumulation of waste in the environment. Moreover, the circular bioeconomy integrates the principles of sustainability and use of biological resources, promoting the shift from the traditional linear model of the bioeconomy, i.e., the take-make-dispose model, to a circular production model (Fig. 10.1). It focuses on the reduction of waste, the efficient use of resources, and the valorization of biological resources through technological development, industrial symbiosis, and the implementation of policies that support these activities. Further, the waste generated in the process are converted into valuable products such as bioenergy, bioplastics, biofertilizers, and biochemicals (Ouro-Salim and Guarnieri 2022).

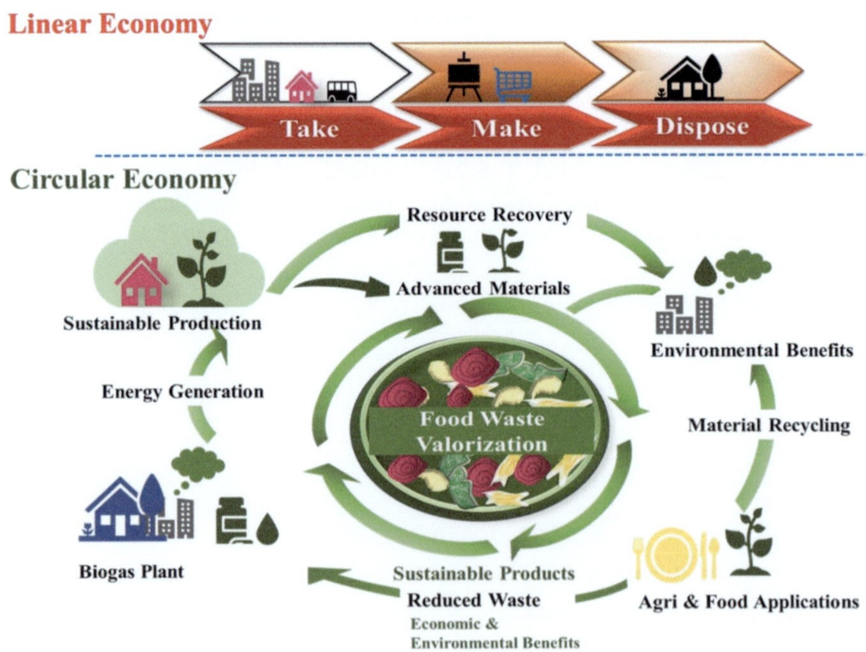

Fig. 10.1 Linear economy vs. circular bioeconomy model for food waste valorization

10.1.2 Principles of Circular Economy vs. Linear Economy

The conventional economic system has largely operated in a linear pattern where resources are exploited, converted into products, utilized, and finally discarded. This take-make-dispose pattern has led to the excessive utilization of resources, increased generation of waste, and intensified environmental degradation (Lavanya et al. 2026). Thus, linear approaches result in increased costs of production and environmental pressures (Kirchherr et al. 2017).

On the other hand, the circular economy proposes a new economic model that aims to eliminate waste through product design, product longevity, and ecosystem services. In the current economic system, generated waste is not discarded but rather utilized as a resource in the creation of new products (Ojha et al. 2020).

10.1.3 Role of Bioresources in Sustainable Development

The renewable biological resources, including plants, animals, and microbes, are the key component of the circular bioeconomy. The circular economy helps in the production of biobased products such as bioenergy and biofertilizers, thereby reducing the dependence on fossil-based products. The inclusion of bioresources in the circular economy helps in the improvement of soil health, conservation of biodiversity, and creation of employment opportunities in rural areas. The circular economy helps in maintaining the ecological balance and promoting the concept of sustainable development, which aligns well with the United Nations Sustainable Development Goals (Geissdoerfer et al. 2017).

Especially in the food waste utilization, peels with pectin content can be used to produce biodegradable packaging, or crop residues can be used to produce carbon nanomaterials (CNMs), which can be used in water filtration and sensing. Additionally, the contribution of bioresources to UN SDGs, such as SDG 2, zero hunger (reduction of food loss and promotion of efficient agriculture); SDG 7, affordable and clean energy (promotion of renewable energy); SDG 12, responsible consumption and production (reduction of waste and enhancement of resource efficiency); SDG 13, climate action (reduction of greenhouse gas emissions through sustainable waste management); and SDG 15, life on land (conservation of biodiversity through organic fertilizers and reduction of pollution) (Tamasiga et al. 2022).

The focus of this chapter is the valorization of food waste and CNMs and their implications toward circular economy concept. It includes an overview of the generation and characteristics of food waste, the importance of circular economy concepts, and the various methods that can be applied for the valorization of food waste to produce CNMs, as well as the recent developments, challenges, and future scope for this emerging field, to provide knowledge to the interested readers, including researchers, scientists, and practitioners in the field of sustainable material development and waste management.

10.2 Food and Agricultural Waste

Food waste is considered one of the prominent challenges across the globe. In recent times, the environmental, social, and economic impacts of food waste are very much critical. The Food and Agriculture Organization has reported that almost one-third of food produced globally, i.e., 1.3 billion tonnes of food, is lost or wasted. This not only represents inefficient utilization of key resources such as water, energy, and labor but also indicates a significant portion of greenhouse gas emissions resulting from the degradation of organic matter during landfilling. In the circular economy perspective, valorization of food waste plays an important role in regulating sustainability. Valorization of food waste for the production of CNMs resolves environmental concerns and further enables utilization in advanced applications (Nath et al. 2023; Silva et al. 2025).

10.2.1 Sources of Food Waste

Particularly in the food processing industries, seeking knowledge about the causes of food waste is considered essential to the development of strategies to address the prevention and effective valorization. Each link in the production chain contributes its distinct role in determining the volume and type of waste, which is the result of the complex interplay between technological limitations, consumer behavior, economic conditions, and policy frameworks (Jeevanandam and Danquah 2025).

10.2.2 Types of Food/Agro-industrial Waste

Food and agricultural waste encompasses organic materials produced across all stages of the food system, including cultivation, processing, distribution, consumption, and final disposal. Although commonly regarded as undesirable by-products, such waste streams constitute substantial biomass resources with potential for sustainable recovery and reuse. Identifying and categorizing these waste are therefore an essential initiative in the development of effective waste management practices and valorization strategies (Padhan et al. 2025).

10.2.2.1 Field and Farm-Level Waste

At the agricultural production stage, a variety of waste generated rarely enters consumer markets but constitutes a substantial biomass pool. Examples include the following:

1. *Crop residues*: After the harvest of crops such as rice, wheat, maize, sugarcane, and sorghum, certain portions of the plant remain in the field as residues. These include straw, stalks, leaves, husks, and cobs.
2. *Harvest losses*: Mechanical harvesting, inadequate harvesting techniques, and pest damage can lead to this type of losses in the field, particularly crops that do not meet quality standards for further processing (Kiaya 2014; Garrido Romero et al. 2024).

10.2.2.2 Processing and Industrial Waste

In general, food processing industries while transforming raw agricultural products into consumables might generate large amounts of organic waste (Davidraj et al. 2024), such as the following:

1. *Fruit and vegetable waste*: Processing of fruits and vegetables for juices, purees, canning, peeling, and slicing leads to the generation of skin, seeds, pomace, cores, and pulp residues, for example, citrus processing yields peels and pulp, tomato processing yields skins and seeds, and potato processing generates peel and cut waste (Tezer et al. 2022; Sun et al. 2018).
2. *Cereal and grain processing wastes*: These wastes primarily consist of lignocellulosic materials (Tatrari et al. 2021), such as cellulose, hemicelluloses, and lignin. The cellulose is abundant in straw, husks, stalks, and bagasse.
3. *Hemicellulose*: It is amorphous in nature, comprising various sugar monomers (e.g., xylose, mannose, arabinose).
4. *Lignin*: It is a complex, irregular polymer of phenylpropanoids that combines with cellulose and hemicellulose, thereby conferring mechanical strength in the cereal crops (Bognár et al. 2025; Mohapatra and Singh 2026).

The milling of cereal grains produces bran, germ, and broken grains, which are nutritionally valuable but often underutilized. The processes involved in breweries and distilleries produce grains with high protein and fiber content. These waste should be processed properly owing to their tendency to cause rapid microbial spoilage, leading to health hazards if left undisposed (Alves et al. 2011).

10.2.2.3 Retail and Household Waste

In the distribution and/or consumption stages, the retail waste is mainly impacted by overstocking and expiry-related issues, which cause food wastage in the retail sector, especially in supermarkets. The damage during the handling of products, as well as the marketing strategies, is another main reason for the generation of retail waste, as consumers end up purchasing products in excess than their needs (Cuffey et al. 2023).

Moreover, restaurants, hotels, and institutional kitchens are also key contributors to food waste, and the important reasons are explained as follows:

1. Overpreparation: Buffet-style restaurants and catering businesses tend to overproduce food in anticipation of demand, which may reduce over time. This creates leftover that is not utilized.
2. Portion sizes: Excessive portion sizes contribute to waste that are not consumed.
3. Menu planning: Inadequate knowledge of inventory levels and inflexible menus may result in ingredients to spoil before cooking.
4. Lack of waste tracking: In many food service establishments, there is very limited tracking of waste, which makes it hard to identify the source of the problem (Hazarika et al. 2025).

Institutions such as hospitals and schools may experience food wastage due to overproduction, unexpected changes in attendance, and food restrictions. Also, meals that are prepared in bulk have no ability to adjust in real time, thereby resulting in food waste (Sadegh Hassani et al. 2023).

10.2.2.4 Postharvest and Transportation Losses

Harvesting may cause food wastage because of pests, diseases, mechanical damage, and denial of nonstandard products. Losses may occur to crops due to environmental stresses such as unpredictable weather, pests, and diseases, inefficient harvesting, and inadequate storage facilities, leading to spoilage before the food can reach the markets. Overproduction and processing losses such as trimming and peeling of food products may also result in additional wastage. The classification of postharvest wastage provides a basis in understanding toward investigating opportunities for the development of value-added products through fermentation, thermochemical processes, enzymatic bioconversion, and novel technologies for the production of materials such as CNMs from food wastages (Kiaya 2014).

10.2.3 Biochemical Characterization

The biochemical content of the food and agricultural waste plays an essential and instrumental role in controlling its environmental interactions as well as assistance with the conversion of the waste materials into valuable products. These waste materials are primarily made up of carbon-based compounds such as carbohydrates like cellulose, hemicellulose, and starch, as well as lignin, proteins, fats, and other secondary metabolites. The identification of such waste materials is significant in determining the best conversion processes such as fermentation, anaerobic digestion, pyrolysis, hydrolysis, and composting in turning the waste to value-added products (Liu et al. 2025).

10.2.3.1 Lignocellulosic Biomass

Generally, lignocellulosic structural polymers usually add up to 60–80% of agricultural residues, for example, the rice straw is normally composed of these 3 lignocellulosic products as follows: 30–40% cellulose, 20–25% hemicellulose, and 15–20% lignin. The proportion of each of these polymers is important in determining the most suitable processing technique. Biochemical approaches such as hydrolysis are appropriate in the case of cellulose and hemicellulose, while lignin is more effectively converted using thermochemical approaches (Silva et al. 2024; Yousuf et al. 2020).

10.2.3.2 Starch-Rich Waste

Certain food waste are rich in starch, which is a natural polymer of glucose made up of amylose and amylopectin. Few examples for high-starch waste includes potato peels and waste from processing, bread and bakery waste, cassava waste, and expired grain products. It can be hydrolyzed to fermentable sugars by enzymatic activity or acids and can be used to produce bioethanol and biodegradable polymers (Kringel et al. 2020; Quilez-Molina et al. 2022).

10.2.3.3 Protein-Rich Waste

Usually protein-rich waste comprises of the discarded biological waste materials that are basically abundant in proteins. Moreover, the residual waste obtained in food processing by-products, agricultural waste, dairy industry, and animal tissues is a potential source of energy. These waste can be converted into biodiesel using transesterification reactions or utilized in anaerobic digestion processes (Zhou et al. 2023; De Schouwer et al. 2019).

10.2.3.4 Lipid- and Fat-Rich Waste

Similarly, lipid- and fat-rich waste consists of surplus materials that are high in fats and oils, particularly used cooking oil in frying operations, dairy by-products, and meat processing waste. The degradation of such organic waste in low-oxygen environments, such as landfills, results in the production of methane gas, which is a potent greenhouse gas. Food waste, in general, contributes around 8–10% of the total greenhouse emissions. Then, there is the way in which far rich animal waste and agricultural waste are handled, leading to an increase in soil microorganisms, which in turn produce nitrous oxide (N_2O) (Garrido-Romero et al. 2024; Diamantis et al. 2021).

10.2.3.5 Minor Components and Bioactive Molecules

Normally, minor components and bioactive molecule waste include waste rich in compounds such as phenolics, pigments, and antioxidants. Fruit peels and plant by-products are recognized as valuable sources of bioactive compounds, including polyphenols, flavonoids, carotenoids, pectin, and essential oils. Most of these are known for their potent antioxidant properties, and they can be recovered for various uses in food and biotechnological, nutraceutical, and pharmaceutical applications. The conventional example of waste recovery is citrus peel, which is known for being used as a source of limonene, a solvent, and a flavoring agent (Blasi et al. 2022; Vukic and Vujasinovic 2023).

10.2.3.6 Ash and Mineral Content

By-products of post-combustion and some of the waste streams are naturally rich in mineral such as calcium, potassium, and silica as they contain inorganic ash particles with high such concentrations of minerals. These minerals can find applications as soil conditioners, and some of them can be selectively removed to enhance the efficiency of thermochemical conversion processes (Puri et al. 2024; Fu et al. 2022).

10.2.4 Environmental Impacts of Untreated Waste

Inadequate management of food and agricultural waste results in real problems for the environment, economy, and society. Improper disposal of organic waste allows the leakage of leachate into water bodies, and this carries with it high levels of biochemical oxygen demand and nutrients that cause eutrophication. Meanwhile, agricultural waste being washed away from long-term stored agricultural residues also contribute nutrients such as nitrogen and phosphorus to water bodies (Sikder et al. 2025).

In agricultural areas, burning crop leftovers to prepare the fields rapidly leads to air pollutants harmful to air quality. Wasted food causes a significant drain on resources as disposing of food wastage results in a substantial loss of water and energy consumption. Ineffective handling of waste contributes to habitat loss, erosion of land, and reduced biodiversity (Kim et al. 2024). As organic waste decomposes, it creates a favorable environment for disease vectors such as flies, rodents, and pathogenic microorganisms, thereby increasing the risk of disease transmission. For example, untreated slaughterhouse waste may harbor hazardous pathogens that can spread to animals, livestock, and humans (Abayomi et al. 2024).

10.3 CNMs from Food Waste

The emergence of carbon nanomaterials such as carbon dots, carbon nanotubes (CNTs), graphene, graphene oxide, and biochar has transformed many science and technological domains due to their outstanding electrical, thermal, mechanical, and chemical properties. In recent years, utilization of food waste as a source of nanomaterials has gained important attention as a green, carbon-rich, and cost-effective supply. This is in line with environmental and economic sustainability, where biomass waste is converted into valuable nanomaterials with huge potential in many fields (Sun et al. 2023; Xia et al. 2022).

10.3.1 Why Food Waste for CNMs?

Food waste is abundant and biodegradable, consisting of a lot of carbon-containing materials such as cellulose, lignin, carbohydrates, proteins, and fats. These are ideal raw materials for carbonization, which can then be chemically or thermally treated (Santagata et al. 2021). On the positive side, there are cheap and plenty raw materials, minimized waste disposal, reduced petroleum usage, and green synthesis methods.

10.3.2 Types of CNMs Derived from Food Waste

The high content of organic carbon in food waste makes these waste valuable sources for the manufacture of a variety of CNMs, which may be of varied shape, structure, and physicochemical characteristics, appropriate for applications in energy storage, catalysis, environmental remediation, and biomedicine. A range of food waste types, including fruit peels such as banana, orange, and pomegranate; vegetable waste such as potato, tomato, and carrot; coffee grounds; eggshells; and sugarcane biomass, have been studied as sources for the production of CNMs (Deng et al. 2016).

10.3.2.1 Carbon Dots (CDs) and Graphene Quantum Dots (GQDs)

These are very small, almost globular pieces of carbon that are less than 10 nanometers in diameter. Their ability to emit their own light on excitation by an external light source is what makes them so unique in bioimaging, sensing, and optoelectronics. GQDs are an exceptional unit of carbon dots that have crystalline domains with graphene-like properties (Khan and Patil 2020).

Food waste such as fruit peels and agricultural residues, which are rich in lignocellulosic biomass, can be converted into CDs and GQDs using mild techniques such as low-temperature hydrothermal carbonization, microwave-based synthesis, or ultrasonic synthesis. These techniques can convert food waste into nanomaterials that are rich in functional groups. The quantum yield of CDs can be improved by hydrothermal synthesis at reduced temperatures (120–250 °C) with the support of catalysts. The fluorescent CDs engineered using food waste, such as orange peels, banana peels, and sugarcane bagasse, possess diverse photoluminescent properties depending on the synthesis conditions (Mahajan et al. 2023).

Carbon dots provide a varied range of unique characteristics, such as tunability of photoluminescence and improved stability. These attributes offer chances for carbon dots to find different purposes in various fields. The capacity of carbon dots to diffuse in water and their biocompatibility make them appropriate for various applications. In addition to these, carbon dots function in photocatalytic reactions and light-emitting devices (LEDs), as well as in the delivery of drugs. The stability of carbon dots makes them suitable for these applications (Khan et al. 2025).

10.3.2.2 Carbon Nanofibers (CNFs)

CNFs are characterized as thin, cylindrical fibers with diameters varying between tens and hundreds of nanometers. Unlike CNTs, these fibers have stacked graphitic structures whose orientation is tilted in relation to the fibers' axis, thus having unique mechanical and electrical properties (Li et al. 2024). The production of CNFs from waste materials derived from food products utilizes a technique known as catalytic pyrolysis, which is equivalent to the synthesis of carbon nanotubes. The synthesis of these fibers employs rich carbon materials such as coffee or "chitin from shellfish, in addition to a metal catalyst and heat. The arrangement of the graphitic structure in these fibers confers strength and conductivity properties. Thus, these fibers are used in the production of nanocomposites, reinforcement, electrode materials, and filters, owing to their high aspect ratios and conductivity, thereby promoting the development of uniform carbon nanostructures (Liang et al. 2024).

10.3.2.3 Graphene and Graphene-Like Sheets

Graphene consists of a hexagonal lattice of carbon atoms, which is a single atom thick. It has excellent electrical properties, mechanical strength, flexibility, and rapid transport of charge carriers. The synthesis of graphene from food waste involves the controlled degradation of the carbon-containing material, which is then aimed toward the self-organization of the carbon atoms into a graphene structure. Graphene nanosheets, as well as graphene, may be formed by heating the biomass-derived carbon at high temperatures in an inert atmosphere; by chemically reducing the graphene oxide, which is produced by the oxidation of biomass char, followed by exfoliation; or through the hydrothermal carbonization of the biomass, followed by heat treatment of the acquired material, which is focused toward rising the degree of graphitization (Uddin et al. 2023).

Carbon-rich food waste, such as used tea leaves, fruit peels, and nut shells, are potential options for the production of graphene, as they can be separated after the initial carbonization step, leading to the production of graphene nanosheets. Consequently, this obtained biomass-derived graphene has the major properties of the conventional material, including high electrical conductivity, wide surface area, increased mechanical modulus, and efficient thermal stability, which can qualify the material for the production of advanced applications, including flexible electronics, energy storage devices, sensors, as well as composites with excellent mechanical properties (Muhmood et al. 2024; Liang et al. 2023).

10.3.2.4 Carbon Nanotubes

CNTs are carbon nanotubes, nanostructures consisting of sp^2-hybridized carbon atoms with hexagonal lattice structure. CNTs can be either single-walled CNTs or multi-walled CNTs based on the number of layers of hexagonal lattices of carbon atoms present. CNTs are mechanically strong because of their cylindrical shape, with high L/D ratios, high surface area, and good electrical conductivity (Sajid et al. 2022).

The conversion of food biomass into CNTs involves high-temperature processes such as catalytic pyrolysis with metal catalysts such as iron, nickel, and cobalt. Carbon-rich food biomass, such as cellulose from food peelings and food scraps, undergoes thermal decomposition at elevated temperatures to form hydrocarbons, which can be adsorbed onto the catalysts to form CNTs. For instance, rice husks and banana peel have been used as precursors to form multi-walled CNTs through pyrolysis. The porosity of the food biomass significantly impacts the formation of CNTs. Thus, obtained CNTs from food waste demonstrate high-quality tensile strength and electrical conductivity, making them suitable for applications in energy storage devices, composite materials, sensors, and environmental catalysts (Kim et al. 2024; Murjani et al. 2022; He et al. 2024).

10.3.2.5 Activated Carbon and Porous Carbon Materials

Activated carbon is a carbon material with a porous framework and a large surface area. These characteristics have led to the widespread use of activated carbon in adsorption, filtration, and catalysis. Porous carbon nanomaterials derived from food waste have the same fundamental properties as activated carbon. However, the main variation lies in the pore sizes and structure. Carbon nanomaterials are synthesized using the carbonization process at 400–900 °C in an inert atmosphere. Then, chemicals are used in the chemical activation process. Finally, the nanomaterials are cleaned and dried. These two nanomaterials have different properties. These properties have led to their use in different applications in energy, environmental remediation, electronic devices, and biomedicine (Nandi et al. 2023).

10.4 Synthesis Methods for CNMs from Food Waste

The transformation of food and agricultural waste into useful carbon-based nanomaterials (CNMs) depends on the manufacturing and controlled routes of processing. Carbon-based nanomaterials such as carbon dots, nanotubes, graphene, biochar, and activated carbon provide a potential opportunity in the field of green engineering with the ability to utilize the existing carbon content in food waste (Aksu and Güzdemir 2025) (Fig. 10.2). These nanomaterials can be synthesized using different physical, chemical, and biological approaches based on the carbon content in food waste.

10.4.1 Hydrothermal Carbonization

HTC is an eco-friendly and widely adopted technique for producing carbon dots and carbon microspheres from food-derived waste. This thermochemical process converts wet biomass into hydrochar by operating in an aqueous environment at moderate temperatures (approximately 180–250 °C) and self-generated pressures ranging from 2 to 10 MPa. By replicating the natural coal formation pathway on a significantly accelerated timescale, HTC enables the conversion of biomass into

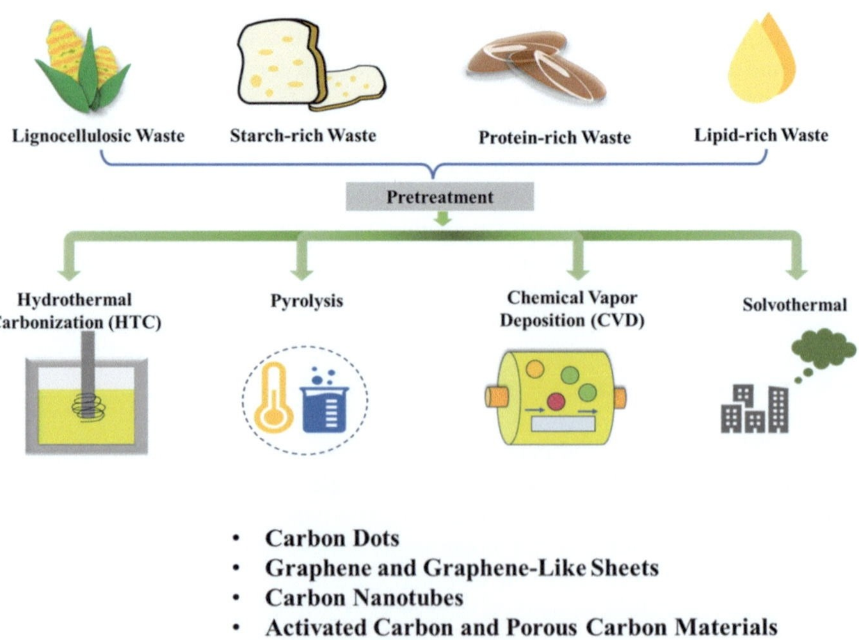

Fig. 10.2 Pathways for conversion of food waste into CNMs

carbon-rich materials within hours rather than geological periods. Under subcritical water conditions, biomass undergoes hydrolysis, dehydration, decarboxylation, and polymerization to form hydrochar with improved carbon content and aromaticity (Chen et al. 2024b). However, scaling up HTC for high-quality CNM synthesis is still under development (Yu et al. 2024).

10.4.2 *Pyrolysis*

Pyrolysis refers to the thermochemical breakdown of organic materials under an oxygen-free environment, generally performed at temperatures ranging from 300 to 900 °C. In this process, major biomass constituents including cellulose, hemicellulose, lignin, proteins, and lipids are decomposed into three primary products: a solid biochar fraction, a liquid bio-oil phase, and a gaseous mixture commonly referred to as syngas. It can be slow or rapid, depending on the heating system. Slow heating favors biochar production; on the other hand, when rapid heating is involved, the production of liquids increases (Rahman et al. 2023).

It occurs through a series of consistent chemical reactions. As the temperature increases, the chemical bonds in the biomass break down, releasing volatile organic compounds and producing a solid residue rich in carbon and aromatics. This modification in chemical composition increases the aromatic and carbon content of the solid, which is appropriate in the production of carbon nanomaterials (CNMs) (Zhang et al. 2024).

10.4.3 *Gasification*

Gasification transforms the organic feedstock into a gas known as syngas, composed of carbon monoxide, hydrogen, and methane, through partial burning with a restricted amount of oxygen or steam at temperatures above 700 °C. Unlike pyrolysis, this method utilizes a small amount of air or steam, thereby allowing the carbon-rich feedstock to convert into gaseous products and a solid residue known as char. The process occurs in four steps: drying, pyrolysis, oxidation, and reduction, as presented by Tezer et al. (2022) and Santos et al. (2022).

The key parameters affecting the gasification process are temperature, with 800–1000 °C promoting efficient gasification at the expense of char; equivalence ratio, which is the ratio of the oxygen supplied; feedstock, with biomass rich in carbon preferred; and pressure, which has the effect of improving the efficiency of the process, as discussed by Tezer et al. (2022).

Advantages and Challenges

Gasification is a method for converting waste into energy and CNMs with high efficiency. It produces clean syngas, which can be used as a source of energy for fuel cells or as a starting material for producing chemicals. The major difficulty is controlling the char structure for producing tailored CNMs and preventing the development of tar. For example, rice husk biochar-derived GO has been used for producing composite membranes for water purification (Santos et al. 2022).

10.4.4 Chemical Vapor Deposition (CVD)

The key principle of a circular economy is the design of turning food waste into carbon nanomaterials (CNMs). The most common methods of synthesis of these materials are hydrothermal carbonization (HTC) and pyrolysis. However, more sophisticated methods such as microwave synthesis and chemical vapor deposition (CVD) produce materials for specialized applications. The choice of synthesis basically depends on economic, scale-up, and environmental considerations. In this regard, upcoming future research should be focused on optimizing these methods for increased efficiency and integrating these methods into decentralized waste management infrastructures so that maximum utilizations of this concept of valorization of food waste can be performed (Heydari Gharahcheshmeh 2025; Sabzi et al. 2023).

10.4.5 Solvothermal and Ultrasonic-Assisted Synthesis

They are fundamentally the advanced versions of the hydrothermal methods that incorporate organic solvents or the use of sound waves to accelerate the carbonization process. Organic solvents, such as ethanol or ethylene glycol, in solvothermal methods assist in the suspension of the precursors and the controlled growth of carbon nanomaterials. The application of sound waves in the ultrasonic techniques facilitates the degradation of the food waste matrices, enhancing the formation of carbon nanomaterials. These advantageous methods have made it possible to produce carbon dots with flexible photoluminescent properties (Li et al. 2024; Ogura et al. 2023).

10.5 Biochemical and Catalytic Conversion Methods for CNMs

Biochemical and catalytic mechanisms convert biomass and waste-derived materials into carbon nanoforms, offering green, environmentally friendly, and sustainable alternatives to conventional high-temperature thermochemical processes. These biochemical techniques utilize enzymes, microbes, and catalysts that convert the complex organic feedstocks into useful carbon nanostructures. Therefore, this chapter discusses the biochemical ways of enzymatic hydrolysis, microbial assistance, and catalytic synthesis of CNMs, their mechanisms, the recent technological advancements, and the challenges that still need to be addressed in this field (Liu et al. 2025).

10.5.1 Enzymatic Hydrolysis

In enzymatic hydrolysis, specific enzymes aim and degrade the complex molecules of biomass, namely, cellulose, hemicellulose, and lignin. This process is considered eco-friendly, owing to its specificity and mild operating temperatures and pH. Thus, this process of breaking down biomass into simpler molecules is a vital step in creating valuable products from biomass. Moreover, enzymatic hydrolysis is highly compatible with catalytic processes that assist the production of carbon nanomaterials. The enzymes used for this process include endoglucanases (slice internal bonds of cellulose chains), *exoglucanases* (remove cellobiose units from chain ends), *β-glucosidases* (hydrolyze cellobiose into glucose), *cellulases* (break β-1,4-glycosidic bonds in cellulose, releasing glucose), *hemicellulases* (target hemicellulose polymers releasing xylose and other sugars), and *ligninases (laccases, peroxidases)* (degrade lignin's complex aromatic network). Enzymatic hydrolysis is preferred as it operates in mild conditions (ambient temperature, neutral pH) and generates fewer by-products compared to conventional chemical hydrolysis (Idowu and Yupanqui 2025).

10.5.2 Microbial-Assisted Synthesis of CNMs

The concept of microbial-assisted production utilizes the natural biological activities of microbes, including bacteria, fungi, and algae, in the production of carbon-based nanomaterials (CNMs) through biotransformation and biomineralization techniques. These systems utilize the natural biological activities of microbes in the conversion of organic substrates or metal-based compounds into CNMs in natural environments, particularly in water-based systems. These systems provide an eco-friendly approach to the conventional, energy-consuming techniques of CNM production (Mohapatra and Singh 2026; Mishra et al. 2023).

Furthermore, microbes participate in the degradation of organic waste or biomass, thereby producing EPS, which contains carbon, and through the biological activities of the microbes, biomass is carbonized to produce CNMs. Microbial synthesis involves several mechanisms, including *biomineralization* (microbes inducing nucleation and growth of carbon-based nanoparticles), *biotransformation* (microbial metabolism modifying the substrates into high carbon compounds), *extracellular synthesis* (microbial enzymes and metabolites facilitating carbon nanomaterial generation) (Mohapatra and Singh 2026; Mishra et al. 2023), *carbon dots and GQDs* (microbial synthesis producing fluorescent carbon dots for bioimaging and sensors), *bio-CNTs and nanofibers* (microbial treatment of waste followed by carbonization leading to nanotube formation), and *green synthesis* (eliminating toxic chemicals and high-energy requirements) (Dong et al. 2014).

10.5.3 Catalysts for CNM Production

Catalysts play an essential role in enhancing efficient transform of biomass and waste into carbon nanomaterials. This is through an enhanced selectivity and quality by dropping energy barriers and controlling carbon nanostructure formation. Both heterogeneous and homogeneous catalysts are used (Chidhambaram et al. 2023;Yao et al. 2022).

10.5.3.1 Types of Catalysts

Metal catalysts, particularly transition metals such as iron (Fe), nickel (Ni), cobalt (Co), and copper (Cu), as well as their alloys, are the most frequently used catalysts in the process of carbon source degradation and of initiation of CNT and graphene formation. These metals may be supported on oxide substrates, for example, aluminum oxide (Al_2O_3) or silicon dioxide (SiO_2), in order to maximize the catalytical activity. As far as the acid-base catalysts are concerned, the acidic catalysts, such as sulfuric acid and zeolites, amplify the process of dehydration, while the basic catalysts, such as alkaline earth metals, increase speed of the polymerization process and cross-linking. Biocatalysts offer an innovative approach to the production of CNM precursors, as enzymes such as laccases and peroxidases enhance lignin degradation, thereby facilitating the formation of CNM precursors (Yao et al. 2022).

10.5.3.2 Role of Catalysts in CNM Formation

Nucleation sites: Metal nanoparticles provide the initial sites toward the collection of carbon atoms for the formation of CNMs.

Graphitization enhancement: Catalysts assist in the reorganization of carbon atoms to form well-ordered graphitic domains.

Control of selectivity: Designing catalysts allows the synthesis of targeted material, such as selecting single-walled or multi-walled carbon nanotubes.

Energy reduction: The use of these catalysts facilitates processing under milder temperature conditions and reduced reaction times relative to non-catalytic approaches. For example, the catalytic pyrolysis of lignocellulosic biomass using Ni-Fe bimetallic catalyst supported on silica for the synthesis of CNMs can further reduce the temperature to 100 °C and increase the yield of single-walled CNTs by 30%. Platinum and cobalt nanoparticles were also explored as catalysts for the production of graphene using the CVD method (Mohamed et al. 2024).

10.5.4 Integration of Biochemical and Catalytic Methods

This integration strategy involves the combination of enzymatic or microbial pretreatment with catalytic conversion to enhance the generation of CNMs. Thus, sugars produced by this approach, through hydrolysis reaction, can be converted to CNMs using thermochemical or catalytic conversion processes to increase the yield of the final products. The enzymatic techniques provide clean carbon sources; moreover, the microbial techniques provide clean biosynthetic ways, and on the other hand, the catalytic techniques provide clean control mechanisms (Li et al. 2019).

10.6 Applications of Food Waste-Derived CNMs

The CNMs derived from food waste address two of the world's critical problems: more smart management of waste and the development of more sustainable and high-performance materials. These materials, including carbon dots, biochar, CNTs, and derivatives of graphene, exhibit remarkable physicochemical properties, including high surface area, chemical stability, and tunability of their functional groups, as well as improved electrical and optical properties. The green and cost-effective synthesis of these materials from wasted food biomaterials offers a great promise for their application in diverse sectors (Chen et al. 2024a; Jayabal et al. 2026; Fahad Halim et al. 2025).

As the global population keeps to rise, further increasing stress on natural resources such as soil and water, and withdrawing the availability of ecosystems, the demand for sustainable agricultural practices and safer food systems has become undoubtedly urgent. For these reasons, the carbon-based nanomaterials such as graphene, carbon nanotubes (CNTs), and carbon quantum dots (CQDs) have been prioritized as possible solutions. Their unique physical and chemical properties are critical to the reduction of postharvest loss and increased agricultural production (Fig. 10.3) (Fahad Halim et al. 2025).

10.6.1 Smart Packaging with Antimicrobial Properties

Microbial contamination-related spoilage is one of the most crucial concerns that the entire food industry is facing today, which results in significant financial losses, reduced shelf life, and significant public health threats from pathogens such as *Salmonella*, *Escherichia coli*, and *Listeria monocytogenes*. Conventional packaging

Fig. 10.3 Applications of food waste-derived CNMs within the circular economy framework

materials such as petroleum-based plastics are known to providing passive protection and inhibit microbial growth and monitor the quality of the product in real time. Hence, there is an imperative requirement to develop advanced packaging solutions that can provide antimicrobial properties, with enhanced barrier properties, and real-time quality monitoring to reduce spoilage and wastage. Consequently, carbon- and nitride-based nanomaterials have great potential as active packaging solutions (Mohamed et al. 2025).

10.6.1.1 CNMs' Antimicrobial Mechanisms

The antimicrobial properties of CNMs are additionally attributed toward the following mechanisms: disruption of the physical integrity of the bacterial cell membrane by the sharp edges of the GO sheets and the needlelike shape of the CNTs, which can penetrate the cell membrane and cause the leakage of the cellular components. The production of ROS is also one of the antimicrobial mechanisms. CNMs, as they can catalyze the production of ROS such as hydroxyl radicals and superoxide anions, can cause oxidative stress that damages the bacterial cell components such as proteins, lipids, and DNA. Electron transfer between the bacterial cell membrane and the CNMs can also impact the respiration and metabolism of the bacteria. The high surface area of the CNMs can also induce bacterial starvation by adsorbing nutrients from the medium (Mohamed et al. 2025).

10.6.1.2 Examples of CNM-Enhanced Smart Packaging

Graphene-based polymer nanocomposites include using biodegradable polymers such as polylactic acid or polyvinyl alcohol with graphene oxide to enhance antimicrobial activity against *E. coli* and *S. aureus*. CNT-based packaging: CNTs dispersed in polymer films display significant antimicrobial activity. Due to their high aspect ratio and electrical conductivity, CNT- based packaging is used to sense gases associated with food spoilage, such as ammonia and ethylene, with the assistance of the sensor platforms that provide real-time information on food freshness. CQDs are fluorescent nanoparticles that sense metabolites associated with microorganisms or chemical changes linked with food spoilage, indicating as non-invasive sensing agents within food packaging films (Mohamed et al. 2025).

10.6.1.3 Antimicrobial Activity and Food Safety

Several studies have proven that packaging materials impregnated with CNMs can reduce microbial contamination effectively. Laboratory tests have revealed that over 90% inhibition of the growth of *Listeria* and *Salmonella* was achieved on food items wrapped with GO nanocomposite packaging materials. CNM-based packaging materials significantly extend the shelf life of perishable commodities such as fruits, vegetables, and meat products by maintaining the level microbes at or below the level that can cause spoilage. If CNM-based sensing of technology is incorporated in the packaging materials, the level of contamination can be detected at the right time (Sadegh Hassani et al. 2023).

10.6.1.4 Sustainability and Biodegradability

Combining CNMs with biodegradable polymers addresses the pollution caused by plastic, thereby developing materials with improved biodegradability without compromising performance. This also aligns with the principles of the circular economy by reducing waste and enabling reuse or composting (Üçüncüoğlu and Korucu 2025).

10.6.2 Nano-fertilizers and Precision Delivery Systems

However, these fertilizers are known to experience low nutrient efficiency. This leads to nutrient loss of about 50–70%, resulting in environmental pollution and increased emissions of greenhouse gases. This increases the cost of production and eventually leads to soil degradation. This highlights the need for the development of nano-fertilizers based on nanotechnology, especially of CNMs, as a revolutionary approach to improve nutrient uptake efficiency with minimal environmental impacts (Davidraj et al. 2024).

10.6.2.1 CNMs as Nutrient Carriers

The CNMs are effective vehicles for the adsorption and delivery of nutrients due to the high surface area that can accommodate the nutrients. The surface of the CNMs can be engineered with carboxyl, hydroxyl, and amino groups to facilitate the binding of nutrient ions such as nitrate, phosphate, potassium, zinc, and iron. The nutrient delivery can be managed with the nano-carriers to equivalent the requirements of the crop to reduce nutrient loss (Davidraj et al. 2024).

10.6.2.2 Types of CNM-Based Nano-fertilizers

Graphene oxide-based fertilizers incorporate the use of graphene oxide sheets, which are able to capture the nutrients such as urea, ammonium, and phosphate ions. These nutrients are held, and they are released gradually, enhancing the accessibility of the nutrients (Yu et al. 2024).

CNT-based fertilizers incorporate the use of functionalized carbon nanotubes that transport micronutrients and macronutrients directly into the rhizosphere or the leaf tissues of the plants. The structure of the nanotubes facilitates the steady release of the nutrients, enhancing the accessibility of the nutrients in the plant system (Yang et al. 2023).

10.6.2.3 Mechanisms of Nutrient Delivery

Basically, CNMs facilitates nutrient supply to plant system through the root, which are absorbed either through the roots or via foliar application into leaves of the plants. This is because of their small size, enabling them to pass through the root's surface and leaves' stomata. It's possible for these nanoparticles to be designed to release nutrients based on specific activators, such as changes in pH levels, moisture levels, and the existence of soil microbes. The nanoparticles also have an impact on soil microbes through their ability to bind to their by-products or by delivering microbes alongside nutrients (Tatrari et al. 2021).

10.6.2.4 Research and Applications

Nitrogen fertilizers based on graphene oxide increased the yield of wheat by 20%. The fertilizers also improved the assimilation of the nutrient by the crop. Zinc fertilizers based on carbon nanotubes increased the content of the nutrient in rice grains and improved the enzymatic antioxidant capacity of the crop. Spraying CQDs on the leaves of the crop increased the content of chlorophyll in the leaves of the tomato crop and improved photosynthetic efficiency. Precision farming techniques such as drones can be used with CNM-based fertilizers (Sun et al. 2018).

10.6.3 Integration and Future Perspectives

The combined use of CNMs in smart packaging and nano-fertilizers suggests a holistic way forward for sustainable food production and distribution. By integrating these technologies, we can reduce waste in post-harvest activities and optimize the efficiency of resource utilization in crop production, addressing global food security concerns. The development of sustainable CNMs from renewable resources, particularly agricultural and food waste, supports circular economy, as waste materials are being utilized as a resource. Green synthesis, which utilizes biomass and food residues as carbon precursors, supports sustainability. Future breakthroughs in these materials could lead to the integration of antimicrobial packaging, nutrient delivery, and environmental sensing into multifunctional CNM-based systems, allowing for real-time monitoring of food quality and soil health (Bhanvase and Barai 2024).

10.7 Energy and Environmental Applications

However, carbon-based nanomaterials such as graphene, carbon nanotubes, and quantum dots are unique because of their outstanding properties, such as their excellent conductivity and large surface area. These materials are seen as having a great potential for energy storage, especially in the development of the latest supercapacitors and batteries, owing to the increased rate of charge transfer and increased energy density. Apart from that, the unique properties of CNMs make them useful in cleaning up the environment, particularly in the treatment of wastewater and CO_2 capture, depending on their surface properties (Kim et al. 2025; Bhatia et al. 2023).

10.7.1 CNMs in Energy Storage

Supercapacitors and batteries are two of the most important components in the field of storing electrical power, as they control the whole setup from gadgets in our hands to electric vehicles and the integration of renewable power sources. Supercapacitors are able to perform well in terms of charging and discharging, as they are able to store power through the adsorption of ions, thus outperforming traditional batteries in terms of performance in fast cycle operations. Carbon-based nanomaterials (CNMs), such as graphene, and carbon nanotubes (CNTs) are two of the best conductors that provide improved conductivity and surface area of these supercapacitors and batteries. Graphene, for instance, has an enormous surface area, with over 2600 m^2 per gram, thus improving the capacitance and energy density of the supercapacitors and batteries, while CNTs improve the power density of the supercapacitors and batteries. These nanomaterials, when combined with metal oxides or conducting polymers, form composite electrodes that provide improved performance, as they are able to exploit the conductivity and the favorable structure of the CNMs (Shivalkar et al. 2021).

Batteries, especially lithium-ion batteries, are at the core of the current energy storage sector. Carbon nanomaterials in the electrodes of batteries improve the efficiency of batteries by allowing the smooth flow of electrons, as well as the expansion and contraction of the electrodes during the discharge cycle of the batteries. Graphene and carbon nanotubes form a flexible framework that helps accommodate active materials with high capacities, like silicon and sulfur, thus improving the battery capacity and charge rates and overall stability. In addition, the doping of these nanomaterials with heteroatoms helps improve the flow of charges as well as the storage of ions. Currently, nanomaterials are being tested in sodium-ion, lithium-sulfur, as well as lithium-air batteries, to address conductivity limitations and capacity degradation in batteries. Because of their lightness, nanomaterials are the key to the next generation of wearable batteries for compact devices (Venkataraman and Sundramoorthy 2026).

CNMs are distinguished by their ability to store energy while functioning as effective agents in the cleaning of the environment, especially in the cleaning of wastewater. The ability of CNMs to address the limitations of traditional wastewater treatment methods is through the effective application of advanced technologies, especially adsorption and membrane separation technologies. Graphene oxide membranes are distinguished by their high permeate flux of water, as well as the ability to filter pollutants, especially due to the presence of a wide variety of oxygen-containing groups that improve their interaction with pollutants. Carbon nanotube membranes are distinguished by their ability to improve the durability of water, especially due to their ability to improve the resistance of water to fouling, thus promoting long- term water use. In addition, the ability of CNMs to improve the efficiency of photocatalytic technologies is by promoting the separation of charge carriers, as well as redox reactions, especially in the removal of pollutants in an environmentally friendly way (Jeevanandam and Danquah 2025).

10.7.2 Environmental Applications

In terms of cleaning up the environment, carbon-based nanomaterials (CNMs) are now at the center of cutting-edge wastewater treatments owing to their unrivaled adsorption ability and robust catalytic activity. These nanomaterials contribute to the removal of a wide range of pollutants, such as organic pollutants, toxic metal ions, as well as harmful microbes, thereby advancing sustainable water purification techniques (Mehta et al. 2025).

10.7.3 Wastewater Treatment

Wastewater generated from both industrial and urban sources may contain toxic metals, synthetic dyes, organic matter, and infectious pathogens, making it a potential threat to both environment and human health. Conventional wastewater treatment technologies have been reported to be less effective, with low removal efficiency, low selectivity, and high operating costs. The sustainability of these technologies has also been questioned. In contrast, carbon nanotube membranes have shown unique properties such as high mechanical strength, chemical stability, and resistance to fouling, making them more suitable for long-term water purification. Furthermore, electrochemical technologies employing CNMs as electrodes have shown great promise for wastewater treatment through enhanced redox reactions. This shows that these nanomaterials have great potential for use in future water purification technologies (Sahoo et al. 2025; Mecha et al. 2025).

The CNMs in adsorption applications are distinguished by their large surface area and porosity. These adsorbents are quite effective in adsorbing dyes; heavy metals such as Pb^{2+}, Hg^{2+}, and Cd^{2+}; and other organic pollutants. By modifying their surface with functional pieces such as oxygen-containing groups, their selectivity and adsorption capacity are enhanced. These adsorbents can be easily recycled and reused multiple times with minimal losses in capacity. To cite some examples, graphene oxide is found to have a high adsorption capacity for methylene blue and rhodamine dyes; CNTs with amine group functionalization are effective in adsorbing heavy metals, and with their magnetic composite forms, separation is easy (Sahoo et al. 2025).

10.7.4 CO_2 Capture and Conversion

The capture and utilization of CO_2 and its conversion into valuable products are one of the key environmental challenges that need to be overcome in the fight against climate change and carbon neutrality. Chemical nanostructured materials (CNMs) have been identified as potential candidates for CO_2 capture owing to their high surface area, tunable pore structures, and functional surface chemistry (Kamaruzaman et al. 2023). For example, membranes based on porous structures of graphene and carbon nanotubes have shown preferential permeation of CO_2 compared to other gases. This allows for the separation of CO_2 from flue gases and other industrial emissions. The key advantage of these capture and utilization technologies is their potential for creating a carbon cycle, similar to the concept of a circular economy, where CO_2 is used as a valuable resource. This has implications for greenhouse gas reduction and carbon capture and utilization for creating valuable chemicals and economic value from CO_2 (Mohamed et al. 2024).

10.8 Process Optimization and Scale-Up Challenges

Process optimization is at the heart of enabling these materials to perform well in energy and environmental applications. By constantly improving the way we make these materials, controlling their shape and size, and cleverly doping them with heteroatoms, we are able to increase surface area, conductivity, and selectivity for the desired species. In terms of energy storage, the way we design the electrode microstructure is critical, as it has to allow for facile ion transport, durability, and both high power and high energy density. However, the principal advancement required is not merely in the laboratory-scale synthesis of these materials, but in the development of scalable processes that preserve material performance while ensuring economic feasibility and sustainability (Zhang et al. 2024).

Achieving the best performance from carbon nanomaterials (CNMs) depends on the material structure, morphology, and surface chemistry. The design of the electrode structure and the pore structure can significantly enhance the performance of the materials with respect to ion transport, the electric path, and the mechanical strength. It is also very important to assess the environmental and health impacts of the CNMs to facilitate the development of the technology (Chand et al. 2025).

10.9 Life Cycle Assessment and Techno-economic Analysis of Waste-to-CNMs Processes

Therefore, the increased environmental challenges, the depletion of fossil fuel resources, and the rising tide of global waste have created the need for novel technologies, such as the conversion of waste into valuable CNMs. However, it is important to understand that, despite the promise of waste as sources of valuable CNMs, it is imperative that the environmental and economic sustainability of these novel material conversion technologies be carefully gauged through life cycle assessment (LCA) and techno-economic analysis (TEA), as presented in the study by Santagata et al. (2021).

10.9.1 Life Cycle Assessment of Waste-to-CNM Processes

This is because the life cycle assessment (LCA) is a comprehensive framework that outlines the environmental impact of the product from the collection of the raw materials (the cradle) to the final disposal of the waste (the grave). By using the LCA to assess the production routes of waste-to-carbon nanomaterials (CNMs), one can identify the environmental advantages or costs of converting food waste, agricultural waste, and industrial waste into carbon-based nanomaterials. This can be achieved by identifying the hotspots in the production processes of waste-to-carbon nanomaterials to make informed decisions to enhance the sustainability of the processes (Singh and Kumar 2025; Abdelmonem et al. 2025).

In defining the goal and scope of life cycle assessment for waste-derived carbon nanomaterials (CNM) pathways, it is important to start with an articulation of purpose and scope. In general, the goal of life cycle assessment of these pathways is focused on determining some of the key environmental impact categories, including greenhouse gas emissions, total energy demand, water consumption, and toxic releases. The scope of life cycle assessment for these pathways includes defining boundaries from collection and transport of raw materials through pretreatment and synthesis of CNMs, post-processing and purification, and finally disposal or recycling (Abdelmonem et al. 2025).

Inventory Analysis: During this stage, an in-depth understanding of the material and energy flows in the entire chain of production is created. However, in the production of carbon nanomaterial (CNM), the focus shifts to the processes that are more energy-intensive, i.e., high-temperature pyrolysis or activation, as well as the processes that involve chemicals, i.e., oxidation and doping, as these processes may be the major contributors to environmental concerns. Also, the production of by-products and waste, including the final distribution of the of CNMs, is taken into account during the analysis (Abdelmonem et al. 2025).

Interpretation and Improvement: The LCA analysis has demonstrated that optimizing the design of a process is essential to reduce both energy consumption and chemical requirements. The integration of renewable energy sources into energy-intensive steps such as pyrolysis can significantly minimize the overall carbon footprint. Moreover, utilizing by-products such as syngas and bio-oil effectively not only improves resource efficiency but also enhances the overall environmental benefit associated with waste-based CNM (Ciccullo et al. 2021).

A complete LCA enables us to make direct comparisons between waste-derived CNMs and their traditional, conventionally produced counterparts, such as graphene derived from graphite. In many cases, these direct lookups show that waste-derived materials reduce environmental impacts because they address waste management problems while also producing useful materials. Yet, there are limitations, mainly related to data gaps, variability in processes, and a lack of data on the performance of these new technologies at an industrial scale (Zhao et al. 2023).

Techno-economic Analysis of Waste-to-CNM Processes

Although the environmental performance is significant, the actual viability depends on the cost involved in the production of CNMs on a larger scale. TEA examines the total cost involved in the profitability of the business, its scalability, and the viability of producing CNMs from waste streams.

The capital costs are one of the major factors involved in the production of CNMs on a larger scale. The production of CNMs involves the use of high-end technology, including reactors for the processes involved in pyrolysis or hydrothermolysis, as well as the costs involved in the downstream processes and the assurance of the quality of the products. The initial costs involved in the production of CNMs are primarily due to the high-temperature furnaces and the need to maintain the processes at high standards. The pilot-scale production costs provide the initial insights regarding the costs involved in the production of CNMs, but the actual production costs are achieved at the economies of scale level (Cox et al. 2010; Subramani et al. 2025).

Operating Costs: These include everything necessary to run the process, including getting the feedstock, powering the process, paying the staff, buying chemicals, keeping the equipment up to date, and dealing with the waste. Since the primary feedstock comes from waste, the initial cost of the material may be very low or even negative, thanks to incentives for the disposal of the material, thus keeping the operating costs low. However, the energy required for the thermochemical reactions and the chemicals for the doping and purification processes remain key costs in the operation (Tropea 2022).

Revenue and Market Dynamics: The revenue or value of CNMs is largely driven by their structural properties, functionality, and demand in various sectors such as electronics, energy storage, materials, and environmental remediation. In order for waste-derived CNMs to become commercially viable, they must become cost-effective and comparable in terms of performance to those produced by conventional manufacturing methods. The production of coproducts such as bio-oil and syngas can contribute to the overall viability of the process by generating additional revenue streams (Chen et al. 2024b).

Economic indicators: They are the most common way of looking at economic performance. The tools of the trade, like net present value, internal rate of return, and the payback period, combine to give us an idea of the profitability of the project and the extent of the investment's attractiveness. Sensitivity analysis is an important aspect as well, as it points us toward the major cost drivers and the places of uncertainty. The availability of feedstock, as well as the of prices of energy, could have a major influence on the economic feasibility of the project (Rațu et al. 2023).

Scale-Up and Commercialization: Scaling up the production of CNMs from the laboratory scale, through pilot-scale, to industrial-scale operations has its own technical as well as economic challenges. Intensification of the process, automation, and modular design of the plants are some of the options that could help reduce the overall cost of production as well as improve the efficiency of the operations. Partnerships with the waste management industry and the agro-industrial sectors are the key to the availability of a reliable feedstock supply. The economic viability of the waste-based CNM technologies could be significantly enhanced by the availability of supportive policies, including financial, carbon credits, and green product schemes (Kefas et al. 2024).

10.10 Policy Frameworks and Stakeholder Engagement

Maintaining robust support policies and broad-based participation by all stakeholders is essential in moving food waste-based CNMs forward in a circular economy. The collaboration between all stakeholders will be essential in unlocking the full potential that CNMs can offer in moving forward with sustainability (Al-Dairi et al. 2022; Tariq and Mushtaq 2023).

10.10.1 Regulatory Guidelines for Nano-products

The rapid development of nanotechnology, particularly carbon-based nanomaterials, has given rise to specific regulatory challenges. The novel physicochemical properties, risks to humans and the environment, and the complex aspects of nanomaterial synthesis all contribute to this problem. To address this, specific regulatory guidelines are being prepared worldwide to address these problems while balancing safety and innovation. Organizations such as the United States Environmental Protection Agency, the European Chemicals Agency, and the US Food and Drug Administration highlight the importance of performing detailed safety assessments on products involving nanotechnology. These include detailed toxicological profiles, exposure scenarios, as well as environmental impact assessments. Due to their small size and highly reactive surfaces, nanomaterials can behave quite differently from conventional substances, thereby requiring specific testing procedures (Li et al. 2015).

The classification and labeling of products also play an important role in their openness, risks, and environmental release. Organizations such as the OECD and ISO help us standardize how we characterize and test the quality of nanomaterials and how we do life cycle analyses. Waste management strategies stress recycling and disposal of waste without contaminating of ecosystems. The involvement of all players, including academia, industry, and government, is crucial for ensuring that we get maximum benefits from these advancements while always of keeping safety first (Sun et al. 2013).

10.10.2 Case Studies of Successful Circular Economy Models

Real-life examples illustrate the integration of the concept of circular economy with the concept of nanotechnology and waste to value, with the support of policies and stakeholders to achieve sustainable outcomes.

10.10.2.1 Industrial Symbiosis in Scandinavian Countries

Sweden and Denmark are examples of countries that have established industrial symbiosis platforms facilitating the transformation of waste from one sector to resources for another, particularly the production of CNMs. This concept is also supported by these countries' policies that promote the reduction of waste and the recycling of products, as well as the development of green technology (Choudhury et al. 2022).

10.10.2.2 Private Sector Circular Initiatives

Globally, several companies are pioneering circular economy approaches by converting food waste into CNMs for applications in packaging, water treatment, and energy storage. These projects are always presented in conjunction with governments and communities to ensure everything is performed within the law and that safety and public trust are maintained (Gunawardene et al. 2022).

10.10.3 Challenges and Future Perspectives

The existing synthesis approaches such as pyrolysis, hydrothermal carbonization, and catalytic transformation are carried out on a laboratory scale. However, if it were scaled up to an industrial level, significant technical challenges still remain; these includes energy efficiency, product quality, and reactor engineering. Continuous flow reactors and modular reactors are promising approaches that may be scaled up. However, more research and investment are necessary in this area (Ji et al. 2025).

The physicochemical properties of carbon nanomaterials are quite unique. However, their potential toxicity and environmental risk are not completely investigated. Therefore, life cycle assessment and toxicology must be an integral part of product development. The regulations are inadequate than the rapid developments in nanotechnology, thereby resulting in high uncertainty (Sudalyandi and Jeyakumar 2022). In addition, there is public concern over the potential environmental and human health risks of nanomaterials. To overcome this, there is a requirement for effective regulations and active stakeholder engagement. This may be beneficial in facilitating global trade and cooperation if international regulations are aligned (Aksu and Güzdemir 2025).

10.11 Conclusion

The prospect of utilizing food waste to produce CNMs is indeed one that has considerable potential to address various global concerns simultaneously, namely, waste management, resource depletion, and sustainable development. The ability to leverage various waste management strategies that involve smart thermochemical and biochemical transformations has significant potential to produce valuable nanomaterials that can be utilized in energy management, agriculture, and environmental systems. The ability to produce CNMs within a sustainable development framework has considerable potential to maximize resource efficiency, minimize environmental impact, and create new economic value chains. However, this approach faces significant challenges including managing feedstock variability, ensuring safety and health implications, remaining economically competitive, and responding to regulatory dynamics. Therefore, the ability to produce CNMs utilizing food waste has considerable potential to advance new knowledge in material management and environmental management that can accompany in a sustainable, carbon-neutral future.

References

Abayomi OO, Yuen GC, Mishra P (eds) (2024) Food waste valorization: emerging trends, techno-economic and environmental considerations. https://doi.org/10.1016/C2022-0-02274-1

Abdelmonem M, Nair RD, Ibrahim MA, Soomro R, Meli AD, Pratiwi AR et al (2025) Synthesis and characterization of agricultural waste-derived CNMs (AWD-CNMs) for tissue engineering application. In: Waste-derived carbon nanostructures: synthesis and applications. Springer Nature Switzerland, Cham, pp 233–270

Aksu M, Güzdemir Ö (2025) Food waste-derived CQDs and their applications in food technology: a critical review. Food Bioprocess Technol 18(8):6753–6778

Al-Dairi M, Pathare PB, Al-Yahyai R, Opara UL (2022) Mechanical damage of fresh produce in postharvest transportation: current status and future prospects. Trends Food Sci Technol 124: 195–207

Alves JO, Zhuo C, Levendis YA, Tenório JA (2011) Catalytic conversion of wastes from the bioethanol production into CNMs. Appl Catal B Environ 106(3–4):433–444

Bhanvase BA, Barai DP (eds) (2024) Bio-derived carbon nanostructures: fundamentals, synthesis and applications. Elsevier

Bhatia L, Jha H, Sarkar T, Sarangi PK (2023) Food waste utilization for reducing carbon footprints towards sustainable and cleaner environment: a review. Int J Environ Res Public Health 20(3): 2318

Blasi F, Tringaniello C, Verducci G, Cossignani L (2022) Bioactive minor components of Italian and Extra-European hemp seed oils. LWT 158:113167

Bognár S, Despotović V, Finčur N, Putnik P, Jovanović D, Hainz K, Šojić Merkulov D (2025) Waste-derived photocatalysts for light-driven water treatment: a comprehensive review of green materials from agri-food, industrial, and E-waste. Separations 12(12):337

Chand S, Kumar R, Kumar K (2025) Plant waste-derived CNMs. In: Waste derived CNMs, vol 1. American Chemical Society, pp 115–148

Chen W, Yin H, Cole I, Houshyar S, Wang L (2024a) Carbon dots derived from non-biomass waste: methods, applications, and future perspectives. Molecules 29(11):2441

Chen Z, Guo Y, Luo L, Liu Z, Miao W, Xia Y (2024b) A critical review of hydrochar based photocatalysts by HTC: synthesis, mechanisms, and applications. Biochar 6(1):74

Chidhambaram N, Kay SJJ, Priyadharshini S, Meenakshi R, Sakthivel P, Dhanbalan S et al (2023) Magnetic nanomaterials as catalysts for syngas production and conversion. Catalysts 13(2):440

Choudhury BJ, Muigai HH, Kalita P, Moholkar VS (2022) Biomass blend derived porous carbon for aqueous supercapacitors with commercial-level mass loadings and enhanced energy density in redox-active electrolyte. Appl Surf Sci 601:154202

Ciccullo F, Cagliano R, Bartezzaghi G, Perego A (2021) Implementing the circular economy paradigm in the agri-food supply chain: the role of food waste prevention technologies. Resour Conserv Recycl 164:105114

Cox J, Giorgi S, Sharp V, Strange K, Wilson DC, Blakey N (2010) Household waste prevention—a review of evidence. Waste Manag Res 28(3):193–219

Cuffey J, Li W, Yu Y, Miao R (2023) Retail food environment and household food waste: an empirical study. Food Policy 117:102457

Davidraj JM, Sathish CI, Benzigar MR, Li Z, Zhang X, Bahadur R et al (2024) Recent advances in food waste-derived nanoporous carbon for energy storage. Sci Technol Adv Mater 25(1): 2357062

De Schouwer F, Claes L, Vandekerkhove A, Verduyckt J, De Vos DE (2019) Protein-rich biomass waste as a resource for future biorefineries: state of the art, challenges, and opportunities. ChemSusChem 12(7):1272–1303

Deng WJ (2016) Turning food waste into biofuel. In recycling of solid waste for biofuels and bio-chemicals. Springer, Singapore, pp 357–379

Diamantis V, Eftaxias A, Stamatelatou K, Noutsopoulos C, Vlachokostas C, Aivasidis A (2021) Bioenergy in the era of circular economy: anaerobic digestion technological solutions to produce biogas from lipid-rich wastes. Renew Energy 168:438–447

Dong Y, Lin J, Chen Y, Fu F, Chi Y, Chen G (2014) GQDs, graphene oxide, CQDs and graphite nanocrystals in coals. Nanoscale 6(13):7410–7415

Fahad Halim AFM, Poinern GEJ, Fawcett D, Sharma R, Surendran S, Rajeshkannan R (2025) Biomass-derived CNMs: synthesis and applications in textile wastewater treatment, sensors, energy storage, and conversion technologies. CleanMat 2(1):4–58

Fu B, Hower JC, Zhang W, Luo G, Hu H, Yao H (2022) A review of rare earth elements and yttrium in coal ash: content, modes of occurrences, combustion behavior, and extraction methods. Prog Energy Combust Sci 88:100954

Garrido-Romero J, D'Souza AB, Hanelt D, Abomohra A (2024) Lipid-rich particles of processed food waste for microalgae harvest through lipid-enriched floating biomat formation. Bioresour Technol 394:130251

Geissdoerfer M, Savaget P, Bocken NM, Hultink EJ (2017) The circular economy–a new sustainability paradigm? J Clean Prod 143:757–768

Gunawardene OH, Gunathilake CA, Vikrant K, Amaraweera SM (2022) Carbon dioxide capture through physical and chemical adsorption using porous carbon materials: a review. Atmosphere 13(3):397

Hazarika U, Xavier KM, Barman P, Goswami P, Ngasotter S, Pathak J (2025) Food waste-derived CNMs. In: Waste derived CNMs, vol 1. American Chemical Society, pp 225–248

He Z, Zhang C, Zhu Z, Yu Y, Zheng C, Wei F (2024) Advances in CNTs and carbon coatings as conductive networks in silicon-based anodes. Adv Funct Mater 34(48):2408285

Heydari Gharahcheshmeh M (2025) Fabrication of conjugated conducting polymers by CVD method. Nanomaterials 15(6):452

Idowu OA, Yupanqui CT (2025) Enzymatic hydrolysis with pepsin enhanced the nutrient compositions of unfractionated soy protein hydrolysate and its cell viability and nitric oxide activities. Food Sci Nutr 13(11):e71147

Jayabal R, Mehar K, Dash I, Sandhwar V, Marupati S, Prabhakar P (2026) Microwave-assisted waste valorization for hydrogen and CNMs: a critical review. Int J Hydrog Energy 198:152762

Jeevanandam J, Danquah MK (2025) Agricultural waste-derived CNMs for biomedical applications. In: Waste-derived carbon nanostructures: synthesis and applications. Springer Nature Switzerland, Cham, pp 213–232

Ji Y, Xue J, Wu J, Fan C, Guo E, Yang H (2025) Plant-derived insoluble protein aggregates: formation mechanisms, influencing factors, solubilization, functional properties, and applications. Food Chem: X 31:103042

Kamaruzaman NA, Jaafar MH, Mohideen M, Fatinathan S (2023) Effects of treated and untreated sludge applications on human health, the environment and other ecological factors. In: Microbial fuel cell (MFC) applications for sludge valorization. Springer Nature Singapore, Singapore, pp 23–42

Kefas EB, Aliyu B, Tashiwa YI (2024) Postharvest losses, causes and mitigation in tomato transportation: a systematic review. J Hortic Postharvest Res 7(3):223–236

Khan ZG, Patil PO (2020) A comprehensive review on carbon dots and GQDs based fluorescent sensor for biothiols. Microchem J 157:105011

Khan R, Qureshi A, Azhar M, Hassan ZU, Gul S, Ahmad S (2025) Recent progress of fluorescent carbon dots and GQDs for biosensors: synthesis of solution methods and their medical applications. J Fluoresc 35(5):2623–2640

Kiaya V (2014) Post-harvest losses and strategies to reduce them. Technical Paper on Postharvest Losses, Action Contre la Faim (ACF) 25(3):1–25

Kim M, Goerzen D, Jena PV, Zeng E, Pasquali M, Meidl RA, Heller DA (2024) Human and environmental safety of CNTs across their life cycle. Nat Rev Mater 9(1):63–81

Kim SW, Ahmadi Y, Kim KH, Lee J (2025) Carbon dots derived from non-biomass waste for a sensing application. Crit Rev Environ Sci Technol 55(8):536–552

Kirchherr J, Reike D, Hekkert M (2017) Conceptualizing the circular economy: an analysis of 114 definitions. Resour Conserv Recycl 127:221–232

Kringel DH, Dias ARG, Zavareze EDR, Gandra EA (2020) Fruit wastes as promising sources of starch: extraction, properties, and applications. Starch-Stärke 72(3–4):1900200

Lavanya P, Priyanga G, Theagarajan R (2026) Avocado seeds (Persea americana). Underutilized Seeds: Sustainable Solutions for Health, Industry, and Environment. Springer Nature, Singapore, pp 39–75

Li X, Rui M, Song J, Shen Z, Zeng H (2015) Carbon and GQDs for optoelectronic and energy devices: a review. Adv Funct Mater 25(31):4929–4947

Li M, Chen T, Gooding JJ, Liu J (2019) Review of carbon and GQDs for sensing. ACS Sens 4(7): 1732–1748

Li W, Guo Y, Zhang X, Dong W, Li X, Yu T, Wang K (2024) Development of self-sensing ultra-high-performance concrete using hybrid carbon black and CNFs. Cem Concr Compos 148: 105466

Liang Y, Zhou Y, Smith RL Jr, Qi X (2023) Ultra-thin highly-wrinkled graphene-like nanosheets for supercapacitor electrodes via 4-nitrocatechol and solvent-induced self-assembly. Carbon 204:495–506

Liang F, Dong H, Dai J, He H, Zhang W, Chen S et al (2024) Fast energy storage of SnS2 anode nanoconfined in hollow porous CNFs for lithium-ion batteries. Adv Sci 11(4):2306711

Liu X, Wang L, Zhang X, Liu Y, Li Y, Li Z et al (2025) Identification and biochemical characterization of a novel porcine circovirus associated with porcine respiratory and diarrheal diseases. Microbiol Spectrum 13(11):e02299–25

Mahajan MR, Nangare SN, Patil PO (2023) Nanosize design of carbon dots, GQDs, and MOFs based sensors for detection of chlorpyrifos in food and water: a review. Microchem J 193: 109056

Mecha CA, Monchari C, Chollom MN, Bakare BF (2025) Applications of solid waste-derived CNMs in water treatment. In: Waste derived CNMs, vol 2, pp 85–101

Mehta S, Kumar R, Kumar S, Kaur GA, Kainthla I, Shandilya M (2025) Applications of solid waste-derived CNMs in catalysis. Waste derived CNMs 2:213–232

Mishra S, Sahoo S, Sahoo PK, Sahoo NK (2023) Recent advancements in the plant and microbial assisted green synthesis of nanomaterials. Mater Today Proc. https://doi.org/10.1016/j.matpr.2023.10.144

Mohamed MA, Adel M, El Nady J (2024) Evaluation of the electrochemical energy storage performance of symmetric supercapacitor devices based on eco-friendly synthesized nitrogen-doped graphene-like derivative electrodes from the perspective of their nanostructural characteristics. Energy Adv 3(12):2947–2964

Mohamed MG, Mansour NA, Mousa AA, Mazrouaa AM (2025) Applications of solid waste-derived CNMs in packaging material. In: Waste derived CNMs, vol 2. American Chemical Society, pp 1–16

Mohapatra M, Singh S (2026) Resource recovery and valorization of food industry waste: paradigm shift of food industry waste into high valued resources through valorisation as an integrated sustainable approach. In: Microbial applications in industrial wastewater treatment. Springer Nature Singapore, Singapore, pp 275–295

Muhmood T, Ahmad I, Haider Z, Haider SK, Shahzadi N, Aftab A et al (2024) Graphene-like graphitic carbon nitride (g-C3N4) as a semiconductor photocatalyst: properties, classification, and defects engineering approaches. Mater Today Sustain 25:100633

Murjani BO, Kadu PS, Bansod M, Vaidya SS, Yadav MD (2022) CNTs in biomedical applications: current status, promises, and challenges. Carbon Lett 32(5):1207–1226

Nandi R, Jha MK, Guchhait SK, Sutradhar D, Yadav S (2023) Impact of KOH activation on rice husk derived porous activated carbon for carbon capture at flue gas alike temperatures with high CO2/N2 selectivity. ACS Omega 8(5):4802–4812

Nath A, Das K, Dhal GC (2023) Global status of agricultural waste-based industries, challenges, and future prospects. In: Agricultural waste to value-added products: technical, economic and sustainable aspects. Springer, Singapore, pp 21–45

Ogura Y, Taniya K, Horie T, Tung KL, Nishiyama S, Komoda Y, Ohmura N (2023) Process intensification of synthesis of metal organic framework particles assisted by ultrasound irradiation. Ultrason Sonochem 96:106443

Ojha S, Bußler S, Schlüter OK (2020) Food waste valorisation and circular economy concepts in insect production and processing. Waste Manag 118:600–609

Ouro-Salim O, Guarnieri P (2022) Circular economy of food waste: a literature review. Environ Qual Manag 32(2):225–242

Padhan SR, Rathore SS, Jat SL, Saini S, Mohanty A, Panigrahi KK et al (2025) Synthesis of CNMs from agro-industrial wastes and their extensive applications. In: Waste-derived carbon nanostructures: synthesis and applications. Springer Nature Switzerland, Cham, pp 71–106

Puri L, Hu Y, Naterer G (2024) Critical review of the role of ash content and composition in biomass pyrolysis. Front Fuels 2:1378361

Quilez-Molina AI, Chandra Paul U, Merino D, Athanassiou A (2022) Composites of thermoplastic starch and lignin-rich agricultural waste for the packaging of fatty foods. ACS Sustain Chem Eng 10(47):15402–15413

Rahman MH, Bhoi PR, Menezes PL (2023) Pyrolysis of waste plastics into fuels and chemicals: a review. Renew Sust Energ Rev 188:113799

Rațu RN, Veleșcu ID, Stoica F, Usturoi A, Arsenoaia VN, Crivei IC et al (2023) Application of agri-food by-products in the food industry. Agriculture 13(8):1559

Sabzi M, Mousavi Anijdan SH, Shamsodin M, Farzam M, Hojjati-Najafabadi A, Feng P et al (2023) A review on sustainable manufacturing of ceramic-based thin films by CVD (CVD): reactions kinetics and the deposition mechanisms. Coatings 13(1):188

Sadegh Hassani S, Daraee M, Navazani S, Rahimi F (2023) Carbon-based quantum dots from food waste: synthesis to application in food safety. https://doi.org/10.1039/9781839169564-00275

Sahoo PK, Das S, Rao YS, Behera AK (2025) Applications of solid waste-derived CNMs in water purification. In: Waste derived CNMs, vol 2. American Chemical Society, pp 55–83

Sajid M, Asif M, Baig N, Kabeer M, Ihsanullah I, Mohammad AW (2022) CNTs-based adsorbents: properties, functionalization, interaction mechanisms, and applications in water purification. J Water Process Eng 47:102815

Santagata R, Ripa M, Genovese A, Ulgiati S (2021) Food waste recovery pathways: challenges and opportunities for an emerging bio-based circular economy. A systematic review and an assessment. J Clean Prod 286:125490

Santos SM, Assis AC, Gomes L, Nobre C, Brito P (2022) Waste gasification technologies: a brief overview. Waste 1(1):140–165. MDPI

Shivalkar S, Gautam PK, Verma A, Maurya K, Sk MP, Samanta SK, Sahoo AK (2021) Autonomous magnetic microbots for environmental remediation developed by organic waste derived carbon dots. J Environ Manag 297:113322

Sikder S, Toha M, Rahman MM (2025) Environmental sustainability and future challenges of waste-derived CNMs. In: Waste derived CNMs, vol 1. American Chemical Society, pp 309–330

Silva TAL, Varão LHR, Pasquini D (2024) Lignocellulosic biomass. In: Handbook of biomass. Springer Nature Singapore, Singapore, pp 105–142

Silva SDO, Mafra AKC, Pelissari FM, Rodrigues de Lemos L, Molina G (2025) Biotechnology in agro-industry: valorization of agricultural wastes, by-products and sustainable practices. Microorganisms 13(8):1789

Singh K, Sawini, Kumar D (2025) Waste-derived CNMs (WD-CNMs): synthesis and characterization. In: Waste-derived carbon nanostructures: synthesis and applications. Springer Nature Switzerland, Cham, pp 1–27

Subramani VB, Lakmini NLM, Don KDA, Yashavant Chougale M, Dubal DP, Moghaddam L (2025) Bio-oil and high-performance carbon materials from biomass–plastic co-liquefaction for sodium-ion capacitors. Available at SSRN: https://ssrn.com/abstract=5624935 or https://doi.org/10.2139/ssrn.5624935

Sudalyandi K, Jeyakumar R (2022) Kinetics and modelling of hydrolysis. In: Biofuel production using anaerobic digestion. Springer Nature Singapore, Singapore, pp 85–97

Sun H, Wu L, Wei W, Qu X (2013) Recent advances in GQDs for sensing. Mater Today 16(11):433–442

Sun H, Zhou Y, Ren J, Qu X (2018) Carbon nanozymes: enzymatic properties, catalytic mechanism, and applications. Angew Chem Int Ed 57(30):9224–9237

Sun C, Han Z, Wang X, Liu B, Li Q, Li H et al (2023) Advanced carbons nanofibers-based electrodes for flexible energy storage devices. Adv Funct Mater 33(52):2305606

Tamasiga P, Miri T, Onyeaka H, Hart A (2022) Food waste and circular economy: challenges and opportunities. Sustainability 14(16):9896

Tariq A, Mushtaq A (2023) Untreated wastewater reasons and causes: a review of most affected areas and cities. Int J Chem Biochem Sci 23(1):121–143

Tatrari G, Karakoti M, Tewari C, Pandey S, Bohra BS, Dandapat A, Sahoo NG (2021) Solid waste-derived CNMs for supercapacitor applications: a recent overview. Mater Adv 2(5):1454–1484

Tezer Ö, Karabağ N, Öngen A, Çolpan CÖ, Ayol A (2022) Biomass gasification for sustainable energy production: a review. Int J Hydrog Energy 47(34):15419–15433

Tropea A (2022) Food waste valorization. Fermentation 8(4):168

Üçüncüoğlu D, Korucu H (2025) From food waste to functional nanomaterials: sustainable production, characterization, and application trends. Int J Agric Environ Food Sci 9 (Special):337–348

Uddin MM, Kabir MH, Ali MA, Hossain MM, Khandaker MU, Mandal S et al (2023) Graphene-like emerging 2D materials: recent progress, challenges and future outlook. RSC Adv 13(47):33336–33375

Venkataraman S, Sundramoorthy AK (2026) Waste-derived CNMs for electrochemical applications: toward a circular and sustainable future. RSC Adv 16(1):733–747

Vukic M, Vujasinovic V (2023) Minor bioactive compounds of cold-pressed oils and their biomedical functions: a brief review. Int J Biomed Healthc 11(4):364–370

Xia D, Yu H, Li H, Huang P, Li Q, Wang Y (2022) Carbon-based and carbon-supported nanomaterials for the catalytic conversion of biomass: a review. Environ Chem Lett 20(3): 1719–1744

Yang X, Chen Y, Zhang C, Duan G, Jiang S (2023) Electrospun CNFs and their reinforced composites: preparation, modification, applications, and perspectives. Compos Part B 249: 110386

Yao D, Li H, Mohan BC, Prabhakar AK, Dai Y, Wang CH (2022) Conversion of waste plastic packings to CNMs: investigation into catalyst material, waste type, and product applications. ACS Sustain Chem Eng 10(3):1125–1136

Yousuf A, Pirozzi D, Sannino F (2020) Fundamentals of lignocellulosic biomass. In: Lignocellulosic biomass to liquid biofuels. Academic Press, pp 1–15

Yu S, He J, Zhang Z, Sun Z, Xie M, Xu Y et al (2024) Towards negative emissions: HTC of biomass for sustainable carbon materials. Adv Mater 36(18):2307412

Zhang M, Qi Y, Zhang W, Wang M, Li J, Lu Y et al (2024) A review on waste tires pyrolysis for energy and material recovery from the optimization perspective. Renew Sust Energ Rev 199: 114531

Zhao G, Liu S, Wang Y, Lopez C, Ong A, Chen X (2023) Reducing food waste from social innovation perspective: a review of measures, research gaps and future directions. Int Food Agribus Manag Rev 26(2):199–224

Zhou J, Li D, Zhang X, Liu C, Chen Y (2023) Valorization of protein-rich waste and its application. Sci Total Environ 901:166141

Chapter 11
Elimination of Polythene Pollution Challenges Through Recycling Approaches

Bapi Ray Sarkar, Nikita Ghosh, Ganesh Dey, Tiyasha Gupta, Indira Gupta, Mrinmoy Basak, and Angila Theengh

11.1 Introduction

Plastic, as an engineering material, has proven advantageous to both developed and developing nations since the invention of the first wholly synthetic plastic, Bakelite, by Leo Baekeland in New York in 1907. Plastic waste management has become a critical environmental issue of the twenty-first century, with worldwide plastic manufacturing totaling 359 million tons in 2018 and an anticipated 12 billion tons expected to be in landfills by 2050. Present waste management procedures yield a mere 19.5% recycling rate for plastics, with the remaining materials being either discarded in landfills (55%) or burnt (25.5%) (Osman et al. 2025). An estimated 100 million tons of plastic are produced worldwide each year, according to a 2014 report by Greenpeace International. This figure raises serious concerns about environmental safety.

The polymer of ethylene gas is called polythene and also known as polythene. It is commonly used in shampoo bottles, grocery bags, and bulletproof jackets. There are various varieties of polythene, such as low-density, medium-density, high-density, and very low-density polythene (Rivard et al. 1995). Numerous industrial applications are made possible by polythene's flexibility, and the present study focuses on applications related to energy and improved fibers and the exploitation

B. R. Sarkar (✉) · N. Ghosh · G. Dey · T. Gupta · I. Gupta
Department of Pharmaceutical Technology, University of North Bengal, Darjeeling, West Bengal, India
e-mail: brs2019@nbu.ac.in

M. Basak
Faculty of Pharmaceutical Science, Assam down town University, Guwahati, Assam, India

A. Theengh
Government Pharmacy College, Gangtok, Sikkim, India

Y. K. Mohanta et al. (eds.), *Bioresource Waste to Carbon Nanomaterial*, Nanotechnology in the Life Sciences, https://doi.org/10.1007/978-981-92-2405-0_11

of unique thermal and optical properties. PE is widely used in consumer goods, packaging, films, containers, pipelines, cables and wires for insulation, and other engineering components (Bredikhin et al. 2017).

The usage of plastic, particularly polythene, is increasing daily. An annual accumulation of 25 million tons of synthetic plastics occurs in the terrestrial environment, as well as on sea beaches. The bulk of all manufactured plastic is made of polythene, which makes up 64% of the total. It is extensively used in the manufacturing of garbage cans, bottles, margarine tubs, water pipes, carry bags, and other throwaway items (Lee et al. 1991). Eventually, rivers, canals, and urban drainage systems allow all of the created plastic waste—including polythene waste—to enter marine water. As a result, it states that beaches were the best places for dumping waste made of polythene (plastic). Polythene waste breaks down at disposal sites through both mechanical and chemical weathering, but it takes a while for it to mineralize and may remain in a microscopic form for a long time (Corcoran et al. 2009). Polythene bags are carelessly disposed of in millions of landfills across the globe, which cover trillions of hectares of land and release extremely toxic leachates from the landfills as well as hazardous methane and carbon dioxide gases during their decomposition stages (Andrady 2015). Sewerage system blockages are a prevalent issue in cities due to plastic waste. If polythene bags are introduced into agricultural fields, they hinder water percolation and adequate soil aeration. Among the effects of polythene bags on agriculture are decreased soil fertility, reduced nitrogen fixation, increased soil nutrient loss, and imbalanced soil flora and fauna (Mohasin et al. 2025).

Recycling polythene is primarily accomplished through mechanical, chemical, and thermal methods; each method involves trade-offs in terms of emissions, product quality, and feedstock tolerance. The conventional approach for recovering polythene waste is mechanical recycling, which aims to preserve the backbone of the polymer while converting waste into secondary resin that can be reused. Through sophisticated sorting and collection, mechanical recycling isolates the various types of PE. It removes impurities before being shredded, cleaned, dried, and re-extruded into pellets or finished goods, with optional compatibilization or the addition of additives (Constantinescu et al. 2022). Chemical recycling allows material recovery that can surpass the purity constraints of mechanical recycling by reconverting polymer chains into monomers, oligomers, or other chemical feedstocks. Catalytic depolymerization, pyrolytic cracking, and hydrocracking, as well as solvolysis (for condensation polymers) and dissolution/reprecipitation techniques that produce monomers or feedstocks for new polymers or fuels, are examples of chemical (tertiary) recycling methods that are appealing when feedstocks mix and contaminate or when bottle-to-bottle quality is required (Thiounn and Smith 2020). The use of heat thermally converts PE into char, oils, and gases (pyrolysis) or syngas (gasification); alternatively, it recovers heat and electricity through controlled combustion. These processes can handle polluted and mixed feedstocks and are usually used when material recovery is impractical (Eze et al. 2021). More environmentally friendly choices for polythene and other plastics are sought through biological recycling and more recent electrochemical, photocatalytic, and AI-enabled sorting techniques (Fig. 11.1).

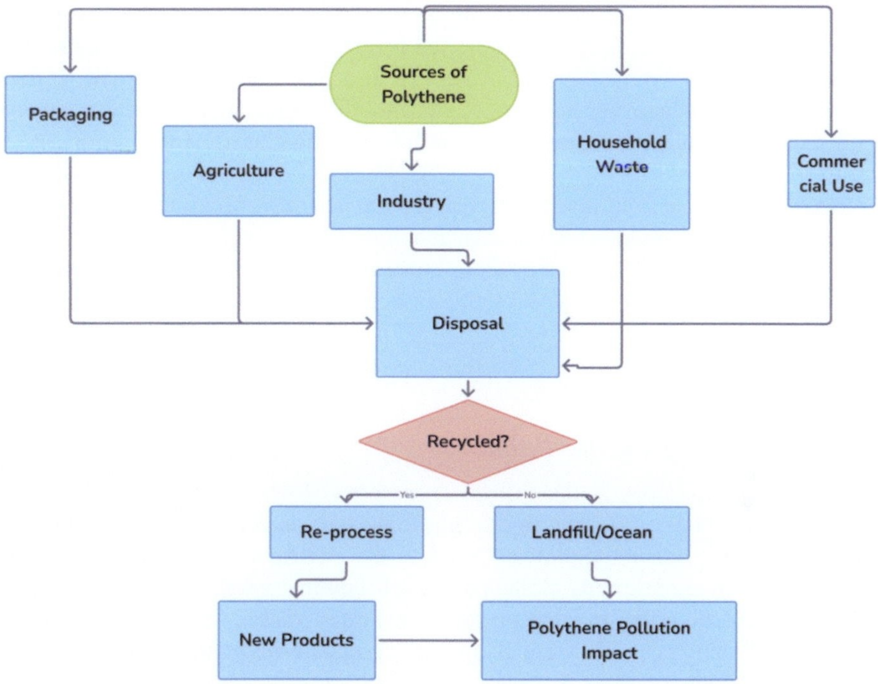

Fig. 11.1 Flowchart of polythene pollution

11.2 Sources and Types of Plastic Pollution

11.2.1 Sources of Plastic Pollution

The manufacturing, use, and improper handling of plastic materials are the leading human-caused causes of plastic pollution. Approximately 60 to 80% of marine trash is made of plastic, most of which enters oceans either directly or indirectly. They can be generally classified as having terrestrial and aquatic origins. It has been observed that global annual plastic production exceeds 359 million tons, worsening the issue (Thushari and Senevirathna 2020). Land-based sources are responsible for the majority of plastic pollution and include mismanaged waste from urban activities, littering landfills, and industrial discharges. Larger amounts of microplastics are discovered near contaminated sites, and plastics are eventually carried in reverse by city runoff, ultimately ending up in the oceans. Compared to product categories such as consumer goods and packaging, plastics used in electronics, transportation, and construction are the least prevalent in aquatic environments (Jambeck et al. 2018). Sea-based sources originate from marine activities including fishing gear (e.g., nets and lines), shipping waste, and offshore industries. Fishery items are particularly dominant in oceanic pollution.

The transport of plastics depends on the size, density, and shape of the plastic. Low-density, thick-walled items are carried by currents from rivers to oceans, and many of them accumulate in sediments, on beaches, and within biota. System dynamic models help to highlight that plastic mobilizes into the marine environment, with most accumulating on the seabed rather than floating (Kaimathuruthy et al. n.d.).

According to research, plastic stays in the environment because it is not biodegradable, which causes bioaccumulation and organisms to form at rates of 0.1–15.03 particles per individual. It poses a risk to human health (e.g., cancer, cardiovascular problems), biodiversity, and ecosystem services like tourism and fisheries (Ford et al. 2022).

11.2.2 Types of Plastic Pollution

Numerous materials, differing in composition, size, and origin, are included in plastic pollution, causing significant harm to the environment. Based on multiple reviews, plastics are primarily classified in the following order according to size and type. Macroplastic refers to large plastic items, often visible waste such as bottles, bags, and finishing gear. They are also found in aquatic environments and require a significant amount of time to break down into smaller particles. Macroplastics dominate specific product categories, such as packaging and fishing items, particularly in rivers and oceans (Rellán et al. 2023; Schwarz et al. 2019). These microplastic particles, which have a size of less than 5 mm, are a major concern due to their widespread presence and capacity to infiltrate food chains. They are further classified into primary microplastic and secondary microplastic. Primary microplastics are internationally manufactured for use in cosmetics and industry or as exfoliating abrasives. Secondary microplastics formed as a result of large plastics breaking down mechanically, photooxidizing, and being exposed to ultraviolet light. Due to their high dispersibility in soil, water, and air, these particles have a significant impact on both terrestrial and aquatic ecosystems (Thushari and Senevirathna 2020; Pilapitiya and Ratnayake 2024). Plastics are also classified as megaplastic (huge, substantial, considerable debris), mesoplastic (intermediate sizes), and nanoplastics (very small, smaller than microplastics, down to nanoscale). Polymer composition plays a crucial role, with polythene and polypropylene being the most widespread across environments due to their low density and extensive use (Schwarz et al. 2019; Pilapitiya and Ratnayake 2024). Plastics are divided into biodegradable (which can break down under certain conditions) and nonbiodegradable (persistent in the environment), with the latter contributing most to long-term pollution. Worldwide, microplastics are found to accumulate at rates of 0.001–140 particles/m^3/m^3 in water and 0.2–8766 particles/m^3/m^3 in sediments, with concentrations being higher in the vicinity of cities. Simply becoming entangled, being consumed, causing starvation, and transferring toxins like heavy metals into living things, these pollutants seriously harm the environment (Napper and Thompson 2020).

11.3 Impact of Polythene Pollution on the Terrestrial Environment

Once regarded as a miracle material, polythene has now turned into a disaster for the environment. Although it is inexpensive, lightweight, and long-lasting, these characteristics make it dangerous because it takes hundreds of years for it to break down. Every year, millions of tons of polythene are produced, and a large portion ends up in landfills, rivers, and oceans. From plastic bags and bottles to packaging materials, these items are clogging our soil, polluting our water, and endangering both marine and terrestrial life. Animals often mistake plastic for food, leading to injury and death. Microplastics pose a major threat to human health and have even entered our food chain. Additionally, burning polythene releases harmful gases that contribute to global warming and air pollution. Our planet will be covered in polythene tomorrow if we do nothing today. Saying no to polythene pollution and yes to a sustainable future is a commitment that safeguards our environment (Bodor et al. 2024; Ullah et al. 2023; De-la-Torre 2020).

11.3.1 Impact of Polythene Pollution on Soil Microbiota

As a synthetic ethylene polymer, polythene is highly resistant to microbial breakdown and is nonbiodegradable. Its growing buildup in soil environments has an adverse effect on microbial communities and soil health. Microplastics and polythene sheets block the flow of water and air, which lowers soil aeration and moisture penetration. It alters the soil's physical composition, decreasing its porosity and water-holding capacity, which has an immediate impact on the habitats of microorganisms. In soils contaminated with polythene, the diversity of beneficial soil microorganisms, such as bacteria, fungi, and actinomycetes, is diminished. Long-term buildup reduces soil fertility by causing a transition from diverse microbe populations to opportunistic or stress-tolerant species (Iqbal et al. 2020). Soil enzymes such as urease, dehydrogenase, and phosphatase, which are crucial for nutrient cycling, show reduced activity under polythene contamination. Plant growth is hindered due to inadequate carbon, phosphorus, and nitrogen cycles. Plasticizers, stabilizers, and dyes found in polythene seep into the ground. These substances are harmful to microorganisms, damaging cells, preventing the activity of enzymes, and sometimes even killing microorganisms. Breakdown of polythene into microplastics provides a new surface for colonization but often favors pathogenic or non-beneficial microbes. Microplastic increases the risk of soil-borne plant diseases and alters the natural microbial balance. Soil microbes are the main agents of organic matter decomposition. Polythene blocks contact between microbes and organic residues, leading to slower humus formation and lower soil fertility. A persistent buildup puts the microbiota in a chronic state of stress. The loss of beneficial microbes, including *Rhizobium*, *Azotobacter*, and mycorrhizal fungi, results in decreased soil productivity and poses a threat to sustainable agriculture. The soil ecosystem is also severely disrupted by polythene contamination, which alters the physical properties of the soil, reduces microbial diversity, and obstructs essential biochemical activities. To preserve soil microbiota, it is necessary to use less plastic, promote biodegradable alternatives, and implement effective waste management practices (Aminiyan et al. 2024).

11.3.2 Impact of Polythene Pollution on Plants

As a nonbiodegradable substance, polythene has emerged as a primary contributor to soil contamination. Its detrimental effects on agricultural productivity and plant growth are readily apparent (Alimi et al. 2018).

1. Soil Blocking: Polythene remains in the soil for years, blocking pores and reducing the natural aeration and water movement. It makes the soil less fertile for plant growth.
2. Nutrient Loss: When soil is covered with polythene, plant roots cannot properly absorb water and essential nutrients. As a result, plants show stunted growth and low productivity.
3. Toxic Effects: With time, polythene breaks down into microplastics and releases toxic chemicals. These harmful substances enter the plant system, disturbing its metabolism and reducing crop quality (Bettas Ardisson et al. 2014).
4. Poor Seed Germination: A soil covered with polythene prevents seeds from getting proper contact with the soil. It reduces seed germination and affects crop cycles.
5. Hindrance to Microorganisms: Soil microorganisms, which are essential for plant growth, are badly affected by polythene. Their activity decreases, resulting in a loss of soil fertility.
6. Long-Term Damage: Continuous accumulation of polythene in soil can make fertile land barren. Forests and natural vegetation also lose their regeneration capacity due to this pollution. Polythene pollution suffocates the soil, poisons the environment, and directly hampers the growth and survival of plants. To protect agriculture and biodiversity, strict measures to reduce, recycle, and replace polythene are urgently needed (Li et al. 2021).

11.3.3 Impact of Polythene Pollution on Animals

Numerous animals misinterpret polythene as food, become stuck in it, or experience harmful side effects. Polythene pollution has consequently become a serious threat to biodiversity, animal health, and ecological balance (Ali et al. 2020). Many animals, especially cows, dogs, birds, fish, and marine animals, mistake polythene bags for food. Once swallowed, the plastic blocks their digestive tract, leading to starvation, internal injuries, or death. Small animals and birds can choke while trying to swallow plastic pieces. Marine creatures often get suffocated when plastic bags wrap around their nose, mouths, or gills (Arthur et al. 2009). Polythene contains harmful chemicals like BPA, phthalates, and heavy metals. These toxins can enter an animal's body, causing hormonal imbalance, reproductive failure, and organ damage. Animals often get entangled in polythene sheets, sacks, or rings, which restrict their movement. It leads to injuries, infections, or drowning in aquatic habitats. Microplastics from polythene enter small organisms (plankton, fish), which are consumed by larger animals. Over time, toxic chemicals accumulate in higher organisms, disturbing the whole food chain, including humans. Polythene litter reduces soil and water quality, making habitats less suitable for animals. Grazing animals (like cattle and goats) in urban areas are especially vulnerable to consuming polythene waste. Polythene pollution is life-threatening to animals. It causes injury, starvation, poisoning, and death while also disturbing the ecological balance (Bradl 2004).

11.3.4 Impact of Polythene Pollution on Insects and Wildlife

Polythene pollution poses a significant environmental threat to insects, as it disrupts natural cycles and harms their vital role in pollination, decomposition, and ecological balance, causing their decline over decades (Garcês and Pires 2024; Andrades et al. 2019).

On Insects When polythene deposits on soil, it interferes with the flow of water and air, decreasing soil moisture and aeration. Polythene damages soil insects such as termites, ants, beetles, and insects associated with earthworms (Avery-Gomm et al. 2013). Polythene prevents organic waste from decomposing properly, reducing food availability for decomposer insects. Insects dependent on flowers and plants may suffer, as plant health is also affected by plastic pollution. As polythene breaks into microplastics, insects unknowingly ingest them. It can damage their digestive systems, reduce energy levels, and shorten their lifespan. In some cases, it interferes with reproduction, lowering population growth. Polythene contains additives and absorbs poisonous pollutants from the environment. When insects come into contact with or ingest contaminated particles, it can affect their nervous system, immunity, and breeding cycles (Codina-García et al. 2013). In rivers, ponds, and wetlands, polythene blocks oxygen exchange. Larvae of dragonflies, mosquitoes, and other aquatic insects may suffocate or die prematurely, disrupting the aquatic food chain. Insect decline due to polythene pollution reduces pollination, leading to poor crop and plant growth, which has a cascading effect on birds, reptiles, and animals that depend on insects as food. Polythene pollution harms insects by destroying their habitats, reducing food availability, causing toxic ingestion, and disrupting reproduction. Since insects are critical for ecosystems, their decline due to plastic waste threatens the balance of nature (Ribeiro-Brasil et al. 2022).

On Wildlife Accumulation of polythene waste in forests, wetlands, rivers, and oceans disrupts natural habitats. It blocks burrows, covers soil, and clogs water bodies, depriving many species of suitable breeding and feeding grounds. Animals such as deer, cows, elephants, turtles, and marine species often mistake polythene bags for food. Once ingested, it blocks their digestive tracts, leading to starvation, internal injury, and, in most cases, death. With time, polythene breaks down into microplastics. These particles carry harmful chemicals that accumulate in the bodies of insects, birds, and wild animals, causing poisoning, hormonal imbalance, and reduced immunity (Akindele and Alimba 2021). Wildlife often gets trapped in discarded polythene materials, restricting movement. Birds, small mammals, and reptiles may die due to suffocation, strangulation, or inability to escape predators. Small organisms consuming polythene particles pass them up the food chain. Larger predators, including birds of prey and carnivorous mammals, are affected, leading to bioaccumulation and biomagnification of harmful substances (Akindele et al. 2020). Continuous exposure to polythene pollution leads to a decline in populations of sensitive species. It weakens ecosystem balance, threatening long-term biodiversity and natural heritage. Polythene pollution poses a grave danger to insects and wildlife by causing habitat loss, ingestion hazards, toxic exposure, and biodiversity decline. To safeguard ecosystems, strict waste management, eco-friendly alternatives, and awareness are essential.

11.4 Impact of Polythene Pollution on the Aquatic Environment

Globally, plastic pollution is a serious environmental problem, particularly for aquatic ecosystems. Fish and amphibians, which are vital to food webs and ecosystem functioning, are in jeopardy due to widespread use and poor waste management, highlighting the physical, chemical, and biological effects.

11.4.1 Impact on Fish and Amphibians

Polythene pollution reaches water bodies through littering and illegal dumping, urban runoff carrying plastic debris, industrial effluents, discarded fishing equipment, wind, and rain transporting plastics from land to water (Saikia and Handique 2024).

11.4.1.1 Impact on Fish

Fish species, spanning from freshwater to marine environments, are profoundly affected by polythene pollution. Fish frequently mistake polythene fragments, microplastics, and fibers for prey, especially small zooplankton or food particles. Ingested plastics can cause intestinal obstruction, resulting in reduced nutrient absorption, internal injuries, and eventual death. Fish feel "full" after ingesting plastics but receive no nutrition, leading to starvation. Sharp plastic fragments can puncture or abrade delicate tissues. Larger polythene debris, such as discarded fishing nets, can entangle fish, restricting movement, causing injuries, or making them vulnerable to predators (Pounds 2024). Polythene can absorb hydrophobic pollutants, such as persistent organic pollutants (POPs), including PCBs (polychlorinated biphenyls) and PAHs (polycyclic aromatic hydrocarbons), as well as heavy metals like lead, cadmium, and mercury, and pesticides and herbicides from agricultural runoff. When fish ingest polythene loaded with these toxins, the chemicals can desorb in the gut, entering the fish's tissues and bioaccumulating. It causes oxidative stress, damaging cells, endocrine disruption affecting growth and reproduction, immune system suppression, increasing disease susceptibility, and genetic mutations and abnormalities (Saikia and Handique 2024).

Behavioral and Physiological Changes

The soil ecosystem is also severely disrupted by polythene contamination, which alters the physical properties of the soil, reduces microbial diversity, and obstructs essential biochemical activities. To preserve soil microbiota, it is essential to reduce plastic use, promote biodegradable alternatives, and implement effective waste management practices.

11.4.1.2 Impact on Amphibians

Amphibians are often found in freshwater and semiaquatic habitats where polythene pollution builds up. Amphibians, which include frogs, salamanders, and newts, are especially sensitive bioindicators of ecosystem health. Polythene debris blocks water flow and reduces oxygen levels, while microplastics accumulate in breeding pools, making habitats unsuitable for egg-laying. Altered water chemistry affects larval development. Amphibians can become trapped or entangled in plastic debris, leading to drowning or starvation. Polythene litter in wetlands reduces access to feeding and hiding spots. Amphibians' permeable skin allows them to absorb pollutants more easily. Research has indicated that tadpoles that consume microplastics experience hormonal disturbances, slower growth rates, and delayed metamorphosis. Polythene-derived chemical contaminants result in deformities (such as limb malformations) and infertility (Wagner et al. 2014; Besseling et al. 2014).

11.4.2 Impact on Invertebrates

Polythene pollution has become a pervasive environmental issue, especially in aquatic and terrestrial ecosystems. While the effects of polythene on larger fauna such as fish and amphibians are extensively studied, its impact on invertebrates—which play crucial roles in ecosystem functioning—has received less attention. Invertebrates, including crustaceans, mollusks, insects, and worms, form the base of many food webs and contribute to nutrient cycling, soil aeration, and water purification. This resource examines the effects of polythene pollution on invertebrate health, with a particular emphasis on growth inhibition and decreased fecundity, two critical markers of ecosystem health and population survival (Cole et al. 2011). Physical effects include ingestion; invertebrates often mistake microplastics or polythene fragments for food. Gut blockage, which is plastic ingestion, can cause physical blockage, reducing nutrient absorption; abrasion, that is, sharp plastic particles can damage gut lining or external tissues; and entanglement, where larger polythene debris can trap or immobilize invertebrates. Polythene adsorbs and concentrates hydrophobic pollutants (POPs, heavy metals). The leaching of additives (plasticizers, flame retardants) from polythene causes toxicity, and ingested plastics release these toxins internally, leading to physiological stress (Cole et al. 2015). Nutrient deficiencies block the digestive tract, toxicity interferes with metabolism, stress shifts energy to detoxification, and behavioral changes lead to reduced eating, movement, and food consumption (Huerta Lwanga et al. 2016).

Impact on Reduced Fecundity Mechanisms leading to reduced fecundity include endocrine disruption, which involves chemicals associated with plastics that interfere with hormone regulation, and an energy allocation shift, where resources prioritize survival over reproduction. Physical damage includes reproductive organs that may be damaged by plastic particles, and behavioral impairment, which deals with altered mating or egg-laying behavior, reduces reproductive success (Rochman et al. 2013).

11.4.3 Impact on Algae Blooms

Polythene pollution, primarily macroplastics, is affecting aquatic ecosystems globally by affecting algae blooms and microbial community dynamics. This study examines how polythene pollution impacts algal proliferation, alters microbial communities, and exacerbates harmful algal blooms (Zhang et al. 2017). One component of surface shading that reduces light penetration and has a particular effect on photosynthetic algae species and substrate availability is floating polythene debris. Microplastics provide algae with surfaces to adhere to and form biofilms, sometimes favoring bloom-producing species. Polythene adsorbs and releases nutrients and contaminants, influencing algal growth, and leaching of additives (plasticizers, stabilizers) may stimulate or inhibit microbial and algal communities.

Microplastics and polythene debris alter microbial community composition, facilitate colonization by specific bacterial taxa associated with algal blooms, and influence competition, predation, and symbiotic relationships within microbial consortia. Disrupted nitrogen and phosphorus cycling due to microbial imbalance, enhanced release of greenhouse gases like methane and nitrous oxide, and impaired degradation of organic matter lead to the accumulation of nutrients fueling blooms.

11.5 Challenges of Polythene Waste Management

Several challenges are faced by countries in managing plastic waste. From implementation, practice, monitoring, enforcement, and other related activities, all of these require a proper framework of policies, concern from the manufacturer to the consumer, adequate recycling methods, and monitoring by local authorities and the Government. Annually, approximately 318 million tons of plastic resin are produced, of which 218 million tons are generated as waste products post-consumption. Globally, only around 15% of the waste is recycled, and ten million tons of plastic waste enter the aquatic ecosystem annually. There is a huge stark difference between plastic production and recycling post-consumption (Ragossnig and Agamuthu 2021). Several regulatory gaps and a lack of public awareness are mentioned below (Pontes et al. 2024).

11.5.1 Overview of Global Regulation

Based on a report by Thomas Lindhqvist, who looked into the environmental costs and product end-of-life cycle, the Swedish Government implemented the extended producer responsibility (EPR) policy in 1990 (Quinn and Sinclair 2006). Lindhqvist defined EPR as "a policy principle that promotes total life cycle and environmental improvements of product systems" (Lindhqvist 2000). According to the EPR policy, it is the financial and physical responsibility of the manufacturer/brand owners for post-consumer products, not the municipalities, thus requiring companies to rebuy post-consumer waste products. Therefore, the strategy allows the enterprises to carefully plan their material selection and design. 80% of EU nations currently have EPR policies in place, thanks to Germany's pioneering work on the Duales System Deutschland, which was developed in partnership with a nonprofit organization (Zhang et al. 2017). Furthermore, in 1994, the European Commission introduced the Packaging and Packaging Waste Directive (PPWD) 94/62/EC to reduce the environmental impact of packaging and harmonize the waste management system (Hänsch and Kinkel 1995). The Directive was amended in 2018 via EU/2018/852, which focused on preventing waste generation and promoting the reuse, recycling, and recovery of packaging waste. The EU also implemented a strategy to manage plastic waste, aiming for all plastic packaging in the European market to be either reusable or economically recyclable by 2030. The PPWD also set recycling targets of at least 50% of plastic packaging waste by 2025 and 55% by 2030. In addition, extended producer responsibility (EPR) schemes must be established for all packaging in the EU (EU-27) by the end of 2024 (Wilkki and Reeve 2021; Lorang et al. 2022). In India, the Environment Protection Act of 1986 focused on the recycling of plastic waste, while the 1999 rule focused on the manufacturing and sale of plastic packaging; however, these rules did not address disposal and other packaging issues. In 2011, the Government introduced the EPR policy in India, and the policy became stronger with the introduction of plastic waste management (PWM) rules, 2016, which applied to all, from manufacturers to local authorities to individuals (Profile 2024).

11.5.2 Central Pollution Control Board

The Central Pollution Control Board (CPCB) is a government body under the Ministry of Environment, Forest, and Climate Change (MOEF&CC) responsible for controlling pollution in India. The board was first made on September 22, 1974, under the Water Act 1974 to prevent water pollution. Later, the board must oversee air pollution under the Air Act 1981 and noise pollution (Chand 2018). The principal technical organization responsible for developing and implementing standards, guidelines, and rules under the Environment Protection Act is the CPCB. To manage the growing problem of plastic waste, the plastic waste management (PWM) rule, 2011, was introduced, assigning the responsibility of managing plastic waste to local authorities, with the state body committee tasked with monitoring it (Bhattacharya et al. 2018). The 2011 rule was revised in 2016, providing more details and extending the rule to rural areas, as well as imports, and also introduced EPR. The new rule also specified the thickness of the plastic bag, reducing it from 40 to 50 microns to facilitate recycling and increase the cost of manufacturing, thereby eliminating free distribution practices. The new rule also instructed the manufacturer/brand owner to strategize plans with the local authorities for the collection back system (Kumar et al. 2018). The PWP 2016 rule was revised and is now known as the Plastic Waste Management (Amendment) Rules 2018. There were three significant changes in the revised rule, like (Bhattacharya et al. 2018; Association, I.P.C 2022) the following: The term "nonrecyclable multilayered plastic" under Section 9 (3) was substituted by "multilayered plastic which is nonrecyclable or non-energy recoverable or with no alternate use." Section 15, which required the shop vendor (providing plastic bags) to register with the local authorities and pay an annual fee, was removed. Manufacturer/brand owners who are operating in more than two states have to register with the CPCB.

These new rules have strengthened plastic waste management; however, implementation remains a challenge.

11.5.3 Gaps in Regulations for Polythene Recycling Approaches

The EPR regulations are too general, often lacking clarity, giving loopholes to manufacturers/brand owners to avoid responsibilities. It is very challenging to monitor the waste management practices of manufacturers and brand owners at the local level. The policies cover plastic waste management for mostly packing materials, whereas complex, contaminated, and low-value polythene is often ignored (Hossain et al. 2022).

There are several gaps in the guidelines for plastic waste management, as the guidelines do not have specifications on the various stages of waste processing, like reduction, recycling, and reusing of plastic waste (Olatayo et al. 2022). In India, the regulations often do not include waste pickers, who are the main workers coming into contact with the polythene waste post-consumption. They are responsible for collecting, sorting, and recycling the majority of the waste.

11.6 Different Recycling Approaches

11.6.1 Chemical Recycling

11.6.1.1 Solvolysis

In this process, plastics are broken down (depolymerized) into their smaller constituent monomer units by using a solvent at relatively low temperature. It is basically a nucleophilic substitution reaction in which the solvent serves as the nucleophile. The main types of solvolysis reactions include hydrolysis, depolymerization using water; glycolysis, depolymerization using glycols; and methanolysis, depolymerization using methanol (Pandey et al. 2023).

Hydrolysis A favorable chemical process where polythene terephthalate (PET) converts into its monomeric molecules, which are terephthalic acid (TA) and ethylene glycol (EG) by depolymerization reaction. Generally, hydrolysis reactions are categorized into three types: neutral, acidic, and alkaline.

In neutral hydrolysis, only water or steam is utilized at elevated temperature (200–399 °C) and high pressures (1–4 MPa). Acidic hydrolysis utilizes strong mineral acids such as HCl, H_2SO_4, and HNO_3. Acidic hydrolysis can occur even under mild conditions, but the reaction rate depends on the temperature. In alkaline hydrolysis, bases such as sodium hydroxide and potassium hydroxide are used in aqueous or alcoholic media. Here, PET is converted into disodium (or dipotassium) terephthalate and ethylene glycol (EG). Acidification of this product gives high-purity terephthalic acid (TA). This method works at lower pressures and produces purer monomers than neutral or acidic hydrolysis (Sinha et al. 2010; Mancini and Zanin 2007).

Glycolysis PET undergoes glycolysis via transesterification, which results in the formation of bis(2-hydroxyethyl) terephthalate (BHET), which acts as an essential intermediate for the synthesis of various polymers. It is also recombined with fresh BHET and utilized as a component for the resynthesis of PET (Pandey et al. 2023).

Methanolysis It is an effective chemical recycling process for PET. It produces ethylene glycol (EG) and dimethyl terephthalate (DMT), both of which are recovered and utilized for the resynthesis of PET. This process breaks down by more than 99% through reaction with methanol at 210 °C in the presence of aluminum triisopropoxide, which acts as a catalyst. The choice of catalyst is essential. It improves both the reaction efficiency and the product selectivity. To increase the rate of depolymerization, lower energy consumption, and reduce the creation of byproducts, a variety of catalysts have been studied recently, including metal acetates, alkoxides, and heterogeneous catalytic systems. These developments increase methanolysis' application in large-scale PET recycling while also enhancing process sustainability (Chanda 2021a).

Aminolysis It is a reaction where amines react with PET to produce terephthalic acid (TPA) and ethylene glycol (EG), for example, PET breaks down with ethanolamine using dibutyltin oxide, which acts as a catalyst. PET reacts with excess ethylenediamine (EDA), producing products such as α, ω-aminooligo(ethylene terephthalamide) (AOET) and bis(2-aminoethyl) terephthalamide (BAET). The method of PET recycling has not been studied much, and more research is needed to explore its efficiency and possible uses (Teotia et al. 2017; Chanda 2021b).

11.6.1.2 Dissolution/Precipitation

It is the most versatile approach for the recycling of plastic. The process applies it to extruded polystyrene. This technique separates a single polymer from multilayer films. Alternative multilayer separation methods have been employed, utilizing a solvent-based microemulsion along with a carboxylic acid for delamination. In contrast, PVC separation involves swelling the polymer with a boiling solvent, followed by treatment with hot water and density-based sifting. Dissolution or precipitation also enables the removal of additives and colorants, producing high-purity resin. Choice of solvent is a vital point. The process uses butanol and steam for PVC recycling. Polystyvert employs cymene (from citrus waste) for PS recovery (Vollmer et al. 2020).

11.6.1.3 Pyrolysis

Recycling industries widely use pyrolysis because it can handle a broad range of raw materials and offers flexibility in processing. In this technique, plastics are subjected to extreme temperatures in the absence of oxygen, causing their long polymer chains to break down into smaller molecules, for instance, polyolefins such as polythene can be decomposed back into similar compounds, sometimes even monomers, when heated to around 1000 °C (Rahimi and García 2017). The process of pyrolysis is classified into three main types. Thermal pyrolysis is considered one of the more economical methods for managing plastic waste. The distribution of products, however, varies with the type of polymer used, for example, when PET is subjected to pyrolysis at 500 °C under a nitrogen atmosphere with a heating rate of 10 °C/min, the process yields approximately 39.9% liquid, 52.1% gas, and 8.0% solid residue. The relatively high portion of gas suggests that PET is particularly suitable for generating gaseous products. In contrast, experiments conducted in the 350–550 °C range have demonstrated that pyrolysis can produce 79 to 85% liquid oil, which exhibits properties comparable to those of conventional fuels. Polystyrene (PS) behaves differently, as it tends to yield a high amount of liquid oil. Studies with virgin PS report oil recoveries of nearly 92–93%, while the presence of impurities or combinations slightly lowers this value to around 82.5% (Marcilla et al. 2009). Thermal cracking of plastics generally requires very high energy input. However, the introduction of catalysts reduces both the reaction temperature and the processing time, making the method more cost-effective and selective. This catalytic pyrolysis pathway is particularly significant for generating valuable products such as liquid fuels and light olefins (C2–C4). Mesoporous materials, such as Pt/h-HZSM-5 and Pt/ITQ-6, have demonstrated strong performance in the hydrocracking of polystyrene. In medium-density polythene (MDPE) conversion studies, it has been found that catalytic efficiency is governed more by the strength of Bronsted and Lewis sites than by the overall acidity of the material (López et al. 2010). Co-pyrolysis of plastics with biomass enhances product yields through synergistic interactions, for instance, lignin co-paralyzes with polythene, generating syngas, hydrogen, and carbon monoxide. Blending polythene terephthalate with biomass sources improves product distribution and oil recovery. In another study, the co-pyrolysis of waste newspaper and high-density polythene at 500 to 800 °C achieved a maximum alcohol and hydrocarbon yield of nearly 86%. When biomass combines with polythene, the suit forms during pyrolysis and reduces to a level below theoretical prediction. Catalyst-assisted co-pyrolysis is shown to provide better conversion efficiency at lower operating temperatures. Depending on the approach, catalyst applies directly during pyrolysis (in situ) or in subsequent stages (ex situ), for instance, corn stalks and polystyrene thread with the HZSM-5 catalyst demonstrated improved product profiles. Likewise, a 1:1 mixture of low-density polythene (LDPE) and cellulose pyrolyzed with zeolite (Si/Al = 25) at 450 to 650 °C enhances the yield of aromatic hydrocarbons while suppressing coke formation. In another case, polypropylene mixed with cellulose, xylene, and lignin in a 1:3 ratio, using K_2CO_3 and $ZnCl_2$ as catalysts at 450 °C, produced pyrolytic oil with improved quality and composition (Liu et al. 2020; Chen et al. 2017).

11.6.1.4 Gasification

Gasification is a thermal process in which solid waste is converted into a combustible gas mixture, rather than being directly burned, unlike incineration, which releases harmful pollutants such as sulfur oxides, nitrogen oxides, dioxins, furans, and heavy metals. Gasification mainly produces syngas, a mixture of carbon monoxide (CO) and hydrogen (H_2). This syngas is utilized for various purposes, including generating clean fuel for electricity, powering transportation, and producing liquid fuels and chemicals.

Steps involve in gasification are drying (moisture is removed from the waste), pyrolysis (the dried waste is heated in limited oxygen, producing volatile gases and solid char), gas formation (the volatile gases convert into syngas along with some tar), and char gasification (the remaining char undergoes further reactions, yielding additional syngas and leaving behind ash). Plastics are highly suitable for gasification due to their high volatile content, which enables them to break down almost entirely into gases during pyrolysis in co-gasification. Plastics are often combined with other feedstocks, such as biomass or municipal solid waste. The interaction between these materials improves overall efficiency and enhances syngas quality (Lopez et al. 2018).

11.6.1.5 Upcycling

Biocollection transforms discarded plastics, such as polythene (PE), into useful organic acids. These acids serve as building blocks for producing high-value materials, such as coatings and solvents. This approach offers a more efficient and sustainable alternative to conventional recycling techniques (Rao and Radha 2025).

11.6.1.6 Hydrocracking

Using different carrier gases during pyrolysis can significantly influence the product distribution. When hydrogen is introduced instead of nitrogen, it promotes hydrocracking reactions that suppress coke formation. It occurs because hydrogen stabilizes coke precursors, allowing the process to run at lower temperatures and thereby extending catalyst lifespan. In addition, the presence of hydrogen favors the production of saturated hydrocarbons (alkanes), which are valuable feedstocks for steam cracking and the subsequent manufacture of polymers such as polythene (PE) and polypropylene (PP).

Reactive gases, such as CO_2, have shown benefits in reducing coke and tar yields, particularly during the pyrolysis of plastics like PVC and PET, while also decreasing the activity of the resulting oils. Hydrogen offers another advantage in dealing with heteroatoms (e.g., chlorine and bromine) commonly found in plastic waste (da Rocha Filho et al. 1993; Soni et al. 2021). However, the formation of corrosive acids poses challenges for reactor durability and material selection (Vollmer et al. 2020).

11.6.1.7 Depolymerization in Supercritical Fluids

Supercritical fluid exhibits a unique combination of properties: They possess densities comparable to liquids while maintaining the diffusivity and energy levels of gases. This dual nature greatly accelerates chemical reactions. In the case of PET, rapid depolymerization can be achieved using supercritical water or methanol, leading to the formation of dimethyl terephthalate (DMT) and ethylene glycol (EG). Using methanol under supercritical conditions yields up to 98% DMT in less than half an hour. The reaction mechanism generally involves an initial stage of random chain growth, producing oligomers, followed by end-chain cleavage that generates the monomeric products.

Enzymatic polymerization: Conventional lipases are capable of degrading aliphatic polyesters but are ineffective against highly stable aromatic polymers, such as PET. However, an enzyme known as TfH, derived from Thermo fibridifusca, has been shown to break down PET into valuable intermediates, including monomers and oligomers. The PET's crystallinity and reaction temperature affect the degradation efficiency of TfH. The thermophilic origin and unique structural characteristics of materials with lower crystallinity, like PET used in beverage bottles, make them more prone to degradation. Compared to standard lipases, TfH is substantially more efficient for PET polymerization because it performs best at approximately 55 °C (Genta et al. 2005; Goto 2009).

11.6.2 Mechanical Recycling

Mechanical recycling is one of the most commonly used processes for plastic recycling; it doesn't alter the polymer chain, but it separates the polymer from contaminants and reprocesses it into granules. It includes various mechanical processes, such as separation, contaminant removal through washing, and size reduction with the aid of grinders and crushers. Subsequently, the process uses heat to extrude the ground, flaked, or pelletized polymer and reform it into products with defined geometries. Waste plastics are recycled to produce plastic through the mechanical recycling process. Afterward, plastic can be recycled into various products, such as doormats and footwear, using either mechanically molded or chemical recycling methods. In place of virgin plastic, using recycled material for production will reduce the demand for virgin polythene and lower waste generation. For reusing plastic waste, mechanical recycling is a cost-effective and technically sound method. However, it has some drawbacks regarding the quality of the materials, as they degrade in terms of fracture properties and heat aging. To upgrade the mechanical and thermal properties of polypropylene, mineral fillers or elastomers are added. However, the products produced through mechanical recycling are of lower quality compared to those made from virgin plastic. In addition, mechanical recycling is responsible for producing microplastic pollution, and the annual emission of microplastics by these methods ranges between 0.014 and 5.8 t/year (Schyns and Shaver 2021; Pfaendner 2010).

11.6.3 Photodegradation

Plastics can undergo degradation or mineralization into carbon dioxide through a process known as photodegradation. The mechanical qualities, including tensile strength, toughness, and elasticity, progressively decrease when exposed to electromagnetic radiation, like as sunshine. Alongside this, their chemical characteristics, such as molecular weight and composition, also change. Under aerobic conditions, the process often involves the generation of highly reactive radical species, such as hydroxyl radicals (-OH), which drive nonselective oxidation reactions. As a result, the plastic material breaks down into smaller fragments and diverse byproducts, including carbon dioxide, organic compounds, as well as microplastic and nanoplastic particles (Bułkowska et al. 2024; Yadav et al. 2026; Fotopoulou and Karapanagioti 2017).

11.6.4 Incineration of Plastics

Approximately 26% of the plastic waste produced worldwide in 2015 was incinerated, often in conjunction with other solid wastes. Industries commonly use this technique because plastics have a high calorific value similar to that of fossil fuels. Alternative approaches include co-processing in cement kilns or metallurgical foundries, where plastic can serve as a partial replacement for traditional fuels, for instance, using plastics in the iron and steel industry instead of coal can lower the carbon footprint by up to 30% and improve energy efficiency thanks to their high hydrogen content. Despite these benefits, incineration poses environmental risks. Inefficient or outdated facilities may release harmful gases such as sulfur dioxide (about 0.23 kg/MWb), nitrogen oxides about 1.5 kg/MWb), and toxic dioxins—particularly when burning chlorine-rich plastics like PVC. Additionally, incomplete combustion can leave microplastics behind in the ash. For every time of municipal solid waste containing about 12% plastic, the residual ash may hold anywhere between 1.9 and 565 microplastic particles per kilogram. On the energy recovery side, modern waste-to-energy plants are able to convert 70 to 80% of the energy into hot water, 20 to 25% into electricity, or about 50 to 60% into both when using cogeneration systems (Nikiema and Asiedu 2022; Devasahayam et al. 2019).

Transforming Plastic Waste to Fuel

Over the past three decades, plastic waste generation has increased rapidly. Various techniques, including landfilling, chemical recycling, and mechanical recycling, are employed to recycle waste plastics. Thermal, co-pyrolysis, and gasification are several conversion techniques that are currently used. Depending on the type of plastic waste, the chosen techniques vary accordingly. The properties, such as calorific value, density, and viscosity, of fuel oils derived from waste plastics are equal to or superior to those of commercial diesel fuel. The conversion of plastic waste into fuel has helped decrease the negative impacts of plastic pollution on the environment and produce an alternative by reducing demand (Chandran et al. 2020).

11.6.5 Bioremediation

Bioremediation, a sustainable and eco friendly method of removing harmful pollutants from the environment, utilizes microorganisms. Generally, the bioremediation process is classified into two categories: in situ and ex situ. In situ is a process of bioremediation that is performed directly at the contaminated site, while the ex situ method requires removal of contaminated materials for treatment elsewhere. There are various bioremediation methods commonly used, such as land farming, bioventing, bioaugmentation, rhizofiltration, composting, bioreactors, and biostimulators. Organisms known as bioremediators actively perform these functions. There are some drawbacks to bioremediation. Not all contaminants can be treated by microorganisms, for example, heavy metals such as cadmium and lead cannot be easily absorbed or degraded biologically (Kensa 2011).

11.7 Future Measures

11.7.1 Public Awareness

Governments, nongovernmental organizations, youth organizations, and grassroots communities are spearheading campaigns to raise awareness about polythene contamination.

11.7.2 Technologies: Advanced Technologies

Conventional methods for treating plastic waste include mechanical recycling, incineration, and landfilling. Due to the advancement of new technologies, the treatment process for plastic waste has changed over time.

11.7.2.1 Pyrolysis Technology

Pyrolysis is a process to treat plastic waste in high temperatures (300–900 °C) in the absence of oxygen or low oxygen concentration (Miandad et al. 2016). Instead of burning the plastic, the process breaks it down into smaller parts and converts it into liquid fuel, gas, and sometimes beneficial chemicals as well. All plastics are composed of monomers, which are small building blocks joined together to form long-chain polymers. Pyrolysis is a reverse process, where it breaks the chain and turns it back into smaller units. This method offers several advantages, such as (Eze et al. 2021; Lee et al. 2025) after every cycle of recycling the plastic becomes weaker, less clear, and less flexible, resulting in a lower-value product. Catalytic pyrolysis yields a higher-quality product. The conventional recycling method requires sorting, washing, and blending before recycling, which increases the cost, whereas the pyrolysis technique does not require the initial process, thus resulting in decreased cost.

11.7.2.2 Microwave-Assisted Pyrolysis Method

Microwave-assisted pyrolysis is a newer method similar to pyrolysis. In this method, instead of using conventional heating, the microwave heating technique is utilized. When microwave heat is absorbed, it causes the molecules of the plastic to vibrate, thereby producing heat within the material. The absorbent material absorbs microwaves and transfers heat through conduction. The microwave can interact with the plastic material in three different ways. It either can be reflected by the conductors or transmitted by the insulators or can be absorbed depending on the type of material's property (Eze et al. 2021). The type and amount of absorbent will determine the even heating of the material. However, plastics themselves cannot absorb microwaves; they are often mixed with materials that can strongly absorb microwaves (e.g., tires, silicon carbide, iron mesh, etc.).

11.7.2.3 Microbial Degradation

Biodegradation refers to the breakdown of organic matter by an organism, typically bacteria, fungi, or algae. In the past, people believed that synthetic plastics could not be biologically degraded; however, recent studies have shown that certain microbes can produce special enzymes to break them down. The biodegradation of plastics happens in several steps (Glaser 2019; Soong et al. 2022; Jaiswal et al. 2020; Purohit et al. 2020) due to several environmental factors such as sunlight, oxygen, temperature, moisture, pressure, and abrasion. They break the large plastic pieces into smaller fragments. Microorganisms that can utilize plastic as a source of carbon attach to the surface of the plastic and form colonies and grow as biofilms. These biofilms weaken the plastic by making pores and cracks. Microbes release enzymes (like lipases, proteases, esterases, depolymerases, etc.) that cut the long polymer chains into smaller molecules (oligomers, dimers, monomers). The small molecules are taken up by microbial cells and used as nutrients (carbon and nitrogen sources) to build biomass. Finally, the degraded material fully converts into simple end products such as carbon dioxide (CO_2), methane (CH_4), nitrogen gas (N_2), and water (H_2O).

11.7.2.4 Gasification of Polythene (PE)

Unlike the pyrolysis method, in the gasification process, the plastic waste is treated at a higher temperature (500–1300 °C) in the presence of steam, oxygen, air, etc. to produce syngas (gas mixture of hydrogen and carbon monoxide) (Saebea et al. 2020). In this process, the waste undergoes pretreatment: it is rinsed, milled, broken down into smaller fragments, and dried. The fragments undergo pyrolysis, where the PE waste is further broken down and produces cracked hydrocarbons. The cracked hydrocarbons are turned into hydrogen and carbon monoxide; this is the reforming of PE waste (Bashir et al. 2025).

11.7.2.5 Chemical Recycling Process

Chemical recycling occurs through the depolymerization of plastic polymers into smaller building blocks using chemicals and catalysts. The end products can be recycled to produce valuable materials, such as newer plastics and chemicals. Newer chemical recycling processes, depending on the chemical used, include glycolysis, alkaline hydrolysis, and methanolysis (Xi et al. 2005). PET is converted into smaller molecules by a transesterification reaction during the glycolysis chemical recycling process. The process creates BHET using ethylene glycol. In the alkaline hydrolysis process, PET is present in an aqueous solution of sodium hydroxide, resulting in the formation of a water-soluble disodium salt form of TPA and ethylene glycol. The addition of HCl will lead to the precipitation of TPA, which can be separated. The resulting purified TPA can be used to make PET bottles, polyester films, and polyester fibers (Aguado et al. 2014).

In the methanolysis chemical reaction process, PET is broken down using methanol at high temperatures (180–280 °C) and high pressures (20–40 atm). The process will lead to the production of DMT, which is as pure as a virgin DMT and ethylene glycol. There are different types of liquid, vapor, and supercritical methanolysis processes. Zinc acetate, calcium phosphate, and aluminum triisopropoxide, among others, can be used in the process as catalysts (Laldinpuii et al. 2022).

11.7.3 Emerging Risk and Compound of Concern

Both the ecosystem and the health of all living things, including humans, are now at risk from microplastics and nanoplastics. Through injection, inhalation, and contact with water, air, and soil, these microparticles and nanoparticles are taken up by the environment and enter the animal or human body. These microscopic particles frequently cause problems with the reproductive system (Urli et al. 2023) and pulmonary and gastrointestinal issues (Yee et al. 2021).

11.8 Conclusion

The issue of plastic pollution, particularly from polythene, is one of the most pressing environmental problems of the twenty-first century. This problem requires a comprehensive solution that utilizes innovative recycling techniques. This study has thoroughly explored the various uses of polythene, its detrimental environmental effects, and potential remedies enabled by advanced recycling technologies. Polythene contamination reaches far beyond simple garbage accumulation. Microplastics have developed and spread, causing widespread pollution in all ecosystems—from agricultural soils to marine habitats. This pollution endangers food security and biodiversity. There is an urgent need for systemic change in plastic waste management. Established effects on human health, land, and marine ecosystems affect the cardiovascular, hepatic, renal, and pulmonary systems. The results of this study indicate different recycling approaches and strong potential in new recycling methods that build upon traditional mechanical recycling. Chemical recycling, such as glycolysis, alkaline hydrolysis, and methanolysis, can break down plastics into reusable building blocks, closing the loop in plastic use. Bioremediation, which utilizes specialized microorganisms to break down polythene with enzymes, offers eco-friendly options that could transform how we manage waste. In order to prevent plastic waste mismanagement, such as ocean disposal and illicit landfill activities, it is imperative to investigate innovative, safe, and economical recycling methods for plastic trash while promoting international cooperation and law. Effectively tackling the problem of plastic trash requires international cooperation.

References

Aguado A et al (2014) Chemical depolymerisation of PET complex waste: hydrolysis vs. glycolysis. J Mater Cycles Waste Manag 16(2):201–210

Akindele EO, Alimba CG (2021) Plastic pollution threat in Africa: current status and implications for aquatic ecosystem health. Environ Sci Pollut Res 28(7):7636–7651

Akindele EO, Ehlers SM, Koop JH (2020) Freshwater insects of different feeding guilds ingest microplastics in two Gulf of Guinea tributaries in Nigeria. Environ Sci Pollut Res 27(26): 33373–33379

Ali U et al (2020) Rice straw, biochar and calcite incorporation enhance nickel (Ni) immobilization in contaminated soil and Ni removal capacity. Chemosphere 244:125418

Alimi OS et al (2018) Microplastics and nanoplastics in aquatic environments: aggregation, deposition, and enhanced contaminant transport. Environ Sci Technol 52(4):1704–1724

Aminiyan MM, Shorafa M, Pourbabaee AA (2024) Mitigating the detrimental impacts of low-and high-density polyethylene microplastics using a novel microbial consortium on a soil-plant system: insights and interactions. Ecotoxicol Environ Saf 283:116805

Andrades R et al (2019) Scavenging as a pathway for plastic ingestion by marine animals. Environ Pollut 248:159–165

Andrady AL (2015) Plastics and environmental sustainability. Wiley

Arthur C, Baker JE, Bamford HA (2009) Proceedings of the international research workshop on the occurrence, effects, and fate of microplastic marine debris., September 9–11, 2008. University of Washington Tacoma, Tacoma

Association IPC (2022) Plastic waste management (amendment) rules 2022. Author. Retrieved from http://coa. delhigovt. nic. in/wps/wcm/connect …

Avery-Gomm S et al (2013) Plastic ingestion in marine-associated bird species from the eastern North Pacific. Mar Pollut Bull 72(1):257–259

Bashir MA et al (2025) Plastic waste gasification for low-carbon hydrogen production: a comprehensive review. Energy Adv 4:330–363

Besseling E et al (2014) Nanoplastic affects growth of S. obliquus and reproduction of D. magna. Environ Sci Technol 48(20):12336–12343

Bettas Ardisson G et al (2014) Biodegradation of plastics in soil and effects on nitrification activity. *A laboratory approach*. Front Microbiol 5:710

Bhattacharya R et al. (2018) Challenges and opportunities: plastic waste management in India

Bodor A et al (2024) Soils in distress: the impacts and ecological risks of (micro) plastic pollution in the terrestrial environment. Ecotoxicol Environ Saf 269:115807

Bradl HB (2004) Adsorption of heavy metal ions on soils and soils constituents. J Colloid Interface Sci 277(1):1–18

Bredikhin P et al (2017) Preparation of basalt incorporated polyethylene composite with enhanced mechanical properties for various applications. MATEC Web Conf.. EDP Sciences

Bułkowska K, Zielińska M, Bułkowski M (2024) Blockchain-based management of recyclable plastic waste. Energies 17(12):2937

Chand P (2018) Environmental protection and regulations in India: role of the central pollution control board. Indian J Public Adm 64(4):645–663

Chanda M (2021a) Chemical aspects of polymer recycling. Adv Ind Eng Polym Res 4(3):133–150

Chanda M (2021b) Advanced industrial and engineering polymer research

Chandran M, Tamilkolundu S, Murugesan C (2020) Conversion of plastic waste to fuel. In: Plastic waste and recycling. Elsevier, pp 385–399

Chen W et al (2017) Fast co-pyrolysis of waste newspaper with high-density polyethylene for high yields of alcohols and hydrocarbons. Waste Manag 67:155–162

Codina-García M et al (2013) Plastic debris in Mediterranean seabirds. Mar Pollut Bull 77(1–2): 220–226

Cole M et al (2011) Microplastics as contaminants in the marine environment: a review. Mar Pollut Bull 62(12):2588–2597

Cole M et al (2015) The impact of polystyrene microplastics on feeding, function and fecundity in the marine copepod Calanus helgolandicus. Environ Sci Technol 49(2):1130–1137

Constantinescu D et al (2022) Technological considerations regarding the mechanical recycling of waste from polyethylene and polypropylene packaging. In: Proceedings of the advanced materials and systems ICAMS, Bucharest, pp 26–28

Corcoran PL, Biesinger MC, Grifi M (2009) Plastics and beaches: a degrading relationship. Mar Pollut Bull 58(1):80–84

da Rocha Filho G, Brodzki D, Djéga-Mariadassou G (1993) Formation of alkanes, alkylcycloalkanes and alkylbenzenes during the catalytic hydrocracking of vegetable oils. Fuel 72(4):543–549

De-la-Torre GE (2020) Microplastics: an emerging threat to food security and human health. J Food Sci Technol 57(5):1601–1608

Devasahayam S, Raju GB, Hussain CM (2019) Utilization and recycling of end of life plastics for sustainable and clean industrial processes including the iron and steel industry. Mater Sci Energy Technol 2(3):634–646

Eze WU et al (2021) Plastics waste management: a review of pyrolysis technology. Clean Technol Recycl 1(1):50–69

Ford HV et al (2022) The fundamental links between climate change and marine plastic pollution. Sci Total Environ 806:150392

Fotopoulou KN, Karapanagioti HK (2017) Degradation of various plastics in the environment. In: Hazardous chemicals associated with plastics in the marine environment. Springer, pp 71–92
Garcês A, Pires I (2024) The detrimental impacts of plastic pollution on wildlife. Res Ecol 6(2): 42–46
Genta M et al (2005) Depolymerization mechanism of poly (ethylene terephthalate) in supercritical methanol. Ind Eng Chem Res 44(11):3894–3900
Glaser JA (2019) Biological degradation of polymers in the environment. In: Andrady AL (ed) Plastics in the environment, vol 1. Wiley, p 13
Goto M (2009) Chemical recycling of plastics using sub-and supercritical fluids. J Supercrit Fluids 47(3):500–507
Hänsch K, Kinkel K (1995) European Parliament and Council Directive 94/62/EC of 20 December 1994 on packaging and packaging waste. J Environ Law 7(2):323–337
Hossain R et al (2022) Plastic waste management in India: challenges, opportunities, and roadmap for circular economy. Sustainability 14(8):4425
Huerta Lwanga E et al (2016) Microplastics in the terrestrial ecosystem: implications for Lumbricus terrestris (Oligochaeta, Lumbricidae). Environ Sci Technol 50(5):2685–2691
Iqbal S et al (2020) Unraveling consequences of soil micro-and nano-plastic pollution on soil-plant system: implications for nitrogen (N) cycling and soil microbial activity. Chemosphere 260: 127578
Jaiswal S, Sharma B, Shukla P (2020) Integrated approaches in microbial degradation of plastics. Environ Technol Innov 17:100567
Jambeck J et al (2018) Challenges and emerging solutions to the land-based plastic waste issue in Africa. Mar Policy 96:256–263
Kaimathuruthy BJ et al. (n.d.) Physical drivers of transport, dispersion and trapping of microplastics in a macrotidal, hyper-turbid fluvial–estuarine system: a modelling approach. Available at SSRN 5817349
Kensa VM (2011) Bioremediation-an overview. J Ind Pollut Control 27(2):161–168
Kumar A et al (2018) Estimation of the generation rate of different types of plastic wastes and possible revenue recovery from informal recycling. Waste Manag 79:781–790
Laldinpuii Z et al (2022) Methanolysis of PET waste using heterogeneous catalyst of bio-waste origin. J Polym Environ 30(4):1600–1614
Lee B et al (1991) Biodegradation of degradable plastic polyethylene by Phanerochaete and Streptomyces species. Appl Environ Microbiol 57(3):678–685
Lee JE et al (2025) Current methods for plastic waste recycling: challenges and opportunities. Chemosphere 370:143978
Li M et al (2021) Impacts of polyethylene microplastics on bioavailability and toxicity of metals in soil. Sci Total Environ 760:144037
Lindhqvist T (2000) Extended producer responsibility in cleaner production: policy principle to promote environmental improvements of product systems. Lund University
Liu X et al (2020) On deconvolution for understanding synergistic effects in co-pyrolysis of pinewood and polypropylene. Appl Energy 279:115811
López A et al (2010) Pyrolysis of municipal plastic wastes: influence of raw material composition. Waste Manag 30(4):620–627
Lopez G et al (2018) Recent advances in the gasification of waste plastics. A critical overview. Renew Sust Energ Rev 82:576–596
Lorang S et al (2022) Achievements and policy trends of extended producer responsibility for plastic packaging waste in Europe. Waste Disposal Sustain Energy 4(2):91–103
Mancini SD, Zanin M (2007) Post consumer pet depolymerization by acid hydrolysis. Polym-Plast Technol Eng 46(2):135–144
Marcilla A, Beltrán M, Navarro R (2009) Thermal and catalytic pyrolysis of polyethylene over HZSM5 and HUSY zeolites in a batch reactor under dynamic conditions. Appl Catal B Environ 86(1–2):78–86

Miandad R et al (2016) Catalytic pyrolysis of plastic waste: a review. Process Saf Environ Prot 102: 822–838
Mohasin M et al (2025) Microplastics in agricultural soils: sources, impacts, and mitigation strategies. Environ Monit Assess 197(6):684
Napper IE, Thompson RC (2020) Plastic debris in the marine environment: history and future challenges. Global Chall 4(6):1900081
Nikiema J, Asiedu Z (2022) A review of the cost and effectiveness of solutions to address plastic pollution. Environ Sci Pollut Res 29(17):24547–24573
Olatayo KI, Mativenga PT, Marnewick AL (2022) Does policy on plastic waste support higher waste management hierarchy options? Recycling 7(3):36
Osman AI et al (2025) Synergistic technologies for a circular economy: upcycling waste plastics and biomass. Front Chem Sci Eng 19(1):2
Pandey KP et al (2023) Practical ways to recycle plastic: current status and future aspects. J Mater Cycles Waste Manag 25(3):1249–1266
Pfaendner R (2010) Additives to upgrade mechanically recycled plastic composites. In: Management, recycling and reuse of waste composites. Elsevier, pp 253–280
Pilapitiya PNT, Ratnayake AS (2024) The world of plastic waste: a review. Cleaner Mater 11: 100220
Pontes LAM et al (2024) The complexities of solid and plastic waste legislation implementation: a comparative study across the European Union, United States, and Brazil. J Mater Cycles Waste Manag 28:3
Pounds AM (2024) The role of fish in food security: multidisciplinary approaches to support Kenyan fishing communities around Lake Victoria
Profile C (2024) Plastic policies In India
Purohit J, Chattopadhyay A, Teli B (2020) Metagenomic exploration of plastic degrading microbes for biotechnological application. Curr Genomics 21(4):253–270
Quinn L, Sinclair AJ (2006) Policy challenges to implementing extended producer responsibility for packaging. Can Public Adm 49(1):60–79
Ragossnig AM, Agamuthu P (2021) Plastic waste: challenges and opportunities. Waste Manag Res 39(5):629–630
Rahimi A, García JM (2017) Chemical recycling of waste plastics for new materials production. Nat Rev Chem 1(6):0046
Rao PM, Radha P (2025) Pioneering approaches to plastic biodegradation and upcycling for sustainability. Environ Monit Assess 198(1):23
Rellán AG et al (2023) Sources, sinks and transformations of plastics in our oceans: review, management strategies and modelling. Sci Total Environ 854:158745
Ribeiro-Brasil DRG et al (2022) The impacts of plastics on aquatic insects. Sci Total Environ 813: 152436
Rivard C et al (1995) Starch esters as biodegradable plastics: effects of ester group chain length and degree of substitution on anaerobic biodegradation. Enzym Microb Technol 17(9):848–852
Rochman CM et al (2013) Ingested plastic transfers hazardous chemicals to fish and induces hepatic stress. Sci Rep 3(1):3263
Saebea D et al (2020) Gasification of plastic waste for synthesis gas production. Energy Rep 6:202–207
Saikia KK, Handique S (2024) Microplastics abundance and potential ecological risk assessment in sediment, water and fish of Deepor Beel—a Ramsar wetland of the Brahmaputra plain, India. Environ Monit Assess 196(10):977
Schwarz AE et al (2019) Sources, transport, and accumulation of different types of plastic litter in aquatic environments: a review study. Mar Pollut Bull 143:92–100
Schyns ZO, Shaver MP (2021) Mechanical recycling of packaging plastics: a review. Macromol Rapid Commun 42(3):2000415
Sinha V, Patel MR, Patel JV (2010) PET waste management by chemical recycling: a review. J Polym Environ 18(1):8–25

Soni VK et al (2021) Thermochemical recycling of waste plastics by pyrolysis: a review. Energy Fuel 35(16):12763–12808
Soong Y-HV, Sobkowicz MJ, Xie D (2022) Recent advances in biological recycling of polyethylene terephthalate (PET) plastic wastes. Bioengineering 9(3):98
Teotia M, Tarannum N, Soni RK (2017) Depolymerization of PET waste to potentially applicable aromatic amides: their characterization and DFT study. J Appl Polym Sci 134(31):45153
Thiounn T, Smith RC (2020) Advances and approaches for chemical recycling of plastic waste. J Polym Sci 58(10):1347–1364
Thushari GGN, Senevirathna JDM (2020) Plastic pollution in the marine environment. Heliyon 6(8):e04709
Ullah R et al (2023) Micro (nano) plastic pollution in terrestrial ecosystem: emphasis on impacts of polystyrene on soil biota, plants, animals, and humans. Environ Monit Assess 195(1):252
Urli S et al (2023) Impact of microplastics and nanoplastics on livestock health: an emerging risk for reproductive efficiency. Animals 13(7):1132
Vollmer I et al (2020) Beyond mechanical recycling: giving new life to plastic waste. Angew Chem Int Ed 59(36):15402–15423
Wagner M et al (2014) Microplastics in freshwater ecosystems: what we know and what we need to know. Environ Sci Eur 26(1):12
Wilkki CM, Reeve N (2021) Communication from the Commission to the European Parliament, the Council, the European Economic and Social Committee and the Committee of the Regions on European Missions European Commission Directorate-General for Research and Innovation Directorate G—Common Policy Centre. European Commission: Brussels
Xi G, Lu M, Sun C (2005) Study on depolymerization of waste polyethylene terephthalate into monomer of bis (2-hydroxyethyl terephthalate). Polym Degrad Stab 87(1):117–120
Yadav E, Pathak A, Poluri KM (2026) Engineered nanocatalysts for degradation and valorisation of micro/nanoplastics. Chem Commun 62:2154
Yee MS-L et al (2021) Impact of microplastics and nanoplastics on human health. Nano 11(2):496
Zhang W et al (2017) Microplastic pollution in the surface waters of the Bohai Sea, *China*. Environ Pollut 231:541–548

Chapter 12
Carbon-Based Nanosensors for Agricultural and Food Waste Threshold Determination

Arabinda Baruah and Salma A. Khanam

12.1 Introduction

Food production and distribution systems around the world are currently facing serious challenges. The global population is steadily increasing, which means that the demand for safe and nutritious food is also rising. According to the Food and Agriculture Organization (FAO), nearly one-third of the food produced globally for human consumption is lost or wasted every year (Hegde et al. 2024). Such losses also waste important resources like water, land, energy, and fertilizers and contribute to greenhouse gas emissions. Therefore, reducing agricultural losses and food waste has become an important global goal.

One effective way to address this problem is through early and accurate threshold determination. In other words, identifying the point at which agricultural systems begin to deteriorate or when food quality starts to decline can help us in minimizing the loss. In agriculture, thresholds may include situations where soil nutrients fall below optimal levels, toxic substances accumulate in soil or plants, or stress markers in plants rise beyond safe limits. In food systems, thresholds may involve changes in pH, the release of spoilage gases, or the presence of microbial metabolites that indicate the loss of freshness or safety. Detecting these limits at an early stage allows timely action, which can help prevent spoilage and reduce food waste. Traditional analytical techniques such as high-performance liquid chromatography (HPLC), gas chromatography-mass spectrometry (GC-MS), and atomic absorption spectroscopy (AAS) are widely used for chemical analysis because they offer high sensitivity and accuracy. However, these methods usually require sophisticated laboratory equipment, trained personnel, and significant operational costs. As a result, they are not

A. Baruah (✉) · S. A. Khanam
Department of Chemistry, Gauhati University, Guwahati, Assam, India
e-mail: arb@gauhati.ac.in

Y. K. Mohanta et al. (eds.), *Bioresource Waste to Carbon Nanomaterial*,
Nanotechnology in the Life Sciences, https://doi.org/10.1007/978-981-92-2405-0_12

suitable for rapid, on-site monitoring in farms, storage facilities, or transportation systems (Sabui et al. 2023). Modern agricultural and food systems, especially those focusing on sustainability and precision farming, need sensing technologies that are portable, affordable, and capable of providing rapid results.

Nanotechnology has emerged as a promising solution to these challenges. Among various nanomaterials, carbon-based nanomaterials have gained considerable attention due to their unique physical and chemical properties. These materials typically possess large surface areas, adjustable surface chemistry, high electrical conductivity, and strong optical properties (Zhang et al. 2023) (Guo et al. 2025). Examples include carbon quantum dots (CQDs), graphene derivatives, carbon nanotubes (CNTs), and biochar-derived nanocarbons. These materials are particularly useful for sensing applications because they can convert very small chemical or biological interactions into measurable signals. Carbon nanosensors usually work through mechanisms such as changes in fluorescence, electrical signals, or visible color, for example, fluorescent carbon quantum dots can show either an increase or a decrease in fluorescence intensity when they interact with specific target molecules. This property allows rapid detection of contaminants or spoilage indicators in food samples (Kumar and Rhim 2024; Zou et al. 2024). Similarly, graphene and CNT-based sensors are widely used in electrochemical detection because they facilitate fast electron transfer and exhibit high electrical conductivity (Guruprasath et al. 2024). These characteristics enable sensors to detect very small amounts of analytes quickly, which is essential for monitoring threshold levels.

In agricultural systems, carbon nanosensors can be used to monitor several important parameters such as soil nutrients (e.g., nitrate and phosphate), heavy metals, plant growth hormones, moisture levels, and gases like carbon dioxide. Some nanocarbon-based sensors have also been developed for pesticide detection. Interestingly, certain sensors are produced from agricultural waste materials such as tea residues, which makes them both sustainable and cost-effective (Sinha and Ray 2024). These sensing tools can help farmers make better decisions regarding irrigation, fertilization, and crop protection, which are key aspects of precision agriculture.

Carbon nanosensors are also highly useful in postharvest monitoring and food storage systems. During food spoilage, microorganisms release gases such as ammonia, hydrogen sulfide, and various volatile organic compounds. Biogenic amines and changes in pH can also indicate food deterioration. By detecting these indicators at an early stage, nanosensors can help determine whether food is still fresh and safe to consume, for instance, nitrogen-doped carbon dots have been successfully used to detect heavy metals and other contaminants at very low concentrations, which is important for maintaining food safety (Mohammed 2025). Early identification of such threshold levels can significantly reduce food waste and improve food quality control.

Another important advantage of carbon nanomaterials is that they can often be synthesized using green and sustainable methods. Many researchers now produce carbon quantum dots and related materials from biomass, plant extracts, or agricultural waste. This approach not only lowers the cost of production but also supports sustainability and circular economy concepts (Aksu and Güzdemir 2025). In addition, biogenic nanomaterials are generally considered to be less toxic and more environmentally friendly, which makes them suitable for applications in agriculture and food systems.

Although carbon nanosensors show great potential, some challenges still need to be addressed before they can be widely used in real-world applications. Agricultural and food samples often contain complex mixtures of chemicals that may interfere with sensor performance. Therefore, further research is needed to improve sensor stability, reproducibility, and large-scale manufacturing methods. In addition, proper field testing and regulatory approvals will be necessary before these technologies can be commercially adopted. Thus, it can be concluded that carbon-based nanosensors represent an emerging and powerful tool for monitoring agricultural systems and determining food quality thresholds. Their high sensitivity, fast response, portability, and adaptability make them highly suitable for modern smart farming and food safety monitoring systems. By enabling early detection of critical changes, these sensors have the potential to reduce food waste, conserve resources, and support sustainable agricultural practices.

12.2 Fundamentals of Carbon-Based Nanomaterials for Sensing

The carbon-based nanomaterials including graphene, CNTs, CQDs, and mesoporous carbon have unique structural and electronic properties due to the presence of C atoms. The grouping of carbon materials is determined by their shape, size, and dimensionality: three-dimensional (carbon nanomaterials), two-dimensional (graphene), zero-dimensional (bulky ball and carbon dots), and one-dimensional (carbon nanotubes and carbon nanofibers). Due to the strong adsorption capability, adjustable surface chemistry, large surface area, improved optical properties, and cheap production of these materials, they have gained much attention in the field of chemical and biological sensing applications (Biswas and Mohanta 2024).

Nanostructured carbon-based materials are widely used in electrochemical sensing applications for monitoring heavy metals, pesticides, and dyes present in the environment. They are also utilized in biomedical sensing platforms like anticancer therapy, diagnosis, and biomolecule delivery along with food analysis for alcoholic beverages, herbicides, and chemical detection (Thakur and Kumar 2023). The distinct properties of nanomaterials allow them for effective interaction with analytes and quick signal generation. Carbon nanotubes mainly single-walled carbon nanotubes (SWCNTs) and multiwalled carbon nanotubes (MWCNTs) having high surface area, mechanical durability, and electronic conductivity have gained much attention in sensing. The single-layered graphite that produces graphene exhibits considerable potential for application in environmental cleanup procedures. Graphene or CNTs can be functionalized with different chemical groups or nanoparticles, thereby improving the sensitivity and selectivity of the materials

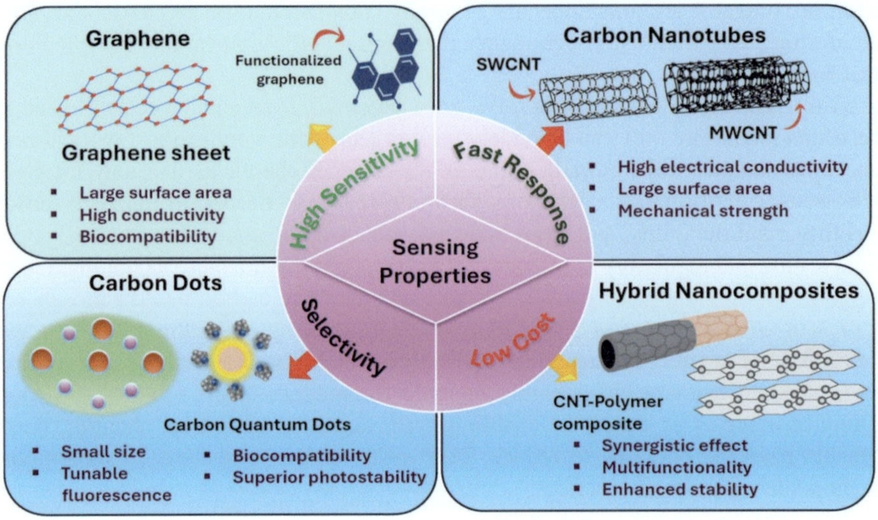

Fig. 12.1 Schematic representation of different types of carbon nanomaterials and their sensing properties

toward the target molecules (Meskher et al. 2024). CQDs are also an emerging class fluorescent carbon nanomaterial that exhibit excellent photoluminescence, high thermal stability, tunable excitation and emission properties, and optical stability. These attributes make CQDs suitable for application in sensor (Mondal et al. 2024). Figure 12.1 schematically represents various types of carbon-based nanomaterials used for sensing contaminants present in food and agricultural products. The basic characteristics of carbon-based nanomaterials that make them appropriate for sensing applications are discussed in the following section.

12.2.1 Structural Features and Physicochemical Properties of Carbon Nanomaterials

The sensing capabilities of carbon nanomaterials are influenced by the distinct structural properties of nanostructure arrangement which arises due to the distinct hybridizations of C atoms. Graphene sheets are formed by the two-dimensional hexagonal arrangement of sp^2 hybridized C atoms. It possesses exceptionally high surface-to-volume ratio, facilitating the molecular adsorption and sensitivity of the material. The cylindrical arrangement of graphene sheets results in the formation of CNTs, namely, SWCNTs or MWCNTs. The large surface area and superior electrical conductivity of CNTs enable them to transfer electron quickly during the sensing process. Also, it helps in the diffusion of analytes due to its cylindrical structure enhancing the interaction between the target species and the sensing surface. Defects like vacancies, edges, and dopant atoms also influence the sensor functionality by acting as active centers for catalysis and adsorption. Additionally, due to good mechanical durability, thermal stability, as well as chemical resistance, they can be used in harsh reaction conditions. The physicochemical characteristics of carbon-based nanomaterials can be influenced by adjusting the size, pore structure, morphology, and heteroatom doping. Doping of metals like boron (B), sulfur (S), or nitrogen (N) to the carbon framework enhances the catalytic activity by altering the electronic density.

12.2.2 Optical and Electronic Properties Relevant to Sensing

The adjustable optical and excellent electrical properties of carbon nanomaterials play a crucial role in optical, electrochemical, and fluorescence-based sensing technologies. Due to the presence of delocalized π-electron systems, they effectively interact with the electromagnetic radiation ensuing effective charge transport. Both graphene and CNTs have high mobility and excellent electrical conductivity. The interaction between analyte molecules and the surface of carbon nanomaterial surfaces can change the electrical resistance (R), current (A), or potential (V) of the system.

Carbon nanomaterials are capable of detecting very low concentration of target molecules since the electrical signals are influenced by the slight changes in analyte concentration. Many carbon nanomaterials have distinct optical responses in addition to electrical characteristics. Very stable, unique, and tunable fluorescent optical properties are exhibited by small-sized CQDs. They show significant photoluminescence and size-dependent emission characteristics. CQDs are also used as biosensors for their ability to emit light if excited. The smaller the size of the dots, the higher is the emission energy. Interaction with analytes can either enhance or quench the fluorescence, providing a useful technique for optical sensing. Similarly, the optical detection sensitivity is increased due to the surface plasmon interactions in carbon-based nanocomposites (Elugoke et al. 2024). Another significant characteristic is their wide absorption ability in the visible, near-infrared, and ultraviolet region making them useful for spectroscopic sensing platforms. High stability of carbon nanomaterials also assurances steady optical responses over prolonged time period. Because of these characteristics, carbon nanomaterials are used in a variety of sensing platforms broadening their range of applications.

12.2.3 Surface Chemistry and Functional Groups

The sensing capacity is significantly influenced by surface modification of carbon nanomaterials. Introduction of functional groups like hydroxyl (-OH), carboxyl (-COOH), carbonyl (>C=O), etc. on the surface of nanomaterials like CNTs, graphene, or CQDs via surface oxidation or chemical treatment provides the active sites to interact with the target molecules. Both covalent (electrostatic interaction) and non-covalent (chemical bond formation) methods are used for functionalization (Liu et al. 2023). With the help of covalent bonding, hydrogen bonding, and electrostatic interactions, these functional groups enable the selective detection of particular molecules. Incorporation of oxygen-containing functional group to hydrophobic nanomaterials makes it dispersible in aqueous media, increased chemical reactivity can be used to change the properties of the surface, and the core of CDs can be altered by functional group manipulation and doping (Alosime 2023). CDs exhibit dichroism due to the presence of the functional group containing sp^2- and sp^3-hybridized C atoms. The hybrid systems designed by adding metal or metal oxide nanoparticles to carbon nanomaterial also improve the sensitivity and selectivity of the material with the help of catalytic activity of metals along with the high conductivity and stability of carbon frameworks. Also, addition of heteroatoms like N, P, or S changes the surface properties of carbon nanomaterials. It provides more active sites and makes it easier for analytes to stick to the surface, which helps in the detection of gases, biomolecules, heavy metals, pesticides, and other pollutants (Zhi et al. 2025).

12.2.4 Signal Transduction Mechanisms in Carbon Nanomaterials

The process of turning the interaction between an analyte and a sensing material into a signal that can be measured is known as signal transduction. Many transduction mechanisms are supported by carbon nanomaterials due their unique electrical, optical, and catalytic properties. In electrochemical transductions like amperometric, potentiometric, or impedimetric sensors, the target molecules come into contact with carbon nanomaterials; some redox reaction takes place on the surface of the electrode. These reactions develop electrical signals like changes in current (A), potential (V), or impedance (I) that can be measured. Carbon nanomaterials with high surface area and electrical conductivity facilitate the rapid electron transfer and enhance the electrochemical sensor ability (Hashem and Abdallah 2024). In chemiresistive sensors, the variation in resistance can be observed due to the concentration of the charge carriers resulting from the adsorption of target molecules onto carbon nanotubes or graphene sheet. This principle is commonly utilized in gas sensing applications (Jo et al. 2023). In fluorescence-based sensor, the interaction of analyte with carbon dots or graphene quantum dots can cause fluorescence quenching, energy transfer, or emission shifts, making it easier to detect biological or chemical species (Dagdag et al. 2024). Carbon nanomaterials serve as the conductive channel in the sensors based on field-effect transistor (FET) devices. Here, analytes bind to the surface of carbon nanomaterials. It changes the local electric field and flow of current, which can be measured. Also, catalytic signal amplification is observed when carbon nanomaterials or enzymes are mixed with metal nanoparticles (Dai et al. 2022). This synergistic effect improves the detection limits and response times.

12.3 Types of Carbon Nanomaterials Used in Nanosensors

12.3.1 *Graphene and Graphene Derivatives (GO, rGO)*

Graphene is a two-dimensional sheet of sp^2-hybridized carbon atoms arranged in a honeycomb lattice. Its high surface area, excellent electrical conductivity, and mechanical strength make it a foundational material for nanosensors that convert chemical interactions into measurable electrical or optical signals. Graphene derivatives such as graphene oxide (GO) and reduced graphene oxide (rGO) provide tunable surface chemistry through oxygen-containing functional groups, enabling robust analyte adsorption and functionalization for improved selectivity. Recent reviews highlight the use of functionalized GO and rGO in electrochemical and biosensing platforms, where hybridization with polymers or metal oxides enhances sensitivity and stability, especially for environmental, biomedical, and food analytes (Mshelia et al. 2025).

12.3.2 *Carbon Nanotubes (SWCNTs and MWCNTs)*

CNTs are cylindrical nanostructures of carbon atoms with exceptionally high aspect ratio, electrical conductivity, and surface area. Single-walled CNTs (SWCNTs) and multiwalled CNTs (MWCNTs) support rapid electron transfer and strong analyte interaction through π-π stacking, making them highly attractive for chemiresistive, electrochemical, and gas sensors. CNTs have also been widely incorporated into nanocomposites with conductive polymers and other matrices, exploiting percolation networks for enhanced charge transport and improved signal transduction in sensor devices. Recent reviews document CNT-based sensors' high sensitivity, deformation flexibility, and applicability across gas, biological, and environmental detection schemes (Kausar and Ahmad 2024).

12.3.3 *Carbon Dots and Graphene Quantum Dots*

CDs and graphene quantum dots (GQDs) are zero-dimensional carbon nanomaterials with sizes typically below 10 nm, exhibiting significant quantum confinement effects. They possess strong photoluminescence, high biocompatibility, tunable surface states, and facile surface functionalization. Because their optical and electronic properties change upon interaction with target molecules, CDs and GQDs are widely used in fluorescence and ratiometric sensors for rapid analyte detection. Recent literature emphasizes advanced CDs for electrochemical and optical sensing, noting their low toxicology and facile synthesis, including green routes from biomass or waste (Kayani et al. 2025).

12.3.4 *Porous Carbon and Activated Carbon Nanostructures*

Porous carbon materials such as activated carbon, carbon aerogels, and templated nanoporous carbons offer high internal surface area and controlled pore architectures that facilitate rapid mass transport and analyte adsorption. These properties improve sensitivity in gas and chemical sensors where diffusion to active sites is critical. Recent research on porous carbon-based sensors underscores their applicability in humidity, volatile organic compound (VOC), and environmental gas detection, and their integration with functional nanostructures supports enhanced signal transduction and analyte selectivity (Panigrahi et al. 2024).

12.3.5 *Hybrid Carbon Nanocomposites*

Hybrid nanocomposites combine carbon nanomaterials with polymers, metal oxides, conducting polymers, or other nanostructures to exploit synergistic effects. These composites often demonstrate enhanced sensitivity, faster response/recovery, and improved stability compared to single-component materials, for example, rGO-polymer composites can deliver high performance in gas sensing, while carbon dots or CNTs integrated with metal oxides enhance electrochemical detection of biomolecules or environmental contaminants. Recent work highlights CQD-CNT-graphene hybrids that achieve significantly improved sensing responses through combined high surface area and efficient charge transfer pathways (Al-Sagheer and Farea 2025).

12.4 Mechanisms of Carbon-Based Sensing

Carbon-based nanosensors work by converting chemical interactions into detectable signals. The most common mechanisms, such as electrochemical, optical, colorimetric, and chemiresistive sensing, are briefly outlined below.

12.4.1 *Electrochemical Sensing*

Electrochemical sensors detect contaminants by measuring changes in electrical signals such as current, voltage, or charge at the sensor surface. The sensor usually contains an electrode coated with a sensing material. When a target substance such as a pesticide, heavy metal ion, or toxin comes in contact with the electrode, it either reacts chemically or binds to the sensing material. This interaction changes the flow of electrons at the electrode surface. As a result, the electrical current or potential of the sensor changes. These changes are measured using a small electronic device and are directly related to the amount of contaminant present in the sample. Electrochemical sensors are very sensitive and can detect very small amounts of chemicals. They are also fast and portable and require only a small sample. Because of these advantages, electrochemical sensing is widely used to detect pesticide residues in fruits and vegetables, heavy metals in soil and irrigation water, and toxins in milk or food products.

12.4.2 Optical Sensing

Optical fluorescence sensors detect contaminants by measuring changes in light emitted from a sensing material. In these sensors, the sensing material is usually a fluorescent substance such as carbon dots, quantum dots, or dye molecules. When light of a certain wavelength shines on the material, it emits light of another wavelength. This emitted light is called fluorescence. When a target contaminant interacts with the sensing material, the intensity of the emitted light changes. In some cases, the fluorescence becomes weaker (called fluorescence quenching). In other cases, the fluorescence becomes stronger. The change in light intensity indicates the presence of the contaminant and can also help estimate its concentration. Fluorescence sensors are highly sensitive and are commonly used to detect pesticides, antibiotics, heavy metals, and toxins in food and agricultural samples. They can also be integrated with small optical devices or smartphones for rapid analysis.

12.4.3 Colorimetric Sensing

Colorimetric sensors detect contaminants through visible color changes. These sensors contain materials that react with specific target molecules and produce a change in color, for example, the sensor may change from colorless to blue, yellow to red, or light to dark when the contaminant is present. One of the biggest advantages of colorimetric sensing is that the result can often be seen with the naked eye without the need for complex instruments. In some cases, the color change can also be measured using a smartphone camera or a simple optical reader to determine the amount of contaminant present. Because they are simple and low-cost, colorimetric sensors are widely used in paper-based test strips and portable kits. These sensors are useful for detecting pesticides, heavy metals, and harmful chemicals in fruits, vegetables, grains, milk, and water used for irrigation.

12.4.4 Chemiresistive Sensing

Chemiresistive sensors work by measuring changes in the electrical resistance of a sensing material. Resistance is the ability of a material to oppose the flow of electric current. In these sensors, the sensing material is placed between two electrodes. When a target gas or chemical molecule comes into contact with the sensing surface, it interacts with the material and changes how easily electrons move through it. This interaction causes the electrical resistance of the sensor to either increase or decrease. By measuring this change, the presence and concentration of the contaminant can be determined. Chemiresistive sensors are simple in design and can respond quickly to chemical species. They are particularly useful for detecting gases such as ammonia, hydrogen sulfide, or other volatile compounds that may be released from spoiled food or contaminated agricultural environments. They can also be used to monitor chemical vapors during storage and food processing.

Overall, these sensing techniques provide fast and sensitive ways to detect harmful substances in agricultural products and food items, helping ensure food safety and quality.

12.5 Detection of Contaminants in Agricultural Systems

Accurate and rapid detection of contaminants in agricultural environments is essential for ensuring crop health, environmental safety, and food quality. Traditional laboratory methods for measuring pesticides, nutrient residues, metals, and microbial pathogens are accurate but often slow, costly, and impractical for on-site monitoring. Carbon-based nanosensors offer potential solutions by providing fast, sensitive, and portable detection of key contaminants. Their sensing mechanisms rely on changes in electrical, optical, or fluorescence signals when target substances interact with carbon nanomaterials such as CQDs, graphene derivatives, and carbon nanotubes. Figure 12.2 shows different types of pesticides, pathogens, and heavy metals present in agricultural systems which can be detected using carbon-based sensors.

12.5.1 Pesticides and Herbicides

Pesticides and herbicides are widely used to control pests and weeds but can become harmful when residues remain in soil or crops above safe thresholds. Carbon-based nanosensors have been developed to detect these residues with high sensitivity. One common approach uses CQDs functionalized to interact selectively with pesticide molecules. CQDs exhibit strong fluorescence that can be *quenched* (reduced in intensity) upon binding with target pesticides. This quenching signal can be

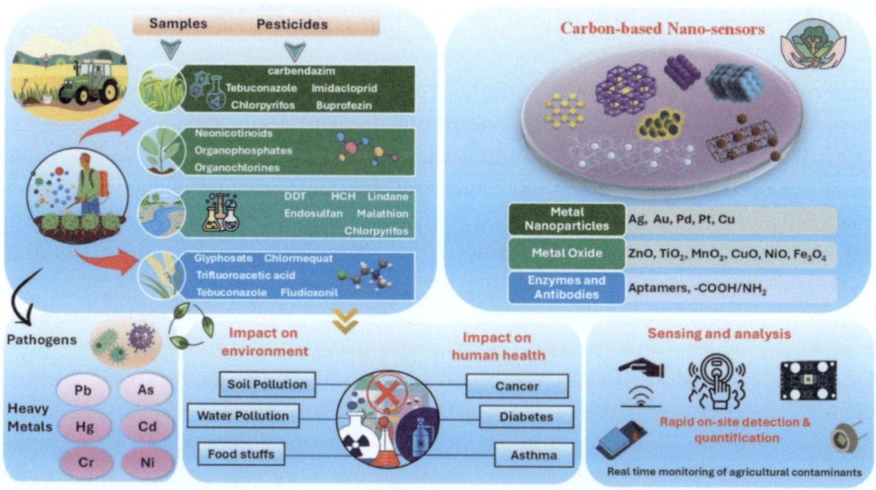

Fig. 12.2 Schematic representation of different types of pesticides, pathogens, and heavy metals present in agricultural systems which can be detected using carbon-based sensors

correlated to concentration, allowing quantification at very low levels, for example, Sharma and colleagues developed biomass-derived CQDs that detect organophosphate pesticides through fluorescence change, with rapid response and good selectivity even in complex agricultural samples (Sharma et al. 2026. Electrochemical nanosensors based on graphene or carbon nanotubes have also been reported for herbicide detection, where adsorption of pesticide molecules onto the carbon surface alters the sensor's current response. These platforms allow field-deployable detection without expensive equipment (Kumar and Rhim 2024; Sinha and Ray 2024).

12.5.2 Fertilizer Residues and Nitrogen Species

Fertilizers rich in nitrogen compounds like nitrate, nitrite, and ammonia are essential for crop growth. However, excess accumulation in soil or runoff water can be harmful to the environment and human health. Rapid detection of these nitrogen species at threshold levels is therefore important. Carbon-based sensors typically detect nitrogen species through electrochemical mechanisms. In these sensors, nitrogen ions interact with the carbon nanomaterial surface, causing measurable changes in current or potential, for instance, graphene and rGO electrodes have been used to monitor nitrate and nitrite ions with high sensitivity. The large surface area and excellent conductivity of graphene enhance ion adsorption and electron transfer, enabling detection at low-concentration thresholds relevant to agricultural guidelines (Zhang et al. 2023). Ammonia, a volatile nitrogen species, can be sensed using porous carbon materials whose resistance changes upon gas adsorption. These changes can be used to quantify ammonia concentrations noninvasively, making them useful for soil and environmental monitoring (Panigrahi et al. 2024).

12.5.3 Heavy Metals in Soil and Irrigation Water

Heavy metals such as cadmium (Cd), lead (Pb), arsenic (As), and mercury (Hg) can accumulate in soils and irrigation water from industrial emissions, fertilizers, and waste disposal. Even at low concentrations, these metals can be toxic to plants and humans. Carbon nanomaterials like CQDs and functionalized graphene have shown excellent performance in detecting heavy metals through fluorescence quenching and electrochemical responses. In fluorescence-based sensing, metal ions bind to functional groups on CQDs, causing a decrease in fluorescence intensity that correlates to metal concentration, for example, nitrogen-doped CQDs have been used to detect lead and cadmium ions at sub-part-per-billion levels, providing a simple optical readout for environmental monitoring (Singh et al. 2025). Electrochemical nanosensors based on graphene or carbon nanotubes modified with metal-binding ligands also show high sensitivity for heavy metal detection. These sensors exploit rapid electron transfer to produce measurable current changes when metal ions are captured on the sensor surface, allowing on-site detection with minimal sample preparation (Li et al. 2025).

12.5.4 Pathogens and Microbial Contaminants

Microbial contamination of soil and irrigation water poses risks to plant health and crop safety. Bacteria such as *E. coli*, *Salmonella*, and *Pseudomonas* spp. can thrive in agricultural environments and be transferred to crops, eventually entering the food chain. Carbon-based nanosensors for pathogens often combine carbon materials with biological recognition elements such as antibodies, aptamers, or peptides that bind specifically to target microbes. When a pathogen binds to the functionalized sensor surface, it induces a measurable change in electrical conductivity or optical signal, for example, studies have reported graphene-based FET sensors functionalized with aptamers that can detect *E. coli* with high specificity at low concentrations (Sabui et al. 2023). Other platforms use CQDs whose fluorescence is modulated in the presence of microbial metabolites or cell surface components, providing a simple optical readout. These sensors have detected bacterial contamination in irrigation water samples with rapid response times, making them promising for real-time water quality monitoring (Kumar and Rhim 2024).

Thus, it is evident that, in agricultural systems, carbon-based nanosensors enable rapid, sensitive, and often on-site detection of a wide range of contaminants including pesticides, nitrogen species, heavy metals, and microbial pathogens. Their mechanisms typically involve measurable changes in electrical current or optical signals (such as fluorescence) when target substances interact with carbon nanomaterials.

12.6 Detection of Food Contaminants and Quality Indicators Using Carbon-Based Nanosensors

Carbon-based nanosensors especially those built from CQDs, graphene derivatives, and hybrid carbon nanocomposites have emerged as powerful platforms for monitoring diverse contaminants and quality markers in food systems. These sensors utilize changes in electrical signals, optical properties, or fluorescence emission when a target analyte interacts with the carbon nanomaterial, enabling threshold-level detection with high precision. Figure 12.3 shows a schematic representation of carbon-based sensors for the detection of pesticides, pathogens, and heavy metals in food items.

12.6.1 Mycotoxins and Natural Toxins

Mycotoxins are toxic secondary metabolites produced by certain fungi that can contaminate cereals, nuts, coffee, and other food crops. Even at very low concentrations, compounds such as aflatoxin B_1 (AFB_1), ochratoxin A (OTA), and zearalenone pose *serious health risks* including liver damage and cancer. Carbon-based nanosensors detect mycotoxins primarily via fluorescence modulation and electrochemical signaling, for example, CQDs can be functionalized with aptamers or molecular recognition elements that bind specifically to mycotoxin molecules. Upon binding, the fluorescence intensity of CQDs changes sharply typically being *quenched* (reduced) or *enhanced* depending on the interaction which serves as a

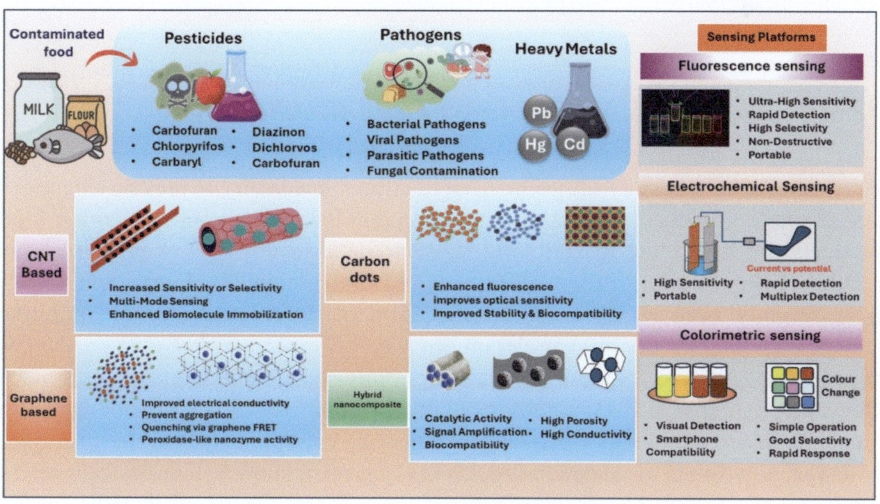

Fig. 12.3 Schematic representation of carbon-based sensors for the detection of pesticides, pathogens, and heavy metals in food items

quantitative signal for the toxin concentration. These fluorescence changes are often based on mechanisms such as the inner filter effect (IFE) or Förster resonance energy transfer (FRET), where the energy transfer between the toxin and CQDs alters optical emission. Advanced platforms also combine carbon nanomaterials with electrochemical detection, where target binding produces measurable current or potential shifts. These methods offer high selectivity and low detection limits, making them suitable for real-time monitoring of food safety in grains, juices, and processed foods.

12.6.2 Heavy Metals in Food Products

Heavy metals such as lead (Pb), cadmium (Cd), and mercury (Hg) can accumulate in food due to environmental pollution, contaminated soil, or improper handling. Ingestion of heavy metals can lead to neurological disorders and organ damage. Carbon nanomaterials, especially nitrogen-doped CQDs and graphene derivatives, can be designed to interact selectively with heavy metal ions. For optical sensing, metal ions often quench the photoluminescence of CQDs by accepting electrons or altering the surface states that influence fluorescence. This change is proportional to metal concentration and can be used to quantify even trace amounts of metals.

Electrochemical sensors incorporating graphene or functionalized CNTs can provide complementary detection, where the binding of metal ions affects electron transfer kinetics and thus modifies current responses in voltametric or impedance measurements. These platforms offer fast, label-free, and sensitive detection that is especially useful for analyzing juices, dairy, and seafood products where heavy metal contamination is a concern.

12.6.3 Antibiotic and Veterinary Drug Residues

Antibiotics and veterinary drugs are used in animal husbandry to prevent infections and promote growth. Residues of these compounds in meat, milk, and other food products can contribute to antibiotic resistance and health risks if present above regulatory thresholds. Carbon-based nanosensors particularly fluorescent CQDs have recently been reported for the detection of different classes of antibiotic residues. Dual-color fluorescent carbon dots can be *covalently linked to antibodies* specific for antibiotic compounds, enabling highly sensitive immunoassay formats. One study demonstrated a dual-color carbon dot immunoassay capable of detecting aminoglycoside antibiotics (such as gentamicin and streptomycin) in meat samples with very low limits of detection (~0.13–0.32 ng/mL), showcasing excellent selectivity and sensitivity. The sensing mechanism in these systems often relies on the change in fluorescence intensity when the target antibiotic displaces or alters the local environment of the carbon dots, enabling quantification. Fluorescence modulation mechanisms like IFE or energy transfer are commonly exploited, and surface functional groups on CQDs can be tuned to improve specificity toward particular drug residues.

12.6.4 Food Spoilage Indicators and Freshness Markers

Food spoilage results from chemical and biological changes in food products such as protein breakdown, lipid oxidation, and microbial activity. *Biogenic amines* (e.g., histamine, putrescine) and shifts in pH or VOCs are well-recognized freshness indicators. Detecting these markers rapidly can prevent consumption of spoiled food and reduce waste. Carbon nanomaterials integrated into smart sensors can detect spoilage markers using several mechanisms, for examples, CQDs can be engineered to respond to specific amines released during spoilage. When amines interact with surface functional groups on the CQDs, the emission spectrum changes, providing a visible or measurable fluorescence shift that correlates with spoilage level. Graphene, rGO, and CNTs can serve as sensitive layers where adsorption of VOCs or amines alters the electrical resistance. Such chemiresistive sensing is applicable in smart packaging where the headspace gas composition changes with freshness. Composite films containing carbon nanomaterials and pH-sensitive dyes can change color when the food's pH shifts due to microbial metabolism, offering a simple visual freshness indicator on packaging. Carbon nanosensors in packaging materials thus help ensure real-time indication of spoilage without requiring laboratory equipment, significantly aiding quality control in supply chains.

12.6.5 Adulterants and Chemical Additives

Food adulteration, the addition of inferior substances or illegal chemicals, undermines food integrity and can pose health risks. Common adulterants include industrial dyes, melamine, and non-permitted additives used to mimic desirable characteristics. Carbon-based nanosensors are being developed for rapid detection of such adulterants, for example, CQDs functionalized with specific ligands can interact with adulterant molecules, resulting in measurable fluorescence quenching or enhancement. The optical signal change often guided by energy transfer or electron transfer mechanisms correlates with adulterant concentration. Graphene-based FET sensors functionalized with aptamers or molecularly imprinted polymers can also selectively detect small adulterant molecules by changes in surface charge and conductivity. These sensors offer rapid, sensitive detection in complex food matrices like milk or fruit juices, where adulteration is common.

12.7 Portable and On-site Detection Platforms

For agricultural and food safety applications, rapid and on-site assessment of contaminants and quality indicators is highly valuable. Traditional laboratory analysis often requires transport of samples, time-consuming preparation, and specialized personnel. In contrast, portable and on-site detection platforms enable real-time monitoring directly in fields, processing units, storage facilities, and distribution chains. Carbon-based nanosensors play a central role in these systems because of their high sensitivity, low cost, ease of integration, and compatibility with flexible substrates. Recent research has focused on integrating carbon nanomaterials into paper devices, smartphone readouts, wearable formats, and IoT-connected sensor networks (Vijaybhai et al. 2025).

12.7.1 Paper-Based Sensors and Test Strips

Paper-based sensors and test strips are among the simplest and most widely deployed portable detection formats. These platforms use cellulose paper as a substrate, onto which carbon nanomaterials (such as carbon quantum dots, graphene oxide, or carbon nanotubes) are immobilized along with recognition elements that interact with target analytes. When a sample (soil extract, food juice, or water) is applied, the interaction between the analyte and carbon nanomaterial produces a color or fluorescence change that can be seen by eye or measured with a simple reader, for example, carbon quantum dots have been integrated into lateral flow test strips for detection of pesticides and heavy metals, where analyte binding causes visible color shifts. These devices are inexpensive, easy to use, and suitable for rapid screening without any laboratory setup (Sun et al. 2024). Paper-based carbon nanomaterial sensors are particularly useful for resource-limited settings and for users with minimal technical training, such as farmers and food handlers.

12.7.2 Smartphone-Assisted and Point-of-Care Devices

Smartphones now serve as powerful portable instruments because of their high-resolution cameras, processing capabilities, and connectivity. Carbon-based nanosensors can be integrated with smartphone interfaces to enable quantitative on-site analysis. In these systems, carbon nanomaterials are used as sensing elements, for example, carbon dots or graphene-based films that produce measurable optical changes when targets bind. The smartphone camera captures the signal (color intensity, fluorescence, or pattern), and a dedicated app or algorithm processes the data to provide concentration readouts. This approach has been used for detecting food contaminants, spoilage gases, and heavy metals with high sensitivity and rapid turnaround, for instance, fluorescent carbon dots embedded in hydrophilic matrices have been paired with smartphone imaging to quantify mycotoxins in cereals; changes in emission intensity captured by the phone correlate directly with toxin level. These platforms are highly suitable for point-of-care use in kitchens, markets, and farm gates (Murugesan et al. 2025).

12.7.3 Wearable and Field-Deployable Sensors

Wearable and field-deployable sensors extend on-site detection capabilities by offering continuous monitoring in real environments. These sensors embed carbon nanomaterials into flexible substrates, textiles, or handheld devices that can be carried into fields or attached to equipment. In agriculture, wearable sensors based on carbon nanotubes or graphene films have been investigated for monitoring gases associated with plant stress or soil conditions. These sensors change resistance or electrical output in response to gas adsorption, enabling farmers to detect thresholds of ethylene, ammonia, or VOCs linked to crop health. Because carbon nanomaterials are mechanically robust and electronically responsive, they can operate continuously over long periods under field conditions without frequent recalibration (Das et al. 2025). Field-deployable handheld devices incorporating carbon nanomaterial sensors can also measure heavy metals or nutrients on location by combining electrochemical platforms with portable electronics, offering immediate feedback for soil and irrigation management.

12.7.4 Integration with Internet-of-Things (IoT) Monitoring

The internet of things (IoT) transforms standalone sensors into connected networks that enable real-time data collection, remote monitoring, and automated decision-making. When carbon-based nanosensors are coupled with IoT platforms, sensor readings can be transmitted wirelessly to cloud servers, dashboards, or mobile apps for continuous tracking. In smart agriculture, networks of soil moisture and nutrient sensors enhanced with carbon nanomaterial electrodes can relay soil status data across fields. These networks allow precision irrigation and fertilization by analyzing trends over time and space, reducing resource waste and environmental impact. Similarly, IoT-connected gas or spoilage sensors in storage facilities can alert managers when food quality falls below threshold markers. Carbon nanomaterials are well suited for IoT applications because they can be fabricated on flexible, low-power electronic boards and integrated with wireless modules (such as Bluetooth or LoRa). Their high electrical sensitivity makes them ideal for miniaturized sensor nodes that require low energy while maintaining accuracy (Nguyen et al. 2025).

12.8 Conclusion and Future Scope

Carbon-based nanosensors have become very useful tools for monitoring contaminants and quality indicators in agriculture and food systems. In this chapter, we discussed different carbon nanomaterials such as graphene and its derivatives, carbon nanotubes, carbon dots, porous carbon, and hybrid nanocomposites. These materials are now being used in portable platforms like paper-based strips, smartphone-assisted devices, wearable sensors, and IoT systems. Together, these technologies show that carbon nanomaterials are not only scientifically interesting but also suitable for practical monitoring in real situations. One important advantage of carbon nanosensors is that their surfaces can be easily modified. Scientists can attach functional groups, polymers, biomolecules, or metal nanoparticles to their surface. This helps the sensors selectively detect substances such as pesticides, heavy metals, mycotoxins, nitrogen compounds, antibiotics, pathogens, and spoilage gases. Because many carbon nanomaterials have high electrical conductivity, strong fluorescence, and large surface area, they can detect very small amounts of contaminants.

Despite these advantages, several challenges still exist. One major challenge is selectivity in real samples. Agricultural soil, water, and food contain many interfering substances such as salts, proteins, and organic matter. These can affect the sensor signal and sometimes produce incorrect results. Therefore, better surface modification and calibration methods are needed to improve sensor accuracy. Another challenge is reproducibility and large-scale production. Many nanosensors are currently produced only in laboratories. However, making carbon nanomaterials with uniform size and properties on an industrial scale is difficult. Small changes during synthesis can affect sensor performance. Standard production methods and strict quality control are therefore necessary. Stability and shelf life are also important concerns. Some sensors contain biological components such as enzymes or antibodies that may lose activity over time. In agricultural environments, sensors must tolerate changes in temperature, humidity, dust, and mechanical stress. Therefore, improving the long-term stability of these sensors is essential.

Regulatory approval is another important issue. For food safety and environmental monitoring, new sensors must be validated against established analytical methods. Regulatory agencies require strong evidence of accuracy and reliability before approving new technologies. Without such validation, nanosensors may only be used for preliminary screening. Cost is also an important factor. Although carbon materials themselves are often inexpensive, the total cost of the sensing device includes fabrication, packaging, electronics, and data processing systems. For widespread adoption, the final devices must remain affordable for farmers and small food producers. Looking ahead, the future of carbon-based nanosensors is very promising. One important research direction is multiplex sensing, where a single device can detect several contaminants at the same time, for example, future sensors may simultaneously measure nitrate, heavy metals, and pesticide residues in a single test. Artificial intelligence and data analysis will also become important. Continuous sensor data can be analyzed using machine learning to predict contamination risks and suggest preventive actions, for example, soil nutrient data combined with weather information could help optimize fertilizer use and reduce environmental pollution.

Finally, progress in this field requires collaboration among scientists from many disciplines, including materials science, chemistry, agriculture, food technology, engineering, and data science. Such collaboration will help transform laboratory research into reliable commercial technologies. In summary, carbon-based nanosensors provide a sensitive, flexible, and portable approach for monitoring contaminants and quality indicators in agriculture and food systems. They allow rapid detection and can be integrated with modern digital technologies. However, challenges related to selectivity, reproducibility, stability, regulatory approval, and large-scale production must still be addressed. With continued research and technological development, carbon-based nanosensors are expected to play an important role in improving agricultural productivity, food safety, and environmental sustainability in the future.

References

Aksu M, Güzdemir Ö (2025) Food waste-derived carbon quantum dots and their applications in food technology: a critical review. Food Bioprocess Technol 18(8):6753–6778

Alosime EM (2023) A review on surface functionalization of carbon nanotubes: methods and applications. Discover Nano 18(1):12

Al-Sagheer L, Farea M (2025) Advanced graphene and polymer nanocomposites for next-generation environmental gas sensing. Diam Relat Mater 152:111991

Biswas K, Mohanta YK (2024) An overview of carbon-based nanomaterials and their derivatives for different sensing applications. In: Nanoscale matter and principles for sensing and labeling applications. Springer, pp 305–325

Dagdag O, Quadri TW, Daoudi W, Berdimurodov E, Kim H (2024) Carbon nanodots: basics, properties, and fundamentals. In: Carbon dots: recent developments and future perspectives. ACS Publications, pp 127–145

Dai C, Liu Y, Wei D (2022) Two-dimensional field-effect transistor sensors: the road toward commercialization. Chem Rev 122(11):10319–10392

Das P, Nath PC, Pandey VK, Singh R, Rustagi S, Shaikh AM, Kovács B (2025) Carbon quantum dots as emerging biosensors for food safety and environmental applications: advances and challenges. Appl Food Res 5:101255

Elugoke SE, Uwaya GE, Quadri TW, Ebenso EE (2024) Carbon quantum dots: basics, properties, and fundamentals. In: Carbon dots: recent developments and future perspectives. ACS Publications, pp 3–42

Guo S, Radecka I, Eissa AM, Ivanov E, Stoeva Z, Tchuenbou-Magaia F (2025) Recent advances in carbon-based sensors for food and medical packaging under transit: a focus on humidity, temperature, mechanical, and multifunctional sensing technologies—a systematic review. Materials 18(8):1862

Guruprasath N, Sankarganesh P, Adeyeye SAO, Babu AS, Parthasarathy V (2024) Review on emerging applications of nanobiosensor in food safety. J Food Sci 89(7):3950–3972

Hashem HM, Abdallah A (2024) A rational study of transduction mechanisms of different materials for all solid contact-ISEs. Sci Rep 14(1):5405

Hegde V, Bhat MP, Lee J-H, Kurkuri MD, Kim CS, Lee K-H (2024) Carbon-based nanomaterials: Multifaceted role in agrochemical recognition, remediation, and release. Nano Today 57: 102388

Jo YM, Jo YK, Lee JH, Jang HW, Hwang IS, Yoo DJ (2023) MOF-based chemiresistive gas sensors: toward new functionalities. Adv Mater 35(43):2206842

Kausar A, Ahmad I (2024) Highpoints of carbon nanotube nanocomposite sensors—a review. E-prime–Adv Electr Eng Electron Energy 7:100419

Kayani KF, Ghafoor D, Mohammed SJ, Shatery OB (2025) Carbon dots: synthesis, sensing mechanisms, and potential applications as promising materials for glucose sensors. Nanoscale Adv 7(1):42–59

Kumar JV, Rhim J-W (2024) Fluorescent carbon quantum dots for food contaminants detection applications. J Environ Chem Eng 12(2):111999

Li R, Yue J, Zhu F, Zhou J, Liu X (2025) Application progress of carbon quantum dot composites in fluorescent detection of food safety. J Food Sci 90(5):e70299

Liu L, Lei L, Zeng K, Wu K, Yang N (2023) Surface chemistry determined electrochemical sensing performance of red phosphorus and single walled carbon nanotube composites. Adv Funct Mater 33(3):2211335

Meskher H, Ragdi T, Thakur AK, Ha S, Khelfaoui I, Sathyamurthy R et al (2024) A review on CNTs-based electrochemical sensors and biosensors: unique properties and potential applications. Crit Rev Anal Chem 54(7):2398–2421

Mohammed AS (2025) Nitrogen-doped carbon dots in food sensing: a review of detection mechanisms and applications. RSC Adv 15(57):48727–48756

Mondal J, Lamba R, Yukta Y, Yadav R, Kumar R, Pani B, Singh B (2024) Advancements in semiconductor quantum dots: expanding frontiers in optoelectronics, analytical sensing, biomedicine, and catalysis. J Mater Chem C 12(28):10330–10389

Mshelia SS, Jain P, Gaur M (2025) Graphene oxide-based smart sensors: synthesis, functionalization strategies, and biosensing applications. Nano-Struct Nano-Objects 43:101532

Murugesan A, Li H, Shoaib M (2025) Recent advances in functionalized carbon quantum dots integrated with metal–organic frameworks: emerging platforms for sensing and food safety applications. Foods 14(12):2060

Nguyen DH, Muthu A, Elsakhawy T, Sheta MH, Abdalla N, El-Ramady H, Prokisch J (2025) Carbon nanodots-based sensors: a promising tool for detecting and monitoring toxic compounds. Nanomaterials 15(10):725

Panigrahi PK, Chandu B, Puvvada N (2024) Recent advances in nanostructured materials for application as gas sensors. ACS Omega 9(3):3092–3122

Sabui P, Mallick S, Singh KR, Natarajan A, Verma R, Singh J, Singh RP (2023) Potentialities of fluorescent carbon nanomaterials as sensor for food analysis. Luminescence 38(7):1047–1063

Sharma N, Pandey T, Pandey V (2026) Biomass-derived carbon quantum dots as sustainable nanosensors for pesticides and toxic metabolites. Environ Sci Nano 13:14

Singh H, Aggarwal V, Kachore A, Bala E, Kumar R, Sharma RK, Verma PK (2025) Carbon dots: an emerging food analysis nanoprobes for detection of contaminants. Food Chem 485:143180

Sinha N, Ray S (2024) Application of carbon quantum dots derived from waste tea for the detection of pesticides in tea: a novel biosensor approach. ACS Omega 9(51):50201–50213

Sun F, Li P, Wu G, He F, Liu S, Shen Y et al (2024) Carbon nanomaterials-based smart dual-mode sensors for colorimetric and fluorescence detection of foodborne hazards. Trends Food Sci Technol 152:104681

Thakur A, Kumar A (2023) Recent trends in nanostructured carbon-based electrochemical sensors for the detection and remediation of persistent toxic substances in real-time analysis. Mater Res Express 10(3):034001

Vijaybhai BM, Singh A, Singh G, Tangra AK, Sankhla MS (2025) Advances in carbon quantum dot-based nanocomposites for targeted detection of food toxins: a focus on safety applications. J Food Compos Anal 148:108321

Zhang C, Lai Q, Chen W, Zhang Y, Mo L, Liu Z (2023) Three-dimensional electrochemical sensors for food safety applications. Biosensors 13(5):529

Zhi S, Dai Q, Wang H, Wu D, Zhao L, Hu C, Dai L (2025) Heteroatom-doped carbon materials for multifunctional noncatalytic applications. ACS Nano 19(33):29860–29897

Zou Y, Shi Y, Wang T, Ji S, Zhang X, Shen T et al (2024) Quantum dots as advanced nanomaterials for food quality and safety applications: a comprehensive review and future perspectives. Compr Rev Food Sci Food Saf 23(3):e13339

Chapter 13
Sustainable Recycling of Dairy-Based Effluents for Carbon Nanomaterial Synthesis: Technologies, Mechanisms, and Applications

Biswabhusan Dash, Himadri Tanaya Behera, Santosh Mahanty, and Abhik Mojumdar

13.1 Introduction

Dairy industry wastewater, which is generally considered to be a major environmental pollutant owing to its high organic content and large volume generated worldwide due to milk production on a large scale, has been found to be used to synthesize carbon-based nanomaterials that are considered to be a valuable resource (Qasim et al. 2023). Instead of being used as wastewater that causes pollution to water resources and contributes to greenhouse emissions, these nutrient-rich wastewater streams are found to be used to synthesize carbon-based nanomaterials such as carbon dots, graphitic structures, and porous carbon materials. By using these organic matter present in dairy wastewater to synthesize carbon-based nanomaterials, not only is the problem of waste management being addressed, but also eco-friendly nanomaterials are being synthesized for various applications (Sonawane and Murthy 2023; Fahad Halim et al. 2025). Such dual benefits from using dairy wastewater to synthesize carbon-based nanomaterials place dairy wastewater in a positive category in the overall framework of circular economy and green nanotechnology.

B. Dash (✉)
Center for Biotechnology, Siksha 'O' Anusandhan (Deemed to be University), Bhubaneswar, Odisha, India
e-mail: biswabdash@soa.ac.in

H. T. Behera · S. Mahanty
Imgenex India Pvt. Ltd., Bhubaneswar, Odisha, India

A. Mojumdar
Biochemical Analysis Team/Center for Research Equipment, Korea Basic Science Institute (KBSI), Chungbuk, Republic of Korea

Y. K. Mohanta et al. (eds.), *Bioresource Waste to Carbon Nanomaterial*,
Nanotechnology in the Life Sciences, https://doi.org/10.1007/978-981-92-2405-0_13

13.1.1 Overview of Daily Wastewater Composition and Global Generation Statistics

Dairy wastewater is known for its difficulty in handling because of the high concentration of organics present in it. These waters have a typical COD range of 1000–6000 mg O2/L and a BOD range of 500–3000 mg/L. Along with these, high concentrations of fats, oils, grease, suspended solids (up to 3000 mg/L), and nutrient waste such as nitrogen and phosphorus from milk residue and cleaning agents are also present in dairy wastewater (Azari et al. 2025; Sinha et al. 2018). Besides these high concentrations of organics and suspended matter, the high volume of wastewater itself creates a problem in handling dairy wastewater. It requires about 2.5 L of water to produce 1 L of milk. Also, 1 ton of dairy product generates 1 cubic meter of wastewater (Kolev Slavov 2017). These values may vary depending on the scale of the factory and the cleaning in place methods used in the industry. These waters are used in the processes of washing, heating, cooling, and specific operations in the dairy industry, making it one of the most water-intensive industries in food production (Kaur 2021).

13.1.2 Environmental and Ecological Impacts of Untreated Dairy Effluents: Conventional Effluent Treatment Technologies and their Limitations

Untreated dairy effluent poses a major threat to the environment. The organic content of dairy effluent is very high. The high organic content of dairy effluent depletes dissolved oxygen at a very rapid rate. The depletion of dissolved oxygen results in a hypoxic environment that is harmful to aquatic life. Excessive nutrients like nitrogen and phosphorus cause eutrophication. pH and temperature changes caused by dairy effluent also affect aquatic life. These changes ultimately affect biodiversity as well as human health. Conventional treatment technologies like physicochemical treatment and biological treatment are inadequate for the treatment of dairy effluent. These technologies are unable to remove COD completely. These technologies are also unable to remove nitrogen completely from dairy effluent. Another drawback of conventional treatment technologies is their high cost. Additionally, these processes generate large amounts of sludge, typically containing 10–25% dry matter (Kaur 2021). Even though some technologies like anaerobic digestion and ultrafiltration are more effective than conventional technologies, like other technologies, these are also limited to some extent. The limitation of these technologies is that only some amount of biogas is generated during anaerobic digestion. These technologies are also limited by fluctuations in the concentration of pollutants as well as by a lot of energy consumption. These are some of the drawbacks that make it imperative to use more efficient technologies for the treatment of dairy effluent (Ashekuzzaman et al. 2019).

13.1.3 *Rationale and Significance of Converting Dairy*

The transformation of dairy effluent into carbon nanoparticles changes the paradigm of wastewater management from disposal to resource recovery and addresses the major shortcomings of the prevailing wastewater management practices. With the use of green synthesis techniques, the carbon-rich nature of dairy effluent containing lactose, proteins, and fats can be converted into nanostructures that possess antioxidant properties and potential biological activities. This method minimizes the environmental impacts while following the principles of the circular economy that considers garbage as an important raw material for the production of high-value products. The functional nanostructures produced have the potential to combat oxidative stress and enhance medical applications, thus illustrating the role of agricultural and industrial waste in the development of nanotechnology. Most importantly, the scalability of the method provides the advantage of being an eco-friendly and sustainable method of waste disposal while offering new opportunities for the design of green materials.

13.2 Chemical and Biochemical Composition Analysis of Dairy Effluents as Carbon Precursors

Dairy effluents have a promising potential in the production of carbon nanomaterials due to the high amount of available carbon content. Dairy waste treatment comprises a high amount of lactose, casein, and whey proteins derived from milk. These wastes also have a high amount of inorganic salts. These ingredients have a high potential in the production of functional carbon structures through hydrothermal carbonization or pyrolysis. This method not only helps in utilizing an untapped waste material but also provides a means of developing green technology in the production of functional carbon structures with various applications in medicine and energy storage (Shivsharan et al. 2013).

13.2.1 *Detailed Profiling of Lactose, Proteins, Fats, and Minor Organics in Dairy Effluents*

Dairy effluent contains a diverse range of organic carbon sources, primarily derived from milk, that can effectively serve as the precursor to producing a variety of nanomaterials. The main components of dairy effluent are lactose, proteins (casein and whey), and fats, along with some lesser components, such as salts, detergents, and additives that modify carbonization processes. Lactose in a range of 20–50 g/L has been reported to be present in whey-rich dairy wastewater. Lactose is also the dominant soluble carbohydrate in dairy waste (accounting for about 3% of the total solids) and is responsible for up to 50% of the total chemical oxygen demand (COD). Due to its high degree of biodegradability, lactose is an excellent candidate for producing a carbon source by hydrothermal carbonization and can provide a porous carbon structure (Ghatawat et al. 2019). Proteins, like casein and whey, normally represent ~3% of the total solids, but throughout the thermal conversion process, they add sulfur and nitrogen functionalities to the product and allow the production of heteroatom-doped carbon with enhanced biocompatibility and antioxidant activity. Collectively, the fat, oil, and grease (FOG) found in dairy effluent can produce between 45 and 450 mg/L of COD but can exceed 1000 mg/L in the processing of butter or ghee and serve as templates for the creation of mesoporous carbon structures. Other organic materials that may be present in lesser amounts (<5% of total COD), including detergents, sanitizers, residual sugars, and phosphorus, can have no chemical contribution to the generation of solid carbon but help enrich surface characteristics by natural heteroatom doping and porogenic effects. Therefore, dairy effluents can be used as a green feedstock to produce functional carbon nanomaterials.

13.2.2 Comparative Carbon Content and Elemental Analysis Versus Other Biomass Wastes

Dairy effluent is a moderately rich carbon source mainly due to organics like lactose and proteins derived from milk. Although the carbon content of dairy effluent is comparable to other agricultural wastes, it is important to note that dairy effluent is different from other wastes due to its moisture content as well as its nutrients. These nutrients are important for doping heteroatoms during carbon nanomaterial synthesis; thus, dairy effluent is a potential candidate for creating functional and green carbon nanomaterials (Carawan et al. 1979; Sinha et al. 2018) (Table 13.1).

The carbon content of dairy effluents is comparable with that of sewage sludge, while the latter is distinguished by the availability of a better source of nitrogen, as it contains a protein fraction. This is a natural source of N-doping, which makes dairy wastes better than lignocellulosic wastes for the production of carbon nanomaterials with better surface properties and better efficiency for specific uses, such as catalysis, antioxidants, and biological interfaces.

Table 13.1 Comparative carbon content and elemental analysis versus other biomass wastes (Oh et al. 2024)

Biomass source	C (wt%)	H (wt%)	O (wt%)	N (wt%)	S (wt%)	Ash (wt%)
Dairy effluents	38–48	6–7	23–29	3–5	0.3–1.0	11–26
Sewage sludge	38.7	6.0	23.1	5.7	0.9	25.6
Food waste	47.9	7.1	29.2	4.2	0.3	11.3
Livestock manure	42.6	6.2	24.4	3.9	0.9	22.0

13.2.3 Challenges and Potential Inhibitors in Dairy Effluent Valorization Processes

The valorization of dairy wastes for the production of carbon nanomaterials by means of hydrothermal carbonization of pyrolysis is a promising method, but it is associated with several challenges that may affect the efficiency of the processes and the overall yields of carbon, as dairy wastes are complex substrates with a high organic load and inhibitory properties, as suggested by Casero-Díaz et al. (2025).

13.2.3.1 Key Challenges

High Organic Load and Acidity

The high concentration of organic matter, with a COD of up to 50 g/L, causes a rapid production of volatile fatty acids, leading to a reduction in pH and thus the creation of acidic environments that may affect microbial and thermal processes. Maintaining a food/microorganism ratio of two to four is a challenge that may cause the collapse of the system, as suggested by Casero-Díaz et al. (2025).

Sludge Formation and Costs

The amount of sludge that is formed is huge, 200–350 kg/500,000 L of milk effluent. The sludge absorbs a lot of toxins. The cost of cleaning up this sludge is very expensive, at times reaching 60% of the total treatment cost (Udourioh et al. 2025).

Process Variability

Variability in pH-, temperature-, retention time-, and waste-to-solvent ratio is a challenge that is encountered during HTC. The variability impacts HTC's efficiency in nutrient distribution and hydrochar production (Udourioh et al. 2025).

13.2.3.2 Potential Inhibitors

Nutrient and Filamentous Growth

Excessive nutrients from proteins can cause filamentous growth of fungi. Excessive nutrients cause inhibition in sequencing batch reactors (SBRs) (Usmani et al. 2022).

Low Protein Concentrations

Low protein concentration from whey requires expensive concentration steps before hydrolysis can be effective (Pires et al. 2021).

Oxygen Depletion and Scale-Up Issues

Excessive BOD and COD cause depletion of oxygen. The depletion of oxygen limits the use of aerobic treatment. The process is also limited by scalability. The process is less feasible for industrial-scale implementation due to scalability challenges (Rocha and Guerra 2020).

13.3 Fundamentals and Classification of Carbon Nanomaterials from Biomass

Carbon nanomaterials derived from biomass, including dairy effluents, are produced through thermal decomposition of organic-rich precursors. Techniques such as pyrolysis (400–900 °C under inert atmosphere), hydrothermal carbonization (HTC, 180–250 °C in aqueous systems), and microwave-assisted activation rearrange C-C bonds into sp^2–/sp^3-hybridized structures with nanoscale dimensions (1–100 nm) (Lesch et al. 2025). These zero- to three-dimensional carbon allotropes naturally inherit heteroatoms (N, O, P, S) from proteins and minerals present in dairy waste, enhancing their biocompatibility and reactive oxygen species (ROS)-scavenging capacity key features for oxidative stress therapies (Fahad Halim et al. 2025).

13.3.1 Carbon Nanotubes (CNTs)

CNTs are cylindrical graphene sheets rolled into seamless tubes, classified as single-walled or muti-walled. Biomass-derived CNTs can be synthesized via catalytic chemical vapor deposition, where volatiles from dairy effluents deposit on Fe/Ni catalysts naturally present in milk minerals. These structures exhibit remarkable mechanical and electrical properties, tensile strength exceeding 100 GPa, conductivity up to 10^6S/cm, and aspect ratios greater than 10^3 making them ideal for conductive scaffolds in biochemical composites. Interestingly, biomass-derived CNTs often show fewer defects due to natural templating effects (Qasim et al. 2023).

13.3.2 Graphene

Graphene consists of single atomic layers of carbon arranged in a honeycomb lattice (0.34-nm interlayer spacing). Biomass routes, including oxidation-exfoliation or flash joule heating of char, yield few layer graphene or graphene oxide (GO). With a theoretical surface area of 2630 m^2/g, thermal conductivity of approximately 5000 W/m·K, and a Young's modulus of about 1 TPa, graphene exhibits exceptional mechanical strength, thermal conductivity, and flexibility. In biomedical contexts, it enables drug delivery membranes, while dairy lactose can facilitate defect-free sheet formation through epitaxial growth during reduction (Zhao et al. 2023).

13.3.3 Carbon Quantum Dots (CQDs)

CQDs are quasi-spherical nanocrystals (2–10 nm) with crystalline cores and functionalized surfaces. Hydrothermal synthesis from citric acid like sugars in dairy effluents produces CQDs with tunable fluorescence (blue to red) via quantum confinement and surface states. They are highly biocompatible, showing less than 5% cell viability loss, and exhibit strong antioxidant activity, with ROS quenching efficiencies up to 90% in DPPH assays (Lesch et al. 2025).

13.3.4 Activated Carbon

Activated carbon forms disordered three-dimensional porous networks (pore sizes 0.5–50 nm) through physical activation (steam/CO_2 at approximate 80 °C) or chemical activation (KOH, $ZnCl_2$). Biomass-derived activated carbons achieve surface areas of 1000–3500 m^2/g and micropore volumes >1 cm^3/g. Casein proteins contribute nitrogen doping (5–10 wt%), boosting capacitance and adsorption selectivity, while dairy fats generate mesopores that improve pollutant capture. This aligns with green synthesis principles, positioning dairy effluent-derived activated carbon as a sustainable material for environmental remediation.

13.4 Synthesis Routes for Carbon Nanomaterials from Dairy Effluents

13.4.1 Physical and Chemical Synthesis Routes for Dairy Effluents

Dairy effluents, enriched with lactose, proteins, and fats, represent sustainable feedstocks for carbon nanomaterial (CNM) production. Their high moisture content and organic richness make them particularly suitable for both physical and chemical synthesis routes, enabling one-pot green conversion (Ritambhara et al. 2018).

The surface area of the biomass-based ACs ranges between 1000 and 3500 m^2/g, while the volume of the micropore is greater than 1 cm^3/g. The casein proteins also act as nitrogen dopants (5–10 wt%), thus enhancing the capacitance and selectivity of the ACs. On the other hand, the dairy fats act as mesopore creators, thus improving the selectivity of the ACs. This meets the green synthesis requirements, thus categorizing dairy effluent-based ACs as green materials for environmental remediation purposes.

13.4.2 Physical and Chemical Synthesis Routes for Dairy Effluents

Dairy effluents, being rich in lactose, proteins, and fats, thus qualify as green materials for the synthesis of CNMs. The presence of moisture in dairy effluents qualifies them as green materials for physical and chemical synthesis routes (Figs. 13.1 and 13.2).

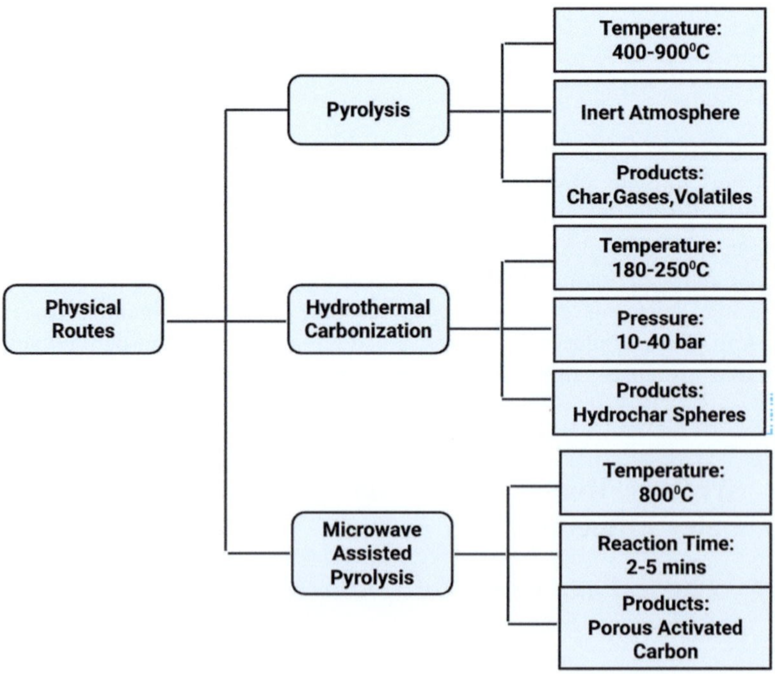

Fig. 13.1 Physical synthesis routes for dairy effluents

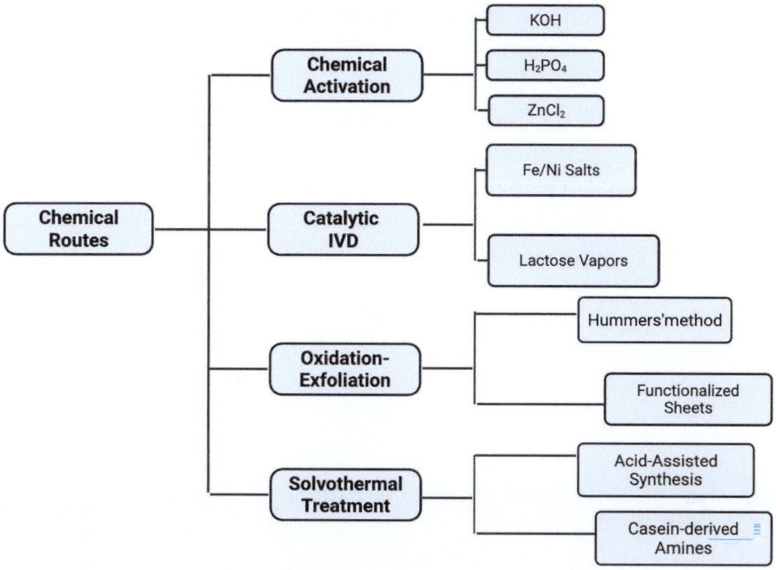

Fig. 13.2 Chemical synthesis routes for dairy effluents

13.5 Thermochemical Conversion of Dairy Effluents into CNMs

Dairy waste is high in lactose (20–50 g/L), protein (5–10 g/L), and lipid content (50–1000 mg/L); therefore, carbon nanomaterials (CNM) can be produced using dairy waste through thermochemical conversion processes. Dehydration, depolymerization, and graphitization reactions will occur and result in the production of either hydrochar or CNTs (20–50 wt.%) (Udourioh et al. 2025).

13.5.1 Pyrolysis

Pyrolysis is a thermal process that takes place at an inert environment and when dairy waste has a temperature between 400° and 900 °C. Lactose undergoes pyrolyzed reaction and produces HMF and furan; lipids produce hydrocarbons and char (30–45%) when used with calcium carbonate and iron, an elemental catalyst to produce carbon nanotubes (CNTs). Microwave-assisted pyrolysis produces CNTs (20–50 nm diameter with a maximum yield of 25 wt.%) at approximately 800 °C with a reaction time ranging between 2 and 5 minutes. Proteins will also assist in producing doped CNTs using about 3% of the weight of the protein used to produce CNTs (Azam et al. 2024).

13.5.2 Hydrothermal Carbonization (HTC)

Hydrothermal carbonization treats wet effluent directly without a pretreatment step (temperature, 180–250 °C; pressure, 10–40 bar; time, 4–24 h). During HTC, the sugars will undergo hydrolysis and dehydration to produce furan intermediates, which will polymerize into roughly 1–10 μm-sized spherical hydrochar particles with a yield of 35%–50%. The Maillard reaction of protein (caseins) and sugar will create heteroatoms (nitrogen and oxygen) in the produced hydrochar. This reaction will produce carbon quantum dots (producing photoluminescent quantum yields of approximately 20%–40%). The lipids in dairy waste will contribute to the creation of mesoporous structures (activated with KOH) with specific surface areas (greater than 2000 m^2/g) (Khalaf et al. 2023).

13.5.3 Gasification and Hybrid Routes

Gasification of dairy manure at 800 °C will produce not only syngas but also CNTs (approximately 2–5 wt.%). Catalysis (nickel) in the ash that results from the dairy processing operation will enhance the gasification reaction. Combining HTC (at 220 °C) with pyrolysis (at 700 °C) will optimize production of fixed carbon (approximately 45%) and decrease tar formation as well (Udourioh et al. 2025).

13.5.4 Process Optimization and Yields

The optimized conditions involve 10–20 wt% biomass, 4–7 pH, and ~12-h retention time. Yields are directly proportional to chemical oxygen demand (COD). Higher fat composition favors mesoporous CNMs. Importantly, dairy-derived minerals can cut down on catalyst usage by 50%–70%, thus enabling more “green” and cost-effective production of CNTs and grapheme (Zhi et al. 2024).

13.6 Nanostructure Formation and Morphology Control in Dairy Effluent-Derived CNM

When dairy waste is processed under heat, its main components, namely, lactose, proteins, and fats, break down to smaller reactive molecules. These fragments then recombine and organize into aromatic clusters through three key stages: nucleation, growth, and graphitization. This bottom-up pathway is what drives the creation of carbon nanomaterial (Islam et al. 2022).

13.6.1 Hydrothermal Carbonization (HTC)

13.6.1.1 Reaction Mechanisms and Process Conditions Under HTC (180–250 °C)

Lactose first hydrolyzes into glucose, which dehydrates into 5-hydroxymethylfurfural (HMF). HMF molecule polymerizes into furan rings that stack together via π-π interactions, forming spherical hydrochar particles (1–10 μm). Proteins undergo Maillard reactions with sugars, introducing nitrogen and oxygen atoms that create core-shell structure with hydrophilic surfaces. The particle's shape and density depend on processing conditions: Longer residence time leads to denser spheres. Acidic pH favors furan-based carbon quantum dots (CQDs) smaller than 10 nm (Khalaf et al. 2023).

13.6.1.2 Post-HTC Pyrolysis and Activation for Porous Carbon Formation

KOH and activating chemicals are used to refine the material being processed. The activation process will etch the carbon materials through a chemical reaction, and as a result of that etching, micropores and mesopores will be developed due to those reactions. By using a 1:2 ratio of activating agent to KOH, a specific surface area of 2500 m^2/g can be developed (Kwapinska et al. 2020).

13.6.1.3 Factors Influencing Morphology and Structural Properties

The use of casein micelles, as well as the use of dairy fats as templating agents, has been shown to successfully produce mesoporosity. The use of microwave assistance will help produce plasma and align CNTs. High graphitization was confirmed by Raman spectroscopy and will be further confirmed via the use of protein doping if this is desired for optical processing. There are many other parameters of consideration that will assist in refining the products of different types of carbon nanomaterials and therefore assist in developing the carbon nanotube and carbon dot process (de Sousa and Monteiro 2025).

13.7 Catalytic and Advanced Physical Techniques for CNM Production from Dairy Effluent

The production of CNMs from dairy effluent not only represents an intelligent method for waste recycling but also has the potential to become an important tool in cost-effective nanotechnology. By exploiting the inherent chemistry of milk, researchers can use catalytic and advanced physical methods for the production of

CNMs. These approaches achieve impressive outcomes, such as 2–5 wt% yields for CNTs and surface areas exceeding 2000 m^2/g for activated carbons (Azari et al. 2025).

13.7.1 Biomass-Derived Catalysts in CNM Formation

Minerals from milk proteins and salts provide catalytic activity during CVD pyrolysis (600–1000 °C), for example, Fe from casein phosphates nucleates CNT tip growth via methane decomposition, producing multi-walled CNTs (20–50 nm diameter) with high graphitization. Ni salts, meanwhile, promote graphene curling. Pretreatments such as effluent concentration (10%–20% solids) enhance catalyst density, improving yields by 30%–50% compared to synthetic catalysts (Lu et al. 2025).

13.7.2 MOF Precursors

Metal organic frameworks such as MIL-88A (Fe) or ZIF-8 can be templated onto dairy-derived carbon substrates via hydrothermal mixing (e.g., Fe $(NO_3)_3$ + fumaric acid on activated char at 110 °C for 6 h). Upon pyrolysis, these MOFs yield F-e/N-doped CNMs. The porous MOF architecture directs uniform CNT forests or hierarchical graphene, while Fe clusters catalyze graphitization. Resulting materials exhibit high surface areas (800–1500 m^2/g) and nitrogen content (5–8 wt%), enhancing conductivity and ROS-scavenging capacity (Azari et al. 2025).

13.7.3 Metal Nanoparticle Catalysts

Nanoparticles supported on dairy-derived ash further boost CNM synthesis. Co-Ni-Mo nanoparticles (5–20 nm, supported on MgO) in pyrolysis CO_2 capture CVD systems increase MWCNT yields to 2.4 wt% while co-producing hydrogen (161 mL/g). $NiFe_2O_4$ nanoparticles accelerate dark fermentation before carbonization, enriching volatile precursors for cleaner CVD feeds. Microwave plasma techniques can also generate Au/Ag nanoparticles in situ, producing carbon quantum dots (CQDs) with photoluminescence quantum yields (PLQY) exceeding 40% (Fahoul et al. 2022).

13.7.4 Advanced Physical Techniques for CNM Synthesis

Physical innovations further refine CNM morphology. Microwave-assisted CVD (approx. 800 °C, 2–5 min) with dairy catalysts induces plasma for rapid CNT

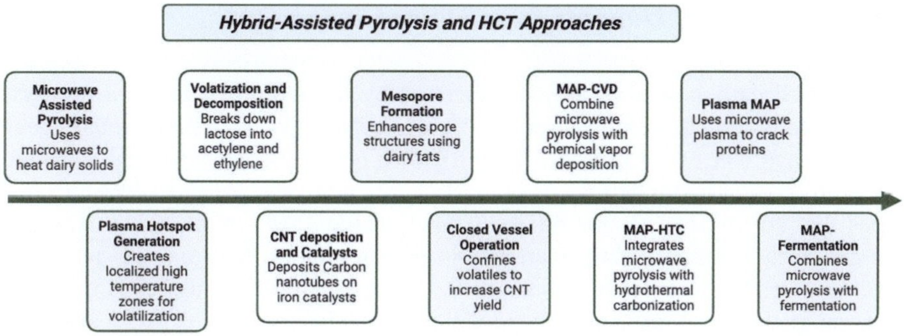

Fig. 13.3 Overview of microwave-assisted pyrolysis and hybrid thermochemical routes for converting dairy waste

alignment, achieving aspect ratios >1000. Flash Joule heating (approx. 5000 K/s) exfoliates graphene sheets directly from hydrochar. Combined cavitation-microwave pretreatments solubilize up to 40% of COD, increasing carbon availability for subsequent synthesis (Adeola et al. 2023).

13.7.5 Microwave-Assisted Pyrolysis and Hybrid Thermochemical Approaches

Microwave-assisted pyrolysis offers rapid and uniform heating, which helps convert into bio-oil, gas, and char with good energy efficiency. Hybrid thermochemical approaches pair microwave heating with techniques like catalytic cracking or hydrothermal treatment to improve conversion and product quality (Adeola et al. 2023). A brief overview of these pathways is shown in Fig. 13.3.

13.8 Role of Dopants and Functionalization in Tailoring CNM Properties in Dairy Effluents

The synthesis of CNMs from dairy effluent is characterized by the presence of heteroatoms and organic groups from milk components (Kumar et al. 2021). These include proteins, lactose, and fats that act as carbon sources as well as dopants such as nitrogen (N), phosphorous (P), sulfur (S), and oxygen (O), which are present in milk. These heteroatoms and organic groups fine-tune the electronic structure and surface reactivity of CNMs. As a result, these materials are characterized by high antioxidant activity, electrical conductivity, and biocompatibility.

13.8.1 Inherent Dopants from Dairy Components

Dairy-based carbon sources are characterized by heteroatoms that act as dopants during hydrothermal carbonization and pyrolysis. Nitrogen from milk proteins such as casein and whey is responsible for 3–10 wt% nitrogen content through Maillard reactions during hydrothermal carbonization or pyrolysis. Nitrogen has different types that are responsible for different functionalities (Ma et al. 2024). Nitrogen present in milk-based carbon materials is responsible for high oxygen reduction reaction (ORR) onset potentials (about 0.9 V), pyridinic N being responsible for electron presence. Nitrogen is responsible for high reactive oxygen-scavenging activity (>80% DPPH), pyrrolic N being responsible for spin density. Nitrogen is responsible for high graphitization. Phosphorous from milk-based carbon materials is responsible for high capacitance through pseudocapacitance effects due to P-O-C bridges. Sulfates are responsible for thiophenic S present on the surface of carbon materials, acting as a selectivity promoter in catalytic reactions. Additionally, oxygen functional groups such as hydroxyl (-OH) and carboxyl (-COOH) from lactose are responsible for high dispersion stability as well as pH-dependent swelling behavior in drug delivery systems (Sen et al. 2023).

13.8.2 Functionalization and co-Doping (N/P/S) Strategies (Kanakaraju et al. 2022; Mozafari and Soroush 2021)

Microwave-assisted pyrolysis at approx. 8000 °C synergistically incorporates N and P, thus creating hierarchical pores. This functionalization method increases superoxide dismutase-like activity by a factor of five. Grafting further improves the post-HTC functionalization with whey-derived amines, thus passivating CQDs and increasing photoluminescence quantum yields. This also reduces cytotoxicity, with cell viability >95% at 100 μg/mL. Targeted functional groups, such as the folate group, may be used for targeted cancer therapy. Metal doping with dairy-derived Fe and Ca complexes is also effective, acting as dopants in CVD, catalyzing CNT growth, and introducing defect states for fluorescence emission at 450–550 nm.

13.8.3 Property Tailoring Effects

The presence of pyridinic N in the lattice shifts the Fermi level, thus conferring n-type conductivity >104 S/m. O/P co-doping at mesoporous edge sites increases adsorption capacity. Graphitic N stabilizes the lattice, thus increasing thermal resistance. Careful control of dopant levels is required, balancing stability with lattice stability. This functionalization method allows for the creation of CNMs with ROS quenching efficiencies between 80% and 95%, thus showing the potential for dairy-derived materials to be used in a variety of applications, replacing conventional nanomaterials with a greener alternative (Kueasook et al. 2025).

13.9 Scale-up Challenges and Reactor Design Considerations for Dairy Effluent-Derived CNMs

The scale-up of dairy effluent-derived CNMs is a far more complex process than initially indicated by small-scale experiments. While small-scale experiments with 100-mL reactors achieve yields >30 wt%, high-quality materials are also produced. However, the scale-up to a 10–100-L pilot plant poses a number of engineering and operational challenges (Khalaf et al. 2023).

13.9.1 Key Scale-up Challenges

HTC and pyrolysis demand hydrochar energy of 22–50 MJ/kg. Heat transfer is low in large-scale reactors. Temperature gradients of ±500 °C between the core and the wall cause irregular CNM structure (Adeola et al. 2023). Variability in effluent composition due to seasonal changes in COD (1–6 g/L) and fats (50–1000 mg/L) also causes instability in the reactor (Lesch et al. 2025). The pressure of the HTC reactor ranges from 10 to 40 bar. Equipment failure and safety hazards arise due to pressure buildup and the presence of corrosive process water. Scale-up also leads to the generation of considerable by-products (65% liquor and 15% solid waste), as well as the production of greenhouse gases. Catalysts also undergo continuous changes. A decrease of 40–60% in the production of CNT occurs due to the continuous operation of the reactor. Catalysts need to be regenerated or replaced frequently due to the continuous changes that occur in the reactor.

13.9.2 Hydrothermal Carbonization Reactors

The cylindrical reactor design is the most common reactor design. The reactor is provided with a PID controller. Temperature (180–2500 °C), residence time (1–24 h), and pH (3–9) are precisely controlled. The reactor also has an internal heating coil. The two-stage hydrolysis carbonization reduces the energy consumption by 20–30% (Khalaf et al. 2023).

The fluidized bed reactor is also an effective reactor. The temperature ranges from 500 to 9000 °C. N2 gas flow is also provided. The reactor is also provided with the facility of using the microwave reactor. The use of SiC hot spots also occurs. The continuous screw reactor is also effective. The reactor is effective for the pyrolysis of wet feed. Autogenous pressure seal occurs in the reactor (Adeola et al. 2023).

The modular reactor design allows the reactor to be tested on a small scale. The reactor throughput ranges from 1 to 5 kg/h. The online monitoring tools also allow the reactor to be monitored. Raman spectroscopy and XRD also allow the reactor to be monitored

13.10 Characterization Techniques of Dairy Effluent-Derived CNMs

13.10.1 Morphological Tools: Scanning Electron Microscopy (SEM), Transmission Electron Microscopy (TEM), and Atomic Force Microscopy (AFM)

The detailed characterization of the form, shape, and size of carbon nanomaterials from dairy effluents is crucial to optimize their characteristics and utilization. At different scales, morphology techniques including scanning electron microscopy, transmission electron microscopy, and atomic force microscopy offer invaluable insights. SEM provides a realistic characterization of the surface topography to examine the overall shape, porosity, and size distribution of the carbon structures prepared (Shah et al. 2021). SEM would also be used to verify the formation of nanostructures from dairy effluent-derived CNMs and to determine the structure of the carbon (spheres, tubes, sheets), as well as the extent of agglomeration, which can affect their catalytic or adsorption performance, among others. Transmission electron microscopy (TEM), especially high-resolution TEM, can provide an even finer resolution by visualizing internal structures, crystallinity, lattice fringes, and defects at atomic level. This process is fundamental for validating CNM-type construction, for example, of carbon nanotubes or layers and materials of graphene, by characterizing their internal structure. TEM can also demonstrate the preparation and transformation of biomass precursors to carbon nanostructures as was done using biochar pore transformation into carbon nanotubes (McKenzie Jr et al. 2025). TEM on dairy effluents can reveal the extent of graphitization, amorphous carbon presence, and interior pore patterns that are essential for the conductivity of the materials when used in CNMs. Atomic force microscopy (AFM) complements SEM and TEM by providing three-dimensional surface topographical mapping, surface roughness measurements, and mechanical characterization for individual nanoparticles. AFM can be especially valuable for the characterization of thin film or isolated nanoparticles and can enable accurate measurements of height, width, and surface texture that can correlate to the surface area and surface reactivity of the material (Enríquez et al. 2014). These morphological tools combined are essential for a comparison between synthesis methods from dairy effluent and carbon nanomaterials in general from the material with carbon nanometer-like morphology (Wang et al. 2025).

13.10.2 Spectroscopic Analyses: Raman Spectroscopy, Fourier-Transform Infrared Spectroscopy (FTIR), and X-Ray Diffraction (XRD) for Structural Information

The spectroscopy and diffraction techniques are the basic tools to unravel the structural properties and the position of the chemical bonds of the carbon nanomaterials of the dairy effluents. The use of Raman spectroscopy is also significant for the characterization of ordered and disordered carbon structures. The use of graphitic (sp^2-hybridized) and amorphous (sp^3-hybridized) carbon structures is also significant. The ratio of the D-band and G-band intensities provides significant insight into the graphitization extent of the carbon material. This is particularly significant for the dairy effluent-derived CNMs because the nature of the organic matter and the synthesis of the CNMs may vary.

The various surface functional groups of the carbon nanomaterials may be determined using Fourier transform infrared spectroscopy (FTIR). The FTIR analysis will be employed to monitor the changes in the functional groups of the carbon nanomaterials at the synthesis or posttreatment stages, for example, the biochars derived from agricultural waste use the FTIR technique to describe the surface structures that are significant for the adsorption of the waste (Hidayah et al. 2022).

On the other hand, crystallographic data, crystal structure, phase purity, and average crystallite size of the carbon nanomaterials can be characterized by X-ray diffraction techniques. As noticed for some of the carbon nanomaterials, the XRD patterns are generally related to the graphite-like structures, and a peak is associated with the interlayer spacing. These techniques, along with the above, provide a full picture of the characteristics of the synthesized carbon nanomaterials, which is necessary for understanding the influence of the synthesis of carbon nanomaterials from dairy effluent through a bio-mediated method.

13.10.3 Chemical Composition and Surface Functional Group Characterization to Determine Adsorption and Catalytic Potential

The chemical composition and characteristics of surface functional groups are significant characteristics that need to be understood and controlled for determining the potential of carbon nanomaterials synthesized from dairy effluent for adsorption and catalytic properties. Elemental analysis, performed by techniques like CHNS/O, is used for determining the chemical composition of the synthesized carbon nanomaterials, which includes the content of carbon, hydrogen, nitrogen, sulfur, and oxygen, providing a broad picture of the purity and elemental ratio of the synthesized carbon nanomaterials. For waste-derived materials in particular, the composition of the initial dairy effluent can affect the elemental composition at the end of the products. But surface-sensitive techniques are the most essential to assess the functionality. X-ray photoelectron spectroscopy is a powerful surface analysis method providing essential information on the elemental composition and chemical state of the outermost layers (Chen and Zhang 2024; Chen et al. 2024). With carbon nanomaterial type, an X-ray photoelectron spectroscopy (XPS) can quickly describe and quantify the numerous oxygen-containing functional groups present such as hydroxyl, carbonyl, carboxyl, and ether groups on the surface. These carboxyl-functionalized graphenes are important as active sites for adsorbing pollutants and as catalyst factors for nanomaterials. Its presence and distribution of the functional groups also affect its affinity for various substances and its activity in chemical reactions; therefore, XPS is indispensable in shaping CNMs for application in diverse fields such as pollutant removal from wastewater (Talafha et al. 2024) or as a catalyst (Mohapatra et al. 2023). Moreover, such methods as Boehm titration have the advantage to measure the acidic and basic functional groups on the surface and contribute related information in terms of the surface chemistry of the material. It can be characterized to specify CNMs with optimized surface values that can efficiently adsorb contaminants including in the treatment of dairy waste or facilitate catalytic activities for obtaining maximum value from a recovery of recycled dairy waste (Mohapatra et al. 2023). This in-depth chemical characterization guarantees that the sustainable syntheses can produce materials with desired performance characteristics.

13.11 Techno-Economic Analysis and Environmental Impact Assessment

13.11.1 Energy, Mass Balances, and Carbon Footprint Associated with Effluent-to-CNM Pathways

With a circular bioeconomy, the transformation of dairy effluents to carbon nanomaterials requires a complete appreciation of the energy and mass levels together with the overall carbon footprint. Conventional CNM synthesis techniques use expensive, nonrenewable precursors, which are energy-consuming as well as high in cost and environmental burden. Nonetheless, valorization of dairy waste into CNMs suggests a viable alternative. Dairy facilities produce significant quantities of wastewater, which if not treated will lead to environmental pollution and greenhouse gas emissions (Patel et al. 2020). For dairy processing effluents, a multilevel environmental analysis of the anaerobic treatment shows that recovered energy (electricity and heat) generated from the anaerobic digestion of wastewater can reduce the farm's overall environmental impacts and carbon footprint by up to 13% compared to baseline scenarios (Stanchev et al. 2020). This demonstrates that the combination of dairy waste treatment and energy recovery, a fundamental process that is usually required for numerous CNM synthesis pathways, already constitutes a significant environmental benefit. The process of converting organic matter (including lactose, proteins, and fats) in its dairy effluent into carbon-rich intermediates and then into CNMs has to be evaluated with consideration of a mass balance. Mass balance for direct dairy effluent-to-CNM pathways is still emerging, and studies on carbon nanomaterials derived from biomass have shown, for example, that flash Joule heating can generate flash graphene from waste of biomass at much lower environmental costs and in some cases even carbon neutrality (Jia et al. 2022). This indicates that efficient technologies allow lowering the waste throughout the process. Life cycle assessment is one crucial instrument to quantify these impacts, evaluating produced goods, from raw material extraction through processing, manufacturing, and disposal. Hence, for the CNMs derived from dairy effluent, the carbon footprint can mainly be determined by the energy source utilized for processing steps (e.g., thermal treatment, purification), the efficiency of carbon conversion from the effluent, as well as the treatment of leftover by-products. These factors have to be optimized for creating the CNM production route that is genuinely sustainable and low-carbon.

13.11.2 Economic Modeling Highlighting Cost Drivers and Market Potentials

The economic feasibility of converting dairy effluents to carbon nanomaterials is an important economic consideration in the industrial environment and relies on economic modeling which captures cost determinants and market opportunities. Traditional production of carbon nanomaterials is expensive because of energy demands and the costly raw materials. A waste product used as a feedstock, on the other hand, is also a cost-effective solution, taking advantage of an inexpensive (or even negative-cost) resource (Ikram et al. 2023). Techno-economic analyses for similar activities, for instance, microalgae extraction and the treatment of dairy effluent, show that capital expenditure and operating expenses components have been identified, of which centrifugation and electricity supply are the key contributions. For CNM synthesis from waste, the main cost drivers are associated with initial production of the prior preparation of dairy effluent in terms of pretreatment (concentration and hydrolysis), energy from the carbonization or synthesis (pyrolysis, hydrothermal carbonization, or flash Joule heating), purification, and downstream functionalization. A modeling framework for nanomaterial synthesis stresses the importance of finding paths to a decrease in environmental footprints and minimum selling price in parallel; some of the strategies also demonstrate MSP decreases by >35% with cellulose nanocrystals (Lan et al. 2024).

Market opportunity of dairy effluent-based CNMs is substantial because of the growing interest in utilizing sustainable materials in the fields of energy storage, environmental remediation, and biomedical devices, for example, waste-derived

CNMs (e.g., flash graphene from biomass) have been proven to be more than tenfold cheaper than regular graphene and are competitively priced for these applications (Jia et al. 2022). This cost-benefit, along with their being eco-friendly, makes them preferable from the conventional CNM materials. Economic feasibility studies of chemical recovery and energy/water production from dairy wastewater treatment plants have demonstrated that resource recovery can enable revenue generation to offset the cost of treatment (Basem et al. 2024). This seems to imply that generating high-value CNMs would add to the economic potential of dairy waste valorization. In addition, the potential for functional carbon materials from waste with tunable properties for individual applications will create new markets. A growing focus on circular economy principles and corporate sustainability goals also means there is a lot of consumers pull for using products from waste streams, meaning CNMs based on dairy effluents will be able to command a premium while upholding environmental obligations.

13.11.3 Integration Within Circular Bioeconomy and Sustainable Waste Management Frameworks

However, the synthesis of carbon nanomaterials derived from dairy effluents, integrated into circular bioeconomy and sustainable waste management frameworks, offers a transformative potential on waste valorization and resource efficiency (Patel et al. 2020) and, since the dairy industry, is one of the primary producers of wastewater and by-products in the world. It is considered a source of tremendous environmental concerns without effective management. Also, at the end of the "waste-to-wealth" continuum of the circular economy, we want to turn these waste streams into valuable products so that the burden of landfill and pollution is reduced (Ikram et al. 2023). The conversion of dairy effluents into high-value CNMs of interest adheres to this paradigm perfectly because this concept treats waste as not a liability but as a crucial resource needed for advanced materials. It requires a paradigm shift away from traditional linear models (take-make-dispose) to what is often called restorative, keeping resources in use for the longest possible time.

Microbial processes to convert dairy wastes to bioproducts constitute a good example of this integration, and they are playing a significant role in boosting the bioeconomy, by generating bioenergy (Shete and Shinkar 2017), for example, the anaerobic digestion of dairy processing effluents is a well-known waste treatment process that can be employed to obtain valuable biogas, in addition to giving rise to the treated effluent that could be exploited to produce the possible CNM precursors and hence lessen the overall environmental impact of the dairy facility. The sustainable development of microalgae from dairy waste to carbon credit, as well as bioremediation and biofuels, indicates the value of biological pathways to valorize effluents into multiple products, one of which is biomass which can be employed as a CNM precursor (Patel et al. 2020). In addition to reducing waste, the application of dairy effluent for the generation of CNM enhances sustainable waste management through the closing of material loops. Through its organic content, instead of disposing nutrient-rich wastewater, products that contain these materials have become processed for energy storage, environmental protection, or biomedicine, and this provides a way to ensure resource independence that reduces dependence on virgin materials (Verma et al. 2021). This approach serves both to deal with the environmental problems associated with dairy waste and also to bring new economic activities, thereby constructing a stronger and more environmentally sound industry ecosystem.

13.12 Future Perspectives and Technological Innovations

13.12.1 Coupling CNM Production with Bioenergy Recovery for Zero-Waste Dairy Models

The sustainable dairy development in the future rests on integrated waste-free milk production system which generates useful feedstock that is a successful business model for zero-waste milk. One of the most important aspects of such integration is the synergistic integration between carbon nanomaterial synthesis and bioenergy recovery processes in dairy sewage. There is a large volume of waste generated by dairy operations, such as manure and wastewater, which represent large environmental problems in particular because of the high organic load as well as the greenhouse gas emissions (Lanjekar and Panwar 2024). Conventional waste management methods increasingly are considered unsustainable, leading to the importance of shift toward circular economy approach. Integrated polygeneration technologies are emerging as an alternative, and they can transform all of the dairy farm refuse, including even very low concentrations of methane, from the barns into electricity, fresh water, and cooling. Such systems are able to significantly reduce the environmental impact of a dairy farm as they convert waste into resources for energy independence and water conservation (Luqman and Al-Ansari 2021). In this environment, the valorization of dairy waste by microbial, more specifically, anaerobic digestion, is crucial. Anaerobic digestion of cheese whey, in particular, which not only removes the effluent, generates biogas, a renewable energy source, and digestate, a high carbon component (Casallas-Ojeda et al. 2021). This neglected digestate is promising as a pre-step for the creation of CNMs. Advanced carbon nanomaterials made from these carbonaceous residues could be further processed for generating highly advanced carbon nanomaterials, so that by processing more deeply the carbonaceous residues to advanced carbon nanomaterials would be synthesized and used to make one of a new process from waste. In this integrated production approach, the intrinsic energy of dairy effluent is recovered through multiple conversion pathways, enabling effective bioenergy generation. The remaining carbon-rich matrix is then transformed into high-value products, ensuring that both energy and material resources are fully utilized. By converting residual carbon into advanced, high-grade commodities, the process maximizes energy recovery while significantly reducing environmental impact (Chaudhary et al. 2023).

13.12.2 Development of Multifunctional CNM Composites for Next-Generation Applications

Multifunctional carbon nanomaterial composites derived from dairy effluents are a technological breakthrough that will translate into future-level applications in multiple industries. Although environmental remediation and biomedicine, as well as other industry, are highly applicable to the wide field of carbon nanocomposites (Veeman et al. 2021), dairy effluent-derived CNMs for the development of carbon nanomaterials are an exploration that has room for future work. High surface-to-volume ratio, excellent mechanical properties, adjustable electrical conductivity, and range of surface chemistry make them appropriate candidates for incorporation in complex composites with advanced characteristics (Talafha et al. 2024). Utilizing waste resources such as dairy effluent to fabricate the CNMs, the resulting composites therefore possess an inherent environmental value, consistent with tenets of circular economy (Ikram et al. 2023).

Future work will concentrate on developing these dairy effluent-derived CNMs as functionalities that can be synergistically combined in a composite structure, for example, the production of bionanocomposites based on the functionalization of carbon nanomaterials with bacterial cells is already significantly promising in nanobioremediation for industrial pollutants, underscoring the contribution that CNMs play for next-generation wastewater treatment (de Carvalho Lima et al. 2024). Think of a composition in which dairy effluent-derived graphene or carbon dots are incorporated into a polymer matrix which gives increased mechanical properties as well as a sensing mechanism for ambient control and continuous drug release for biomedical implants. In addition, incorporation with conductive polymers could lead to versatile electronics, biocompatible sensors, and even energy-harvesting devices, which may exploit the electronic characteristics of CNMs and the adaptiveness of polymers alike (Veeman et al. 2021). The problem to overcome is to control the synthetic parameters of dairy effluent-derived CNMs for accurate structural and chemical properties to promote their effective integration of the solid components into complex composite structures. Developments in this direction will require advanced material characterization, computational modeling, and manufacturing technologies to exploit the power of such sustainable, waste-derived materials in a truly multifunctional development in the next generation of high-yield technology (Varadharajan et al. 2024). Consider an example in which the composition includes dairy effluent-derived graphene or carbon dots in a polymer that provides improved mechanical properties and an ambient control and continuous drug delivery sensing feature in biomedical devices. Moreover, integration with conductive polymers can result in multifunctional electronics, sensors, and even energy-harvesting devices, which can take advantage of the electronic properties of CNMs and their adaptability, similar to polymers. However, there is an issue that has to be addressed, that is, how to manage the synthetic conditions of dairy effluent-derived CNMs in terms of structural and chemical properties in order to effectively integrate these solid materials into complex composite materials. Advances in this aspect will require advanced characterization, modeling, and manufacturing technologies that can take advantage of the full potential of these sustainable materials in a multifunctional development in the next generation of high-yield technology, as described in Varadharajan et al. (2024).

Declaration of Third-Party Material This manuscript does not contain any third-party material that requires permission for reproduction.

References

Adeola AO, Duarte MP, Naccache R (2023) Microwave-assisted synthesis of carbon-based nanomaterials from biobased resources for water treatment applications: emerging trends and prospects. Front Carbon 2:1220021

Ashekuzzaman S, Forrestal P, Richards K, Fenton O (2019) Dairy industry derived wastewater treatment sludge: generation, type and characterization of nutrients and metals for agricultural reuse. J Clean Prod 230:1266–1275

Azam MZ, Ashraf M, Aslam Z, Kamal MS, Aslam U (2024) Combustion and pyrolysis of dairy waste: a kinetic analysis and prediction of experimental data through artificial neural network (ANN). Therm Sci Eng Prog 53:102746

Azari P, Hasheminejad H, Zarean Mousaabadi K (2025) Synthesis and characterization of MIL-88 A (Fe)/C composite for treatment of dairy factory's wastewater by enhanced electro-Fenton method. Sci Rep 15(1):8503

Basem A, Abdeldayem M, Jasim DJ, Nabi H (2024) Exploring the efficiency of employing Fe3O4-MWCNT nanofluids in a heat sink equipped with circular micro pin-fins. Int J Thermofluids 24: 100928

Carawan RE, Jones V, Hansen A (1979) Wastewater characterization in a multiproduct dairy. J Dairy Sci 62(8):1243–1251

Casallas-Ojeda M, Torres-Guevara LE, Caicedo-Concha DM, Gómez MF (2021) Opportunities for waste to energy in the milk production industry: perspectives for the circular economy. Sustainability 13(22):12892

Casero-Díaz T, Castro-Barros C, Mauricio-Iglesias M, Carballa M (2025) From dairy waste to butyric acid: scaling the valorization process. J Environ Manag 380:124934

Chaudhary AK, Verma S, Chauhan R (2023) Design of a low-cost, environment friendly perovskite solar cell with synergic effect of graphene oxide-based HTL and CH3NH3GeI3 as ETL. Eng Res Express 5(3):035039

Chen C, Zhang J (2024) Enhancing leukemia treatment: the role of combined therapies based on amino acid starvation. Cancer 16(6):1171

Chen S, Xie Y, Ma K, Wei Z, Ran X, Fu X, Zhang C, Zhao C (2024) Electrospun nanofibrous membranes meet antibacterial nanomaterials: from preparation strategies to biomedical applications. Bioact Mater 42:478–518 de Carvalho Lima EN, Souza LHM, Aguiar EM, Octaviano ALM, Justo JF, Piqueira JRC (2024) Can bacteria and carbon-based nanomaterials revolutionize nanoremediation strategies for industrial effluents? Front Nanotechnol 6:1389107

de Sousa MS, Monteiro ALSA (2025) An overview of current application and trends of nanotechnological approaches in the dairy industry: from. In: Nanobiotechnology in agri-food sector. Springer Nature, p 349

Enríquez E, De la Rubia M, Del Campo A, Rubio-Marcos F, Fernandez J (2014) Characterization of carbon nanoparticles in thin-film nanocomposites by confocal Raman microscopy. J Phys Chem C 118(19):10488–10494

Fahad Halim A, Poinern GEJ, Fawcett D, Sharma R, Surendran S, Rajeshkannan R (2025) Biomass-derived carbon nanomaterials: synthesis and applications in textile wastewater treatment, sensors, energy storage, and conversion technologies. CleanMat 2(1):4–58

Fahoul N, Sayadi MH, Rezaei MR, Homaeigohar S (2022) Nickel ferrite nanoparticles catalyzed dark fermentation of dairy wastewater for biohydrogen production. Bioresour Technol Rep 19: 101153

Hidayah N, Putri V, Elma M, Syauqiah I, Amenia A, Putra D, Akbar H, Rahma A (2022) The functionalization of pyrolyzed palm empty fruit bunches-based membranes adsorbent by Fourier-transform infrared spectroscopy. In: IOP conference series: materials science and engineering. IOP Publishing

Ikram R, Mohamed Jan B, Nagy PB, Szabo T (2023) Recycling waste sources into nanocomposites of graphene materials: overview from an energy-focused perspective. Nanotechnol Rev 12(1): 20220512

Islam SU, Shalla AH, Khan SA (2022) Handbook of biomass valorization for industrial applications. Wiley

Jia M, Bai H, Liu N, Hao L, He P, Fan Z, Liu J, Niu R, Gong J, Tang T (2022) Upcycling waste polyethylene into carbon nanomaterial via a carbon-grown-on-carbon strategy. Macromol Rapid Commun 43(18):2100835

Kanakaraju D, anak Kutiang FD, Lim YC, Goh PS (2022) Recent progress of Ag/TiO_2 photocatalyst for wastewater treatment: doping, co-doping, and green materials functionalization. Appl Mater Today 27:101500

Kaur N (2021) Different treatment techniques of dairy wastewater. Groundw Sustain Dev 14: 100640

Khalaf N, Shi W, Fenton O, Kwapinski W, Leahy J (2023) Hydrothermal carbonization (HTC) of dairy waste: effect of temperature and initial acidity on the composition and quality of solid and liquid products. Open Res Eur 2:83

Kolev Slavov A (2017) General characteristics and treatment possibilities of dairy wastewater–a review. Food Technol Biotechnol 55(1):14–28

Kueasook R, Wang P, Chen H, He P, Zeng Z, Xu X, Li L (2025) Solvent-free one-step simple synthesis of N, O-doped microporous carbon using K_2CO_3 as an activation agent and their application to CO_2 capture: synergistic effect of pore structure and nitrogen–oxygen functional groups. Chem Eng Sci 311:121615

Kumar A, Kaladharan K, Tseng F-G (2021) Nanomaterials: surface functionalization, modification, and applications. In: Nanomaterials and their biomedical applications. Springer, pp 405–438

Kwapinska M, Horvat A, Liu Y, Leahy JJ (2020) Pilot scale pyrolysis of activated sludge waste from milk processing factory. Waste Biomass Valoriz 11(6):2887–2903

Lan K, Wang HS-H, Lee T, de Assis CA, Venditti RA, Zhu Y, Yao Y (2024) A modeling framework to identify environmentally greener and lower-cost pathways of nanomaterials. Green Chem 26(6):3466–3478

Lanjekar PR, Panwar NL (2024) A review on hydrogen production from biomass and commercialization assessment through technology readiness levels (TRLs). Bioenergy Res 17(2): 912–931

Lesch R, Visser ED, Seroka NS, Khotseng L (2025) Biomass-derived carbon-based nanomaterials: current research, trends, and challenges. J Renew Mater 13:1935–1977

Lu J, Veksha A, Poh YS, Loh WJA, Lisak G (2025) Catalytic synthesis of biogenic multi-walled carbon nanotubes and hydrogen from manure. Chem Eng J 503:158613
Luqman M, Al-Ansari T (2021) A novel solution towards zero waste in dairy farms: a thermodynamic study of an integrated polygeneration approach. Energy Convers Manag 230:113753
Ma Y, Yao Y, Deng Z, Zeng C, Liu Y, Ma J, Zhang Z (2024) Hydrothermal N-doping, magnetization and ball milling co-functionalized sludge biochar design and its selective adsorption of trace concentration sulfamethoxazole from waters. Chemosphere 363:142855
McKenzie KR Jr, Banek NA, Wagner MJ (2025) Sustainable conversion of biomass to multiwalled carbon nanotubes and carbon nanochains. Materials 18(5):1022
Mohapatra L, Cheon D, Yoo SH (2023) Carbon-based nanomaterials for catalytic wastewater treatment: a review. Molecules 28(4):1805
Mozafari M, Soroush M (2021) Surface functionalization of MXenes. Mater Adv 2(22):7277–7307
Oh DY, Kim D, Park KY (2024) A comprehensive comparative study on microwave-assisted pyrolysis products derived from raw and digested organic waste, with emphasis on sewage sludge, food waste, and livestock manure. Heliyon 10(8):e29618
Patel DK, Kim H-B, Dutta SD, Ganguly K, Lim K-T (2020) Carbon nanotubes-based nanomaterials and their agricultural and biotechnological applications. Materials 13(7):1679
Pires AF, Marnotes NG, Rubio OD, Garcia AC, Pereira CD (2021) Dairy by-products: a review on the valorization of whey and second cheese whey. Foods 10(5):1067
Qasim M, Clarkson AN, Hinkley SF (2023) Green synthesis of carbon nanoparticles (CNPs) from biomass for biomedical applications. Int J Mol Sci 24(2):1023
Ritambhara Z, Vijayaraghavalu S, Prasad HK, Kumar M (2018) Treatment and recycling of wastewater from dairy industry. In: Advances in biological treatment of industrial waste water and their recycling for a sustainable future. Springer, pp 91–115
Rocha JM, Guerra A (2020) On the valorization of lactose and its derivatives from cheese whey as a dairy industry by-product: an overview. Eur Food Res Technol 246(11):2161–2174
Sen P, Bhattacharya P, Mukherjee G, Ganguly J, Marik B, Thapliyal D, Verma S, Verros GD, Chauhan MS, Arya RK (2023) Advancements in doping strategies for enhanced photocatalysts and adsorbents in environmental remediation. Technologies 11(5):144
Shah SS, Qasem MAA, Berni R, Del Casino C, Cai G, Contal S, Ahmad I, Siddiqui KS, Gatti E, Predieri S (2021) Physico-chemical properties and toxicological effects on plant and algal models of carbon nanosheets from a nettle fibre clone. Sci Rep 11(1):6945
Shete B, Shinkar N (2017) Anaerobic digestion of dairy industry waste water-biogas evolution-a review. Int J Appl Environ Sci 12(6):1117–1130
Shivsharan VS, Kulkarni S, Wani M (2013) Physicochemical characterization of dairy effluents. Int J Life Sci Pharma Res 2(2):182–191
Sinha S, Srivastava A, Mehrotra T, Singh R (2018) A review on the dairy industry waste water characteristics, its impact on environment and treatment possibilities. In: Emerging issues in ecology and environmental science: case studies from India. Springer, pp 73–84
Sonawane AV, Murthy Z (2023) Dairy industry wastewater treatment by MOF and 2D nanomaterial engineered PVDF membranes based aerobic MBR: membrane fouling mitigation and stability study. Process Saf Environ Prot 171:680–693
Stanchev P, Vasilaki V, Egas D, Colon J, Ponsá S, Katsou E (2020) Multilevel environmental assessment of the anaerobic treatment of dairy processing effluents in the context of circular economy. J Clean Prod 261:121139
Talafha A, Abu-Dalo M, Albiss BA, Al-Nabulsi A (2024) Design and fabrication of nanocomposite filters for dairy industry wastewater treatment. Research Square
Udourioh GA, Solomon MM, Okolie JA (2025) A review of the valorization of dairy industry wastes through thermochemical, biological, and integrated processes for value-added products. Food Sci Anim Resour 45(2):375
Usmani Z, Sharma M, Gaffey J, Sharma M, Dewhurst RJ, Moreau B, Newbold J, Clark W, Thakur VK, Gupta VK (2022) Valorization of dairy waste and by-products through microbial bioprocesses. Bioresour Technol 346:126444

Varadharajan S, Vasanthan KS, Mathur V, Hariperumal N, Mazumder N (2024) Green synthesis and multifaceted applications: challenges and innovations in carbon dot nanocomposites. Discov Nano 19(1):205

Veeman D, Shree MV, Sureshkumar P, Jagadeesha T, Natrayan L, Ravichandran M, Paramasivam P (2021) Sustainable development of carbon nanocomposites: synthesis and classification for environmental remediation. J Nanomater 2021(1):5840645

Verma R, Maji PK, Sarkar S (2021) Comprehensive investigation of the mechanism for Cr (VI) removal from contaminated water using coconut husk as a biosorbent. J Clean Prod 314: 128117

Wang J, Han J-Q, Yan Y, Yu M-N, Feng Q-Y, Xie L-H (2025) Nanomechanical characterization via atomic force microscopy. ACS Appl Mater Interfaces 17(40):55689–55705

Zhao C, Kang J, Li Y, Wang Y, Tang X, Jiang Z (2023) Carbon-based stimuli-responsive nanomaterials: classification and application. Cyborg Bionic Syst 4:0022

Zhi Y, Xu D, Jiang G, Yang W, Chen Z, Duan P, Zhang J (2024) A review of hydrothermal carbonization of municipal sludge: process conditions, physicochemical properties, methods coupling, energy balances and life cycle analyses. Fuel Process Technol 254:107943

Chapter 14
Molecular Toxicity and Mechanisms Involved in Recycled Carbon Nanomaterial Synthesis and Bioapplications: In Vitro and In Vivo

Liza Handique, Saurav Chakraborty, Tuhina Das, Pranaba Nanda Bhattacharyya, Samim Dullah, Bhaskar Sarma, and Krishna Giri

14.1 Introduction

Carbon nanomaterials (CNMs) are nanoscale carbon-based allotropes or derivatives, typically possessing at least one dimension below 100 nm, whose physicochemical properties differ from bulk carbon due to quantum confinement, high surface area, and tunable surface chemistry (Diez-Pascual 2022). Their key features include exceptional electrical conductivity, mechanical strength, chemical stability, and ease of covalent or non-covalent functionalization. These attributes confer a high attraction for CNMs in applications such as electronics, energy storage and conversion, sensing, biomedicine, composites, and environmental remediation (Jovic et al. 2020; Speranza 2021). From a molecular toxicity perspective, CNMs are relevant in two principal ways: (a) as engineered carriers that improve targeted drug delivery

L. Handique
Department of Botany, Jagannath Barooah University, Jorhat, Assam, India

S. Chakraborty · T. Das
Department of Biotechnology, JIS University, Kolkata, West Bengal, India

P. N. Bhattacharyya
Department of Botany, N.N. Saikia College, Titabor, Jorhat, Assam, India

S. Dullah
Department of Zoology and DBT-IBT Hub, N.N. Saikia College, Titabor, Jorhat, Assam, India

B. Sarma
Department of Botany, Dhemaji College, Dhemaji, Assam, India

K. Giri (✉)
Centre of Excellence on Sustainable Land Management, Indian Council of Forestry Research and Education, Dehradun, Uttarakhand, India

Y. K. Mohanta et al. (eds.), *Bioresource Waste to Carbon Nanomaterial*, Nanotechnology in the Life Sciences, https://doi.org/10.1007/978-981-92-2405-0_14

and reduce systemic toxicity and (b) as active materials capable of adsorbing, neutralizing, or catalytically degrading toxic species such as reactive oxygen species (ROS) and environmental pollutants (Diez-Pascual 2022).

CNMs are mostly classified according to effective dimensional and structural organization. Zero-dimensional (0D) CNMs include fullerenes, carbon quantum dots (CQDs), nanodiamonds, and carbon nano-onions, where all dimensions are confined to the nanoscale. One-dimensional (1D) CNMs possess high aspect ratios, with carbon nanotubes (CNTs), carbon nanofibers, and nanothreads as representative examples. CNTs consist of rolled graphene sheets forming single-walled or multi-walled cylindrical structures with nanometer-scale diameters and micron-scale lengths. Two-dimensional (2D) CNMs exhibit nanoscale thickness but extended lateral dimensions; graphene, graphene oxide (GO), reduced graphene oxide (rGO), and graphene nanoribbons are the most prominent examples. Three-dimensional (3D) CNMs comprise interconnected porous or hierarchical architectures such as carbon aerogels, porous carbons, and aggregated nanodiamond networks (Jovic et al. 2020; Gárate-Vélez et al. 2025). An alternative classification is based on bonding and allotropy, including graphitic (sp^2-rich), diamond-like (sp^3-rich), and amorphous carbons. Variations in atomic arrangement strongly influence electronic behavior, surface reactivity, mechanical strength, and biological interactions, thereby dictating application potential (Gárate-Vélez et al. 2025).

14.1.1 Synthesis and Functionalization, of CNMs

Carbon nanomaterials (CNMs) are synthesized via material-specific routes, with chemical vapor deposition (CVD) dominating the production of high-quality carbon nanotubes and graphene, while exfoliation and oxidative methods are used for graphene derivatives. Fullerenes are typically generated by arc discharge or laser ablation. Surface functionalization—through covalent or non-covalent strategies—is generally required to enhance dispersibility, biocompatibility, and interfacial compatibility. However, scalable control over chirality, defect density, and layer number remains a key bottleneck for industrial implementation (Diez-Pascual 2022). The synthesis pathways of CNMs have been depicted in Fig 14.1.

14.1.2 Key Properties and Why They Matter

The exceptional properties of carbon nanomaterials (CNMs) stem from carbon's versatile bonding and delocalized π-electron systems. Strong sp^2 C–C bonding confers extraordinary mechanical strength to graphene and carbon nanotubes (CNTs), while the same electronic structure underlies their high electrical and thermal conductivity. Large specific surface areas, high aspect ratios, nanoscale confinement, and chemical tunability enable tailored interactions with biomolecules,

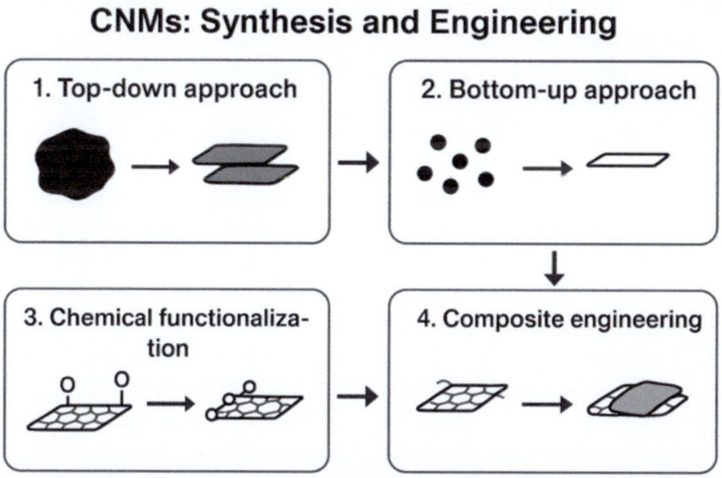

Fig. 14.1 Synthesis pathway of recycled carbon materials. (Created by authors)

polymers, and inorganic phases, supporting multifunctional platforms for catalysis, sensing, adsorption, and drug delivery (Jovic et al. 2020; Diez-Pascual 2022). These combined properties support the wide-ranging applications of carbon nanomaterials (CNMs) across electronics and sensing technologies, energy storage and conversion systems (including batteries, supercapacitors, and solar cells), catalysis, biomedical imaging and therapy, structural composites, and environmental remediation. In energy-related applications, CNMs function as highly efficient electrodes and catalyst supports, for instance, carbon nanohorns have been shown to improve the performance of platinum-based electrochemical catalysts (Baig et al. 2021). In the biomedical field, their nanoscale size and tunable surface functionalization facilitate targeted drug and gene delivery, imaging applications, and therapeutic interventions (Sridharan et al. 2022). Additionally, graphene and carbon nanotubes (CNTs) enhance the mechanical strength of lightweight polymer and metal composites while maintaining low density. Their large surface area, adsorption capacity, redox properties, and adaptable surface chemistry also make them highly effective in environmental applications, including pollutant sensing, water treatment, and selective biological interactions (Speranza 2021).

14.1.3 Recycled Carbon Nanomaterials and Sustainability

Recycling of carbon-based waste into CNMs is increasingly important for sustainable nanotechnology and circular economy strategies. Biomass wastes, plastic residues, composite materials, and other carbon-rich solid wastes can be converted into functional nanocarbons via pyrolysis, activation, hydrothermal treatment, electrochemical exfoliation, or emerging methods such as flash Joule heating. Biomass-derived carbons typically yield porous activated carbons or nanostructured carbons, while plastic waste can be upcycled into graphene-like materials, CNTs, and hybrid nanocarbons (Bedia et al. 2018; Le 2025).

Recycled carbon nanomaterials (CNMs) often retain high specific surface area, tunable porosity, and favorable electrochemical conductivity, which makes them highly effective for applications such as adsorption, heterogeneous catalysis, environmental remediation, and energy storage systems including batteries and supercapacitors. These properties arise from their hierarchical pore structures, large active surface sites, and modifiable surface functional groups that enhance adsorption capacity, catalytic efficiency, and charge transfer performance. Recent studies have demonstrated that waste-derived or recycled carbon materials can exhibit comparable or even superior performance to conventionally synthesized nanocarbons in environmental and energy applications. However, several limitations still constrain large-scale adoption, including feedstock heterogeneity, impurity removal, inconsistent structural properties, and challenges in reproducible synthesis and industrial scale-up. Continuous advancements in catalyst-assisted conversion technologies, activation strategies, heteroatom doping, and process optimization are improving the structural control, functionality, and economic feasibility of recycled CNMs. Consequently, recycling-based synthesis strategies contribute not only to waste valorization and circular economy approaches but also to expanding access to high-value carbon nanomaterials for advanced technological applications (Le 2025; Jiang et al. 2024; Chandran et al. 2023; Mohapatra et al. 2023; Nezafat et al. 2024; Zhang et al. 2025). Overall, recycling strategies not only reduce environmental burden but also expand access to high-value carbon nanomaterials for technological applications.

14.1.4 Recycled and Derived Carbon Nanomaterials

14.1.4.1 Feedstocks and Strategies

Recycled and derived carbon nanomaterials (CNMs) are broadly classified by feedstock origin, which governs carbon yield, porosity, surface chemistry, and functional performance. Key feedstocks include biomass wastes, plastic and polymer wastes, metal-organic framework (MOF) precursors, and other solid or composite wastes. Across all routes, precise process control and activation are essential to achieve high surface area, graphitization, and application-relevant properties (Bedia et al. 2018; Le 2025).

14.1.4.2 Biomass-Derived Carbon Nanomaterials

Biomass and agroforestry wastes (e.g., agricultural residues, shells, husks, and lignin-rich by-products) serve as renewable, low-cost precursors for porous carbons. Processing typically involves inert atmosphere pyrolysis/carbonization to form biochar, followed by physical or chemical activation (commonly KOH or CO_2) to generate high-surface-area activated carbons. These materials can be tailored into 0D–3D architectures, including carbon spheres, nanofibers, nanosheets, and hierarchical porous networks. Activation and heteroatom doping enhance porosity, electrical conductivity, and electrochemical performance, enabling applications in adsorption, catalysis, and energy storage; however, feedstock heterogeneity and scale-up remain key challenges (Bedia et al. 2018; Yan et al. 2025).

14.1.4.3 Plastic Waste-Derived Carbon Nanomaterials

Plastic and polymer wastes constitute abundant carbon-rich feedstocks that align with circular economy strategies when upcycled into CNMs. Pyrolysis and catalytic carbonization convert waste plastics into activated carbons, graphene-like sheets, carbon nanotubes (CNTs), and hybrid graphitic nanostructures. Catalyst- and template-assisted approaches improve graphitization and morphology control, while emerging methods such as flash Joule heating enable rapid, energy-efficient conversion. Across techniques, polymer chains thermally decompose into aromatic intermediates that reorganize into sp^2-rich carbon frameworks. Key challenges include contaminant removal, reproducibility, morphology control, and scalability for producing high-quality graphene and CNTs (Le 2025; Jiang et al. 2024).

14.1.4.4 MOF-Derived and Other Solid Waste-Derived Carbons

Metal-organic frameworks serve as sacrificial precursors for porous carbons and carbon-metal composites, where carbonization transforms organic linkers into conductive porous carbon and metal nodes into catalytic phases or removable species. This strategy enables valorization of spent or waste MOFs into advanced functional materials. Additional unconventional feedstocks include industrial carbonaceous wastes, paper, textile residues, and composite wastes such as CFRPs, which can be converted into nanocarbons via thermal treatment or electrochemical exfoliation. Despite their versatility, these routes face challenges related to mixed material contamination, purification, and quality control compared with pristine synthesis pathways (Shahzadi et al. 2024; Jiang et al. 2024).

Feedstock Influence and Scale-Up Considerations

Feedstock selection critically determines porosity development, surface functional groups, defect density, and overall performance of recycled CNMs. Activation is indispensable, as nonactivated chars generally exhibit poor adsorption and

Table 14.1 Methods of recycling CNMs

Feedstock type	Key conversion method(s)	Example product(s)	Key challenge(s)	References
Plastic waste	Pyrolysis + catalyst/template, flash Joule heating, arc discharge	Graphene nanosheets, CNTs, 1D/2D nanocarbons	Graphene quality, scale-up, energy cost	Le (2025); Jiang et al. (2024)
Biomass waste	Carbonization + activation (chemical/physical), hydrothermal, electrospinning precursor	Porous carbons (0D-3D), CNFs, nanosheets	Feedstock variation, control of structure, graphitization	Bedia et al. (2018); Yan et al. (2025)
Solid/ composite waste	Electrochemical exfoliation, thermal conversion	Graphene oxide, recycled nanocarbons	Mixed composition contamination, purity, separation	Shahzadi et al. (2024); Jiang et al. (2024)

electrochemical behavior. Although laboratory-scale synthesis is well established, industrial deployment remains limited by feedstock heterogeneity, process optimization, cost-effectiveness, and quality consistency. Advances in catalyst design, activation strategies, and scalable conversion technologies are therefore essential for the sustainable production of high-value recycled CNMs (Bedia et al. 2018; Le 2025). The different methods of recycling CNMs are listed in Table 14.1.

14.2 Recycling Strategies and Effects on Properties

Bioresource wastes are increasingly converted into carbon nanomaterials (CNMs) through thermal, chemical, mechanical, physical/top-down, and biological/green approaches that control porosity, surface chemistry, defect density, and particle size. Recycling routes mainly include pyrolysis and activation, hydrothermal/solvothermal, microwave carbonization, chemical vapor deposition (CVD), mechanical/top-down fragmentation, and green/biogenic synthesis, each producing distinct carbon products with tailored physicochemical effects. Reviews highlight that integrating these methods with waste streams enables sustainable CNM production with tunable properties for energy, environmental, and sensing applications (Cheng et al. 2024; Kang et al. 2020; Mehta et al. 2025). Table 14.2 shows the recycling strategies along with the relevant details.

14.2.1 Sources of Waste CNMs

Recycling wastes into CNMs supports a circular economy by turning abundant waste streams into high-value carbon nanostructures such as carbon nanotubes (CNTs), graphene-like materials, carbon nanofibers, and carbon quantum dots.

Table 14.2 Recycling strategies, approach, products, and the material effects of CNMs

Sl. no.	Strategy	Typical feedstock	Typical processing approach	Common product types	Principal material effects	References
1.	Pyrolysis and activation	Crop residues, husks, shells	Slow/fast pyrolysis	Biochar, activated carbon, nano-biochar	Increase porosity and stable aromatic carbon	Tomczyk et al. (2020)
2.	Hydrothermal/ solvothermal	Fruit/vegetable wastes, lignocellulosic slurries	Autogenous pressure, aqueous heating	Carbon quantum dots (CDQs), hydrothermal dots	Surface functionalization and small particle size with oxygen/N groups	Ferjani et al. (2023)
3.	Microwave-assisted carbonization	Shells, peels, agro-residues	Rapid dielectric heating	Nanostructured carbons, CQDs	Rapid carbonization, often smaller or more uniform particles	Duran-Jiménez et al. (2024)
4.	Chemical vapor deposition (CDV)	Bio-derived gases or precursors	High-temperature gas phase growth	High-quality graphene	Produces low-defect, high-conductivity graphene	Manawi et al. (2018)
5.	Top-down mechanical/ball milling	Cellulosic film, waste graphite	Exfoliation, milling, sonication	Graphene-like flakes, CNTs	Reduces lateral size, increases edge defects	Asif and Saha (2023)
6.	Green/biogenic synthesis	Plant extract, agriculture extract	Stabilizer in low-toxicity routes	CDQs, carbon dots, and supported metal NPs	Surface capping, heteroatom doping, benign synthesis	Chahal et al. (2021)

Table 14.3 Sources and their derivative CNMs

Waste source	Types of waste	Typical processing approach	Produced CNMs	References
Agricultural waste	Sugarcane bagasse, corn stalks, crop residues	Pyrolysis to produce carbon-rich gaseous effluents	CNTs, carbon nanostructures	Mohapatra et al. (2023); Nezafat et al. (2024)
Plastic waste	Polyethylene (PE), polypropylene (PP), postconsumer plastic	Pyrolysis, chemical vapor deposition (CVD)	CNTs, graphene-like carbon	Mohapatra et al. (2023); Nezafat et al. (2024)
Industrial waste	Manufacturing by-products, chemical residues	High-temperature conversion, catalytic processes	CNTs, carbon nanofibers	Mohapatra et al. (2023); Nezafat et al. (2024)
Construction waste	Building offcuts, residual composite materials	Thermal conversion, catalytic treatment	Amorphous carbon, carbon microstructures	Mohapatra et al. (2023); Nezafat et al. (2024)
Environmental impact consideration	–	Assessment of bioavailability, ecotoxicity	–	Mohapatra et al. (2023); Nezafat et al. (2024)

Feedstocks include agricultural residues (e.g., sugarcane bagasse, crop stalks), plastic wastes (e.g., PE, PP), industrial by-products, and construction and composite wastes. Thermal decomposition and catalytic processes convert these inputs into CNMs suitable for energy, sensing, and environmental applications while also reducing landfill burden (Mehta et al. 2025). Table 14.3 depicts the waste sources from which different CNMs are derived.

While the conversion of waste into CNMs presents numerous benefits, it is essential to consider the potential environmental risks associated with nanomaterials specially concerning their bioavailability and toxicity in the ecosystems (Suman and Pei 2022).

14.2.2 Recycling Techniques (Thermal, Chemical, Mechanical, Plasma/Laser)

Recycling techniques of waste CNMs play a crucial role in waste management system, with various methods tailored to different materials and desired outcomes. The important techniques include thermal, chemical, mechanical, and plasma recycling, each with unique processes, applications and examples. Below is an overview of these techniques (Cheng et al. 2024; Mehta et al. 2025). Fig. 14.2 and Table 14.4 depicts these recycling techniques in detail:

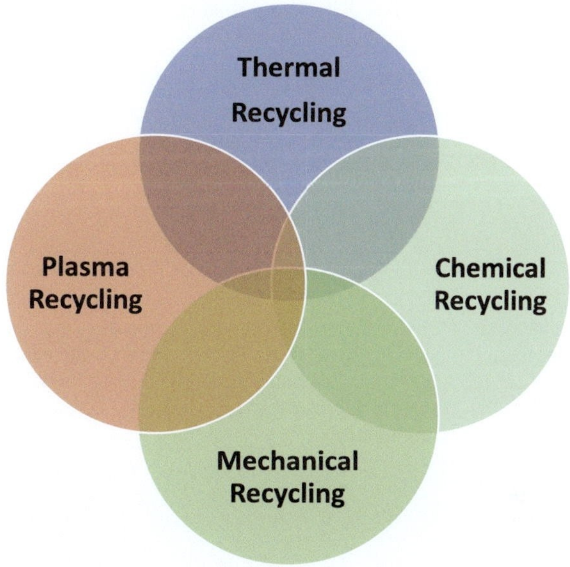

Fig. 14.2 Overview of the mentioned recycling techniques. (Created by authors)

14.2.3 Impact of Recycling on Physicochemical Properties

Recycling influences the physicochemical properties of polymers and derived carbon materials, for example, thermal and mechanical recycling of plastics such as polyethylene (PE) and polypropylene (PP) leads to chain scission, reduced molecular weight, changes in crystallinity, and alterations in mechanical performance like tensile strength and modulus. These effects arise from thermal and oxidative degradation during processing and are key considerations for material quality and reuse in high-value applications (Zhang et al. 2024).

14.3 Molecular Toxicity and Mechanisms

Molecular toxicity refers to adverse effects arising from interactions between exogenous or endogenous agents and biological macromolecules, including DNA, proteins, lipids, and cellular organelles. Toxic responses are initiated by a molecular initiating event (MIE)—such as covalent binding, oxidative damage, or receptor-mediated activation—that triggers downstream biochemical cascades leading to cytotoxicity, genotoxicity, inflammation, or organ dysfunction. Environmental and endogenous toxicants, including heavy metals (e.g., Hg, Cd) and polycyclic aromatic hydrocarbons (PAHs), act primarily through oxidative stress, epigenetic alterations, and protein dysfunction; their effects are often modulated by metabolic activation as well as genetic and nutritional factors (Klaassen 2018/2024;

Table 14.4 Overview of recycling strategies

Recycling technique	Process	Applications	Advantages/ challenges	References
Thermal recycling	Uses high temperatures to decompose materials, converting them into energy or reusable carbon-rich products	Organic waste, thermoset composites, energy recovery	Advantage: reduces waste volume, enables energy recovery Challenge: may generate emissions if not properly controlled	Kusano and Kusano (2024)
Chemical recycling	Breaks down polymers and composites via chemical reactions, recovering fibers and components	Composite materials, polymer waste, recovering valuable fibers	Advantage: can restore materials and recover high-value components Limitation: mechanical properties may degrade after processing	Mehta et al. (2025)
Mechanical recycling	Grinding, shredding, or crushing materials to create recyclates or fillers	Plastics, composites, filler material for new products	Advantage: simple and low cost Challenge: output quality varies; recycled fibers often lower grade	Mehta et al. (2025)
Plasma recycling	Uses ionized gases (plasma) to break down waste at lower temperatures through plasma-assisted reactions	Electronic waste (e-waste), metal recovery, potential wide use in waste processing	Advantage: effective for complex waste, lower temperature processes Future need: improvements required for efficiency and scalability	Kusano and Kusano (2024)

Ajsuvakova et al. 2020; Adegbola and Adetutu 2024; Haidar et al. 2023; Qu and Zheng 2024).

14.3.1 Mechanisms of Molecular Toxicity

Oxidative stress is a core mechanism of molecular toxicity, whereby toxicants including metal nanoparticles, pesticides, and environmental pollutants induce excessive reactive oxygen species (ROS; O_2^-, ·OH, H_2O_2), overwhelming redox homeostasis and causing lipid peroxidation, protein carbonylation, and oxidative DNA damage, leading to mitochondrial dysfunction, ATP depletion, and apoptotic or necrotic cell death (Liu et al. 2025a, b; Sies et al. 2024; Forman et al. 2024). Genotoxicity frequently co-occurs, as agents such as benzo[a]pyrene, aflatoxin B1, and engineered nanomaterials induce DNA adducts, cross-links, and chromosomal aberrations that disrupt replication and transcription, increasing mutagenic and carcinogenic risk; DNA adducts therefore serve as key molecular epidemiological biomarkers (Phillips and Arlt 2024; Wild and Turner 2024).

Additional pathways include covalent protein modification and enzyme inhibition by electrophilic toxicants and heavy metals (Hg^{2+}, Cd^{2+}), membrane disruption by amphiphilic chemicals and nanoparticles that alters fluidity and ion transport, and mitochondrial toxicity via electron transport chain inhibition and permeability transition, exemplified by rotenone-mediated complex I inhibition (Liebler 2023; Klaassen 2018/2024; Nel et al. 2024; Park and Park 2024; Fadeel et al. 2024; Fromenty and Roden 2024; Wang et al. 2024; Green and Llambi 2023; Galluzzi et al. 2024). In parallel, inflammatory and immune-mediated toxicity arises through NF-κB, toll-like receptor, and NLRP3 inflammasome activation, driving cytokine release, chronic inflammation, fibrosis, and tissue injury (Lawrence 2024; Kelley et al. 2024).

A detailed mechanistic understanding of molecular toxicity forms the foundation for drug safety evaluation, safe-by-design development of nanomaterials, and environmental risk assessment. It guides the clinical management of DNA-damaging chemotherapeutic agents such as cisplatin and doxorubicin, supports the optimization of nanomaterial surface properties and biodistribution profiles, and enables the use of molecular biomarkers—including 8-hydroxy-2′-deoxyguanosine (8-OHdG), metallothioneins, and lipid peroxidation products—for monitoring oxidative and genotoxic stress (Galluzzi et al. 2024; Minotti et al. 2024; Fadeel et al. 2024; Abegunde 2025; Valavanidis et al. 2023; Liu et al. 2025a, b).

14.3.2 Factors Affecting Molecular Toxicity

Molecular toxicity is influenced by both the intrinsic properties of toxicants and host-related biological factors that govern interactions with cellular components such as DNA, proteins, lipids, and organelles. These factors collectively determine the intensity, mechanism, and biological outcome of toxic exposure (Kowalska et al. 2024).

14.3.2.1 Chemical Structure and Reactivity

The intrinsic chemical structure of a molecule, including its functional groups, electrophilicity, hydrophobicity, and surface charge, determines its ability to cross biological membranes, bind macromolecules, and form covalent adducts.

Electrophilic chemicals, aromatic compounds, aldehydes, and halogenated substances generally exhibit higher molecular reactivity and toxicity (Liebler 2023; Zanger and Schwab 2024).

14.3.2.2 Dose and Duration of Exposure

Toxicity is fundamentally dose-dependent. Acute high-dose exposures may overwhelm cellular defense systems, whereas chronic low-dose exposures can result in cumulative oxidative damage, DNA lesions, endocrine disruption, and long-term disease risk (Klaassen 2018/2024).

14.3.2.3 Route of Exposure

The route of exposure—including inhalation, ingestion, dermal absorption, or injection significantly—influences toxicokinetics and target organ toxicity. Inhaled nanoparticles primarily affect respiratory tissues, whereas ingested chemicals undergo hepatic first-pass metabolism, which can either detoxify or bioactivate toxicants (Fadeel et al. 2024).

14.3.2.4 ADME (Absorption, Distribution, Metabolism, and Excretion)

Pharmacokinetic mechanisms determine the levels of toxicants that ultimately reach and accumulate in target tissues. Among these processes, metabolism—especially that mediated by cytochrome P450 enzymes—is pivotal, as it can either promote detoxification and clearance or produce reactive intermediates that increase toxic potential (Zanger and Schwab 2024; Harmut, Jaeschke and Ramachandran 2024).

14.3.2.5 Biotransformation and Reactive Metabolites

Some compounds exhibit toxicity only after metabolic activation, a process known as bioactivation. Reactive metabolites such as epoxides, quinones, and ROS can covalently bind to DNA and proteins, triggering mutagenesis and cytotoxicity. Acetaminophen-induced hepatotoxicity via its reactive metabolite N-acetyl-p-benzoquinone imine (NAPQI) represents a classic example (Jaeschke and Ramachandran 2024).

14.3.2.6 Cellular Antioxidant Capacity

Cells possess antioxidant defense systems, including glutathione (GSH), superoxide dismutase (SOD), and catalase, to neutralize ROS. Depletion or inhibition of these systems by toxicants results in uncontrolled oxidative stress, leading to lipid peroxidation, mitochondrial dysfunction, and DNA damage (Forman et al. 2024; Sies et al. 2024).

14.3.2.7 Genetic Factors and Individual Susceptibility

Genetic polymorphisms in drug-metabolizing enzymes, transporters, and DNA repair pathways significantly influence individual susceptibility to toxicants. Variations in mitochondrial DNA further modulate sensitivity to oxidative stress. Age also plays a critical role, as infants possess immature metabolic systems, while elderly individuals exhibit reduced detoxification and antioxidant capacity (Galluzzi et al. 2024; Barker 2024).

14.3.2.8 Nutritional Status

Nutritional deficiencies in antioxidants such as vitamins C and E, selenium, or amino acids required for glutathione synthesis compromise cellular defense mechanisms. Conversely, high-fat diets can enhance the absorption of lipophilic toxicants and exacerbate oxidative stress (Ursini et al. 2024).

14.3.2.9 Chemical Interactions and Co-exposure

Concurrent exposure to multiple chemicals can produce synergistic, additive, or antagonistic effects. Alcohol consumption, for example, induces CYP2E1 activity and increases the toxicity of several drugs by promoting reactive metabolite formation (Cederbaum 2023; Silva et al. 2024).

14.3.2.10 Environmental Conditions

Environmental factors such as temperature, pH, humidity, and ultraviolet radiation influence chemical stability and biological reactivity. Heat stress enhances metabolic ROS production, thereby amplifying toxic effects (Sies et al. 2024).

14.3.2.11 Tissue-Specific Susceptibility

Certain tissues exhibit heightened vulnerability due to high metabolic activity (liver), direct exposure (lungs), or limited regenerative capacity (brain). Organ-specific expression of metabolic enzymes and transporters explains selective toxicity, such as paracetamol-induced hepatotoxicity and neurotoxicity of environmental pollutants (Lucchini et al. 2024).

14.3.2.12 Immune Status and Inflammation

Preexisting inflammation sensitizes tissues to toxicants by increasing cytokine levels and ROS. Toxins may further amplify immune activation, leading to chronic inflammation or hypersensitivity responses (Gao et al. 2024).

14.3.3 Cellular Uptake, Biodistribution, and ROS Generation

Cellular uptake critically influences nanoparticle and toxicant biodistribution, organelle targeting, and downstream molecular toxicity. Uptake occurs primarily via endocytosis (clathrin-mediated, caveolae-mediated, macropinocytosis, scavenger receptor pathways), with additional contributions from passive diffusion for very small or hydrophobic species. The dominant route depends on particle size, surface charge, and cell type, and smaller, highly reactive particles generally show greater internalization and toxicity. Uptake mechanisms strongly influence subcellular fate—endocytic pathways often target lysosomes and mitochondria, enhancing reactive oxygen species (ROS) production, lysosomal membrane destabilization, and apoptosis pathways, whereas passive diffusion may lead directly to cytoplasmic exposure and DNA interactions (Khan and Jia 2023). Overall, increased uptake correlates with elevated ROS and oxidative stress, which both mediate and amplify nanomaterial toxicity across cell types.

Understanding how toxic molecules and nanoparticles enter cells is critical for predicting exposure levels and downstream toxic effects. Cellular uptake mechanisms determine intracellular dose, subcellular localization, and activation of specific toxicity pathways, including organelle damage and cell death. Endocytosis represents the primary uptake route for most toxicants and nanoparticles and includes clathrin-mediated, caveolae-mediated, macropinocytosis, and scavenger receptor-mediated pathways, with the preferred route governed by particle size, surface chemistry, and cell type (Liu and Tang 2020; Choi et al. 2017). Passive diffusion occurs mainly for small, hydrophobic molecules and some nanoparticles, particularly when active transport is limited or at low temperatures (Ha et al. 2021). Protein-facilitated uptake, such as albumin-mediated transport, further modulates intracellular accumulation and toxicity for certain compounds.

The mode of cellular entry critically influences subcellular distribution and organelle targeting, thereby affecting mitochondrial function, lysosomal stability, oxidative stress, and apoptosis (Liu and Tang 2020; Ha et al. 2021; Quevedo et al. 2021; Docampo et al. 2025). Receptor-mediated internalization, including scavenger and heparan sulfate receptor pathways, is often linked to enhanced toxicity due to efficient delivery to sensitive intracellular compartments (Męczyńska-Wielgosz et al. 2025; Zhang et al. 2020; Dutta et al. 2019), whereas passive uptake may bypass endosomal processing, resulting in direct cytoplasmic exposure and DNA damage. Uptake and toxicity are highly cell-type-dependent, with phagocytes predominantly using phagocytosis and non-phagocytic cells relying on endocytic or receptor-mediated pathways (Męczyńska-Wielgosz et al. 2025; Zhang et al. 2020; Cui et al. 2017). Additionally, particle size, surface charge, and surface modifications strongly regulate uptake efficiency and cytotoxicity, with smaller and more reactive particles generally exhibiting higher toxic potential (Choi et al. 2017; Quevedo et al. 2021; Męczyńska-Wielgosz et al. 2025).

14.3.4 DNA Damage, Apoptosis, Epigenetic Changes, and Immune Responses

Excessive reactive oxygen species (ROS) generated during nanoparticle internalization induce oxidative damage to DNA, proteins, and lipids, resulting in genotoxic lesions such as strand breaks, base modifications, and DNA cross-links. Persistent DNA damage can overwhelm repair systems, activating checkpoint arrest and apoptosis or promoting mutation accumulation (Shi and Dansen 2020; Valavanidis et al. 2023). Nanotoxic exposure is also closely linked to epigenetic dysregulation, including alterations in DNA methylation and histone modifications that sustain aberrant gene expression even after exposure ceases (Li and Ji 2021). In parallel, damaged macromolecules and ROS function as danger signals that activate innate immune sensors, including inflammasomes and the cGAS-STING pathway, triggering NF-κB- and interferon-dependent inflammatory responses that contribute to chronic toxicity (Khan and Jia 2023; Andrade 2022).

Molecular toxicity therefore impacts cellular genetic integrity through an interconnected triad of genotoxicity, epigenetic reprogramming, and immune activation. Reactive metabolic by-products, such as aldehydes and ROS, damage DNA and proteins, eliciting cellular stress responses that drive carcinogenesis and other chronic diseases (Morellato et al. 2020; Moretton and Loizou 2020; Lee et al. 2020). DNA repair pathways, including the Fanconi anemia pathway, play a critical protective role in maintaining genomic stability under toxic stress (Morellato et al. 2020; Moretton and Loizou 2020). However, many toxicants induce persistent epigenetic alterations—encompassing DNA methylation changes, histone modifications, and noncoding RNA dysregulation—that promote long-term gene expression changes, carcinogenesis, and chronic disease susceptibility (Qu and Zheng 2024; Parfett and Desaulniers 2017).

Mitochondrial dysfunction is another hallmark of molecular toxicity, leading to impaired energy metabolism, excessive ROS generation, and activation of intrinsic apoptotic pathways. Concurrently, endoplasmic reticulum stress and lysosomal destabilization further contribute to apoptosis and autophagy, amplifying cellular injury (Koyama et al. 2024). Genotoxic effects arise when toxins such as polycyclic aromatic hydrocarbons, heavy metals, ionizing radiation, alkylating agents, or ROS induce DNA strand breaks, adducts, cross-links, and replication stress. Failure of the DNA repair-pathways, including nucleotide excision repair, base excision repair, or homologous recombination, leads to mutation accumulation, chromosomal instability, and malignant transformation, for example, benzo[a]pyrene metabolites form bulky DNA adducts, while ROS generate 8-oxoguanine lesions that promote GC → TA transversions (Galluzzi et al. 2024; Barker 2024).

Environmental toxins promote disease not only through DNA damage but also by inducing persistent epigenetic changes that alter gene regulation. Activation of innate immune pathways sustains inflammation and oxidative stress, reinforcing these epigenetic disruptions and increasing the risk of genomic instability, cell death, or cancer (Panahi et al. 2016; Liu et al. 2025a, b; Kumar et al. 2023).

Environmental toxicants can modify DNA methylation, histone modifications, and noncoding RNA expression, thereby altering gene regulation and promoting carcinogenesis.

Moreover, pollutants often activate innate immune pathways and inflammatory signaling cascades that generate reactive oxygen species, leading to oxidative stress and long-term epigenetic reprogramming.

Sustained inflammation and ROS production further damage DNA and impair repair mechanisms, contributing to genomic instability and increased risk of cancer and other chronic diseases.

14.3.5 Key Signaling Pathways

Nanoparticle-induced stress engages several interconnected signaling networks. NF-κB and MAPK pathways integrate oxidative and inflammatory signals, modulating expression of cytokines and stress response genes, while the tumor suppressor p53 regulates DNA damage responses, cell cycle arrest, and apoptosis (Khan and Jia 2023; Shi and Dansen 2020). ROS often serve as both damage mediators and signal transducers, activating NF-κB through redox-sensitive modifications and associating with inflammatory gene expression (Nigam 2025). Persistent NF-κB activity fosters pro-inflammatory and survival gene expression, whereas p53 activation in response to oxidative DNA damage promotes repair or apoptotic outcomes depending on damage severity (Khan and Jia 2023; Shi and Dansen 2020). Together, these pathways form a network where oxidative stress, genotoxic stress, and immune signaling converge, determining cell fate and long-term consequences of exposure to toxic substances and nanomaterials.

14.4 Toxicity Assessment Approaches

Assessment of molecular toxicity relies on complementary in vitro and in vivo approaches. In vitro assays offer high-throughput, mechanistic insight and ethical advantages, whereas in vivo studies capture systemic, organ-level responses.

Integration of both is essential for accurate hazard identification, risk assessment, and regulatory decision-making (OECD 2023; Hartung and Leist 2024).

In vitro molecular toxicity testing is commonly organized into core classes: cytotoxicity/viability, genotoxicity/DNA damage, oxidative stress/redox imbalance, apoptosis and regulated cell death, mechanistic biomarker and high-content approaches, and advanced 3D or multicellular models. Regulatory genotoxicity assessment typically employs a core battery combining a bacterial mutagenicity test (Ames) with at least one mammalian cell-based genotoxicity assay, such as the micronucleus or comet assay (OECD 2023).

14.4.1 In Vitro Studies

In vitro methodologies constitute the foundation of contemporary molecular toxicology, enabling mechanistic insight while reducing reliance on animal experimentation. These assays evaluate biochemical, cellular, and genomic perturbations triggered by xenobiotics, thereby offering high-resolution readouts that reflect toxicity pathways such as oxidative stress, mitochondrial dysfunction, DNA damage, membrane disruption, and apoptosis. The following section comparatively evaluates widely used in vitro assays, highlighting their scientific basis, methodological workflow, analytical capacity, advantages, and limitations.

14.4.1.1 Cytotoxicity and Cell Viability Assays

Cytotoxicity assays assess cell survival through measurements of metabolic activity, membrane integrity, or intracellular ATP. Tetrazolium-based assays (MTT, MTS, WST) quantify mitochondrial dehydrogenase activity, while LDH release assays detect membrane damage. Dye exclusion methods (e.g., trypan blue) and ATP-based luminescence provide complementary viability endpoints (Asati et al. 2023).

These assays are simple, inexpensive, and widely standardized (e.g., ISO 10993-5), making them suitable for high-throughput screening. However, metabolic readouts may not directly reflect cell number, and nanomaterials or redox-active compounds may interfere with assay chemistry, necessitating orthogonal validation using multiple endpoints (Drasler et al. 2024).

14.4.1.2 Genotoxicity and DNA Damage Assays

The micronucleus (MN) assay detects both clastogenic and aneugenic events by scoring extranuclear chromatin bodies formed during defective mitosis. The use of cytochalasin B ensures analysis of dividing cells, and additional staining (e.g., CREST or FISH) can distinguish chromosome loss from breakage (OECD 2023).

Despite strong regulatory acceptance, cytotoxicity and physicochemical artefacts may confound interpretation.

The γH2AX assay detects phosphorylation of histone H2AX at DNA double-strand break sites and serves as a sensitive early marker of DNA damage response activation. While highly amenable to automation and high-throughput formats, γH2AX does not always indicate persistent damage and is not yet fully standardized for regulatory use (Kopp et al. 2023).

The comet assay (single-cell gel electrophoresis) quantifies DNA strand breaks at single-cell resolution and is among the most sensitive genotoxicity assays. High-throughput variants such as CometChip improve reproducibility and scalability, including applications in 3D cultures. However, detected damage may be transient and subject to DNA repair, requiring careful time-course interpretation (OECD 2023; Ge et al. 2024).

14.4.1.3 Oxidative Stress and Redox Assays

Oxidative stress assays quantify intracellular ROS/RNS levels or downstream oxidative damage. Fluorescent probes such as DCFH-DA and DHE measure general ROS and superoxide, respectively, while enzymatic assays assess antioxidant defenses (GSH/GSSG, SOD, catalase) and lipid peroxidation products (MDA/TBARS) (Murphy et al. 2023).

These assays provide key mechanistic insight but suffer from probe specificity issues, transient ROS lifetimes, and susceptibility to artefacts, making stringent controls and multiparametric approaches essential (Murphy et al. 2023).

14.4.1.4 Apoptosis and Regulated Cell Death Assays

Apoptosis is defined by hallmark features including phosphatidylserine externalization, caspase activation, loss of mitochondrial membrane potential (ΔΨm), and DNA fragmentation. Techniques such as Annexin V/PI staining enable differentiation between early apoptosis and late apoptosis or necrosis, while caspase activity assays and mitochondrial potential-sensitive dyes (e.g., JC-1, TMRE, TMRM) provide deeper mechanistic insight (Galluzzi et al. 2024). Given that apoptosis is a temporally dynamic process, careful selection of exposure durations and appropriate controls is essential. Employing multiparametric assay combinations further enhances the ability to distinguish regulated cell death from nonspecific cytotoxic effects.

14.4.1.5 High-Content and Mechanistic Screening

High-content screening (HCS) combines automated microscopy with multiparametric fluorescent imaging to concurrently evaluate endpoints such as reactive oxygen species (ROS) production, mitochondrial impairment, alterations in nuclear morphology, and cell death. Complementary transcriptomic and toxicogenomic analyses extend this capability by identifying pathway-level disturbances, thereby supporting mechanism-driven prioritization and read-across strategies (Hartung and Leist 2024). Despite their strong predictive value and depth of information, these approaches depend on sophisticated imaging systems, robust computational analysis pipelines, and rigorous biological validation to minimize the risk of misinterpretation or overanalysis.

14.4.1.6 Advanced 3D Models, Organoids, and Co-cultures

3D cellculture models—including spheroids, organoids, and co-culture systems—more accurately mimic in vivo tissue organization, metabolic functionality, and cell-cell communication compared with conventional two-dimensional monolayer cultures, for instance, liver spheroids and organoids exhibit greater sensitivity to metabolically activated genotoxic agents and provide improved prediction of human-relevant toxicity thresholds (Ge et al. 2024). Although these models offer enhanced physiological relevance, they are technically demanding, generally lower in throughput, and often require re-optimization of assays to ensure adequate probe penetration and reliable imaging In-vitro assays for detection of molecular toxicity have been depicted in Table 14.5.

14.4.2 In Vivo Studies

In vivo molecular toxicity studies are essential to (a) confirm whether molecular perturbations observed in vitro manifest in an intact organism, (b) detect systemic and tissue-specific genotoxicity or oxidative damage, (c) measure persistent DNA lesions and mutations, and (d) provide data for regulatory risk assessment that incorporate absorption, distribution, metabolism, and excretion (ADME) as well as toxicokinetics. Accordingly, in vivo assays remain a critical confirmatory tier within integrated testing strategies, particularly when tissue-restricted exposure, metabolic activation, or whole-organism responses are relevant (OECD 2023; Kirkland and Fowler 2024).

Regulatory in vivo genotoxicity assessment relies primarily on the mammalian erythrocyte micronucleus test, the in vivo alkaline comet assay, and transgenic rodent (TGR) mutation assays, which together address chromosomal damage, DNA strand breaks, and gene mutations across tissues and life stages (OECD TG 474, 489, 488).

Table 14.5 In vitro assays for detection of molecular toxicity

Assay type	Principle	Key endpoint	Strengths	Limitations	Key references
MTT/WST	Tetrazolium reduction	Viability	Simple, high throughput	Metabolic interference	Asati et al. (2023)
LDH	Enzyme release	Membrane damage	Nondestructive	Late apoptosis overlap	Asati et al. (2023)
ROS (DCFH-DA)	Probe oxidation	Oxidative stress	Sensitive	Low specificity	Murphy et al. (2023)
$\Delta\Psi m$ (JC-1/TMRE)	Mitochondrial dye uptake	Early apoptosis	Mechanistic	Dye artefacts	Galluzzi et al. (2024)
Comet	DNA migration	DNA damage	Highly sensitive	Repair confounding	OECD (2023)
Micronucleus	Chromosomal loss/breakage	Genotoxicity	Regulatory acceptance	Cytotoxicity effects	OECD (2023)
Annexin V/PI	PS exposure/membrane loss	Apoptosis	Multiparametric	Timing critical	Galluzzi et al. (2024)
High-content imaging	Multiparametric imaging	Mechanistic profiling	Predictive	Cost, data complexity	Hartung and Leist (2024)
3D organoids	Multicellular architecture	Human relevance	Better translation	Lower throughput	Ge et al. (2024)

14.4.2.1 Mammalian Erythrocyte Micronucleus Test

The in vivo micronucleus (MN) assay identifies micronuclei in polychromatic erythrocytes that arise from chromosomal fragments or whole chromosomes failing to segregate properly during erythropoiesis. It is typically performed in rodents after single or repeated dosing, with bone marrow or peripheral blood collected at defined post-exposure time points in accordance with OECD Test Guideline 474. The incidence of micronucleated polychromatic erythrocytes (MN-PCEs) is determined using light microscopy, flow cytometry, or automated imaging systems (OECD 2023). Recognized as a standard regulatory method for detecting clastogenic and aneugenic effects, the assay is well-established, reproducible, and supported by extensive historical control datasets. However, because it depends on active erythroblast proliferation, appropriate timing of sample collection is essential. Additionally, excessive bone marrow toxicity or insufficient systemic exposure may complicate data interpretation (Kirkland and Fowler 2024).

14.4.2.2 In Vivo Alkaline Comet Assay

The alkaline comet assay measures DNA strand breaks and alkali-labile sites in single cells isolated from target tissues such as the liver, stomach, lung, or blood. DNA migration parameters (e.g., %DNA in tail, tail moment) quantitatively reflect damage. OECD TG 489 provides harmonized guidance on study design, tissue selection, sampling times, controls, and reporting (OECD 2023).

The assay is highly sensitive and enables tissue-specific assessment of genotoxicity, thereby reflecting in vivo ADME and toxicokinetic influences. Enzyme-modified variants (e.g., Fpg-comet) permit detection of oxidized DNA bases, and automated scoring platforms improve throughput and reproducibility. Limitations include detection of largely transient lesions and sensitivity to sampling time relative to DNA repair kinetics, necessitating strict standardization (Azqueta et al. 2023).

14.4.2.3 Transgenic Rodent Mutation Assays

Transgenic rodent (TGR) assays employ animals carrying recoverable reporter genes (e.g., lacZ, cII) to directly quantify mutation frequency and spectra in somatic and germ cells following exposure. OECD TG 488 outlines dosing regimens, tissue sampling, and analytical workflows, commonly using 28-day repeat-dose designs (OECD 2023).

TGR assays provide direct measurement of in vivo gene mutations across tissues and are particularly valuable for germ cell mutagenicity and mechanistic interpretation through mutation spectra analysis. Their major limitations are higher cost, longer timelines, and the need for specialized molecular and sequencing infrastructure (Kirkland and Fowler 2024).

14.4.2.4 DNA Adduct Analysis

DNA adduct analysis directly detects covalent binding of reactive metabolites to DNA using sensitive analytical approaches such as ^{32}P postlabelling or LC-MS/MS. These methods are applied when metabolic activation and covalent DNA interaction are suspected contributors to mutagenicity or carcinogenicity. They provide unambiguous mechanistic evidence of DNA reactivity and adduct identity but require advanced instrumentation and careful interpretation in the context of adduct repair and persistence (Guengerich 2024).

14.4.2.5 Biomarkers of Oxidative DNA Damage

Quantification of 8-hydroxy-2′-deoxyguanosine (8-OHdG/8-oxo-dG) in tissues, plasma, or urine is widely used to assess oxidative DNA damage in vivo. Measurements by LC-MS/MS, HPLC-ECD, or validated immunoassays support evaluation of oxidative stress-mediated modes of action and allow noninvasive sampling via urine. However, artefactual oxidation during sample processing and limited lesion specificity require stringent methodological controls (Valavanidis et al. 2023).

14.4.2.6 In Vivo γH2AX Detection

γH2AX immunohistochemistry or flow cytometry detects phosphorylation of histone H2AX at DNA double-strand break sites and is increasingly applied to map tissue-localized DNA damage responses. The assay offers high sensitivity and spatial resolution and can be combined with histopathology and cell-type markers. As γH2AX represents an early and repairable signal, appropriate timing, validated antibodies, and rigorous controls are essential for interpretation (Audebert et al. 2023).

14.4.2.7 Chromosomal Aberration Analysis

Classical metaphase chromosomal aberration analysis in bone marrow or other proliferative tissues provides direct visualization of structural chromosomal changes. Although largely superseded by the MN test for routine screening, it remains useful as a complementary mechanistic tool. Its limitations include low throughput, labor intensity, and dependence on actively dividing cells (OECD 2023).

14.4.2.8 Clinical Chemistry, Histopathology, and Mechanistic Biomarkers

Standard toxicological endpoints—including clinical chemistry, organ weights, and histopathology—combined with molecular biomarkers (e.g., cytokines, oxidative stress enzymes, gene expression) provide essential context for molecular lesions detected in vivo. Integration of these data supports mode-of-action and adverse outcome pathway analyses, although distinguishing adaptive from adverse molecular responses requires weight-of-evidence evaluation (Krewski et al. 2024).

14.4.2.9 In Vivo Imaging, Reporter Models, and Omics

Noninvasive imaging modalities and transgenic reporter models enable longitudinal monitoring of pathway activation and tissue distribution, reducing animal use and supporting time-course analyses. In parallel, transcriptomic, proteomic, and metabolomic profiling of tissues and biofluids reveals pathway perturbations and early biomarkers of effect. While highly informative, these approaches demand specialized instrumentation, bioinformatics expertise, and careful control of biological variability (Hartung and Leist 2024).

14.4.2.10 Study Design Considerations

For all in vivo molecular endpoints, demonstration of systemic or tissue exposure is critical, as negative results without exposure are uninterpretable. Time-sensitive endpoints such as comet and γH2AX assays must align sampling with expected damage and repair kinetics. Excessive systemic toxicity should be avoided to prevent secondary genotoxic effects, and study designs should incorporate multiple dose levels, appropriate controls, historical databases where available, and validated analytical methods. In vivo assays are most effective when applied as confirmatory components within integrated testing strategies that build upon in vitro and NAM-based evidence (OECD 2023). These in-vivo assays have been properly shown in Table 14.6.

14.5 Mitigation Strategies and Bioapplications

Carbonaceous nanomaterials (CNMs) have emerged as versatile platforms in biomedicine, environmental remediation, and sensing owing to their high surface area, tunable surface chemistry, electrical conductivity, and relative biocompatibility. Nevertheless, concerns regarding their potential toxicity—particularly long-term

Table 14.6 In vivo assays for detection of molecular toxicity

Method	Endpoint	When to use	Strengths	Limitations	Representative guidance/reviews
In vivo micronucleus (OECD TG 474)	Chromosomal damage (MN-PCE)	Regulatory genotoxicity screen; confirm clastogenic/aneugenic effects	Regulatory acceptance, flow cytometry automation	Bone marrow accessibility, confounded by bone marrow cytotoxicity	OECD TG 474
In vivo comet (OECD TG 489)	DNA strand breaks and alkali-labile sites	Tissue-specific DNA damage, short-term exposures	Sensitive, multiple tissues, reflects ADME	Transient damage, needs timing optimization	OECD TG 489, Frontiers method review
Transgenic rodent (TG 488)	Gene mutation frequency and spectrum	Detect somatic and germ cell mutations after repeat dosing	Direct mutation measurement, tissue sampling flexibility	Resource-intensive, longer timelines	OECD TG 488
DNA adduct analysis	Covalent DNA lesions	Suspected reactive metabolite formation	Mechanistic evidence for covalent binding	Technical sensitivity, adduct repair complicates interpretation	Analytical method papers (Guengerich 2024)
8-OHdG (tissue/urine)	Oxidative DNA damage biomarker	Oxidative stress-mediated toxicity	Noninvasive (urine), mechanistic	Artefacts during processing, specificity concerns	Valavanidis et al. (2023)
γH2AX IHC/flow	DNA DSB response (foci)	Spatial mapping of DSBs in tissues	Sensitive, spatial resolution	Early/repairable signal; antibody validation needed	Audebert et al. (2023)
Omics (in vivo)	Pathway perturbations	Mechanistic profiling, biomarker discovery	Comprehensive mechanistic insight	High complexity, requires bioinformatics	Hartung and Leist (2024)

bioaccumulation, oxidative stress, and environmental persistence—necessitate careful material design and risk mitigation strategies (Zheng et al. 2022; Ma et al. 2024).

14.5.1 Reducing Toxicity

Mitigating nanotoxicity is central to the safe deployment of CNMs, especially for biomedical and environmental applications. Among the most effective strategies are surface modification, green synthesis, and dose optimization, which collectively enhance biocompatibility while preserving functional performance.

Surface functionalization through polymeric or chemical coatings—such as polyethylene glycol (PEG), carboxyl (–COOH), or hydroxyl groups—reduces nonspecific protein adsorption, limits cellular membrane disruption, and improves circulation stability, thereby significantly lowering cytotoxicity and inflammatory responses (Zheng et al. 2022; Tegafaw et al. 2025). Similar approaches applied to carbon dots and quantum dots retain optical properties while suppressing ROS-mediated toxicity.

Green synthesis routes using plant extracts or biomass-derived precursors further reduce toxicity by avoiding hazardous solvents and reagents, producing CNMs with benign surface chemistries and improved biological tolerance (Cheng et al. 2024). Dose optimization remains critical, as nanoparticle toxicity is inherently dose-dependent; careful control of exposure concentrations allows therapeutic efficacy without exceeding toxicity thresholds (Ma et al. 2024). Despite these advances, long-term safety and biodistribution profiles still require systematic evaluation.

14.5.2 Bioapplications

Carbon nanomaterials (CNMs) are increasingly integrated into biomedical and environmental technologies owing to their multifunctionality and tunable physicochemical properties. In drug delivery, CNMs enable targeted and controlled release, enhancing therapeutic efficacy while reducing systemic toxicity (Zheng et al. 2022). Lignin-derived carbons, polymer-carbon nanocomposites, and graphene-based carriers provide high drug and gene-loading capacities with versatile surface functionalization (Cheng et al. 2024).

CNMs and metal-carbon nanocomposites exhibit pronounced antimicrobial and antiviral activity via mechanisms including reactive oxygen species (ROS) generation, membrane disruption, and surface-mediated viral inactivation. Graphene oxide and carbon-based composites are particularly promising for antimicrobial coatings, antiviral surfaces, and protective interfaces (Li and Ji 2021). In biosensing, the high electrical conductivity and surface sensitivity of CNMs enable ultrasensitive detection of biomolecules, pathogens, and disease biomarkers, exemplified by graphene-based sensors for viral diagnostics.

In tissue engineering, CNMs reinforce scaffolds and hydrogels, improving mechanical strength and promoting cell adhesion, proliferation, and differentiation, with applications in bone, skin, and neural tissue regeneration (Zheng et al. 2022). Beyond biomedicine, recycled CNMs derived from plastic and biomass wastes—including graphene, carbon nanotubes (CNTs), and carbon dots—are increasingly applied in environmental remediation, where high adsorption capacity and tunable surface chemistry enable efficient pollutant removal and mitigation of molecular toxicity. Notably, recent studies show that recycled CNMs can achieve performance comparable to conventionally synthesized counterparts, supporting both sustainability and functional efficacy (Cheng et al. 2024).

14.5.3 Molecular Toxicity Pathways Targeted by CNMs

Molecular toxicity typically results from exposure to small-molecule pollutants such as heavy metals, pesticides, dyes, endocrine disruptors, and polycyclic aromatic hydrocarbons, which induce oxidative stress, inflammatory signaling, genotoxicity, and membrane destabilization. Central to these processes is excessive reactive oxygen species (ROS) generation, driving lipid peroxidation, protein oxidation, mitochondrial dysfunction, and DNA damage (Xu et al. 2024). Concurrent activation of inflammatory pathways, particularly NF-κB, promotes pro-inflammatory cytokine release, while genotoxic agents induce direct DNA lesions or ROS-mediated strand breaks (Garriga et al. 2020). These interconnected pathways define the core mechanisms of molecular toxicity that recycled CNMs are designed to interrupt through pollutant adsorption, sequestration, and redox modulation which are shown in Fig. 14.3.

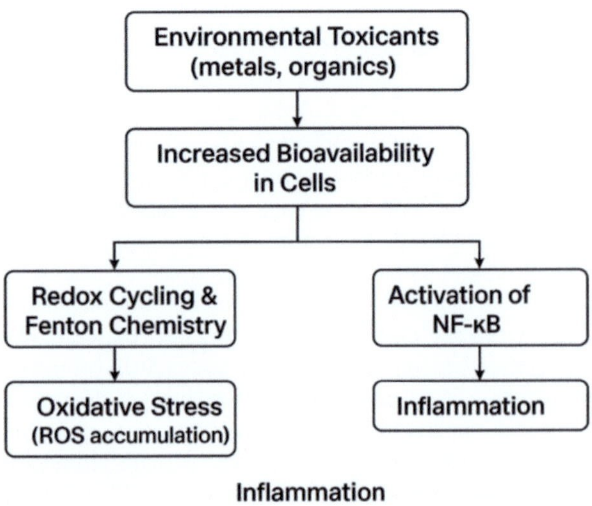

Fig. 14.3 Molecular toxicity pathway flow. (Created by authors)

14.5.4 Mechanisms by Which Recycled CNMs Mitigate Molecular Toxicity

14.5.4.1 Adsorption and Sequestration of Toxicants

The most direct mechanism through which CNMs mitigate molecular toxicity is the adsorption of pollutants onto high-surface-area carbon matrices. GO, rGO, and functionalized CNTs exhibit a rich distribution of surface groups (carboxyl, hydroxyl, epoxide) that facilitate electrostatic attraction, cation-π interactions, complexation, and π-π stacking with metals and organic toxins. For heavy metals such as Pb^{2+}, Cd^{2+}, and Cu^{2+}, GO-based materials demonstrate high affinity, attributed to their oxygenated functional groups that coordinate with metal ions. Magnetic-assisted GO adsorption systems have shown efficient heavy metal removal in water with predictable kinetic behavior (Melchor-Durán et al. 2024). By reducing free metal ion concentrations, CNMs prevent the catalytic initiation of Fenton-type redox cycling, thereby suppressing ROS generation. These have been shown in Fig. 14.4.

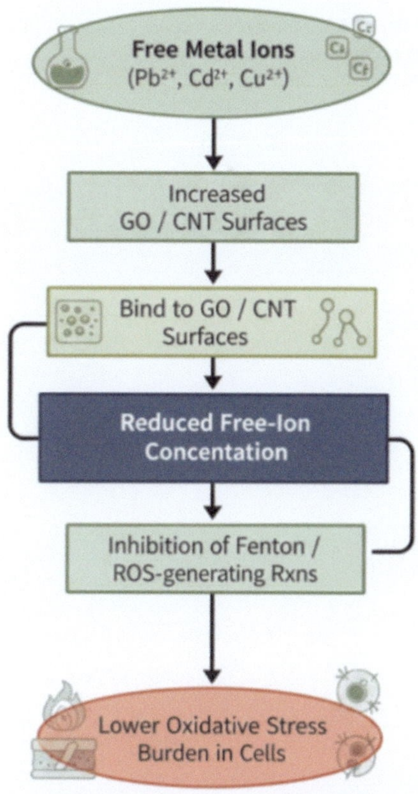

Fig. 14.4 Adsorption as a molecular toxicity interruption point. (Created by authors)

Thus, adsorption functions as a preemptive barrier that blocks toxicants before cellular uptake occurs.

14.5.4.2 Catalytic Degradation and Detoxification

Beyond adsorption, recycled CNMs frequently act as catalytic scaffolds for pollutant degradation. When hybridized with catalytic agents such as TiO_2, Fe_3O_4, Co_3O_4, or peroxymonosulfate activators, CNMs enhance electron transfer, increase active surface sites, and stabilize catalytic centers. This facilitates the oxidative breakdown of persistent organic contaminants into less harmful intermediates or fully mineralized products. Review literature on catalytic wastewater treatment emphasizes the critical role of carbon-based supports in Fenton-like and photocatalytic reactions (Mohapatra et al. 2023). By degrading toxins at the molecular level, CNM-supported catalysts ensure that toxicants never reach concentrations capable of triggering oxidative or inflammatory cellular pathways Fig. 14.5 shows the catalytic degradation flow by the CNMs.

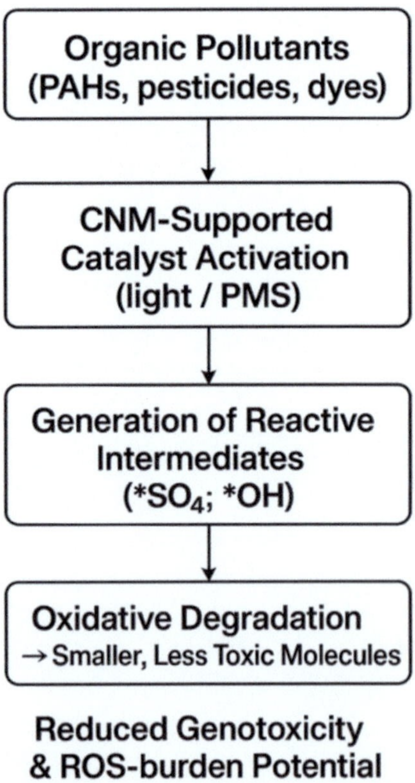

Fig. 14.5 Catalytic detoxification cascade. (Created by authors)

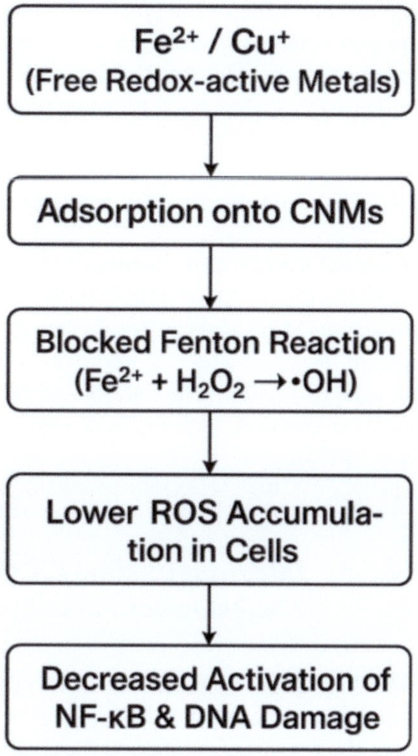

Fig. 14.6 CNM-mediated ROS pathway suppression. (Created by authors)

14.5.4.3 Suppression of ROS-Generating Pathways

A significant portion of molecular toxicity arises from redox-active metals that catalyze ROS formation. CNMs mitigate this by binding these metals and preventing them from participating in the Fenton and Haber-Weiss reactions. This pathway-level mitigation has substantial downstream effects: lipid peroxidation decreases, mitochondrial stress is alleviated, and inflammatory gene expression is reduced. Studies of nanoparticle-driven oxidative stress consistently indicate that limiting redox-active metal availability is an effective means of reducing ROS-mediated cellular injury which are depicted in Fig. 14.6.

By interrupting ROS generation at its origin, CNMs safeguard multiple downstream biological structures.

14.5.4.4 Immobilization and Prevention of Secondary Exposure

An essential aspect of CNM deployment is immobilization within membranes, polymer matrices, or magnetically retrievable composites. Immobilization minimizes the risk of CNM release into the environment while simultaneously allowing captured toxicants to be physically removed. Magnetic GO- and rGO-based systems are particularly effective, offering easy recovery, reusability, and reduced ecotoxicological risk (Melchor-Durán et al. 2024). This ensures long-term stability of remediation systems and prevents secondary contamination cycles.

14.5.4.5 Enhancement of Microbial Detoxification

Recycled CNMs also augment biodegradation processes by acting as conductive scaffolds that enhance extracellular electron transfer in pollutant-degrading microorganisms. This synergy accelerates the breakdown of persistent organic pollutants, lowers their half-life, and improves total system detoxification efficiency. Reviews of catalytic and biological remediation strategies note that CNMs substantially increase electron mobility and catalytic turnover in hybrid systems (Mohapatra et al. 2023).

14.5.5 Integration: How CNMs Intersect with and Suppress Toxicological Pathways

Based on all the aforementioned ways of acting of recycled CNMs in the biological pathways, we can infer the integration of the actions of CNMs and their influence over the toxicological pathways. Given the close interplay among oxidative stress, DNA damage, and inflammatory pathways, the multifaceted interactions of carbon nanomaterials (CNMs) can significantly reshape and potentially mitigate the overall molecular toxicity profile. It can be thus concluded that when mapped onto classical toxicology frameworks, recycled CNMs intervene at multiple crucial junctures. The integrative mechanistic methods of CNM mediated capture and degradation of toxicants have been portrayed in Fig. 14.7.

14.6 Conclusion

Recycled carbon materials bridge waste valorization and advanced nanotechnology, offering tunable physicochemical properties suitable for applications in energy storage, catalysis, environmental remediation, and biomedicine. However, their increasing complexity and nanoscale reactivity necessitate careful evaluation of molecular toxicity mechanisms. Oxidative stress, genotoxicity, and inflammation represent interconnected pathways through which carbon nanomaterials may influence biological systems. Surface functionalization, defect density, residual contaminants, and particle size critically modulate cellular uptake, mitochondrial responses, and DNA integrity, underscoring the importance of structure-toxicity relationships.

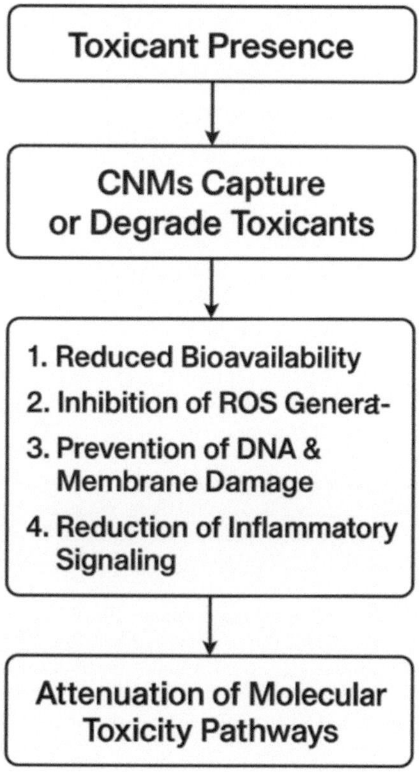

Fig. 14.7 Integrated mechanistic diagram depicting overview of attenuation of CNM-mediated capture and degradation of toxicants. (Created by authors)

The integration of mechanistic toxicology—including ROS profiling, micronucleus assays, mitochondrial membrane potential assessment, transcriptomic analyses, and advanced 3D culture models—provides a predictive framework for safety evaluation. Importantly, recycled CNMs can also be engineered to mitigate toxicity through controlled surface chemistry, purification, and green synthesis strategies, aligning with safe-by-design principles.

Future progress in this field will depend on harmonizing scalable green synthesis methods with rigorous toxicological validation and life cycle sustainability assessments. By embedding molecular toxicity considerations into material development pipelines, recycled carbon nanomaterials can advance as both environmentally sustainable and biologically responsible components of next-generation technologies.

References

Abegunde OO (2025) Physicochemical determinants of nanomaterial toxicity: implications for safe-by-design strategies. NanoImpact 29:100461

Adegbola P, Adetutu A (2024) Genetic and epigenetic modulations in toxicity by heavy metals and PAHs. Toxicol Rep 12:502–519

Ajsuvakova O, Tinkov A, Aschner M et al (2020) Sulfhydryl groups as targets of mercury toxicity. Coord Chem Rev 417:213343

Andrade B (2022) The relationship between reactive oxygen species and the cGAS/STING pathway during inflammaging. PMC

Asati A, Santra S, Kaittanis C (2023) Assay interference and orthogonal strategies in cytotoxicity testing of nanomaterials. Small 19(12):e2206207

Asif FC, Saha GC (2023) Graphene-like carbon structure synthesis from biomass pyrolysis: a critical review on feedstock–process–properties relationship. C 9(1):31

Audebert M, Riu A, Jacques C, Cravedi J-P (2023) γH2AX as a genotoxicity biomarker in vivo. Mutagenesis 38(3):161–173

Azqueta A, Dusinska M, Collins AR (2023) The comet assay in in vivo genotoxicity testing: advances and applications. Front Toxicol 5:1189436

Baig N, Kammakakam I, Falath W (2021) Nanomaterials: a review of synthesis methods, properties, recent progress, and challenges. Mater Adv 2(6):1821–1871

Barker DJP (2024) Developmental origins of health, disease, and toxic susceptibility. J Intern Med 295(1):3–15

Bedia J, Rosas JM, Rodríguez-Mirasol J, Cordero T (2018) Preparation and characterization of activated carbons from biomass wastes and their applications. Waste Manag 78:82–94

Cederbaum AI (2023) Alcohol metabolism and oxidative stress in liver toxicity. Clin Liver Dis 27(3):445–460

Chahal S, Macairan JR, Yousefi N, Tufenkji N, Naccache R (2021) Green synthesis of carbon dots and their applications. RSC Adv 11:25354–25363

Chandran DG, Muruganandam L, Biswas R (2023) A review on adsorption of heavy metals from wastewater using carbon nanotube and graphene-based nanomaterials. Environ Sci Pollut Res 30:110010–110046

Cheng Y, Wang J, Fang C, Du Y, Su J, Chen J, Zhang Y (2024) Recent progresses in pyrolysis of plastic packaging wastes and biomass materials for conversion of high-value carbons: a review. Polymers 16(8):1066

Choi K, Riviere J, Monteiro-Riviere N (2017) Protein corona modulation of hepatocyte uptake and molecular mechanisms of gold nanoparticle toxicity. Nanotoxicology 11:64–75

Cui X, Wan B, Yang Y, Ren X, Guo L (2017) Length effects on the dynamic process of cellular uptake and exocytosis of single-walled carbon nanotubes in murine macrophage cells. Sci Rep 7:1518

Diez-Pascual AM (2022) Carbon-based nanomaterials. Int J Mol Sci 23(15):8294

Docampo E, García-Fernández J, Mármol I et al (2025) Mechanistic insights into sphingomyelin nanoemulsions as drug delivery systems for non-small cell lung cancer therapy. Pharmaceutics 17:461

Drasler B, Sayre P, Steinhäuser KG, Petri-Fink A (2024) Interference of engineered nanomaterials with in vitro toxicity assays. Part Fibre Toxicol 21:6

Duran-Jiménez G, Rodriguez J, Stevens L, Kostas ET (2024) Microwave pyrolysis of waste biomass and synthesis of micro-mesoporous activated carbons. Chem Eng J 488:150926

Dutta S, Finn T, Kuhn A, Abrams B, Raskatov J (2019) Chirality dependence of amyloid β cellular uptake and a new mechanistic perspective. ChemBioChem 20:1023–1026

Fadeel B, Farcal L, Hardy B, Vázquez-Campos S (2024) Advanced tools for the safety assessment of nanomaterials. Nat Nanotechnol 19:111–124

Ferjani H, Abdalla S, Oyewo OA, Onwudiwe DC (2023) Facile synthesis of carbon dots by the hydrothermal carbonization of avocado peels and evaluation of the photocatalytic property. Inorg Chem Commun 160:111866
Forman HJ, Zhang H, Rinna A (2024) Glutathione: overview of its protective roles, measurement, and regulation. Mol Asp Med 94:101224
Fromenty B, Roden M (2024) Mitochondrial dysfunction in drug-induced toxicity. J Hepatol 80(1): 213–226
Galluzzi L, Vitale I, Aaronson SA et al (2024) Molecular mechanisms of regulated cell death and therapeutic implications. Nat Rev Mol Cell Biol 25:285–304
Gao Y, Chen Y, Liu J, Zhang H (2024) Inflammation-associated susceptibility to chemical toxicity: roles of cytokines, oxidative stress, and immune dysregulation. Toxicol Appl Pharmacol 478: 116787
Gárate-Vélez A, Martínez-Huitle CA, Brillas E (2025) Advanced carbon nanomaterials: structure, properties, and environmental applications. Carbon 219:118030
Garriga R, Herrero-Continente T, Palos M et al (2020) Toxicity of carbon nanomaterials and their potential application as drug delivery systems. Nanomaterials 10(8):1617
Green DR, Llambi F (2023) Cell death signaling pathways: apoptosis and necrosis. Cold Spring Harb Perspect Biol 15(9):a041139
Guengerich FP (2024) DNA adduct formation and biological consequences. Chem Res Toxicol 37(1):1–14
Ha Y, Wang X, Liljestrand H, Maynard J, Katz L (2021) Elucidating the mechanism of cellular uptake of fullerene nanoparticles. Environ Eng Res 27:200658
Haidar Z, Fatema K, Shoily S, Sajib A (2023) Disease-associated metabolic pathways affected by heavy metals. Toxicol Rep 10:554–570
Hartmut J, Anup R (2024) Acetaminophen Hepatotoxicity: Paradigm for Understanding Mechanisms of Drug-Induced Liver Injury Annual Review of Pathology: Mechanisms of Disease 19(1):453–478. 10.1146/pathmechdis.2024.19.issue-1. https://doi.org/10.1146/annurev-pathmechdis-051122-094016
Hartung T, Leist M (2024) Next-generation toxicity testing strategies. Nat Rev Toxicol 4:79–96
Jiang L, Zhang Y, Wang H, Liu X (2024) Waste-derived carbon materials for energy and environmental applications. Chem Eng J 475:146569
Jovic M, Vuković G, Babić B (2020) Graphene and carbon nanomaterials: properties and applications. Mater Today Chem 16:100252
Kang C, Huang Y, Yang H, Yan XF, Chen ZP (2020) A review of carbon dots produced from biomass wastes. Nanomaterials 10(11):2316
Kelley N, Jeltema D, Duan Y, He Y (2024) The NLRP3 inflammasome: mechanisms of activation and regulation. J Leukoc Biol 115(1):27–40
Khan A, Jia Z (2023) Recent insights into uptake, toxicity, and molecular targets of microplastics and nanoplastics relevant to human health impacts. iScience 26:106061
Kirkland D, Fowler P (2024) In vivo genotoxicity testing: current practice and future perspectives. Mutagenesis 39(1):1–14
Klaassen CD (2018/2024) Casarett & Doull's toxicology: the basic science of poisons. McGraw-Hill Education
Kopp B, Khoury L, Audebert M (2023) Validation and applications of the γH2AX assay in genotoxicity testing. Mutagenesis 38(4):243–257
Kowalska A, Winiarska K, Szymanski P, Nowak M (2024) Analyzing molecular determinants of nanodrugs' cytotoxic effects. Int J Mol Sci 26(14):6687
Koyama H, Kamogashira T, Yamasoba T (2024) Heavy metal exposure: molecular pathways, clinical implications, and protective strategies. Antioxidants (Basel) 13:76
Krewski D, Andersen ME, Mantus E, Zeise L (2024) Toxicity testing in the 21st century. Environ Health Perspect 132(3):036001
Kumar V et al (2023) Nutriepigenomics in environmental-associated oxidative stress. Antioxidants (Basel) 12(3):771

Kusano Y, Kusano K (2024) Plasma-based waste treatment: mechanisms and prospects. J Clean Prod 412:1–15
Lawrence T (2024) The nuclear factor NF-κB pathway in inflammation. Cold Spring Harb Perspect Biol 16(2):a041321
Le HM (2025) Upcycling plastic waste into graphene and carbon nanomaterials: processes and challenges. Renew Sustain Energy Rev 189:114032
Lee N, Spears M, Carlisle A, Kim D (2020) Endogenous toxic metabolites and implications in cancer therapy. Oncogene 39:5709–5720
Li J, Ji Z (2021) Epigenetic effects of nanomaterials and nanoparticles: DNA methylation and histone modification changes following exposure. J Nanobiotechnol 19:2
Liebler DC (2023) Protein damage by reactive electrophiles: targets and toxicological consequences. Chem Res Toxicol 36(9):1321–1334
Liu N, Tang M (2020) Toxic effects and involved molecular pathways of nanoparticles on cells and subcellular organelles. J Appl Toxicol 40:16–36
Liu Y, Wang J, Zhang X, Chen Z (2025a) Oxidative stress: signaling pathways, biological functions, and disease implications. Antioxidants (Basel) 14(3):211
Liu YY, Ohashi Y, Ushijima T (2025b) How chronic inflammation fuels carcinogenesis as an environmental epimutagen. Discov Oncol 16:1150
Lucchini RG, Dorman DC, Elder A, Veronesi B (2012) Neurological impacts from inhalation of pollutants and the nose–brain connection. NeuroToxicology 33(4):838–841. https://doi.org/10.1016/j.neuro.2011.12.001
Ma X, Tian Y, Yang R et al (2024) Nanotechnology in healthcare, and its safety and environmental risks. J Nanobiotechnol 22(1):715
Manawi YM, Ihsanullah, Samara A et al (2018) A review of carbon nanomaterials' synthesis via the chemical vapor deposition (CVD) method. Materials 11(5):822
Męczyńska-Wielgosz S, Sikorska K et al (2025) Uptake and toxicity of polystyrene NPs in three human cell lines. Int J Mol Sci 26:4783
Mehta J, Dilbaghi N, Deep A et al (2025) Plastic waste upcycling into carbon nanomaterials in circular economy. Carbon 234:119969
Melchor-Durán AP, Moreno-Virgen MR et al (2024) Heavy metal removal from water using graphene oxide in magnetic-assisted adsorption systems. Separations 11(10):294
Minotti G, Menna P, Salvatorelli E, Cairo G, Gianni L (2024) Anthracycline cardiotoxicity: molecular mechanisms and biomarkers. J Clin Oncol 42(6):589–601
Mohapatra L, Cheon D, Yoo SH (2023) Carbon-based nanomaterials for catalytic wastewater treatment: a review. Molecules 28(4):1805
Morellato A, Umansky C, Pontel L (2020) The toxic side of one-carbon metabolism and epigenetics. Redox Biol 40:101850
Moretton A, Loizou J (2020) Interplay between cellular metabolism and the DNA damage response in cancer. Cancer 12:2051
Murphy MP, Holmström KM, Hartley RC et al (2023) Measuring reactive oxygen species and oxidative damage in cells and tissues. Nat Rev Methods Primers 3:61
Nel AE, Xia T, Meng H, Li N (2024) Nanomaterial toxicity: role of oxidative stress and protein corona formation. Nat Nanotechnol 19:98–110
Nezafat Z, Dong Y, Nasrollahzadeh M, Shafiei N, Gharoubi H, Javanshir S (2024) Recent progress in energy conversion and storage of agricultural waste-derived carbon nanomaterials. Green Chem 26:10687–10717. https://doi.org/10.1039/D3GC04332K
Nigam M (2025) Reactive oxygen species: a double-edged sword in NF-κB signaling modulation. MDPI
OECD (2023) OECD guidelines for the testing of chemicals: genetic toxicology. OECD Publishing
Panahi Y, Beiraghdar F, Amirhamzeh A, Poursaleh Z, Saadat A, Sahebkar A (2016) Environmental toxicant exposure and cancer: the role of epigenetic changes and protection by phytochemicals. Curr Pharm Des 22(2):130–140

Parfett C, Desaulniers D (2017) A Tox21 approach to altered epigenetic landscapes. Int J Mol Sci 18:1179

Park EJ, Park K (2024) Metal nanoparticles and their toxicological effects. Arch Toxicol 98:325–351

Phillips DH, Arlt VM (2024) Genotoxicity: damage to DNA and its consequences. Nat Rev Mol Cell Biol 25:321–340

Qu F, Zheng W (2024) Cadmium exposure: mechanisms and pathways. Toxics 12:388

Quevedo A, Lynch I, Valsami-Jones E (2021) Mechanisms of silver nanoparticle uptake by embryonic zebrafish cells. Nanomaterials 11:2699

Shahzadi S, Ahmad M, Hussain S (2024) MOF-derived porous carbons and carbon–metal composites: synthesis and applications. Carbon 218:118021

Shi T, Dansen TB (2020) Reactive oxygen species induced p53 activation: DNA damage, redox signaling, or both? Antioxid Redox Signal 33(12):839–859

Sies H, Berndt C, Jones DP (2024) Oxidative stress. Annu Rev Biochem 93:715–748

Silva E, Rajapakse N, Kortenkamp A (2024) Something from nothing: the potential of chemical mixtures to produce combined effects. Environ Health Perspect 132(2):26001

Speranza G (2021) Carbon nanomaterials: synthesis, functionalization and sensing applications. Nanomaterials 11(4):967

Sridharan R, Monisha B, Kumar PS et al (2022) Carbon nanomaterials and its applications in pharmaceuticals: a brief review. Chemosphere 294:133731

Suman TY, Pei D-S (2022) Nanomaterial waste management. In: Nanomaterials recycling. Elsevier, pp 21–36

Tegafaw T, Mulugeta E, Zhao D et al (2025) Surface modification, toxicity, and applications of carbon dots for cancer theranosis. Nanomaterials 15(11):781

Tomczyk A, Sokołowska Z, Boguta P (2020) Biochar physicochemical properties: pyrolysis temperature and feedstock kind effects. Rev Environ Sci Biotechnol 19:191–215

Ursini F, Maiorino M, Forman HJ (2024) Redox homeostasis and the role of nutrition in oxidative stress. Free Radic Biol Med 206:230–243

Valavanidis A, Vlachogianni T, Fiotakis C (2023) 8-Hydroxy-2′-deoxyguanosine (8-OHdG): a critical biomarker of oxidative stress and carcinogenesis. J Environ Sci Health C 41(3):173–198

Wang Y, Fang J, Leonard SS, Rao KM (2024) Cadmium-induced mitochondrial dysfunction and oxidative stress. Toxicology 489:153430

Wild CP, Turner PC (2024) The toxicology of aflatoxins as a basis for public health decisions. Mutat Res/Rev Mutat Res 819:111771

Xu YY, Wang M, Gong X et al (2024) Carbon-based nanomaterials cause toxicity by oxidative stress to the liver and brain in Sprague–Dawley rats. Nucl Sci Tech 35:109

Yan X, Li Y, Zhang J, Zhao D (2025) Biomass-derived porous carbons: structure control, properties, and energy applications. Energy Storage Mater 63:102991

Zanger UM, Schwab M (2024) Cytochrome P450 enzymes in drug metabolism: regulation and clinical significance. Pharmacol Ther 252:108406

Zhang X, Wesén E, Kumar R et al (2020) Correlation between cellular uptake and cytotoxicity of fragmented α-synuclein amyloid fibrils. ACS Chem Neurosci 11:233–241

Zhang W, Shen J, Guo X et al (2024) Comprehensive investigation into the impact of degradation of recycled polyethylene and recycled polypropylene on thermo-mechanical characteristics and thermal stability of blends. Molecules 29(18):4499

Zhang Y, Li H, Chen X, Wang J, Liu Q (2025) Nanostructured carbon materials derived from biomass waste for electrocatalytic hydrogen production. Int J Hydrog Energy 137:1191–1203

Zheng Y, Tian L, Ouyang H et al (2022) Carbon nanomaterials for drug delivery and tissue engineering. J Nanobiotechnol 20:445

Chapter 15
Bio-Based Carbon Nanostructure-Aided Photocatalysts for Clean Water in the Environmental System

Rajmoni Basumatary and Anjalu Ramchiary

15.1 Introduction

With an ever-growing population and increasing demand for goods, every natural resource has been subjected to exhaustive use, including water resources. Water covers about 70% of the planet's surface; however, only 2.5% of it is fresh water, of which only 1.2% is available in the form of surface water bodies accessible to humans and animals within the ecosystem (Porowski 2019). Improper use of water resources has led to the contamination of existing, easily accessible freshwater sources. The daily activities performed by modern civilization have now caused global contamination of water sources, affecting both terrestrial and aquatic life (Schweitzer and Noblet 2018). This contaminated water also affects agriculture when used for irrigation, poisoning crops and compromising nutrient purity. Consumption of such agricultural produce by humans and animals can lead to bioaccumulation of toxic substances, which may result in adverse health outcomes, including impaired development, endocrine disruption, and carcinogenesis. Therefore, water pollution directly undermines the safety of agricultural products and poses a serious threat to public health and societal well-being. Furthermore, as the agricultural sector is leading globally in water use, which is about 70% of total fresh water for irrigation purposes, it also participates in producing major water pollution through the use of fertilizers, agrochemicals, and other contaminants (Ingrao et al. 2023; Wato et al. 2020). Therefore, maintaining water purity is required both before and after its use in agriculture, which, when successfully achieved, will generate and preserve a circular bio-economy (illustrated in Fig. 15.1).

R. Basumatary · A. Ramchiary (✉)
Department of Physics, Bodoland University, Kokrajhar, Assam, India
e-mail: anjalu1417@buniv.edu.in

Y. K. Mohanta et al. (eds.), *Bioresource Waste to Carbon Nanomaterial*,
Nanotechnology in the Life Sciences, https://doi.org/10.1007/978-981-92-2405-0_15

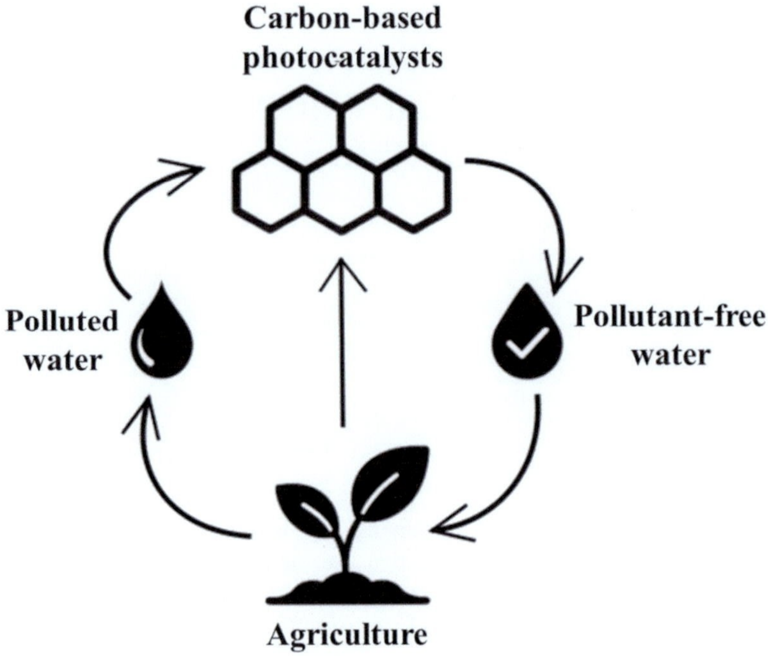

Fig. 15.1 Schematic illustration of carbon-based photocatalysts for sustainable water remediation, enabling pollutant removal and safe water reuse in agriculture within an eco-friendly, closed-loop purification system

In this context, advanced oxidation processes such as photocatalysis for water treatment purposes can provide a sustainable strategy. Photocatalytic systems utilize light as their primary energy source, enabling the generation of highly reactive species capable of mineralizing pollutants into harmless end products (Ahmed and Haider 2018). Since solar irradiation is abundant and renewable, photocatalysis presents a promising low-energy strategy for large-scale water treatment applications. Photocatalytic materials can be strategically engineered to target individual contaminants or simultaneously remove multiple classes of pollutants, depending on their structural design and operational configuration. Furthermore, to achieve a higher degree of water purity, the photocatalytic treatment can be integrated with simple pre-treatment steps to eliminate easily separable impurities, including suspended debris, oils, plastics, and sediments (Salman et al. 2019). The finer filtration in the molecular range to oxidize or reduce particles that are invisible but harmful to agriculture and interconnected to all can be done using photocatalysis.

However, the degradation efficiency of the photocatalyst depends on its intrinsic properties, such as its surface area, stability, charge-carrier dynamics, reproducibility, and reaction kinetics. Consequently, extensive research efforts are focused on designing next-generation photocatalysts with enhanced activity, durability, and environmental compatibility (Djurišić et al. 2020). In this aspect, researchers are also interested in photocatalysts made from bio-based carbon nanostructures such as TiO_2 on graphene, TiO_2 on carbon nanotubes (CNT), TiO_2 on graphitic carbon nitride (g-C_3N_4), ZnO/g-C_3N_4, and Ag-CuO/g-C_3N_4 (Nasir et al. 2022; Osman et al. 2019; Lin et al. 2022). The raw materials required for the synthesis of carbon-based photocatalysts can be easily obtained in abundance from agricultural wastes that are simply discarded, which later turn into garbage and pollutants (Son et al. 2021). The obtained carbon from agricultural waste can serve as a host semiconductor for developing a photocatalyst or as a co-catalyst. These photocatalysts can benefit from their low-dimensional nanostructures that can be deliberately designed. Their reduced size alters features such as pore distribution, chemical reactivity, surface functional groups, mechanical strength, electrical and thermal behavior, as well as optical responses (Speranza 2021). Therefore, it is important to know the different synthesis methods for obtaining carbon nanostructures from agricultural wastes, along with their specific properties, and the techniques to characterize them.

15.2 Fundamentals of Photocatalysis for Water Purification

15.2.1 Principles of Photocatalysis and Charge-Carrier Dynamics

Photocatalysis is a process that involves two entities, photons and a catalyst, in which the reactants undergo reaction at an accelerated rate to yield final products. Many semiconductor materials have been reported to act as photocatalysts under both UV and visible light. Studies have also found that modification of semiconductors with metals and nonmetals can reduce their bandgap energy, improve light response, and enhance the charge separation efficiency. The photogenerated charge carriers can undergo redox reactions with the surrounding molecules, such as oxygen and water, to produce reactive oxygen species. The reactive oxygen species, in turn interacts with the pollutants dissolved in water and undergo chemical transformations. The oxidation and reduction reactions take place in multiple stages, and the pollutants are degraded into less harmful products or completely mineralized into CO_2 and H_2O (Ahmed and Haider 2018). Mostly, wide semiconductors such as TiO_2 and ZnO are used as a host semiconductor chosen for their relatively superior stability, nontoxicity, and photo-corrosion resistance, in comparison to other semiconductors (Pavlović et al. 2024). When semiconductor photocatalysts are irradiated with photons possessing energy equal to or greater than their bandgap, electrons are promoted from the valence band to the conduction band, concurrently generating positively charged holes in the valence band. However, only a portion of these photogenerated charge carriers remains active for photocatalytic reactions. Many excited electrons rapidly lose energy and relax to lower-energy states or recombine with holes through intrinsic defect states or impurity levels within the semiconductor lattice. In some cases, recombination may even occur during the excitation process itself. Consequently, without strategic material engineering to suppress recombination pathways, only a limited fraction of charge carriers contributes effectively to photocatalytic activity. Furthermore, as the sunlight spectrum consists of 5% UV and 44% visible light energy, the UV-active only wide bandgap semiconductors such as TiO_2 with a 3.2 eV bandgap and ZnO with a 3.5 eV suffer from inefficient light energy harnessing capability (Pavlović et al. 2024). Therefore, to improve these limitations, strategies such as doping, defect engineering, forming heterojunctions, and sensitizations are implemented. Doping can be done with metals such as Ag, Au, Ni, Cu, Co, Fe, and Pd and with nonmetals such as B, C, N, and S (Khaki et al. 2017). Both doping and defects lower the bandgap energy from the UV region to the visible light range. They can improve the photocatalytic performance provided there is zero or less contribution into forming unwanted energy levels acting as recombination centers. Another strategy is heterojunction systems, where two or more semiconductors are combined to create interfacial charge transfer pathways that can be classified into three categories, type I, type II, and type III, based on the relative band-edge alignment of the constituent materials (Djurišić et al. 2020). Such engineered heterojunctions typically enhance photocatalytic performance by enabling more efficient charge separation, reducing electron-hole recombination, and improving photostability, for example, integrating a less stable but highly active semiconductor with a robust material like TiO_2 can leverage the stability of the latter while retaining the superior catalytic activity of the former. Well-known systems include Cu_2O-TiO_2, ZnO-TiO_2, and SnO_2-TiO_2, as well as hybrid junctions incorporating carbon-based nanostructures such as carbon dots, carbon nanorods, and other bio-derived carbon materials, which further improve charge mobility and light harvesting (Djurišić et al. 2020; Kadem et al. 2023). Again, sensitization employs functionalization or layering the host semiconductor photocatalyst with functional groups or compounds or with nanostructures that aid in light-harnessing capability by acting as a co-catalyst or an antenna or reactor, consequently boosting the performance of the modified photocatalyst. A few notable examples include dye sensitization or plasmonic sensitization. Using carbon-based support such as graphene, CNT, graphene oxide (GO), and reduced graphene oxide (rGO) also falls under this strategy for optimizing photocatalytic performance. The two-dimensional large surface area and porous structures allow for larger available active sites for photocatalytic reactions.

Essentially, carbon structures obtained from agricultural waste products can be used for developing photocatalysts with a lower bandgap for visible light sensitiveness, such as by C-doping, or using modified semiconducting CNTs or photocatalysts having less recombination and abundant charge carriers, such as with incorporation of graphene, GO, rGO, and CNT.

15.3 Bio-Based Carbon Nanostructures: Properties and Advantages

Some of the known zero-dimensional carbon nanostructures are carbon dots (C-dots), graphene quantum dots (GQDs), carbonized polymer dots (CPDs), fullerene C_{60}, and nanodiamond-like particles. C-dots are spherical particles with a diameter less than 10 nm and can be completely or partially graphitic. They are further categorized as graphene dots, graphite dots, amorphous carbon dots, and polymerlike dots (Kang and Lee 2019). C-dots are renowned for their photoelectrochemical properties, exhibiting strong and tunable photoluminescence (PL). They possess excellent dispersibility in water, low toxicity, biocompatibility, and biodegradability and can be obtained using low-cost synthesis techniques, from naturally abundant precursors (Kang and Lee 2019). Additionally, C-dots are endowed with good surface properties that support diverse functional groups applicable for a range of catalytic applications (Kang and Lee 2019). C-dots enhance solar energy harvesting in semiconductor devices by acting as photosensitizers and can also enhance charge separation by serving as electron mediators (Kang and Lee 2019). When C-dots are applied directly in biological and agricultural environment, they can interact to promote seed germination, increase root elongation, and enhance nutrient uptake and enzyme performance for improving plant's health. Treating rice plants with C-dots has been shown to enhance rice yield by 14.8% with better resistance to pathogens, owing to the nuclear entry of C-dots, DNA relaxation, and upregulation of defense-related genes such as thionin (Kang and Lee 2019). Fullerenes such as C_{60}, considered as 0D, are the closed shell networks of carbon atoms in football-like structure. Owing to their highly conjugated sp^2 carbon network and hollow geometry, fullerenes display unique electronic and surface properties such as the ability to scavenge free radicals like superoxide, methyl radicals, and hydroxyl species (Ahmad et al. 2022). Water-soluble and biocompatible fullerene derivatives can protect cells from oxidative damage by stabilizing reactive oxygen species, making them suitable for biological and materials applications (Ahmad et al. 2022).

One-dimensional (1D) carbon nanostructures are mainly obtained in the form of single-walled carbon nanotubes (SWCNTs), multi-walled carbon nanotubes (MWCNTs), and carbon nanofibers. An SWCNTs is made up of a single graphene sheet rolled into a cylindrical tube with a diameter of approximately 1.4 nm (Gupta et al. 2019), whereas MWCNTs possess sheet rolled into multiple concentric cylindrical tubes having an interlayer spacing of ~3.4 Å and outermost diameters in the 10–20 nm range (Gupta et al. 2019). Lengthwise, a CNT can be hundreds of micrometers to centimeters long (Gupta et al. 2019). The structure of SWCNTs is characterized by the chiral vector (m, n), which distinguishes between the different ways in which a graphene sheet can be rolled into a cylindrical form. This chiral vector determines the tube diameter, chirality, and electronic character (Gupta et al. 2019). Armchair (m, m) tubes display an armchair-like atomic pattern, zigzag (m, 0) tubes show zigzag atomic arrangements, and all other combinations yield chiral nanotubes with helical hexagonal patterns (Gupta et al. 2019). Their electronic properties depend sensitively on (m, n): SWCNTs behave as metals, semiconductors, or small-gap semiconductors based on the rule relating (n-m) to multiples of 3 (Gupta et al. 2019). The sp^2 bonding and delocalized π-electron network allow nanotubes to exhibit remarkable charge transport, including ballistic conduction and extremely high current densities up to 10^9 A/cm^2 (Uhlig et al. 2018). Metallic CNTs possess Fermi velocities comparable to metals and mean free paths approaching 1 μm, yet their resistivity remains modest due to low density of states. Mechanical properties of CNTs rank among the strongest in nature, with Young's modulus around 1.2 TPa and tensile strength over many times greater than steel, enabling large, elastic deformation before failure (Gupta et al. 2019). However, achieving their full technological potential requires synthesis control, as electrical transport is highly sensitive to adsorbed impurities, chirality distribution, and metal-CNT contact resistance (Gupta et al. 2019). MWCNTs show either Russian doll-type concentric shells or parchment-like rolled graphene structures, and CNT diameters generally increase with synthesis temperature (Díez-Pascual 2021). Their hydrophobic, oleophilic, and high-surface-area nature gives CNTs significant selectivity in separations and fluid interactions (Díez-Pascual 2021). CNT sponges, representing a form of 3D CNT assembly, consist of interconnected CNT networks with high porosity, hydrophobicity, and strong affinity toward organic solvents (Kausar 2020). These sponges undergo large reversible deformations and serve as scaffolds for composites when combined with polymers or inorganic materials (Kausar 2020).

Two-dimensional (2D) carbon materials are dominated by graphene, a single sheet of sp^2-bonded carbon atoms arranged in a hexagonal lattice. Graphene is the fundamental building block of other carbon nanostructures: it can be wrapped into 0D fullerenes, rolled into 1D nanotubes, or stacked into 3D graphite (Clancy et al. 2018). Single-layer graphene (SLG) consists of one atomic layer, bilayer and few-layer graphenes comprise two layers and three to ten layers, respectively, and different stacking configurations, AA, AB (Bernal), and ABC (rhombohedral), modify electronic structure (Clancy et al. 2018). Graphene's strong in-plane σ-bonds form the rigid honeycomb network, while out-of-plane π-orbitals determine interlayer interactions. SLG behaves as a semimetal with zero bandgap, featuring linear dispersion at the K and K′ points where charge carriers act as massless Dirac fermions, enabling exceptionally high mobility and conductivity (Speranza 2021). Graphene absorbs ~2.3% of incident white light despite being one atom thick and is transparent from UV to IR wavelengths. Its mechanical strength reaches ~42 N/m with a Young's modulus near 1 TPa (Speranza 2021). In bilayer graphene, the energy-wavevector dispersion near the Dirac point (a point in its electronic band structure where the conduction band and valence band meet) becomes parabolic, indicating quasiparticles that behave as massive Dirac fermion with a finite effective mass. Despite this change in dispersion, the system remains gapless, and a bandgap appears only when the material is modified chemically or subjected to external fields. When the layers are increased to a few-layered graphene, the properties of metallicity increases (Urade et al. 2023). While graphene nanoribbons (GNRs), quasi-1D strips with armchair or zigzag edges, exhibit edge-dependent electronic behavior; armchair GNRs can undergo metal-to-semiconductor transitions under small strain, and zigzag ribbons maintain more robust conduction (Chen et al. 2020).

15.4 Synthesis Strategies and Physicochemical Characterization

Carbon nanostructures can be synthesized from any agricultural waste such as sugarcane bagasse, banana peel, watermelon peel, orange peel, and pineapple leaf, provided they serve as a good carbon source (Baweja and Jeet 2019; Kang et al. 2020). The first step is always drying the waste and grinding them into fine powder to get particles of uniform size. These obtained carbon precursors can then be used to synthesize the nanomaterials using well-known techniques such as pyrolysis, hydrothermal, and microwave irradiation as well as through other advanced synthesis strategies. In pyrolysis, biomass-derived precursors taken from agricultural waste is first converted into carbon product (carbonization) by using high-temperature heating. Depending upon the requirement, pyrolysis is done in a vacuum or an inert atmosphere such as to prevent oxidation and surface contamination. Additionally, concentrated acids or alkalis can be used before or along the process for easier decomposition of organic polysaccharides to extract carbon structure from the precursor matrix (Hashem et al. 2025). This method allows precise control over the nanostructure of carbon dots by adjusting the synthesis temperature, time, and pH. Additional treatments can be incorporated after pyrolysis to obtain 2D or 3D structures such as chemical oxidation and Hummers' method, where the graphitic carbon is exfoliated for obtaining graphene oxide.

In a typical pyrolysis, the carbon-rich wastes are calcined in a high-temperature tubular furnace, typically above 500 °C. The reaction is conducted within a quartz tube under vacuum or in an inert environment to avoid oxidative degradation of the carbon framework. During heating, an inert gas such as nitrogen or carbon dioxide is continuously purged through the system to maintain an oxygen-free atmosphere and facilitate uniform carbonization (Jirimali et al. 2022). The final resulting solid product will subsequently be transformed into nanoscale carbon structures through posttreatment with concentrated acids (e.g., HNO_3, H_2SO_4, etc.) or strong alkalis (e.g., KOH, NaOH, etc.) (Shrestha et al. 2019). This posttreatment selectively removes amorphous regions, disrupts carbon-carbon linkages, and breaks down the bulk carbon into smaller domains. These chemical treatment enables the formation of nanostructures ranging from porous carbon frameworks to nanosheets and quantum dot-sized carbon fragments (Jirimali et al. 2022).

Spray pyrolysis offers a rapid conversion pathway for liquid as well as slurry waste precursors by directly atomizing them into a high-temperature furnace. Further improvement in carbon nanostructure uniformity and better control in particle size can be achieved using ultrasonic-assisted spray pyrolysis. It uses high-frequency acoustic energy to generate uniformly sized microdroplet to nanodroplets, which leads to more consistent nucleation and refined morphology of the carbon nanostructures. Alternatively, the same goal can be achieved by putting the aqueous mixtures, such as containing carbohydrates and sodium bicarbonate, under ultrasonication prior to spraying into a preheated tubular reactor (Jirimali et al. 2022). The atomized droplets undergo rapid vaporization again while entering the hot, inert environment, and the remaining solid residues thermally decompose to form carbon nanostructure. In this process, gases are formed from bicarbonate derivatives that create internal microbubbles, which act as pore-forming agents and significantly influence the resulting nanostructure morphology. Afterward, the carbon product is typically purified, washed to remove inorganic by-products, and may undergo secondary thermal treatment to further enhance structural order and yield high-quality carbon nanomaterials suitable for advanced photocatalytic applications (Jirimali et al. 2022).

Hydrothermal and solvothermal processes are very convenient techniques for producing carbon nanomaterials, including C-dots, from biomass wastes. Without any initial drying requirement, organic carbon sources such as fruit peels, leaves, or food residues are dispersed in water or a particular solvent and sealed inside a Teflon-lined stainless steel autoclave. This autoclave is usually transferred to a furnace that provides high temperature and develops high pressure inside, during which the precursor undergoes hydrothermal carbonization, breaking down into nanoscale carbon structures with surface functional groups that tailor their optical properties. This method is attractive due to its low cost, simplicity, and environmental compatibility. Numerous waste-derived precursors, including citrus peels, pomelo rind, bamboo leaves, and food-processing residues, have been successfully converted into fluorescent C-dots using this method with good quantum yields and suitability for sensing, photocatalysis, and optoelectronic applications (Kang et al. 2020). This method allows control over structural features of the resulting carbon nanomaterials through the variable reaction parameters, such as temperature and reaction time. The reaction pathway can sometime lead to the formation of hydrochar, which occurs when the biomass undergoes sequential breakdown into various intermediate hydrocarbons to behave as intrinsic acid catalysts during hydrothermal decomposition. Hydrochar formed during hydrothermal processing serves as a versatile precursor for advanced carbon structures. Through pyrolysis, chemical activation, and high-temperature calcination, it can be converted into nanoscale carbon materials such as carbon dots, porous nanocarbons, and graphene-like nanosheets (Jirimali et al. 2022).

The microwave irradiation method is an efficient approach for synthesizing carbon materials, as it enables rapid and uniform heating with low-energy consumption. This method allows precise control over temperature and reaction time and adjustable irradiation frequencies. Microwave irradiation enables the formation of a broad carbon nanostructures with tunable porosity and morphology. Micro-cellulose derived from agricultural waste is widely used as a precursor for producing porous carbons using a microwave irradiation process. In one of the studies, a solution of microcrystalline cellulose and polymer was irradiated and then annealed under microwave to yield a microporous carbon sponge (Liu et al. 2018). In another study, waste palm biomass was powdered, carbonized at various temperatures, mixed with KOH, and subsequently activated under microwave irradiation (Liu et al. 2019). Oil palm waste, after drying and grinding, was also subjected to microwave pyrolysis (500–700 W, 2.45 MHz, vacuum, 25 min), with temperature monitored by a type K thermocouple, producing highly porous carbon (Liew et al.

2018). Even domestic microwave units have been successfully applied to produce N-doped carbon dots from D-glucose and urea (Thara and Mathew 2024). The final carbon structure was decided by parameters such as irradiation time, frequency, power, and the impregnation ratio between precursor and activating agent (Ukanwa et al. 2019). Fluorescent C-dots have also been synthesized from natural sesame seeds using microwave pyrolysis. Setting the power at 800 W, the seeds were carbonized rapidly to release oil and leave solidified residue. Subsequent centrifugation was done to remove agglomerates at 8000 rpm, and finally green-emissive C-dots were obtained after lyophilization (Roshni and Divya 2017). Comparative studies also revealed that activated carbons derived through microwave irradiation display similar properties to those produced via conventional pyrolysis, which was observed taking sesame husk shells as carbon precursor (Sharif et al. 2018).

There are other synthesis techniques (illustrated in Fig. 15.2), such as plasma and laser-assisted carbonization, catalytic chemical vapor deposition employing biomass-derived vapors (Devi et al. 2023), molten salt-mediated carbonization, electrochemical oxidation or fragmentation routes for generating nanoscale carbon species, templating strategies for morphology-directed synthesis, and emerging green synthesis techniques that utilize benign solvents, plant-derived reducing agents, or low-energy reaction pathways. These techniques, although they differ in approaches for obtaining the product, they all follow one working principle: biomass is first subjected to dehydration and polymerization processes, which then go through aromatization and partial graphitization to finally form ordered or semi-ordered carbon domains at the nanoscale (Libra et al. 2011). The process that is capable of driving these chemical transitions while simultaneously enabling structural confinement or controlled nucleation can produce carbon nanostructures such

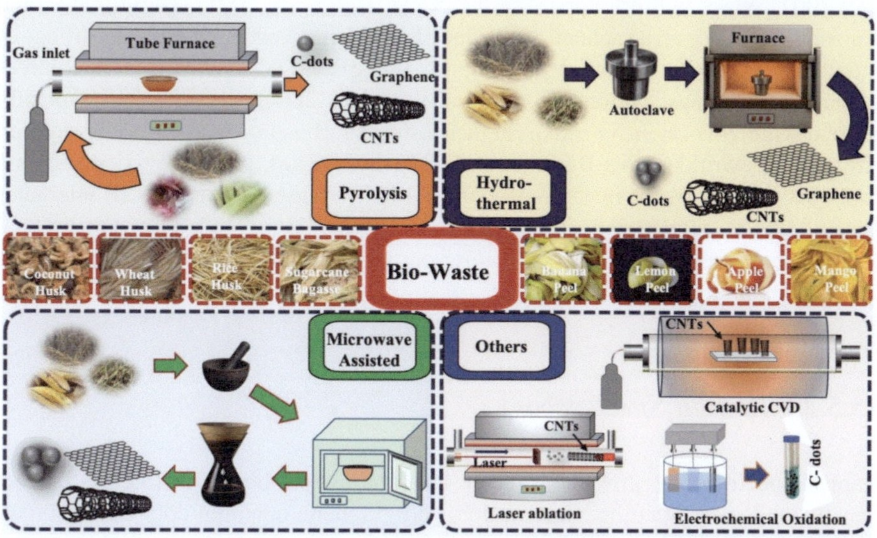

Fig. 15.2 Schematic illustration of various synthesis methods of carbon nanostructures from biowastes

as C-dots, nanofibers, graphene-like nanosheets, and hierarchical porous carbons. Thus, this wide range of methodologies represents alternative but mechanistically equivalent routes for achieving the core transformation of bio-residues into functional carbon nanostructures.

15.5 Structural, Morphological, and Optoelectronic Analysis

15.5.1 X-Ray Diffraction (XRD)

XRD is widely employed to evaluate the crystallinity and structural ordering of carbon nanomaterials. Carbon dots generally display a broad diffraction peak around $2\theta \approx 20$–$25°$ corresponding to the (002) plane, indicative of their predominantly amorphous nature (Luo et al. 2022). Graphene dots are also amorphous with broad peaks between 10 and 15° range and a significant (002) reflection near 25–30°, reflecting the presence of small graphitic domains (Soltani et al. 2025). For CNTs and MWCNTs, a distinct (002) peak at approximately 26° along with weaker (100)/(101) reflections confirms the graphitic arrangement of cylindrical graphene layers (Theodorakopoulos et al. 2024). It was also observed that the intensity of the (002) XRD peak of CNTs increases when the tube axis of CNTs is aligned more in-plane with the substrate surface. The vertical arrangement of nanotubes (with 002 plane parallel to the tube axis and symmetric axis between the incident and scattered beam) allows only the Bragg diffraction from the roof of nanotube to be collected by the detector; therefore, if all the CNTs in the sample are vertically arranged in an array, then (002) peaks will not appear in the XRD (Cao et al. 2001). Two-dimensional nanosheets, such as graphene or reduced graphene oxide, exhibit a strong or broadened (002) reflection depending on the extent of oxidation, while three-dimensional porous carbons or randomly stacked 2D graphene layers typically present wide, low-intensity peaks characteristic of turbostratic and highly disordered carbon frameworks (Love-Baker et al. 2024). In contrast, graphene oxide shows a pronounced (001) peak in the range of 10°–15°, arising from expanded interlayer spacing due to oxygen-containing functional groups, along with a weaker residual (002) feature around 25°–30°, indicating partial graphitic character (Surekha et al. 2020; Soltani et al. 2025).

15.5.2 Raman Spectroscopy

Raman spectroscopy provides crucial information on bonding structure, defect density, and graphitization. Carbon dots and graphene quantum dots exhibit broad D and G bands, with a high I_D/I_G ratio reflecting edge-rich domains and structural disorder (Thang et al. 2021). The D band peaks are also known as the disorder band arising due to scattering from sp^2 carbon-containing defects. G band peaks are observed between 1500 and 1600 cm^{-1}, which result from the presence of the graphite sheet, also known as the graphite or tangential band (Saheeda et al. 2019). However, these quantum dots can also show a lower I_D/I_G ratio when there is an increased graphitization caused by the synthesis techniques or heat treatment in higher temperatures (Dager et al. 2019; Tan et al. 2019). CNTs display the characteristic D, G, and G′ bands, and single-walled CNTs additionally show radial breathing modes (RBMs) at around 100–500 cm^{-1}, which provide direct insight into tube diameter (Berrezueta-Palacios et al. 2024). Multi-walled CNTs show strong G bands associated with multiple graphitic walls and do show RBM modes. In 2D nanosheets, a low I_D/I_G ratio indicates high graphitic quality, while defect-rich sheets show a pronounced D band (Silva Filho et al. 2020). On the other hand, rGO shows much shorter and broader G and D bands than graphene, indicating higher electrical conductivity of rGO than graphene (Balaji Mohan et al. 2016). The higher intensity of D bands with respect to G bands indicates an increase in the amount of disordered phase in rGOs (Silva Filho et al. 2020). Three-dimensional nanostructured carbons typically exhibit high I_D/I_G ratios due to abundant defects and cross-linked amorphous carbon regions.

15.5.3 Scanning Electron Microscopy (SEM)

SEM is primarily widely used to assess surface morphology and microscale structural features. Individual carbon dots and graphene quantum dots are too small to be clearly resolved by SEM, but they often appear as aggregated clusters on substrates (Mansuriya and Altintas 2021). CNTs and MWCNTs exhibit identifiable tubular morphologies whose diameter, length, and degree of entanglement can be visualized, along with any residual catalytic particles. The surface roughness on CNT walls might qualitatively indicate the presence of attached functional groups, particularly oxygen-containing moieties, which modify the inherent smoothness of the graphitic surface (Jun et al. 2019). Two-dimensional carbon nanosheets display characteristic wrinkles, folds, and stacking layers (Xiao et al. 2024; Siburian et al. 2018). Nanosheets are found to be more corrugated at higher temperatures and eventually form larger structures by interweaving (Zhu et al. 2018). In contrast, clean separate nanosheets are found to form at particularly low temperatures. Cross-sectional SEM imaging of the sample reveals the layered structure clearly (Ding et al. 2018). Bulk or three-dimensional porous carbons show interconnected pore networks and hierarchical structures.

15.5.4 Transmission Electron Microscopy (TEM)

TEM and high-resolution TEM (HRTEM) provide detailed visualization of carbon nanostructures at the nanoscale level. C-dots are typically observed as quasi-spherical nanoparticles, generally within the size 2–10 nm range, exhibiting distinct graphitic lattice fringes with interplanar spacings of ~0.21–0.24 nm (Dager et al. 2019). It was observed that heat-treated activated carbon precursors produce carbon quantum dots (CQDs) of larger size (Tan et al. 2019). CQDs obtained from natural biomass or synthetic precursors primarily exhibit amorphous carbon structures. In some instances, the carbon particles are entirely amorphous, or partially ordered, or a graphitic core is encapsulated by an amorphous outer layer. This core-shell contrast arises due to variations in precursor composition and synthesis conditions and is commonly observed in TEM and HRTEM analyses of carbon nanodots. Graphene quantum dots (GQDs) exhibit well-defined crystalline regions that are characteristic of graphene, with lattice fringes corresponding to specific graphitic planes. GQDs generally fall within the size 3.5–6.5 nm, with an average diameter of ~5 nm. HRTEM imaging typically reveals lattice spacings of ~0.24 nm, attributed to the in-plane (100) crystallographic spacing of graphene. This spacing aligns with the basal-plane distance of bulk graphite and reflect the underlying hexagonal lattice structure (Thang et al. 2021). Carbon nanotubes (CNTs) display their characteristic hollow tubular morphology under TEM, while HRTEM enables precise measurement of inner diameter, wall thickness, and the graphitic interlayer spacing of ~0.34 nm. A particular annealing temperature is required to obtain tubes of more uniform structure (Gupta et al. 2019). MWCNTs are characterized by their multiple concentric graphene layers, and the number and regularity can be directly visualized using TEM and HRTEM. Two-dimensional carbon nanosheets appear as thin, semitransparent layers with discernible wrinkles, folds, and edge imperfections. By contrast, bulk or three-dimensional porous nanocarbons exhibit interconnected pore networks, embedded nanocrystalline regions, and variations in framework thickness (Gupta et al. 2019).

15.5.5 Photoluminescence (PL)

Photoluminescence (PL) is highly sensitive to the core characteristics of C-dots, including their type, size, degree of crystallinity, and surface chemistry, particularly oxidation level and heteroatom doping (Jović et al. 2020). C-dots produced by different routes may vary significantly in composition and chemical structure, which in turn leads to distinct PL behavior. C-dots have been shown to exhibit PL bands particularly at 305, 355, 410, 445, and 500 nm, corresponding to intrinsic C (4.1 eV), graphitic N (3.5 eV), pyridinic N (3.0 eV), amino N (2.8 eV), and C=O (2.5 eV)-related electronic levels (Zhao et al. 2020). Excitation wavelength-dependent PL emission may result from relaxation of excited electrons into these different energy states, modulated by the heterogeneity of surface functional groups (Zhao et al. 2020). In nitrogen and sulfur co-doped C-dots, the emission wavelength red shifts with increasing excitation energy, while the PL intensity is further influenced by pH. This behavior has been attributed to variations in particle size and the diversity of functional groups on the surface of doped C-dots (Zhao et al. 2020). For GQDs, the PL emission can originate from intrinsic electronic transitions of $C\pi^* \rightarrow C\pi$ states at sp^2 carbon sites, with the transition energy depending on the size of the GQDs. The PL can also be affected by edge-state contributions. The PL spectra display a broad emission band whose overall shape remains invariant with synthesis temperature, although the intensity varies substantially (Thang et al. 2021). When the emission maximum is found independent of excitation wavelength, the PL of the synthesized GQDs is dominated by intrinsic-state emission (Thang et al. 2021). For graphene oxide (GO), the PL peaks shift toward shorter wavelengths and become increasingly narrow as GO undergoes reduction (say photothermal) to form rGO, with the degree of reduction governed by the exposure time (Ding et al. 2018). Semiconducting SWCNTs with small bandgaps of approximately 1 eV exhibit near-infrared PL, with emission spanning 800–2000 nm, a region that overlaps with the biological tissue transparency window; consequently, they are well suited for bioimaging applications (Narang and Pundir 2018). In CNTs, each PL peak corresponds to a specific nanotube structure and is closely correlated with the nanotube diameter (Liu et al. 2013). In RGO-CNT hybrid systems, the PL intensity evolves systematically with annealing temperature: it increases up to 650 °C and then decreases at higher temperatures. The hybrid sample with an optimal CNT concentration is essential for maximizing PL performance (Liu et al. 2013).

15.5.6 Fourier Transform Infrared Spectroscopy (FTIR)

FTIR spectroscopy is widely employed to probe surface functionalities and chemical modifications in carbon nanostructures. In carbon dots and graphene quantum dots, abundant oxygen-containing groups such as hydroxyl, carbonyl, and carboxyl moieties govern dispersibility and optical properties. Typical C-dot spectra exhibit bands at 3400, 3175, 1720, 1595, 1400, 1335, and 1245 cm^{-1}, assigned to –OH and N–H stretching in NH_4^+, C=O vibrations of ketone/aldehyde groups, C=C stretching from polyaromatic domains, $-COO^-$ modes from carboxylates, C–N stretching, and C–O vibrations of epoxy units, respectively (Tan et al. 2019). Notably, increasing the pyrolysis temperature of the precursor has little effect on the nature of CQD surface functional groups (Tan et al. 2019). The C=C signal reflects a graphitic core, whereas the OH, C=O, and C–H bending bands indicate that these oxygenated and hydrogenated functionalities remain attached to the CQD surface (Dager et al. 2019). FTIR spectra obtained from GQDs typically show a broad O–H band near 3382 cm^{-1}, confirming hydroxyl-based surface functionalization. Along with this, signals observed at 2926 and 2864 cm^{-1} correspond to aliphatic C–H stretching, while the intense peak at 1695 cm^{-1} arises from C=O stretching corresponding to carboxyl groups. Moreover, the signals that have been observed at 1429 and 1262 cm^{-1} reflect C–H bending and C–O stretching, respectively (Soltani et al. 2025). In contrast, CNTs and MWCNTs exhibit weak

FTIR peaks unless chemically oxidized. However, FTIR signals can appear upon oxidation, due to O–H, C=O, and C–O groups, where the spectrum is strongly influenced by its diameter and tube symmetry. With SWCNTs having typical diameters ranging from 0.4 to 2.0 nm and MWCNTs 5–1000 nm, the multiple concentric walls in MWCNTs produce broad composite bands, while SWCNTs show broadened C=C modes due to geometric polydispersity (Țucureanu et al. 2016). When the ends of CNTs are opened, such as by strong oxidants, carboxyl functionalities are developed at defect sites, which can also be removed again after heating above 600 °C. It has been observed that initially an attenuation of the C=C band occurs, which is then followed by the growth of a new peak near 1800–1700 cm^{-1} that corresponds to C=O stretching, generated by oxidative processes. The intensity of this peak indicates the degree of carboxylation, with increasing intensity confirming the formation of CNT-COOH (Țucureanu et al. 2016). FTIR spectra of two-dimensional graphene oxide or reduced graphene oxide typically display epoxy (C–O–C), hydroxyl (C–OH), and carbonyl bands. But when the graphene is pristine, it will only exhibit weak absorptions in FTIR spectra (Azizah et al. 2021; You et al. 2017). In three-dimensional porous carbons, surface functionalities are observed depending upon the activation method, which frequently incorporates oxygenated or heteroatom-containing groups.

15.5.7 X-Ray Photoelectron Spectroscopy (XPS)

XPS is a powerful tool for identifying both the elemental and the chemical bonding environment of carbon nanostructures. In C-dots and GQDs, the C 1s peak typically splits into multiple components representing sp^2 C=C, sp^3 C–C, C–O, and C=O species, often accompanied by detectable signals from doped heteroatoms such as N, S, or P. CNTs and MWCNTs generally show a dominant sp^2 carbon peak, though sp^3 hybridization associated with structural defects and traces of metal catalysts also appears. For two-dimensional nanosheets, the binding energy shift provides quantitative insights into oxygen functionalities, doping levels, and the degree of reduction. In bulk carbons, nitrogen species (pyridinic, pyrrolic, graphitic N), oxygenated groups, and variable sp^2/sp^3 carbon ratios are frequently observed. XPS spectra of CQDs typically exhibit peaks at binding energies 285, 532, and 401 eV, which are related to C 1s, O 1s, and N 1s signals. High-resolution C 1s fitting usually resolves four main peaks of C=C at 284.8 eV, C–O at 286.2 eV, C=O at 287.8 eV, and O=C–OH at 288.7 eV (Tan et al. 2019). For graphene oxide, XPS clearly verifies deoxygenation during reduction. The C 1s region of pristine GO can be separated into aromatic C=C (284.8 eV), C–O (286.2 eV), C=O (287.4 eV), and C(=O)OH (288.8 eV) (Liu et al. 2013). As reduction progresses, the fraction of sp^2 carbon increases, while signals corresponding to oxygen-containing functionalities diminish significantly (Ding et al. 2018).

15.6 Degradation of Agrochemicals, Pesticides, and Organic Pollutants

15.6.1 *Photocatalytic Pathways for Common Pollutants*

Extensive research has demonstrated that carbon nanomaterial-based photocatalysts exhibit superior efficiency in degrading pesticides, agricultural contaminants, and a broad spectrum of organic pollutants, for example, Abdullah et al. have prepared nitrogen-doped $BiVO_4$ (NBVO) and $CrFe_2O_4$ (CF) heterojunction photocatalyst supported on CNT and have observed the highest photocatalytic degradation efficiency in comparison to its counterparts without CNT (Alsubaie et al. 2025). In their study, the performance was evaluated against the degradation of insecticides chlorpyrifos and triazophos, resulting in 88% and 85% removal of them in 160 min. The NBVO/CF nanoparticles and CNTs in the nanocomposite exhibit efficient surface contact due to the enhanced surface area of nanoparticles and surface functionalities of CNTs. A close contact between the spherical NBVO/CF nanoparticles and the carbon nanotubes enhances the stability of the fabricated nanocomposite in the photocatalytic processes. Similarly, another study showcased the improved performance of MWCNT@g-C_3N_4@Ag photocatalyst, for the preparation of which MWCNT was first synthesized from PET bottle waste (Dehghan et al. 2025). Here, the incorporation of MWCNT and Ag-doping improved the performance of g-C_3N_4. This photocatalyst achieved 99.89% removal efficiency of MB in wastewater, with a 0.52-g photocatalyst loading per liter under 231-min reaction time. It was found that enhanced production of superoxide radicals ($\cdot O_2^-$) took place due to the presence of MWCNTs which enabled the photogenerated electrons to migrate from the conduction band (CB) of g-C_3N_4 to the MWCNTs, prohibiting electron-hole pair recombination. Sun et al. evaluated the total ammonia nitrogen (TAN) elimination from wastewater using the prepared Ag/g-C_3N_4-CNT catalyst. They compared the performance of g-C_3N_4 and Ag/g-C_3N_4-CNT for TAN removal and demonstrated that TAN degradation significantly increased from 42.5% to 88.2% with the Ag/g-C_3N_4-CNT catalyst (Sun et al. 2021). Again, several studies have demonstrated that combining TiO_2 with CNTs yields a synergistic photocatalyst with enhanced performance for applications in various types of wastewater treatment (Kamil et al. 2018). It is now known that CNTs suppress electron-hole recombination in TiO_2, can function as photosensitizers that inject electrons into TiO_2, and may introduce defect-related electronic states that further promote photocatalytic activity. CNTs can establish a Schottky barrier at the TiO_2/CNT interface, creating a space-charge region. Because TiO_2 is an n-type semiconductor, its photogenerated electrons readily transfer to CNTs to equilibrate the Fermi levels. The resulting accumulation of holes in TiO_2 then migrates to the surface, where they react with adsorbed water or hydroxide ions to generate reactive $\cdot$OH radicals, which can degrade pollutants in wastewater (Sharma et al. 2021). Agricultural wastewater may also carry harmful bacteria that impede crop and plant health, and photocatalysts can be employed to treat such contamination, for instance, Ag nanoparticle-modified TiO_2-CNT heterostructures synthesized via a deposition-hydrothermal route showed strong visible light antibacterial activity against *E. coli* and *S. aureus* (Osman et al. 2019). The nanocomposites achieved rapid bacterial inactivation under visible light irradiation ($\lambda > 450$ nm, ~60 mW cm^{-2}), as reported through agar well diffusion assays. Control experiments confirmed that visible light alone had no bactericidal effect and Ag^+ ion release was negligible. These results suggest that the antibacterial activity primarily arises from a photocatalytic mechanism involving efficient charge separation within the TiO_2-CNT heterostructure and the subsequent generation of reactive oxygen species (ROS), such as superoxide ($O_2\cdot^-$) and hydroxyl ($\cdot OH$) radicals, which disrupt cellular processes and lead to cell inactivation. Similarly, carbon-based graphene oxide/graphitic carbon nitride was reported as an effective photocatalyst eliminating 97.9% of *E. coli* after 120 min visible light irradiation (Sun et al. 2017) Analysis in this study revealed that introduction of GO contributed to efficient charge separation reducing the recombination of electron-hole pairs formed in g-C_3N_4 to generate more h^+, thus directly improving the bactericidal ability of GO/g-C_3N_4. Two composite materials integrating carbon quantum dots, MOF-808, and iron (Fe-CDs/MOF-808 and Fe-CDs@MOF-808) were engineered for organophosphate pesticide degradation. Fe-CDs/MOF-808 enabled rapid catalytic hydrolysis of paraoxon, achieving complete decomposition within 5 min at a catalyst dose of 0.5 mg·mL^{-1}. Fe-CDs@MOF-808 was examined for light-driven oxidation of parathion, reaching 98.1% degradation in 10 min and full removal after 30 min (da Silva Júnior et al. 2024).

In an agricultural wastewater study, graphene quantum dot-ZnO (GQDs-ZnO) nanocomposites were tested for their ability to remove commercial glyphosate, with the 0.2-M GQD sample showing the strongest catalytic response. This composite showed a degradation rate constant of 8.64×10^{-4} min^{-1}, which is 23% higher than the bare ZnO (7.02×10^{-4} min^{-1}). The improvement is attributed to the influence of GQDs during ZnO formation, where they function as nucleation sites that modify crystal size and dislocation density. These structural changes yield a more crystalline material, which effectively limits electron-hole recombination and prolongs charge-carrier lifetimes at the surface, thereby enhance the photocatalytic behavior of the GQDs-ZnO composite (Phophayu et al. 2020). CuO- and Cu_2O-based photocatalysts integrated with reduced graphene oxide (rGO) have been extensively reported for the visible light-driven degradation of organic dyes such as methylene blue (MB), methyl orange (MO), and rhodamine B (RhB) (Mehmood et al. 2024). While pristine CuO/Cu_2O exhibits a narrow band-gap suitable for visible light absorption, its photocatalytic efficiency is limited by rapid electron-hole recombination. Numerous studies report that incorporation of rGO enhances photocatalytic activity by reducing the effective bandgap (~1.1–1.6 eV), increasing surface area, and promoting efficient electron transfer, thereby suppressing charge-carrier recombination. Enhanced adsorption of aromatic dye molecules on rGO sheets—attributed to strong interactions between the dye molecules and the graphene surface—further improves photocatalytic performance. Under visible light irradiation, rGO functions as an electron mediator, facilitating charge separation and enabling the generation of reactive species such as electrons, holes, and hydroxyl radicals, which collectively drive the efficient degradation of MB, MO, and RhB. As a result, CuO/Cu_2O-rGO-based binary and ternary heterostructures consistently exhibit high degradation efficiencies (>85–95%), demonstrating their strong potential for sustainable wastewater treatment. In another report, C-dots obtained from papaya peel waste were integrated with reduced graphene oxide (rGO) via a one-step hydrothermal method to form rGO/CDs nanocomposites for methylene blue removal (Shahraki et al. 2023). The composition with a 2:1 rGO/CD mass ratio delivered 87% degradation of the dye in 135 min. The enhanced activity stems from the role of rGO sheets as efficient electron acceptors, which facilitate rapid transfer and separation of photoexcited electrons generated by the CDs. The same nanocomposites also exhibited stronger antibacterial action than individual CDs or rGO. When tested against *Bacillus subtilis* (Gram-positive) and *Pseudomonas aeruginosa* (Gram-negative), the rGO/CDs material produced the largest inhibition zones, with a more pronounced effect on the Gram-positive strain (Shahraki et al. 2023). This demonstrates its potential as both an effective photocatalyst and antimicrobial agent. In another report, Boruah et al. reported a graphene-based nanocomposite functionalized with polydopamine and magnetite for simazine detection and photocatalytic breakdown (Boruah et al. 2021). Under solar irradiation, the material achieved 97% removal within 50 min and also showed high activity toward four triazine herbicides simetone (89%), atrazine (87%), ametryn (95%), and prometryn (92%). The catalyst maintained strong performance over repeated operations, retaining 76% efficiency after 12 cycles, highlighting its potential for reusable pesticide treatment in water (da Silva Júnior et al. 2024). Besides the well-structured C-dots, graphene dots, CNTs, or graphene or rGO, several carbon-based photocatalysts such as activated carbon, C-doped photocatalysts, hydrothermal carbonation carbon, biomass carbon, and biochar carbon have been highlighted in studies for treating agricultural and other wastewater with improved photocatalytic performances either as a standalone material or as co-catalysts (Saleh et al. 2023; Liu et al. 2022; Hu et al. 2017; Zhang et al. 2024).

15.7 Conclusion

The carbon-based photocatalysts are a rapidly developing class of photocatalysts that can perform multifunctional environmental roles, including pollutant degradation, water sterilization, and nutrient delivery. This creates a closed-loop agricultural cycle in which contaminated water produced during farming can be treated and reused. Since the carbon-based materials can be regenerated every season from biomass residues, their limited structural stability during repeated photocatalytic cycles is less problematic than for expensive metal-based catalysts. However, for carbon-based material to function effectively as a photocatalyst, at least one component must behave as a semiconductor, possessing a bandgap that allows photoexcitation and subsequent charge-carrier generation. Therefore, in all-carbon composite photocatalysts, the synthesis route must achieve highly reproducible electrical characteristics, whether semiconducting or metallic, since slight variations in chirality, defects, or reduction level can completely alter their conductivity. The materials such as g-C_3N_4, graphene oxide, and elemental doped carbons (by N, O, B, S) demonstrate more reliable semiconducting behavior and thus offer a more stable foundation for designing carbon-based photocatalytic systems, especially when derived from agricultural or food waste streams.

15.8 Future Perspectives

Future development will rely on advancing defect engineering, heteroatom tuning, and hybridization strategies for optimization of visible light absorption, charge separation, and catalytic selectivity. Scalable green synthesis routes, improved mechanistic understanding of carbon-based catalyst-light interactions, and integration with renewable waste-to-carbon conversion technologies will define the next stage of progress. With these strategic advancements, carbon-based photocatalysts hold a strong promise for sustainable environmental remediation and affordable decentralized water treatment systems.

Acknowledgments The author acknowledges the Anusandhan National Research Foundation (ANRF-DST), New Delhi, India, for the financial support through the project Ref. No: EEQ/2023/000212.

References

Ahmad H, Khan RA, Koo BH, Alsalme A (2022) Systematic study of physicochemical and electrochemical properties of carbon nanomaterials. RSC Adv 12(24):15593–15600

Ahmed SN, Haider W (2018) Heterogeneous photocatalysis and its potential applications in water and wastewater treatment: a review. Nanotechnology 29(34):342001

Alsubaie AS, Altaf S, Aldhafeeri TR, Amin MA, Hessien MM, Rafiq U, Warsi MF (2025) Fabrication of nitrogen-doped $BiVO_4/CrFe_2O_4$@CNTs heterojunction for enhanced photocatalytic degradation of pesticides. Phys B Condens Matter 715:417592

Azizah N, Fahira N, Dirfan Z, Zharvan V, Sujiono EH (2021) Comparative study and characterization of reduced graphene oxide (rgo) and porous reduced graphene oxide (p-rgo) based on coconut shell waste. J Nano- Electron Phys 13(6). https://doi.org/10.21272/jnep.13(6).06017

Balaji Mohan V, Debes Bhattacharyya M, Liu D, Jayaraman K (2016) Improvements in electronic structure and properties of graphene derivatives. Adv Mater Lett 7(6):421–429

Baweja H, Jeet K (2019) Economical and green synthesis of graphene and carbon quantum dots from agricultural waste. Mater Res Express 6(8):0850g8

Berrezueta-Palacios C, Nakar D, Wroblewska A, Garrity O, Li H, Shadmi N, Flavel BS, Joselevich E, Reich S, Gordeev G (2024) Collective radial breathing modes in homogeneous nanotube bundles. Carbon 224:119010

Boruah PK, Darabdhara G, Das MR (2021) Polydopamine functionalized graphene sheets decorated with magnetic metal oxide nanoparticles as efficient nanozyme for the detection and degradation of harmful Triazine pesticides. Chemosphere 268:129328

Cao A, Xu C, Liang J, Wu D, Wei B (2001) X-ray diffraction characterization on the alignment degree of carbon nanotubes. Chem Phys Lett 344(1–2):13–17

Chen Z, Narita A, Müllen K (2020) Graphene nanoribbons: on-surface synthesis and integration into electronic devices. Adv Mater 32(45):2001893

Clancy AJ, Bayazit MK, Hodge SA, Skipper NT, Howard CA, Shaffer MS (2018) Charged carbon nanomaterials: redox chemistries of fullerenes, carbon nanotubes, and graphenes. Chem Rev 118(16):7363–7408

da Silva Júnior AH, Müller JDOM, Oliveira CRSD, de Noni Junior A, Tewo RK, Mhike W, da Silva A, Mapossa AB, Sundararaj U (2024) New insights into materials for pesticide and other agricultural pollutant remediation. Materials 17(14):3478

Dager A, Uchida T, Maekawa T, Tachibana M (2019) Synthesis and characterization of monodisperse carbon quantum dots from fennel seeds: photoluminescence analysis using machine learning. Sci Rep 9(1):14004

Dehghan P, Abbasi M, Azari A, Mofarahi M, Nowrouzi M, Dibaj M, Akrami M (2025) Green and sustainable synthesis of MWCNT@g-C_3N_4@Ag photocatalyst from PET for efficient wastewater treatment. Sci Rep 15(1):10601

Devi M, Wang H, Moon S, Sharma S, Strauss V (2023) Laser-carbonization–a powerful tool for micro-fabrication of patterned electronic carbons. Adv Mater 35(38):2211054

Díez-Pascual AM (2021) Carbon-based nanomaterials. Int J Mol Sci 22(14):7726

Ding JJ, Chen HX, Feng DQ, Fu HW (2018) Investigation on photoluminescence emission of (reduced) graphene oxide paper. IOP Conf Ser Mater Sci Eng 292(1):012097

Djurišić AB, He Y, Ng A (2020) Visible-light photocatalysts: prospects and challenges. APL Mater 8:030903

Gupta N, Gupta SM, Sharma SK (2019) Carbon nanotubes: synthesis, properties and engineering applications. Carbon Lett 29(5):419–447

Hashem HM, El-Maghrabey M, Borg H, Magdy G, Belal F, El-Shaheny R (2025) Room-temperature upcycling of agricultural waste to carbon dots and its application as a selective fluorescence nanoprobe for iron (III) analysis in diverse samples. Talanta Open 12:100484

Hu Z, Shen Z, Yu JC (2017) Converting carbohydrates to carbon-based photocatalysts for environmental treatment. Environ Sci Technol 51(12):7076–7083

Ingrao C, Strippoli R, Lagioia G, Huisingh D (2023) Water scarcity in agriculture: an overview of causes, impacts and approaches for reducing the risks. Heliyon 9(8):e18507

Jirimali H, Singh J, Boddula R, Lee JK, Singh V (2022) Nano-structured carbon: its synthesis from renewable agricultural sources and important applications. Materials 15(11):3969

Jović D, Jaćević V, Kuča K, Borišev I, Mrdjanovic J, Petrovic D, Seke M, Djordjevic A (2020) The puzzling potential of carbon nanomaterials: general properties, application, and toxicity. Nanomaterials 10(8):1508

Jun LY, Yon LS, Mubarak NM, Yeo KS, Khalid M, Bing CH (2019) Comparison of drying method on acid-functionalized multi-walled carbon nanotube and their application for dye removal. IOP Conf Ser Mater Sci Eng 495(1):012057

Kadem AJ, Tan ZM, Suntharam NM, Pung SY, Ramakrishnan S (2023) Synthesis of CuO, ZnO and SnO_2 coupled TiO_2 photocatalyst particles for enhanced photodegradation of rhodamine B dye. Bull Chem React Eng Catal 18(3):506–520

Kamil AM, Mohammed HT, Balakit AA, Hussein FH, Bahnemann DW, El-Hiti GA (2018) Synthesis, characterization and photocatalytic activity of carbon nanotube/titanium dioxide nanocomposites. Arab J Sci Eng 43(1):199–210

Kang Z, Lee ST (2019) Carbon dots: advances in nanocarbon applications. Nanoscale 11(41): 19214–19224

Kang C, Huang Y, Yang H, Yan XF, Chen ZP (2020) A review of carbon dots produced from biomass wastes. Nanomaterials 10(11):2316

Kausar A (2020) Overview on nanocarbon sponges in polymeric nanocomposite. Mater Res Innov 24(5):309–320

Khaki MRD, Shafeeyan MS, Raman AAA, Daud WMAW (2017) Application of doped photocatalysts for organic pollutant degradation: a review. J Environ Manag 198:78–94

Libra JA, Ro KS, Kammann C, Funke A, Berge ND, Neubauer Y, Titirici MM, Fühner C, Bens O, Kern J (2011) Hydrothermal carbonization of biomass residuals: a comparative review of the chemistry, processes and applications of wet and dry pyrolysis. Biofuels 2(1):71–106

Liew RK, Nam WL, Chong MY, Phang XY, Su MH, Yek PNY, Ma NL, Cheng CK, Chong CT, Lam SS (2018) Oil palm waste: an abundant and promising feedstock for microwave pyrolysis conversion into good quality biochar with potential multi-applications. Process Saf Environ Prot 115:57–69

Lin H, Xiao Y, Geng A, Bi H, Xu X, Xu X, Zhu J (2022) Research progress on graphitic carbon nitride/metal oxide composites: synthesis and photocatalytic applications. Int J Mol Sci 23(21): 12979

Liu F, Cao Y, Yi M, Xie L, Huang W, Tang N, Zhong W, Du Y (2013) Thermostability, photoluminescence, and electrical properties of reduced graphene oxide–carbon nanotube hybrid materials. Crystals 3(1):28–37

Liu Y, Liu S, Li Z, Ma M, Wang B (2018) A microwave-synthesized mesoporous carbon sponge as an efficient adsorbent for Cr(VI) removal. RSC Adv 8:7892–7898

Liu C, Chen W, Liab M, Hong S, Li W, Pan M, Wu Q, Mei C (2019) Rapid microwave activation of waste palm into hierarchical porous carbons for supercapacitors. RSC Adv 9:19441–19449

Liu D, Li C, Zhao C, Zhao Q, Niu T, Pan L, Xu P, Zhang F, Wu W, Ni T (2022) Facile synthesis of three-dimensional hollow porous carbon doped polymeric carbon nitride with highly efficient photocatalytic performance. Chem Eng J 438:135623

Love-Baker CA, Harrell TM, Vautard F, Klett J, Li X (2024) Analysis of the turbostratic structures in PAN-based carbon fibers with wide-angle x-ray diffraction. Carbon 224:119037

Luo H, Lari L, Kim H, Hérou S, Tanase LC, Lazarov VK, Titirici MM (2022) Structural evolution of carbon dots during low temperature pyrolysis. Nanoscale 14(3):910–918

Mansuriya BD, Altintas Z (2021) Carbon dots: classification, properties, synthesis, characterization, and applications in health care—an updated review (2018–2021). Nanomaterials 11(10):2525

Mehmood A, Sandhu ZA, Iqbal MJ, Manzoor A, Elahi M, Liyaqat S, Nadeem I, Zainab SA, Bibi A (2024) Recent developments in applications of G-CuO nanocomposites for photocatalytic dye removal. Phys Scr 100(1):012001

Narang J, Pundir CS (eds) (2018) Current and future developments in nanomaterials and carbon nanotubes: introduction to carbon nanomaterials. Bentham Science Publishers

Nasir A, Khalid S, Yasin T, Mazare A (2022) A review on the progress and future of TiO_2/graphene photocatalysts. Energies 15(17):6248

Osman G, Mohamed MM, Khairou KS (2019) Photocatalytic bacterial disinfection using Ag°/Ag^{+1} immobilized on CNT modified TiO_2 nanomaterials. J Pure Appl Microbiol 13(2):767–778

Pavlović J, Tušar NN, Rajić N (2024) The synthesis and photocatalytic efficacy of distinct nano-oxides in the breakdown of organic contaminants. Catalysts 14(11):771

Phophayu S, Pimpang P, Wongrerkdee S, Sujinnapram S, Wongrerkdee S (2020) Modified graphene quantum dots-zinc oxide nanocomposites for photocatalytic degradation of organic dyes and commercial herbicide. J Reinf Plast Compos 39(3–4):81–94

Porowski A (2019) Mineral and thermal waters. In: LaMoreaux J (ed) Environmental geology, Encyclopedia of sustainability science and technology series. Springer, New York, pp 149–181

Roshni V, Divya O (2017) One-step microwave-assisted green synthesis of luminescent N-doped carbon dots from sesame seeds. Curr Sci 112:385–390

Saheeda P, Sabira K, Joseph J, Jayaleksmi S (2019) On the intriguing emission characteristics of size tunable carbon dots derived from functionalized multi-walled carbon nanotubes. Mater Chem Phys 225:8–15

Saleh TS, Badawi AK, Salama RS, Mostafa MMM (2023) Design and development of novel composites containing nickel ferrites supported on activated carbon derived from agricultural wastes and its application in water remediation. Materials 16(6):2170

Salman M, Jahan S, Kanwal S, Mansoor F (2019) Recent advances in the application of silica nanostructures for highly improved water treatment: a review. Environ Sci Pollut Res 26(21): 21065–21084

Schweitzer L, Noblet J (2018) Water contamination and pollution. In: Török B, Dransfield T (eds) Green chemistry. Elsevier, pp 261–290

Shahraki HS, Bushra R, Shakeel N, Ahmad A, Ahmad M, Ritzoulis C (2023) Papaya peel waste carbon dots/reduced graphene oxide nanocomposite: from photocatalytic decomposition of methylene blue to antimicrobial activity. J Bioresour Bioprod 8(2):162–175

Sharif YM, Saka C, Baytar O, Sahin Ö (2018) Preparation and characterization of activated carbon from sesame seed shells. Anal Lett 51:2733–2746

Sharma HK, Sharma SK, Vemula K, Koirala AR, Yadav HM, Singh BP (2021) CNT facilitated interfacial charge transfer of TiO2 nanocomposite for controlling the electron-hole recombination. Solid State Sci 112:106492

Shrestha L, Thapa M, Shrestha R, Maji S, Pradhananga R, Ariga K (2019) Rice husk-derived high surface area nanoporous carbon materials with excellent iodine and methylene blue adsorption properties. J Carbon Res 5(1):10

Siburian R, Sihotang H, Raja SL, Supeno M, Simanjuntak C (2018) New route to synthesize of graphene nano sheets. Orient J Chem 34(1):182

Silva Filho JC, Venancio EC, Silva SC, Takiishi H, Martinez LG, Antunes RA (2020) A thermal method for obtention of 2 to 3 reduced graphene oxide layers from graphene oxide. SN Appl Sci 2(8):1450

Soltani M, Ahmadzadeh N, Rajabi S, Besharati N, Khatamian N, Homayouni Tabrizi M (2025) Efficacy of graphene quantum dot-hyaluronic acid nanocomposites containing quinoline for target therapy against cancer cells. Sci Rep 15(1):8494

Son BT, Long NV, Hang NTN (2021) The development of biomass-derived carbon-based photocatalysts for the visible-light-driven photodegradation of pollutants: a comprehensive review. RSC Adv 11(49):30574–30596

Speranza G (2021) Carbon nanomaterials: synthesis, functionalization and sensing applications. Nanomaterials 11(4):967

Sun L, Du T, Hu C, Chen J, Lu J, Lu Z, Han H (2017) Antibacterial activity of graphene oxide/g-C_3N_4 composite through photocatalytic disinfection under visible light. ACS Sustain Chem Eng 5(10):8693–8701

Sun J, Huang Y, Li H, Qiu Y, Zhang J, Hu K, Li J, Ding Y (2021) Silver nanoparticle-loaded graphitic carbon nitride/multiwall carbon nanotube composite with improved denitrification to nitrogen gas for the photocatalytic removal of aqueous ammonia nitrogen. Environ Technol Innovation 24:101815

Surekha G, Krishnaiah KV, Ravi N, Suvarna RP (2020) FTIR, Raman and XRD analysis of graphene oxide films prepared by modified hummers method. J Phys Conf Ser 495(1): 012012. IOP Publishing

Tan C, Zuo S, Zhao Y, Shen B (2019) Preparation of multicolored carbon quantum dots using $HNO_3/HClO_4$ oxidation of graphitized carbon. J Mater Res 34(20):3428–3438

Thang PN, Hung LX, Thuan DN, Yen NH, Hien NTT, Hanh VTH, Khang NC, Laverdant J, Nga PT (2021) Temperature-dependent Raman investigation and photoluminescence of graphene quantum dots with and without nitrogen-doping. J Mater Sci 56(8):4979–4990

Thara CR, Mathew B (2024) Microwave synthesized N-doped carbon dots for dual mode detection of Hg (II) ion and degradation of malachite green dye. Talanta 268(1):125278

Theodorakopoulos GV, Karousos DS, Benra J, Forero S, Hammerstein R, Sapalidis AA, Katsaros FK, Schubert T, Favvas EP (2024) Well-established carbon nanomaterials: modification, characterization and dispersion in different solvents. J Mater Sci 59(8):3339–3362

Țucureanu V, Matei A, Avram AM (2016) FTIR spectroscopy for carbon family study. Crit Rev Anal Chem 46(6):502–520

Uhlig B, Dhavamani A, Nagy N, Lilienthal K, Liske R, Ramos R, Dijon J, Okuno H, Kalita D, Lee J, Georgiev V, Asenov A, Amoroso S, Wang L, Koenemann F, Gotsmann B, Goncalves G, Chen B, Liang J, Pandey R, Chen R, Todri-Sanial A (2018) Challenges and progress on carbon nanotube integration for beol interconnects. In: 2018 IEEE international interconnect technology conference (IITC), pp 16–18, IEEE

Ukanwa KS, Patchigolla K, Sakrabani R, Anthony E, Mandavgane S (2019) A review of chemicals to produce activated carbon from agricultural waste biomass. Sustainability 11(22):6204

Urade AR, Lahiri I, Suresh KS (2023) Graphene properties, synthesis and applications: a review. JOM 75(3):614–630

Wato T, Amare M, Bonga E, Demand BBO, Coalition BBR (2020) The agricultural water pollution and its minimization strategies—a review. J Res Dev Manag 64:10–22

Xiao G, Li H, Yu Z, Niu H, Yao Y (2024) Highly thermoconductive, strong graphene-based composite films by eliminating nanosheets wrinkles. Nano-Micro Lett 16(1):17

You Y, Deng J, Tan X, Gorjizadeh N, Yoshimura M, Smith SC, Sahajwalla V, Joshi RK (2017) On the mechanism of gas adsorption for pristine, defective and functionalized graphene. Phys Chem Chem Phys 19(8):6051–6056

Zhang C, Ge-Zhang S, Wang Y, Mu H (2024) A wooden carbon-based photocatalyst for water treatment. Int J Mol Sci 25(9):4743

Zhao Q, Song W, Zhao B, Yang B (2020) Spectroscopic studies of the optical properties of carbon dots: recent advances and future prospects. Mater Chem Front 4(2):472–488

Zhu Z, Zhan L, Shih TM, Wan W, Lu J, Huang J, Guo S, Zhou Y, Cai W (2018) Critical annealing temperature for stacking orientation of bilayer graphene. Small 14(39):1802498

MIX
Papier aus verantwortungsvollen Quellen
Paper from responsible sources
FSC® C105338

If you have any concerns about our products,
you can contact us on
ProductSafety@springernature.com

In case Publisher is established outside the EU,
the EU authorized representative is:
Springer Nature Customer Service Center GmbH
Europaplatz 3, 69115 Heidelberg, Germany

Printed by Libri Plureos GmbH
in Hamburg, Germany